BARRON'S

AP®

BIOLOGY

PREMIUM

WITH 5 PRACTICE TESTS

SEVENTH EDITION

Deborah T. Goldberg, M.S.

Former AP Biology Teacher
Lawrence High School
Cedarhurst, New York

About the Author

Deborah Goldberg is passionate about Biology. She earned her B.S. and M.S. degrees in Biology at Long Island University. For fourteen years, she did cell research at NYU Medical Center and New York Medical College. For the following twenty-two years, she taught Biology, Chemistry, and Forensic Science at Lawrence High School in Cedarhurst, New York.

Dedication

I dedicate this book to my children, Michael and Sara Boilen, and to my husband, Howard Blue.

Acknowledgments

I wish to thank—

—My husband, Howard Blue, for his constant love and support
—My sister, Rachel, for her mastery of the English language and her eagerness to share it
—Samantha Karasik, my editor, for her expert guidance
—My students at Lawrence High School on Long Island, New York, who made teaching the best job in the world
—Evaluators, who made some great recommendations
—Stephen Traphagen, for his expert advice, suggestions, and feedback throughout this revision

Published by Kaplan, Inc., d/b/a Barron's Educational Series
750 Third Avenue
New York, NY 10017
www.barronseduc.com

ISBN: 978-1-4380-1172-1

10 9 8 7 6 5 4 3 2

Contents

The owl appears next to topics that provide helpful background information and illustrative examples. See page 4 for more details.

As you review the content in this book and work toward earning that **5** on your AP BIOLOGY exam, here are five things that you **MUST** know above everything else.

1

The process of evolution drives the diversity and unity of life.

- Darwin's theory of natural selection is evolution's major driving mechanism.
- Evolutionary theory is supported by evidence from many scientific disciplines.
- Phylogenetic trees and cladograms graphically represent evolutionary history.
- The genetic code is universal because all living things descend from a common ancestor.
- Evolution explains how all life is so similar and also why so much diversity exists.

2

Biological systems utilize free energy and molecular building blocks to grow, to reproduce, and to maintain dynamic homeostasis.

- Reciprocal processes of cellular respiration and photosynthesis cycle H_2O, O_2, and CO_2.
- Surface-to-volume ratios affect the capacity of a biological system to obtain resources and eliminate wastes.
- Selectively permeable plasma membranes regulate the movement of molecules across them, and maintain internal environments that differ from external ones.
- Negative and positive feedback mechanisms maintain dynamic homeostasis.

3

Molecules, cells, and organs coordinate activities for the fitness of the organism as a whole.

- Cells communicate by generating, transmitting, and receiving chemical signals.
- Signal transduction pathways link signal reception with cellular response.
- Nervous systems sense, transmit, and integrate information.
- Cells of the immune and endocrine systems interact in complex ways.

4

Living systems store, retrieve, transmit, and respond to information that is essential to life processes.

- Genetic information is passed from parent to offspring via DNA with a high degree of accuracy, although some mutations do occur.
- DNA directs the production of polypeptides at the ribosome by an elaborate process.
- The cell cycle is complex with highly regulated checkpoints.
- Most traits derive from gene interactions that are more complex than what Mendel described.
- The expression of genes is controlled by cell signaling, transcription factors, alternative splicing of pre-RNA, and environmental factors.

5

Biological systems interact, and these systems possess complex properties.

- Populations, communities, and ecosystems interact and respond to changes in the environment.
- Mathematical operations can be used to quantify interactions among living things in the environment.
- Interactions between living organisms and their environments result in the recycling of matter and the movement of energy through food chains.

PART 1: INTRODUCTION

The **introduction** to this book provides an overview of the AP Biology course and exam (including the Big Ideas, Enduring Understandings, and Science Practices). It also includes:

- A list of the features of this test preparation guide
- A preview of what the exam questions look like
- A detailed description of the exam format
- Instructions for tackling questions that involve math
- Information about how the exam is graded
- Specific tips for achieving success on both sections of the exam

This introduction will not only prepare you for what to expect in your AP Biology course and on the exam, but it will also set you on your path to understanding biology and developing the ability to think like a scientist.

There's a lot to learn and review when preparing for this exam, but do not stress! This book will provide you with the review and practice you need to fully prepare yourself for this test!

1 About the AP Biology Exam

The AP Biology course and exam are organized around a few underlying principles called the **Big Ideas**. These ideas encompass the core scientific principles, theories, and processes that govern living things and biological systems. For each of the Big Ideas, **Enduring Understandings** incorporate core concepts that are necessary to support the Big Idea. The Big Ideas and Enduring Understandings are highlighted throughout Part 2 (Chapters 2 through 13), which focuses on subject area review topics that are likely to be tested on the exam. **Illustrative Examples**, which are covered throughout Part 3 (Chapters 14 through 18), are also important to review. While they are not required content as per the official College Board curriculum, they do help reinforce concepts that will be tested and they serve as excellent examples for you to cite when writing your answers to the free-response questions. In addition, your teacher may use illustrative examples to help get his or her point across. As a result, your teacher may teach different illustrative examples than another AP Biology teacher would. Finally, the six **Science Practices** incorporate important skills that students are expected to develop throughout their AP Biology course. They were presented by the College Board in the latest AP Biology curriculum, and teachers are expected to integrate them into the course. These six Science Practices will be tested on both the multiple-choice questions of Section I of the exam and the free-response questions of Section II of the exam.

NOTE

Throughout this book (particularly in the chapters in Part 2), you will see references (such as "ENE-1") to three pedagogical classifications. These three classifications are the Big Ideas, Enduring Understandings, and Science Practices. Focus on the review content noted next to these classifications; don't spend time memorizing the classification names and numbers. In addition, although they have not been called out throughout the text (so as not to provide an overwhelming number of pedagogical classifications), all of the review and practice material in this book (particularly Chapters 2 to 13) still covers the Essential Knowledge and Learning Objective classifications of the latest AP Biology curriculum. This book has been organized to help you focus on your understanding of Biology. For the complete list of those Essential Knowledge and Learning Objective classifications, refer to the "AP Biology Course and Exam Description" from the College Board: ***https://apcentral.collegeboard.org/pdf/ap-biology-course-and-exam-description .pdf?course=ap-biology.***

Below is the list of four Big Ideas followed by the six Science Practices:

BIG IDEA 1 (EVO): The process of **evolution** drives the diversity and unity of life.

BIG IDEA 2 (ENE): Biological systems use **energy** and molecular building blocks to grow, reproduce, and maintain dynamic homeostasis.

BIG IDEA 3 (IST): Living systems store, retrieve, transmit, and respond to **information** essential to life processes.

BIG IDEA 4 (SYI): Biological systems interact, and these systems and their **interactions** exhibit complex properties.

SCIENCE PRACTICE 1: Describe and explain biological concepts, processes, and models.

SCIENCE PRACTICE 2: Analyze drawings, graphs, and charts that show biological processes and concepts.

SCIENCE PRACTICE 3: State the null hypothesis, or predict the results of an experiment, and identify testable questions based on an observation.

SCIENCE PRACTICE 4: Construct a graph or chart from a data table.

SCIENCE PRACTICE 5: Use mathematical calculations appropriately, and use and interpret confidence intervals and/or error bars.

SCIENCE PRACTICE 6: Support a scientific claim with evidence from biological principles and concepts.

WHAT'S IN THIS UPDATED REVIEW BOOK?

1. **EVERY TOPIC IN THE REVISED AP BIOLOGY CURRICULUM** (covered in Part 2) is aligned with the Big Ideas, Enduring Understandings, and Science Practices. These categories are identified on the first page of each chapter in this part of the book as well as at the end of the answer explanations to these end-of-chapter practice questions.
2. **ILLUSTRATIVE EXAMPLE TOPICS IDENTIFIED WITH AN OWL ICON** are in the margins for easy identification. These topics (which are primarily found in Part 3) are included for two reasons:

   - Some students never learned this background information because their school does not offer a pre-AP Biology course.
   - This supplemental information can be cited when writing answers to free-response questions on the AP exam.
3. **HUNDREDS OF PRACTICE QUESTIONS** are designed to reinforce the basics of every topic as you review the material.
4. **CHAPTER 19: "INVESTIGATIVE LABS"** features a review of the **13 inquiry-based labs** that fulfill the 6 Science Practices.
5. **CHAPTER 20: "HOW THE COLLEGE BOARD GRADES YOUR ANSWERS TO THE FREE-RESPONSE QUESTIONS"** will help you get the maximum points on your answers to the free-response questions on the exam.
6. **TWO PRACTICE EXAMS AT THE END OF THE BOOK** mirror actual questions on the AP exam, and detailed answer explanations are provided for all questions.
7. **THREE PRACTICE EXAMS ONLINE** also mirror actual questions on the AP exam, provide detailed answer explanations for all questions, and allow you to take the exam in practice (untimed) mode or timed mode. Refer to the card at the beginning of this book, which provides instructions for accessing these online exams.

This text has everything you need in a review book to help you get a 5 on the AP Biology exam!

WHAT ARE THE AP EXAM QUESTIONS LIKE?

Sample question:

Hermaphrodites are animals that serve as both male and female by producing both sperm and eggs. Hermaphrodites are often animals that are fixed to a surface (sessile) and are less often motile (free-moving) animals.

Which of the following statements is the best explanation for this phenomenon?

(A) As in all examples of asexual reproduction, large numbers of offspring can be produced.
(B) This is a novel adaptation that evolved to meet the challenge of finding a mate without having the ability to move.
(C) Both the male and the female can produce offspring simultaneously.
(D) Hermaphroditism is a degenerate form of sexual reproduction and is found in primitive animals such as parasites.

Often, basic information is provided in the stem of the question.

The answer is choice B. This question style:

- Does not focus on rote memory.
- Focuses on reasoning and analysis, evaluation, synthesis, and application; not only on vocabulary.
- Contains basic information in the body of the question so that students do not have to recall details, but can demonstrate an understanding of a concept. In this case, the concept being tested is evolutionary adaptation, not reproduction.
- Contains only four answers from which to choose.
- Takes time to read, think about, and answer. That's why there are only 60 multiple-choice questions on the AP Biology exam.

EXAM FORMAT

The AP Biology exam is approximately 3 hours in length and is made up of two sections.

<table>
<tr><th colspan="4">Section I</th></tr>
<tr><th>Question Type</th><th>Number of Questions</th><th>Timing</th><th>Exam Weighting</th></tr>
<tr><td>Multiple-Choice</td><td>60</td><td>90 minutes</td><td>50%</td></tr>
<tr><th colspan="4">Section II</th></tr>
<tr><th>Question Type</th><th>Number of Questions</th><th>Timing</th><th>Exam Weighting</th></tr>
<tr><td>Long Free-Response</td><td>2</td><td rowspan="2">90 minutes*</td><td rowspan="2">50%</td></tr>
<tr><td>Short Free-Response</td><td>4</td></tr>
</table>

*Note that we recommend using the first 10 minutes of Section II as a "reading period" to organize your thoughts. If you do so, you must monitor the length of the reading period yourself; the exam proctor will not do that for you. However, you may also choose to begin writing immediately.

Math

Math is included on the exam. A table of equations and formulas, including standard deviation or chi-square, is part of the exam (see Appendix C). In previous exams, a Hardy-Weinberg question might have required that you take the square root of 16 to find the value of q. Of course, you calculated that answer in your head. Now, you might actually have to use a calculator to find the value of q when q^2 is 23. No big deal!

TIP

Bring a calculator to the exam. A four-function, scientific, or graphing calculator is allowed on both sections of this exam.

GRADES ON THE EXAM

Advanced placement and/or college credit is awarded by the college or university you will attend. Different institutions observe different guidelines about awarding AP credit. Success on the AP exam may allow you to take a more advanced course and bypass an introductory course, or it might qualify you for 8 credits of advanced standing and tuition credit. The best source of specific up-to-date information about an institution's policy is its catalog or website.

Exams are graded on a scale from 1 to 5, with 5 being the best. The total raw score on the exam is translated to the AP's 5-point scale.

AP Grade	Qualification
5	Extremely Well Qualified
4	Well Qualified
3	Qualified
2	Possibly Qualified
1	No Recommendation

Here are the grade distributions for all the students who sat for the AP Biology exam in May 2019. These numbers tend to be consistent from year to year.

AP Grade	Students Scoring that Grade
5	7.02%
4	22.01%
3	35.35%
2	26.71%
1	8.89%

TIPS FOR SECTION I

Section I consists of 60 multiple-choice questions. It takes 90 minutes to complete.

Be Neat

Improperly erased pencil marks can cause the machine to misgrade your paper. On the other hand, you may write or draw anywhere in the question booklet.

Pace Yourself

Every multiple-choice question is worth the same number of points. Skip lengthy or difficult questions at first; go back to them if you have time. Bring a watch and budget your time.

Answer Every Question

Your score on the multiple-choice questions is based on the number of questions that you answer correctly. There is no penalty for incorrect answers or for leaving an answer blank. So always guess—even if you don't know the answer.

TIPS FOR SECTION II

Section II consists of 6 questions. Questions 1 and 2 are long free-response questions that should take about 25 minutes each to answer. Questions 3–6 are short free-response questions that should take about 8 to 10 minutes each to answer.

Just as an Olympic athlete must anticipate what the judges want to see, you must be prepared to give the exam readers what they want to read. If you can do that on the AP exam, you will get a high score.

Here are things the readers **do not** particularly care about:

Spelling
Penmanship (unless they cannot read the paper)
Grammar
Wrong information—**You do not get points off for incorrect statements, unless you contradict yourself.**

Here are the things the graders **do** care about:

The answer must be in essay form, not an outline.
Label the parts of the question that you are answering in each section of your response.
The readers want to see lots of correct information that answers the question asked—so write, write, and write!

You Do Not Lose Points for Giving Incorrect Information

You begin with zero points, and you gain points as you make correct statements that answer the question. The reader is like the person who stands at the entrance to a concert and uses a clicker to count the number of people entering. Every time you state a correct piece of information that answers the question, you get a click; that is, you get a point.

However, **there is one exception**! If you contradict yourself, **you wipe away all points that you earned in that one part.** Here is an example: A 4-point part of a question asks you to "Describe how ATP is produced during cellular respiration." As part of your answer, you state that "most ATP is produced by oxidative phosphorylation," which is correct. However, toward the end of your response, you then state that "most ATP is generated by the citric acid cycle," which is *not* correct. Thus, you have contradicted yourself within your response. To the reader, this demonstrates that you may not have a real understanding of the process of cellular respiration. Therefore, in this scenario, you would have lost all of the points that you had earned on this part of the question.

Watch the Point Count!

On the AP Biology exam, each free-response question will state the point value next to each part of the question. For example, a long free-response question may be divided into four parts, where Part A is worth 1–2 points, Part B is worth 3–4 points, Part C is worth 1–3 points, and Part D is worth 2–4 points. If you spend 25 minutes writing a 10-page masterpiece on Part A and have no time left for Parts B, C, and D, the maximum that you could earn for that question is only 2 points. Therefore, don't waste time writing too much; just be sure to **answer every part of each question completely and correctly** in order to maximize your points.

Bring a Watch and Budget Your Time

You have 90 minutes total for Section II. The exam proctors will **not** announce when it is time to move from one question to another. You must monitor the time.

Pay Attention to Special Words

Read the question and determine what you must do: "**Describe**," "**Explain**," "**Compare**," or "**Contrast**." Be especially mindful of the word *or*. If you are asked to discuss "***either*** the nitrogen cycle ***or*** the carbon cycle," answer one. Ignore the other! The reader grades **only** the first answer written.

Organize Your Thoughts

Before you begin answering the questions in Section II of the exam, you may want to spend the first 10 minutes (of the 90 minutes total that are allotted for Section II) thinking, analyzing, making notes, and generally preparing to write your responses. Although you are allowed to begin writing immediately, this is unwise. The most successful students take the first 10 minutes to organize their thoughts.

Brainstorm and write down all **key words** you can think of that relate to each topic. Then, look over the key words, eliminate the ones that are not relevant, and prioritize the ones you will be writing about. Present your ideas, in order, *from the general to the particular*. After you've spent the first 10 minutes brainstorming, begin to write your responses.

Do Not Leave Out Basic Material

Many students think that a college-level response should contain only the most complex ideas. This is incorrect. Include everything you can think of that is related to the topic and that answers the question. Remember, you are trying to accumulate points by presenting relevant, correct statements.

Do Not Contradict Yourself

No points will be given if you give contradictory information. For example, you will receive no credit if you state, "The Calvin cycle occurs in the stroma of chloroplasts," and you also write, "The Calvin cycle occurs in the grana of chloroplasts."

Label Your Answers

Number each response and label all parts, such as 1a, 1b, 1c, and 1d. For readability, leave at least one line between your responses. If the reader cannot find your answer, you cannot get any credit.

Include Drawings If You Want

Drawings must be *titled* and *labeled* and must be near the text they relate to. However, you may *not* use drawings (or an outline) instead of writing a response.

Do Not Write Formal Literary-Style Essays

You do not need to include an introduction, a body, and a conclusion. Doing so is not expected and may take up too much time. Jump right in and answer the question.

Answer the Question, Then Move On

You do not have to include every piece of information about the topic to get full credit. Usually, each free-response question is very broad, and there are plenty of ways to get full credit. Remember the reader with the clicker, so just write, write, write!

PART 2: SUBJECT AREA REVIEW

Subject area review consists of all the subjects that are included in the latest AP Biology required curriculum, as per the College Board. These subjects, which are covered in this section of the book (Chapters 2 through 13), are Biochemistry; The Cell; Cell Cycle; Energy, Metabolism, and Enzymes; Cell Respiration; Photosynthesis; Heredity; The Molecular Basis of Inheritance; Biological Diversity; Evolution; Ecology; and Animal Behavior. These are the topic areas that you will most likely be tested on during your exam, and thus you should review all of these chapters closely to make sure that you fully grasp each concept before moving on to the next one.

The chapters in this part of the book revolve around the four Big Ideas of AP Biology:

- **EVO** (Evolution)
- **ENE** (Energetics)
- **IST** (Information Storage and Transmission)
- **SYI** (Systems Interactions)

The first page of each chapter identifies which Big Ideas will be covered in that particular chapter. In addition, each chapter also identifies the Enduring Understandings and/or Science Practices that will be covered throughout that chapter (in the form of sidebars, with helpful hints to remember, as well as end-of-chapter practice questions and answer explanations, to reinforce each concept from that chapter). As you review each chapter, be sure to ask yourself how each topic relates to topics that were covered in other chapters. For example, you can easily relate "cell structure" to "evolution" because all eukaryotic cells have internal membranes, which means that they descended from an ancient ancestral cell that also had internal membranes.

Remember, a lot of what you'll review in these chapters was likely covered in your AP Biology course, but now is the time to make sure that you've mastered each of these topics (and that you understand the connections between topics) before you take the exam. Best of luck with your review!

Biochemistry 2

- → ATOMIC STRUCTURE
- → BONDING
- → MOLECULES—POLAR AND NONPOLAR
- → HYDROPHOBIC AND HYDROPHILIC SUBSTANCES
- → PROPERTIES OF WATER
- → pH
- → ISOMERS
- → ORGANIC COMPOUNDS
- → PROTEIN FOLDING

Big Ideas: ENE, IST & SYI

Enduring Understandings: ENE-1; IST-1; SYI-1

For the complete list of Big Ideas and Enduring Understandings refer to the "AP Biology Course and Exam Description" from the College Board: *https://apcentral.collegeboard.org/pdf/ap-biology-course-and-exam-description.pdf?course=ap-biology.*

INTRODUCTION

You may think, "This is a biology course, not a chemistry course. Why am I studying chemistry?" Well, the reason you have to know some chemistry is because cells are sacs of chemicals. Biochemistry affects every aspect of our lives. The fact that sweating cools the skin is a function of the strong forces of attraction between water molecules. Your body maintains your blood at one critical pH because of the bicarbonate buffering system. To prevent heart attacks, you must begin with an understanding about the structure of fatty acids. Mad cow disease is caused by a misfolded protein. Here is a review of basic biochemistry.

ATOMIC STRUCTURE

Atoms are the building blocks of all **matter**. Atoms consist of subatomic particles: **protons**, **neutrons**, and **electrons**.

An atom in the elemental state always has a neutral charge because the number of protons (+) equals the number of electrons (–). Electron configuration is important because it determines how a particular atom will react with atoms of other elements. If all the electrons in an atom are in the lowest available energy levels, the atom is said to be in the **ground state**. When an atom absorbs energy, its electrons move to a higher energy level, and the atom is said to be in the **excited state**. For example, when a chlorophyll molecule in a photosynthetic plant cell absorbs light energy, the molecule becomes excited and electrons are boosted to a higher energy level. When an excited electron loses energy, it falls back to a shell closer to the nucleus.

Isotopes are atoms of one element that vary only in the number of neutrons in the nucleus. In nature, an element occurs as a mixture of isotopes. *Chemically, all isotopes of the same element are identical because they have the same number of electrons in the same configuration.* For example: carbon-12 and carbon-14 are isotopes of each other and are chemically identical. Some isotopes, like carbon-14, are radioactive and decay at a known rate called the **half-life** (see page 243). Knowing the half-life enables us to accurately measure the age of fossils or to estimate the age of the earth, a process known as radiometric dating. **Radioisotopes** (radioactive isotopes) are useful in many other ways. For example, **radioactive iodine** (I-131) can be used both to diagnose and to treat certain diseases of the thyroid gland. Additionally, a **tracer** such as radioactive carbon can be incorporated into a molecule and used to trace the path of carbon dioxide in a metabolic pathway.

BONDING

REMEMBER

Bond formation releases energy.

A bond is formed when two atomic nuclei attract the same electron(s). *Energy is released when a bond is formed because atoms acquire a more stable configuration by completing their outer shell. Energy must be supplied to break a bond.* There are two main types of bonds: **ionic** and **covalent**.

Ionic bonds result from the **transfer** of electrons. An atom that gains electrons becomes an **anion** (a **n**egative **ion**). An atom that loses an electron becomes a **cation**, a positive ion. The Na^+, Ca^{++}, and Cl^- ions are all necessary for normal nerve function.

Covalent bonds form when atoms **share** electrons. The resulting structure is called a **molecule**. A single covalent bond (–) results when two atoms share a pair of electrons. A double covalent bond (=) results when two atoms share two pairs of electrons. A triple covalent bond (≡) results when two atoms share three pairs of electrons. If electrons are shared *equally* between two identical atoms, the bond is a **nonpolar bond**. This is the type of bond found in **diatomic molecules**, such as H_2 (H–H) and O_2 (O=O). If electrons are shared *unequally*, the bond is referred to as a **polar covalent bond**. This is the case between any two different atoms, such as between atoms of carbon and oxygen in CO_2.

This bond is nonpolar or balanced:	H–H ↑	This bond is polar or unbalanced:	C–H ↑

MOLECULES—POLAR AND NONPOLAR

When two or more atoms form a bond, the entire resulting molecule is either **nonpolar** (balanced or symmetrical) or **polar** (unbalanced). CO_2 forms a linear molecule that is symmetrical or balanced and therefore nonpolar. It looks like this: O=C=O. H_2O is asymmetrical and a highly polar molecule. See Figure 2.1.

HYDROPHOBIC AND HYDROPHILIC SUBSTANCES

Hydrophilic means "water loving." Substances that are polar such as hydrochloric acid (HCl), or that carry a charge like the hydronium ion (H_3O^+), or that are ionic like table salt (NaCl) will dissolve in water. Since so many substances dissolve in water, water is known as the "universal solvent."

Hydrophobic means "water hating" and applies to nonpolar substances, which are miscible with or will dissolve in lipids. Salad dressing separates upon standing because oil (hydrophobic) and vinegar (hydrophilic, acetic acid solution in water) are not miscible.

Since the plasma membrane is a phospholipid bilayer, only nonpolar substances can readily dissolve through the plasma membrane. Large polar molecules must travel across a membrane in special hydrophilic (protein) channels.

PROPERTIES OF WATER

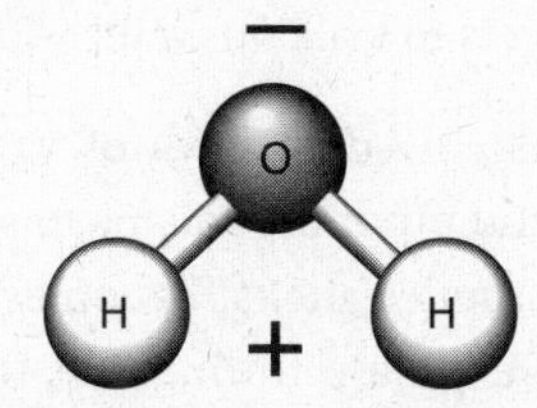

Figure 2.1 H_2O: A Polar Molecule

Because the oxygen atom in a molecule of **water** exerts a greater pull on the shared electrons than do the hydrogen atoms, one side of the molecule has a slightly negative charge and the other side has a slightly positive charge. The molecule is therefore asymmetrical and **highly polar**. See Figure 2.1. In addition, the positive hydrogen of one molecule is attracted to the negative oxygen of an adjacent molecule. The two molecules are held together by this **hydrogen bonding**. The strong hydrogen attraction that water molecules have for each other is responsible for the special characteristics of water that are important for life on Earth. See Figure 2.2, which shows the hydrogen atom of one molecule of water attracted to the oxygen atom of an adjacent molecule.

> **SYI-1**
> Living systems depend on the properties of water that result from its polarity and hydrogen bonding.

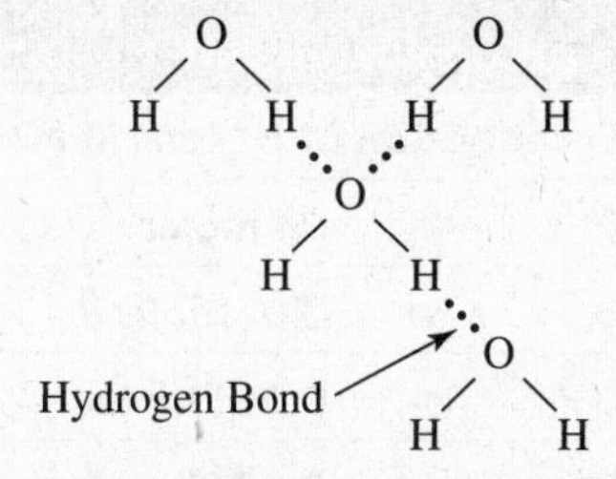

Figure 2.2 Water Molecules with Hydrogen Bonds

- **WATER HAS A HIGH SPECIFIC HEAT.** *Specific heat is the amount of heat a substance must absorb to increase 1 gram of the substance by 1°C.* Because water has a high specific heat, large bodies of water resist changes in temperature and provide a stable environmental temperature for the organisms that live in them. Furthermore, they moderate the climate of the nearby land. High specific heat is responsible for the fact that the marine biome has the most stable temperatures of any biome.
- **WATER HAS A HIGH HEAT OF VAPORIZATION.** Evaporating water requires the absorption of a relatively great amount of heat, so evaporation of sweat significantly cools the body surface.
- **WATER IS THE UNIVERSAL SOLVENT.** Because water is a highly polar molecule, it dissolves all polar and ionic substances.

- **WATER EXHIBITS STRONG COHESION TENSION.** This means that molecules of water tend to attract one another, which results in several biological phenomena.
  1. Water moves up a tall tree from the roots to the leaves without the expenditure of energy by what is referred to as **transpirational-pull cohesion tension**. As one molecule of water is lost from the leaf by transpiration, another molecule is drawn in at the roots.
  2. **Capillary action** results from the combined forces of cohesion and adhesion, attraction of unlike substances.
  3. **Surface tension** allows insects to walk on water without breaking the surface.
- **ICE IS LESS DENSE THAN WATER.** In a deep body of water, floating ice insulates the liquid water below it, allowing fish and other organisms to survive beneath the frozen surface during winter. In the spring, the ice melts, becomes denser, and sinks to the bottom of the lake, causing water to circulate throughout the lake. Oxygen from the surface is returned to the depths, and nutrients released by the activities of bottom-dwelling bacteria during winter are carried to the upper layers of the lake. This cycling of the nutrients in the lake is known as the **spring overturn** and is an important part of the life cycle of a lake.

pH

pH is a measure of the acidity and alkalinity of a solution. Anything with a pH of less than 7 is an acid, and anything with a pH value greater than 7 is alkaline or basic. A pH of 7 is neutral. *The value of the pH is the negative log of the hydrogen ion concentration in moles per liter.* See Table 2.1, which shows pH values compared with molarity.

Table 2.1

pH Compared with Molarity

pH	Concentration of H^+ ions in Moles per Liter		
1	1×10^{-1}	=	0.1 molar
2	1×10^{-2}	=	0.01 molar
3	1×10^{-3}	=	0.001 molar
4	1×10^{-4}	=	0.0001 molar
7	1×10^{-7}	=	0.0000001 molar
13	1×10^{-13}	=	0.0000000000001 molar

A substance with a pH of 3 has 1.0×10^{-3} or 0.001 mole per liter of hydrogen ions in solution, while a substance of pH 4 has a H^+ concentration of 1.0×10^{-4} or 0.0001 mole per liter of hydrogen ions in solution. Therefore, a solution of pH 3 is 10 times more acidic than a solution with a pH of 4. A solution with a pH of 6 is 1,000 times more acidic than a solution with a pH of 9; see Figure 2.3.

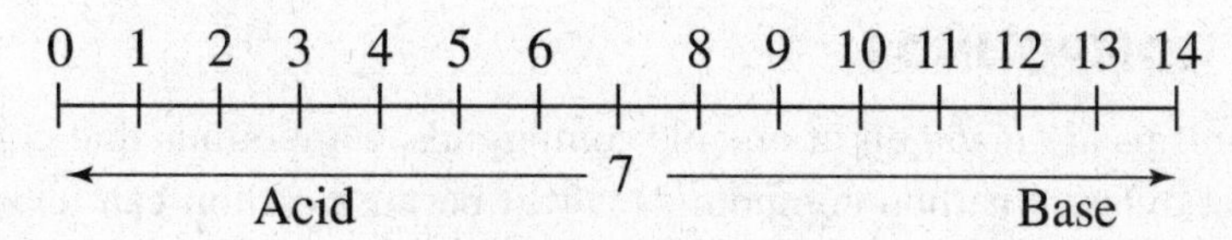

Figure 2.3 pH Scale

IT'S TRICKY

As the H+ concentration increases, the pH decreases.

The pH of some common substances is as follows:

Stomach acid	2
Human blood	7.4
Acid rain	1.5–5.4

The internal pH of most living cells is close to 7. Even a slight change can be harmful. Biological systems regulate their pH through the presence of **buffers**, substances that resist changes in pH. A buffer works by either absorbing excess hydrogen ions or donating hydrogen ions when there are too few. The most important buffer in human blood is the **bicarbonate ion**.

H^+ donor (acid)	Response to rise in pH	H^+ acceptor (base)		Hydrogen ion
H_2CO_3	$\rightleftharpoons$	HCO_3^-	+	H^+
Carbonic acid	Response to drop in pH	Bicarbonate ion		

ISOMERS

Isomers are organic compounds that have the same molecular formula but different structures. Because their structures are different, isomers have different properties. There are three types of isomers: **structural isomers**, ***cis-trans*** **isomers**, and **enantiomers**.

Structural isomers differ in the arrangement of their atoms. ***Cis-trans*** **isomers** differ only in spatial arrangement around double bonds, which are not flexible like single bonds are. **Enantiomers** are molecules that are mirror images of each other. The mirror images are called the L- (left-handed) and D- (right-handed) versions. Knowledge of enantiomers is important in the pharmaceutical industry because the two mirror images may not be equally effective. For example, L-dopa is a drug used in the effective treatment of Parkinson's disease. However, D-dopa, its enantiomer, is biologically inactive and useless in the treatment of the disease. For some reason that we do not understand, all the amino acids in cells are left-handed. See Figure 2.4.

MAKING CONNECTIONS
Here's an example of how a change in structure causes a change in function.

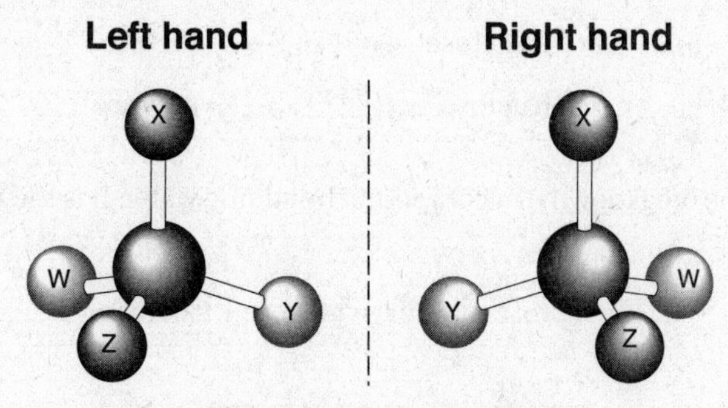

Figure 2.4 Mirror Image

ORGANIC COMPOUNDS

All living organisms are made up of organic compounds, compounds that contain carbon. The number of different carbon compounds is vast because carbon can form single, double, or triple covalent bonds. It can form molecules that are ring shaped, branched, or long chains. There are four classes of organic compounds: **carbohydrates**, **lipids**, **proteins**, and **nucleic acids**.

Carbohydrates

The body uses carbohydrates for fuel and as building materials. Carbohydrates consist of three elements: carbon, hydrogen, and oxygen. The ratio of the number of hydrogen atoms to the number of oxygen atoms in all carbohydrates is always 2 to 1. The empirical formula for all carbohydrates is C_nH_2O. The body uses carbohydrates for quick energy; 1 gram of any carbohydrate will release 4 calories when burned in a calorimeter. Dietary sources include rice, pasta, bread, cookies, and candy. There are three classes of carbohydrates you should know: monosaccharides, disaccharides, and polysaccharides.

Monosaccharides have a chemical formula of $C_6H_{12}O_6$. Three examples are **glucose**, **galactose**, and **fructose**, which are isomers of each other. The structural formula of glucose is shown in Figure 2.5. Notice the conventional numbering of the carbons in the rings. The numbering begins to the right of the oxygen.

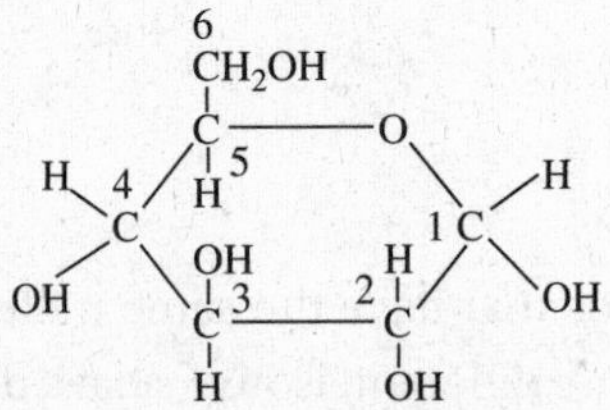

Figure 2.5 Glucose

Disaccharides have the chemical formula $C_{12}H_{22}O_{11}$. They consist of two monosaccharides joined together, with the release of one molecule of water, by the process known as **dehydration synthesis** or **condensation**. Here are the three condensation reactions of monosaccharides that produce the three disaccharides.

monosaccharide	+	**monosaccharide**	→	**disaccharide**	+	**water**
$C_6H_{12}O_6$	+	$C_6H_{12}O_6$	→	$C_{12}H_{22}O_{11}$	+	H_2O
glucose	+	glucose	→	**maltose**	+	water
glucose	+	galactose	→	**lactose**	+	water
glucose	+	fructose	→	**sucrose**	+	water

Hydrolysis is the breakdown of a compound by adding water. It is the reverse of condensation synthesis.

sucrose + water → glucose + fructose

Polysaccharides are macromolecules. They are polymers of carbohydrates, and are formed as many monosaccharides join together by dehydration reactions. There are four important polysaccharides, as shown in Table 2.2.

Table 2.2

Structural and Storage Polysaccharides		
	Structural	**Storage of Energy**
Found in plants:	Cellulose	Starch
	Makes up plant cell walls	Two forms are amylose and amylopectin
Found in animals:	Chitin	Glycogen
	Makes up the exoskeleton in arthropods (and cell walls in mushrooms)	"Animal starch." In humans, this is stored in liver and skeletal muscle

Lipids

Lipids are a diverse class of organic compounds that include **fats**, **oils**, **waxes**, and **steroids**. They are grouped together because they are all hydrophobic, meaning that they are not soluble in water. Structurally, most lipids consist of 1 glycerol and 3 fatty acids; see Figure 2.6.

REMEMBER

The cell membrane consists of phospholipids—lipid molecules with a phosphate attached.

```
     3 Fatty Acids            Glycerol
    H   H   H   H       O       H
    |   |   |   |       ||      |
H—C—C—C—C····C—O—C—H
    |   |   |   |               |
    H   H   H   H               |
                                |
    H   H   H   H       O       |
    |   |   |   |       ||      |
H—C—C—C—C····C—O—C—H
    |   |   |   |               |
    H   H   H   H               |
                                |
    H   H   H   H       O       |
    |   |   |   |       ||      |
H—C—C—C—C····C—O—C—H
    |   |   |   |               |
    H   H   H   H               H
```

Figure 2.6 Lipid

Glycerol is an alcohol and exists only as shown in Figure 2.7.

```
     H
     |
  H—C—OH
     |
  H—C—OH
     |
  H—C—OH
     |
     H
```

Figure 2.7 Glycerol

A **fatty acid** is a hydrocarbon chain with a carboxyl group at one end. Fatty acids exist in two varieties, **saturated** and **unsaturated**, as shown in Figure 2.8.

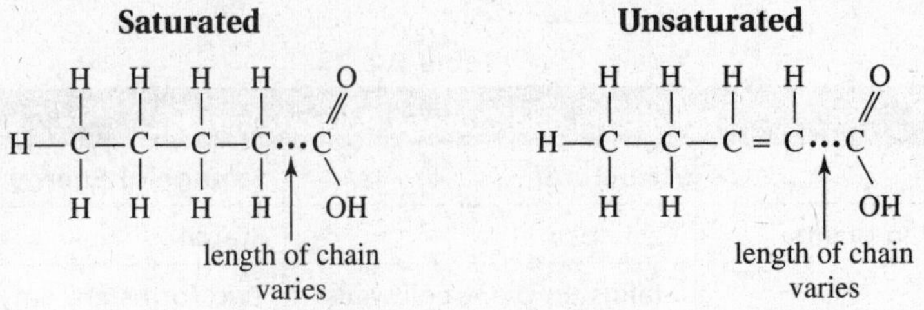

Saturated

```
    H   H   H   H     O
    |   |   |   |    //
H—C—C—C—C•••C
    |   |   |   |     \
    H   H   H   H     OH
```

length of chain varies

Unsaturated

```
    H   H   H   H     O
    |   |   |   |    //
H—C—C—C = C•••C
    |   |              \
    H   H              OH
```

length of chain varies

Figure 2.8 Fatty Acids

In general, saturated fats come from animals, are solid at room temperature, and when ingested in large quantities, are linked to heart disease. An example of a saturated fat is butter. Saturated fatty acids contain only *single bonds* between carbon atoms. Unsaturated fatty acids, in general, are extracted from plants, are liquid at room temperature, and are considered to be healthy dietary fats. Unsaturated fatty acids have at least one *double bond* formed by the removal of hydrogen atoms in the carbon skeleton. As a result, they hold fewer hydrogen atoms than saturated fatty acids. One exception is the group of tropical oils such as coconut and palm oil that are saturated, somewhat solid at room temperature, and are as unhealthy as are fats extracted from animals.

Steroids are lipids that do not have the same general structure as other lipids. Instead, they consist of four fused carbon rings. Figure 2.9 shows the steroid cholesterol. Other examples of steroids are testosterone and estradiol.

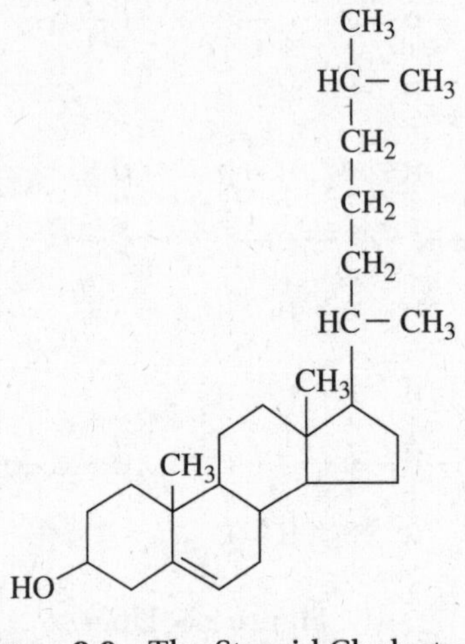

CH_3
HC— CH_3
CH_2
CH_2
CH_2
HC— CH_3
CH_3
CH_3
HO

Figure 2.9 The Steroid Cholesterol

Lipids serve many functions.

- **ENERGY STORAGE:** One gram of any lipid will release 9 calories per gram when burned in a calorimeter and produce twice the energy of 1 gram of carbohydrates or 1 gram of proteins.
- **STRUCTURAL:** Phospholipids (a lipid in which a phosphate group replaces one fatty acid) are a major component of the cell membrane. One steroid, cholesterol, serves as an important component of the plasma membrane of animal cells.
- **ENDOCRINE:** Some steroids are hormones.

Phospholipids

Phospholipids are modified lipids. They consist of only two fatty acids attached to the glycerol backbone, forming two *hydrophobic "tails."* The third hydroxyl group of the glycerol attaches to a phosphate group, which is charged and therefore hydrophilic. This *hydrophilic "head"* attracts water. When phospholipids are added to water, they self-assemble into a double-layered structure called a "bilayer." The phosphate is on the outside and the hydrophobic tails are on the inside, shielded from water. This phospholipid bilayer is the structural basis of all plasma membranes. It forms the boundary between the inside of the cell and the external environment. (See Figure 2.10 and Figure 3.8 of the plasma membrane.)

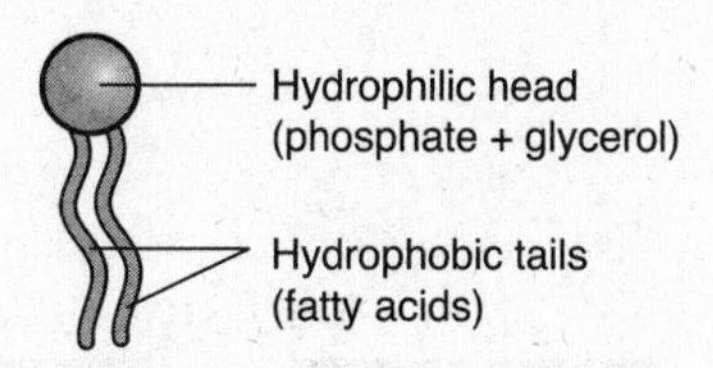

Figure 2.10 Phospholipid

Proteins

Proteins are complex, unbranched macromolecules that carry out many functions in the body including:

- Growth and repair
- Signaling from one cell to another
- Regulation: Hormones such as insulin lower blood sugar
- Enzymatic activity: Catalyzing chemical reactions
- Movement: Actin and myosin are protein fibers that are responsible for muscle contractions

Dietary sources of proteins include fish, poultry, meat, and certain plants like beans and peanuts. One gram of protein burned in a calorimeter releases 4 calories. Proteins consist of the elements S, P, C, O, H, and N. They are **polymers** or **polypeptides** consisting of units called **amino acids**, which are joined by **peptide bonds.**

Amino acids consist of a **carboxyl group**, an **amine group**, and a **variable (R)** all attached to a central asymmetric carbon atom. The R group, also called the side chain or variable, differs with each amino acid. The R groups are categorized by their chemical properties: hydrophobic, hydrophilic, acidic, or basic. The interactions among these R groups ultimately determine the overall structure and function of a protein. With only 20 different amino acids, cells can build thousands of different proteins; see Figure 2.11.

See Figure 2.12, which shows two amino acids combining by dehydration synthesis to form a **dipeptide**, a molecule consisting of two amino acids connected by one peptide bond. In the process, one molecule of water is released.

Each protein has a unique shape or **conformation** that determines what job it performs and how it functions. There are four levels of protein structure that are responsible for a protein's unique conformation. They are **primary**, **secondary**, **tertiary**, and **quaternary structures**; see Figure 2.13.

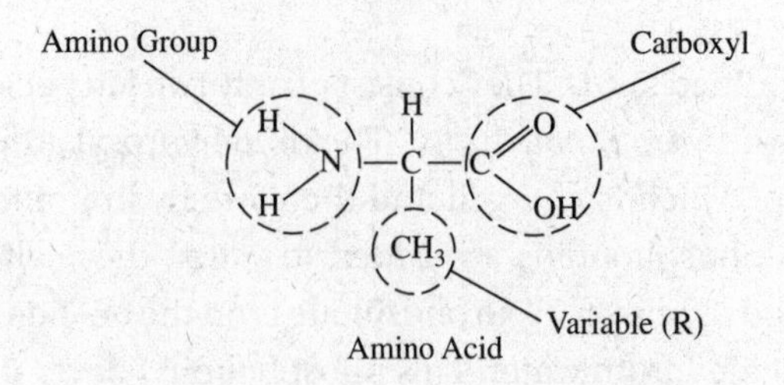

Figure 2.11 Amino Acid

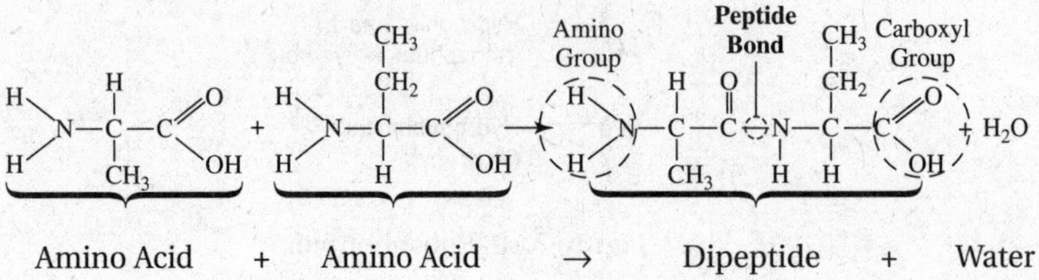

$$\underbrace{H_2N\text{-}CH(CH_3)\text{-}COOH}_{\text{Amino Acid}} + \underbrace{H_2N\text{-}CH(CH_2CH_3)\text{-}COOH}_{\text{Amino Acid}} \rightarrow \underbrace{H_2N\text{-}CH(CH_3)\text{-}CO\text{-}NH\text{-}CH(CH_2CH_3)\text{-}COOH}_{\text{Dipeptide}} + \underbrace{H_2O}_{\text{Water}}$$

Amino Group — Peptide Bond — Carboxyl Group

Figure 2.12 Formation of a Dipeptide by Dehydration Synthesis

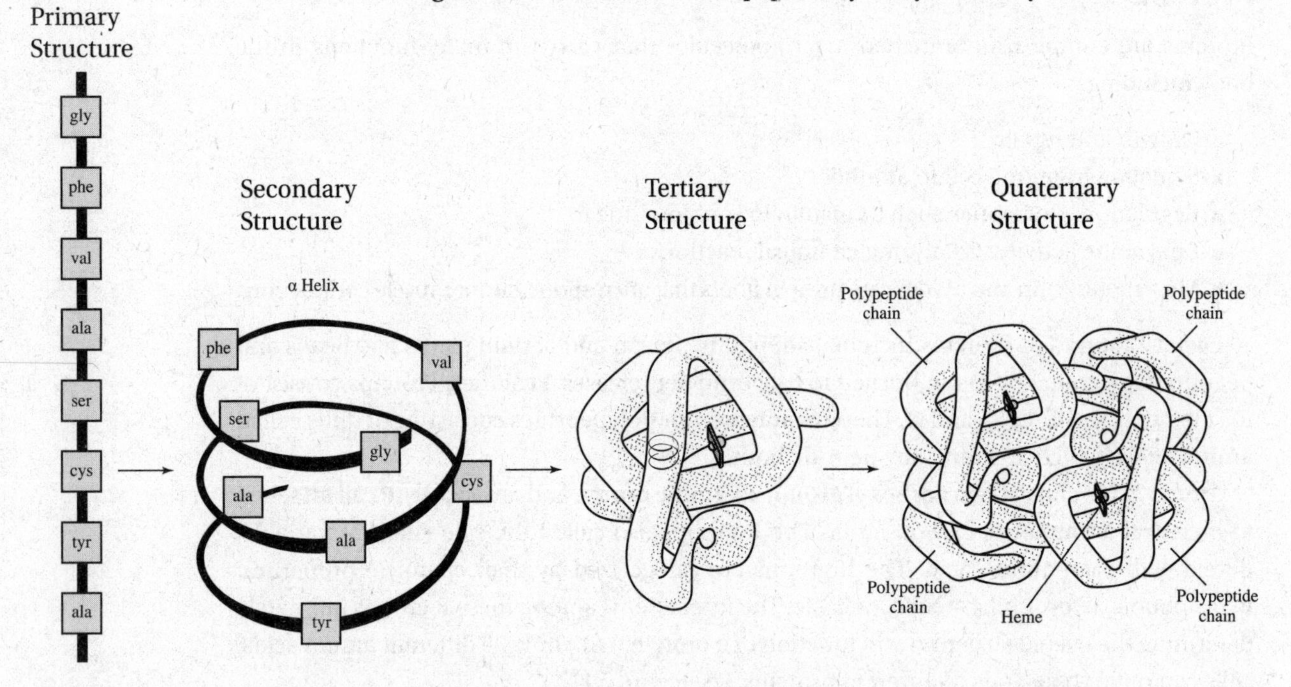

Figure 2.13 Four Levels of Protein Structure of Hemoglobin

The **primary structure** of a protein refers to the unique linear sequence of amino acids. The slightest change in the amino acid sequence of a protein can have major consequences. Such is the case with sickle cell anemia, a life-threatening condition that results from a substitution of one amino acid (valine) for another (glutamic acid) in a molecule of hemoglobin.

> **SYI-I**
> **Explain the connection between the sequence of the subcomponents of a polymer and its properties. Sickle cell anemia is a good example.**

While working in the 1940s and 1950s, Fred Sanger was the first to sequence a protein. That protein was **insulin**, and he received the Nobel Prize for his work.

The **secondary structure** of a protein results from **hydrogen bonding** within the polypeptide molecule. It refers to how the polypeptide coils or folds into two distinct shapes: an **alpha helix** or a **beta-pleated sheet**; see Figure 2.14.

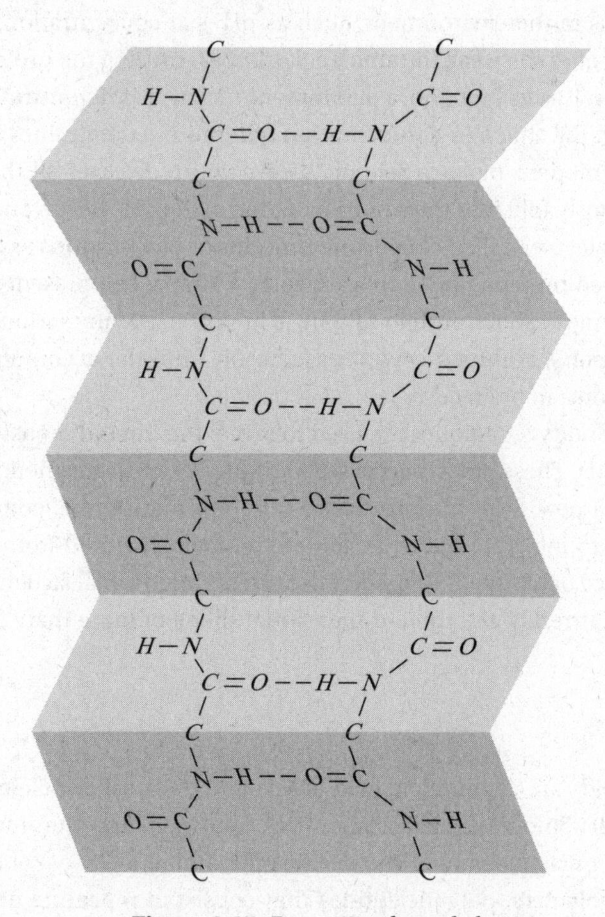

Figure 2.14 Beta- (β)-pleated sheet

Proteins that exhibit either alpha helix or beta-pleated sheet or both are called **fibrous proteins**. Examples of fibrous proteins are wool, claws, beaks, reptile scales, collagen, and ligaments. The protein that makes up human hair, **keratin**, is composed mostly of alpha helixes, while silk and spider webs consist of proteins made of beta-pleated sheets.

Tertiary structure is the intricate three-dimensional shape or conformation of a protein that is superimposed on its secondary structure. Tertiary structure determines the protein's **specificity**. The following intramolecular factors contribute to the tertiary structure:

- Hydrogen bonding between R groups of amino acids
- Ionic bonding between R groups
- Hydrophobic interactions
- Van der Waals interactions
- Disulfide bonds between cysteine amino acids

Quaternary structure refers to proteins that consist of more than one polypeptide chain. Hemoglobin exhibits quaternary structure because it consists of four polypeptide chains, each one forming a heme group.

PROTEIN FOLDING

REMEMBER

A change in structure causes a change in function.

Under normal cellular conditions, the primary structure of a protein determines how it folds into its particular three-dimensional shape. Protein structure also depends on physical and chemical conditions in the environment, such as pH, salt concentration, and temperature. Adverse conditions alter the weak intramolecular forces, causing the protein to lose its characteristic shape as well as its function, a phenomenon known as **denaturation** or **denaturing**.

The concept that the shape or **conformation** of a protein determines how it functions is a basic concept of modern biology. Scientists have yet to discover all the rules about how proteins spontaneously fold into their proper conformation. However, one important recent discovery is that molecules called **chaperone proteins** or **chaperonins** assist in folding other proteins. A misfolded protein, one with an incorrect shape, can present a serious problem for a cell. Learning more about protein folding is important. Many serious diseases, such as Alzheimer's, Parkinson's, and mad cow disease, result from the accumulation of misfolded proteins, called **prions**, in brain cells.

Three complementary techniques are used to reveal the three-dimensional shape or conformation of proteins. These are **X-ray crystallography**, nuclear magnetic resonance (NMR) spectroscopy, and a new field, **bioinformatics**. Bioinformatics uses computers and mathematical modeling to integrate the huge volume of data generated from the analysis of an amino acid sequence of a protein to predict the three-dimensional structure of the resulting protein molecule. Currently, the three-dimensional shape of more than 20,000 proteins has been determined.

Nucleic Acids

The two nucleic acids are **ribonucleic acid (RNA)** and **deoxyribonucleic acid (DNA)**. They encode all hereditary information. Through RNA intermediates, the information encoded in the sequence of nucleotides in DNA specifies the amino acid sequences of all proteins. Nucleic acids are polymers, polynucleotides that consist of repeating units called **nucleotides**. A nucleotide consists of a **phosphate (P)**, a **5-carbon sugar—deoxyribose** or **ribose**, and a **nitrogen base—adenine (A)**, **cytosine (C)**, **guanine (G)**, or either **thymine (T)** (in DNA) or **uracil (U)** (in RNA). The carbon atoms of deoxyribose are numbered from 1 to 5. DNA consists of two strands that run in opposite directions. One strand runs 5′ to 3′, and the other runs 3′ to 5′. See Figure 2.15. Nucleic acids are also discussed in Chapter 8.

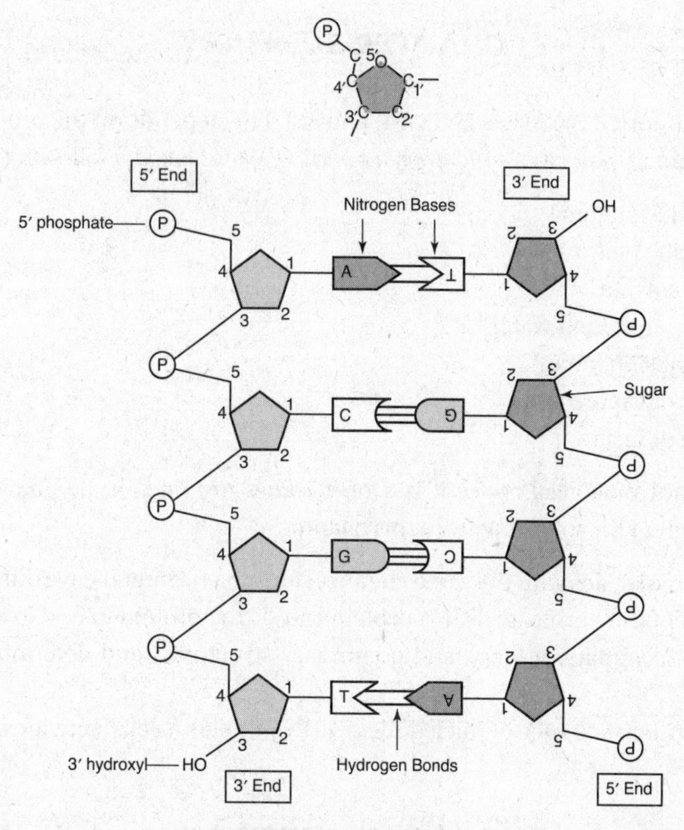

Figure 2.15 DNA

Functional Groups

The components of organic molecules that are most often involved in chemical reactions are known as **functional groups**. These groups are attached to the carbon skeleton, replacing one or more hydrogen atoms that would be present in a hydrocarbon. For example, the difference between testosterone and estradiol is the functional group attached to the carbon skeleton. Each functional group behaves in a consistent fashion from one organic molecule to another. Table 2.3 shows some common functional groups.

Table 2.3

Functional Groups in Organic Compounds

Group	Formula	Name of Compound
Amino	$R-NH_2$	Amine
Carboxyl	$R-C(=O)OH$	Carboxyl (acid)
Hydroxyl	$R-OH$	Alcohol
Phosphate	$R-O-P(=O)(O^-)-O^-$	Phosphate

CHAPTER SUMMARY

This chapter supported Big Ideas: ENE, IST, and SYI. Life depends on the properties of water, which result from its *polarity* and *hydrogen bonding*. Water has the following properties:

- Strong cohesion force
- High specific heat
- Universal solvent
- Ice is less dense than water
- Strong adhesion force
- High heat of vaporization
- High heat of fusion

The all-important biological concept is—*form relates to function*. Be able to give specific examples of each of the following from this chapter:

- Different molecules with the same chemical formula (isomers) have different functions.
- The sequence of amino acids in a protein causes the protein to fold in a particular way (primary, secondary, tertiary, and quaternary structures) and determine the protein's function.
- The particular sequence of nucleotides in DNA directs a cell's activities.

PRACTICE QUESTIONS

1. Which of the following claims about a particular molecule is correct and gives the correct justification for your answer?

   (A) CH_4 is a nonpolar molecule because the bonds between carbon and the four hydrogen atoms are polar.
   (B) O_2 is a polar molecule because the pull on electrons shared between the oxygen atoms is identical.
   (C) CO_2 is a nonpolar molecule because oxygen exerts greater pull on the shared electrons than does carbon.
   (D) H_2O is a polar molecule because part of the molecule is partially positively charged while the remaining part is partially negatively charged.

2. Clothes washing detergent must be able to remove grease and oil stains from clothing while also dissolving and washing away in the rinse water. Which of the following is the correct structure of the detergent molecules and the correct reason for it?

   (A) Detergent molecules are polar and are therefore able to dissolve both grease and water.
   (B) Detergent molecules are nonpolar and are therefore able to dissolve in both grease and water.
   (C) One end of a detergent molecule is nonpolar in order to dissolve oily stains, while the other end is polar to dissolve in the rinse water.
   (D) One end of a detergent molecule is polar in order to dissolve oily stains, while the other end is nonpolar to dissolve in the rinse water.

3. Proteins are formed in a ribosome as amino acids are strung together in a particular sequence determined by instructions from DNA. Here is a diagram of an experiment to investigate protein production and function.

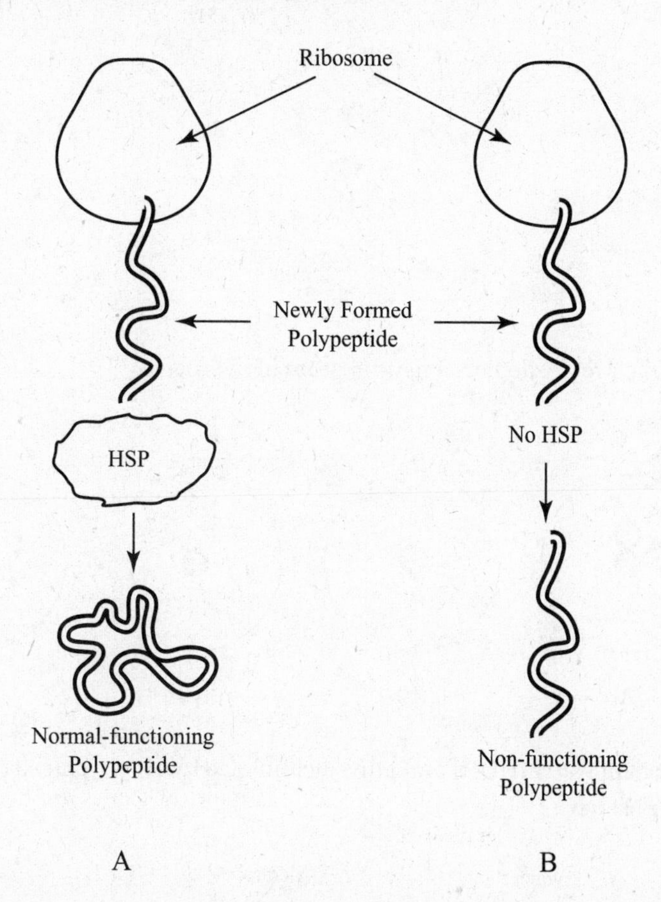

In both trials, newly formed polypeptide is released from the ribosome. In Experiment A, the newly formed polypeptide comes in contact with a heat shock protein (HSP). In Experiment B, the newly formed polypeptide is not exposed to HSP. Both trials are identical except for the presence or absence of HSP.

What is the best explanation for what is demonstrated by this experiment?

(A) HSP is critical to normal ribosome function.
(B) Proper folding of a polypeptide is critical for a protein to function properly.
(C) HSP can cause mutations in DNA, with the result that a normal protein does not form.
(D) Proteins can denature in excessive heat.

4. Which of the following pairs of base sequences could form a short stretch of DNA?

(A) 5′-TTAA-3′ with 3′-AATT-5′
(B) 5′-ACTG-3′ with 3′-ACTG-5′
(C) 5′-ATTC-3′ with 3′-AUUG-5′
(D) 5′-TTAA-3′ with 5′-AATT-3′

5. Here is a sketch of two amino acids—the building blocks of polypeptides. When they combine into a dipeptide, certain atoms are removed. Which atoms are removed?

```
                                    CH3
                                    |
H      H     O       H              CH2    O
  \    |    //         \            |     //
   N — C — C      +     N — C — C          →
  /    |    \          /    |    \
H      CH3   OH      H      H     OH
```

(A) OHO
(B) HOH
(C) NH_2
(D) CH_3

6. Which model correctly depicts a neutral atom of hydrogen?

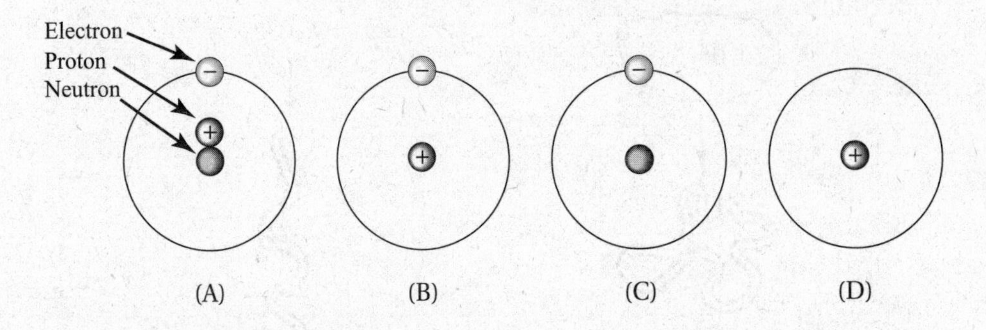

7. One drop each of water (HOH) and ethyl alcohol (C_2H_5OH) are placed onto a glass slide. See the sketch.

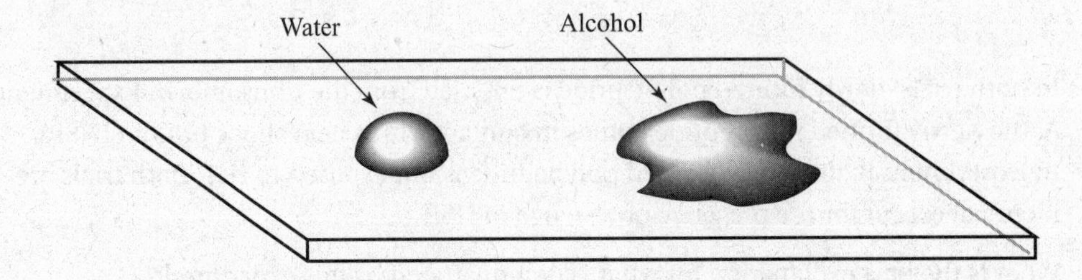

What is the best explanation for the difference in appearance of these two liquids?

(A) Water molecules exhibit greater cohesion tension than do those of ethyl alcohol.
(B) Water molecules exhibit less cohesion tension than do those of ethyl alcohol.
(C) Water molecules exhibit greater adhesion tension than do those of ethyl alcohol.
(D) There is greater hydrogen bonding *within* a molecule of water than *within* a molecule of ethyl alcohol.

8. In 1957 after rigorous testing, a pharmaceutical company in West Germany introduced a new drug to the market as a sedative and sleeping drug for pregnant women. It was called thalidomide and was very effective. Doctors in 46 countries prescribed it to millions of women. Tragically, the drug was responsible for thousands of babies being born with missing or abnormal arms, hands, legs, or feet to mothers who had taken the drug while pregnant. The product was subsequently removed from the market even though the pharmaceutical company proved that the drug contained pure thalidomide with *no* contaminants. Which of the following statements best explains the most likely problem with the drug?

   (A) Some mutations do not show up for several generations.
   (B) The drug contained contaminants that caused mutations in the growing fetuses.
   (C) Enzyme activity in the stomachs of women who took the drug altered the conformation of the drug, making it toxic.
   (D) The drug contained molecules that were isomers and mirror images of each other. One version worked as expected, while the other caused birth defects.

9. Isotopes are atoms of a given element that have the same number of protons but vary in the number of neutrons. The element carbon has three isotopes. Carbon-12 accounts for 99% of all carbon in nature. Which of the following is correct about isotopes of carbon?

   (A) They are all radioactive.
   (B) They contain the same number of neutrons but a different number of protons.
   (C) They contain the same number of electrons but are chemically different because the number of neutrons is different.
   (D) They are chemically identical because they have the same number of electrons.

10. Refer to the figure below.

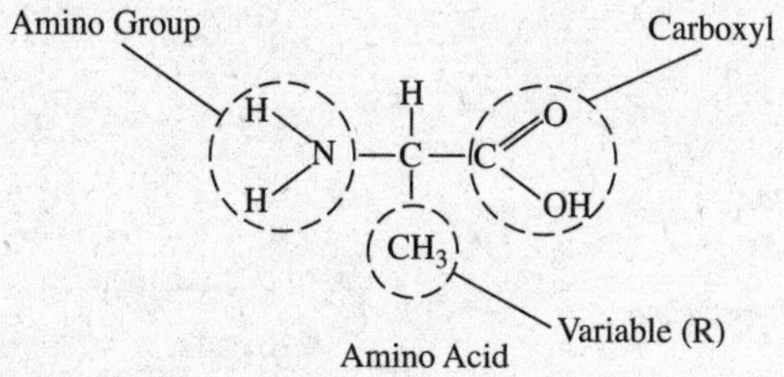

   How are these two amino acids attached together to form a dipeptide?

   (A) carbon atom to carbon atom
   (B) carbon atom to amino group
   (C) amino group to amino group
   (D) amino group to carboxyl group

11. The pH of human blood is normally 7.4. Which of the following correctly explains what that means?

   (A) Blood is slightly acidic.
   (B) Blood carries more OH^- ions than H^+ ions.
   (C) Blood carries more H^+ ions than OH^- ions.
   (D) Blood pH can fluctuate around 7.4 throughout the day.

12. The molecular formula for glucose is $C_6H_{12}O_6$. Which of the following is the molecular formula for an isomer of glucose?

(A) CHO
(B) $C_n (H_2O)$
(C) $C_{12}H_{24}O_{12}$
(D) $C_6H_{12}O_6$

13. Which of the following is stored in the human liver to be used later as a source of energy?

(A) glycine
(B) glycogen
(C) glycerol
(D) glucagon

14. Specific heat is defined as the amount of heat that must be absorbed or lost for 1 g of that substance to change its temperature by 1°C. Specific heat can also be thought of as a measure of how well a substance resists changing its temperature when it absorbs or releases heat. The specific heat of water is high compared to that of other materials. Therefore, a large body of water like an ocean exerts an effect on the local climate.

Which of the following is most likely true about the temperature for Long Beach on this day, given the data on the following map of coastal California?

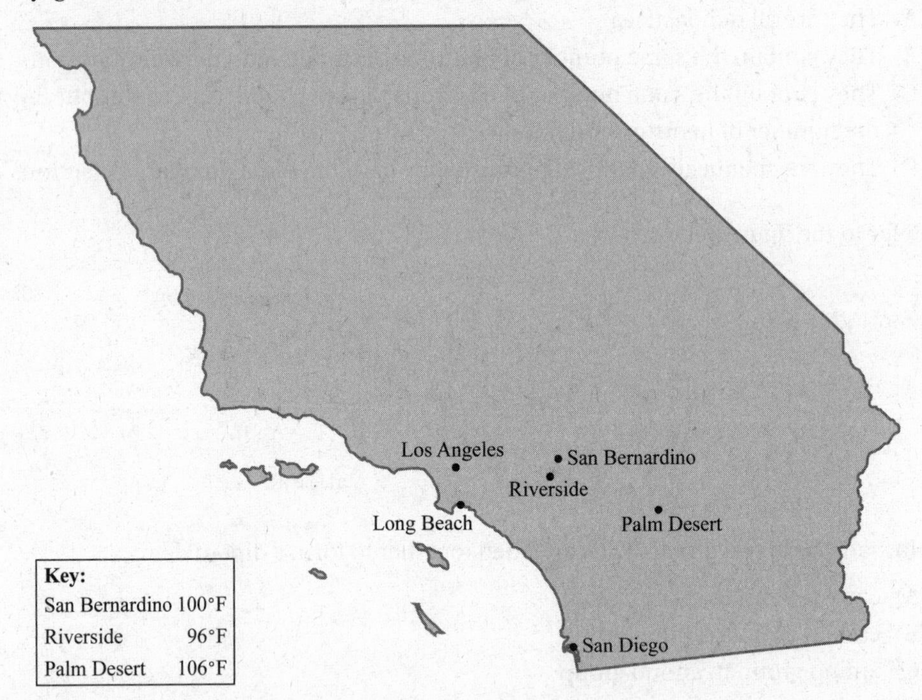

(A) It would be about the same as the temperature for Palm Desert because they are at about the same latitude.
(B) It would be considerably lower than the temperature for Palm Desert because Long Beach is in a depression in the earth.
(C) It would be considerably lower than the temperature for Palm Desert because Long Beach is cooled by the Pacific Ocean.
(D) It cannot be determined by guessing; it must be measured.

Answers Explained

1. **(D)** A *molecule* with an uneven or a lopsided distribution of charges is considered a polar molecule. A molecule where the charges are balanced and distributed equally is nonpolar. Although each C–H bond within a molecule of CH_4 is polar, the arrangement of the 4Hs around the lone carbon is balanced, making the molecule nonpolar. The bonds between oxygen atoms in O_2 are nonpolar, and the overall molecule is also balanced and nonpolar. A molecule of CO_2 contains polar bonds between the carbon and oxygen atoms, but the arrangement of C and Os (O=C=O) is balanced. Therefore the molecule is balanced and nonpolar. Water is a strongly polar molecule because oxygen attracts the hydrogen atoms such that one side of the molecule is partially positive while the other side is partially negative. See Figure 2.1. (SYI-1)

2. **(C)** Oily stains will dissolve only in nonpolar solvents. Each molecule of laundry detergent is constructed so that one end of the molecule is nonpolar and will dissolve the stain. The other end of the molecule is polar and will be soluble in the rinse water. This is better living through chemistry! (SYI-1)

3. **(B)** Notice that the protein in Experiment A is folded, but the protein in Experiment B remains unchanged and unfolded. The heat shock protein must be responsible for the folding that enabled the protein to function. HSP is acting like a *chaperone protein.* In both cases, the ribosome produces a protein. There is nothing wrong with the ribosome. HSP does not come in contact with the DNA in this experiment and therefore does not cause mutations. Although it is correct that heat can denature an enzyme, no heat is applied in this experiment. (SYI-1)

4. **(A)** DNA strands are complementary to each other. The two strands of the helix run opposite each other; one runs 3′ to 5′, and the other runs 5′ to 3′. A bonds with T; C bonds with G. Eliminate choice B because the bases are identical in the two strands, not complementary. Eliminate choice C because U (Uracil) is found only in RNA, not DNA. Eliminate choice D because the two strands both run 5′ to 3′. (IST-1)

5. **(B)** Monomers are combined by dehydration synthesis or condensation, in which a molecule of water is removed. (SYI-1)

6. **(B)** An atom of hydrogen consists of one proton (+) in the nucleus and 1 electron (–) in orbit around the nucleus. (SYI-1)

7. **(A)** The drop of water beads up, while the drop of alcohol does not. The reason is that the molecules of water are strongly attracted to each other by hydrogen bonding and also because water molecules are strongly polar—the negative side of one water molecule attracts the positive side of an adjacent water molecule. These strong intermolecular attractions result in water having strong cohesion tension. Molecules of alcohol do not have such strong attractions for each other. Choice D is incorrect because attraction is between molecules, not within them. (SYI-1)

8. **(D)** This scenario was the case with the drug thalidomide. The terrible outcome did not take several generations to surface. The stem of the question states that there were no contaminants in the drug. Only a small percentage (thousands out of millions) of babies were born with birth defects. Therefore, choice C is probably not reasonable. (SYI-1)

9. **(D)** Since they have the same number of electrons, isotopes of the same element are chemically identical. Only some of the isotopes of carbon (such as C-14) are radioactive. Isotopes vary only in the number of neutrons. (SYI-1)

10. **(D)** The carbon from one carboxyl group from one amino acid combines with the nitrogen of the amino group of the other amino acid. (SYI-1)

11. **(B)** Blood pH is always 7.4. The fact that it does not vary is an example of homeostasis. A pH of 7.4 is slightly basic or alkaline. That means it has more OH^- ions than H^+ ions. (ENE-1)

12. **(D)** Isomers have the same number of atoms of each element; they vary only in the arrangement of those elements. Choice B is not correct because that is an empirical formula, not a molecular formula. However, choice B is the empirical formula for any monosaccharide. (SYI-1)

13. **(B)** Glycogen is a polysaccharide that stores sugars in the liver and skeletal muscle. Glucose is not stored; it is produced and used up constantly. Glycerol and fatty acids make up lipids. Glucagon is a hormone that is responsible for breaking down glycogen into glucose. Glycine is the simplest amino acid. (SYI-1)

14. **(C)** The high specific heat of the water of the Pacific Ocean exerts an effect on land temperatures. It moderates them. (SYI-1)

The Cell

3

→ STRUCTURE AND FUNCTION OF THE CELL
→ TRANSPORT INTO AND OUT OF THE CELL
→ CELL COMMUNICATION
→ APOPTOSIS—PROGRAMMED CELL DEATH

Big Ideas: EVO, ENE, IST & SYI
Enduring Understandings: EVO-1 & EVO-2; ENE-1, ENE-2 & ENE-3; IST-3; SYI-1
Science Practices: 2 & 3

For the complete list of Big Ideas, Enduring Understandings, and Science Practices, refer to the "AP Biology Course and Exam Description" from the College Board: *https://apcentral.collegeboard.org/pdf/ap-biology-course-and-exam-description.pdf?course=ap-biology.*

INTRODUCTION

All organisms on Earth are believed to have descended from a common ancestral cell about three and a half billion years ago. According to the **theory of endosymbiosis**, eukaryotic cells emerged when mitochondria and chloroplasts, once free-living prokaryotes, took up permanent residence inside other larger cells, about one and a half billion years ago. Here was the advent of the radically more complex **eukaryotic cell** with internal membranes that compartmentalized the cell and led to the rapid evolution of multicelled organisms.

Modern cell theory states that all organisms are composed of cells and that all cells arise from preexisting cells. Most animal and plant cells have diameters between 10 and 100 μm (microns or micrometers), although many, like human red blood cells with a diameter of only 8 μm, are smaller.

All cells share certain characteristics. They are all enclosed by a protective and selective barrier called a plasma membrane. They all contain a semifluid substance called **cytosol** in which subcellular components are suspended. Finally, all cells contain ribosomes and genetic material in the form of DNA.

There are two types of cells: prokaryotes and eukaryotes. Prokaryotes are simple cells that contain no nuclei or other internal membranes. All bacteria are prokaryotic cells. Instead of a nucleus, they have a nucleoid region, which is a non-membrane-bound region where

EVO-1 & ENE-2
Be able to describe the difference between prokaryotic and eukaryotic cells.

the chromosome is located. Eukaryotic cells have a nucleus bound by a double membrane. Eukaryotic cells also have organelles and internal membranes that compartmentalize each cell so that complex chemical reactions can be carried out efficiently in separate regions of a cell. All cells of the human body are eukaryotic cells; see Figures 3.1, 3.5 and Table 3.1.

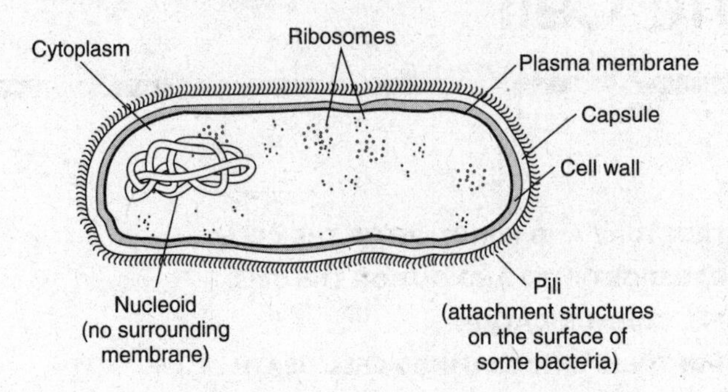

Figure 3.1 Typical Prokaryotic Cell

Table 3.1

Comparison of Cell Types	
Prokaryotes	**Eukaryotes**
No internal membranes: no nuclear membrane, ER, mitochondria, vacuoles, or other organelles	Contain distinct membrane-bound organelles
Circular, naked DNA	DNA wrapped with histone proteins into chromosomes
Ribosomes are very small	Ribosomes are larger
Metabolism is anaerobic or aerobic	Metabolism is aerobic
Cytoskeleton absent	Cytoskeleton present
Mainly unicellular	Mainly multicellular with differentiation of cell types
Cells are very small: 1–10 μm	Cells are larger: 10–100 μm

STRUCTURE AND FUNCTION OF THE CELL

A major theme in biology (and therefore a common free-response question topic) is that *function dictates form* and vice versa. As a result, one would expect that all cells do not look alike. And they do not. The nerve cell, whose purpose is to send electrical impulses, is long and spindly. Cells that store fat are rounded, large, and distended. Cells that make up a tough peach pit resemble square building blocks. Figure 3.2 shows a sketch of different cell types, each with a different overall appearance suited for each different function.

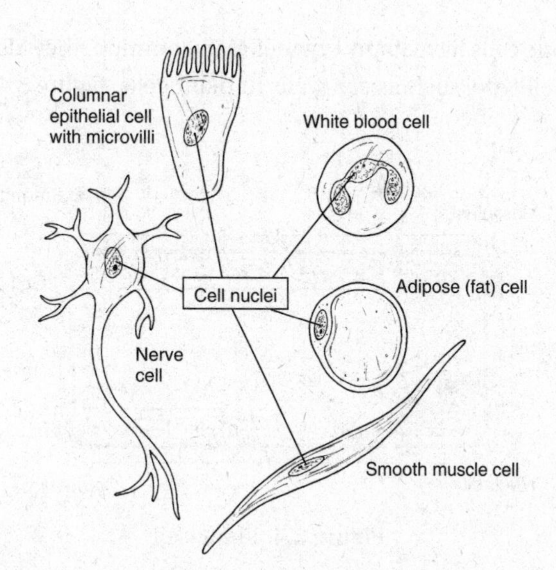

Figure 3.2 Five Different Cell Types

KEEP THIS IN MIND

Form and function go together.

Look for examples as you read this book.

Why Cells Are Small

Although various cells have different characteristics, they are all tiny. This is because the *surface area* of the cell membrane must be able to accommodate the metabolic needs, based on the *volume*, of the cytoplasm.

> **ENE-1**
> Be able to use calculated surface area-to-volume ratios to predict which cell might eliminate wastes or procure nutrients faster.

Compare the two "cells" in Figure 3.3. The ratio of surface area of the cell membrane compared to the total volume of the 1 mm cube (smaller cell) is 6 to 1. The ratio of surface area to volume of the 5 mm cube (larger cell) is only slightly more than 1 to 1. Clearly, the cell membrane of the smaller cell will be faster and more efficient at supplying needed materials and removing waste than the cell membrane of the larger cell. This is the reason why multicelled organisms consist of millions of tiny cells carrying out specialized functions rather than one gigantic all-purpose cell. In fact, as a cell grows and the volume of the cytoplasm becomes too great for the area of the cell membrane, a pathway is triggered and the cell divides in two.

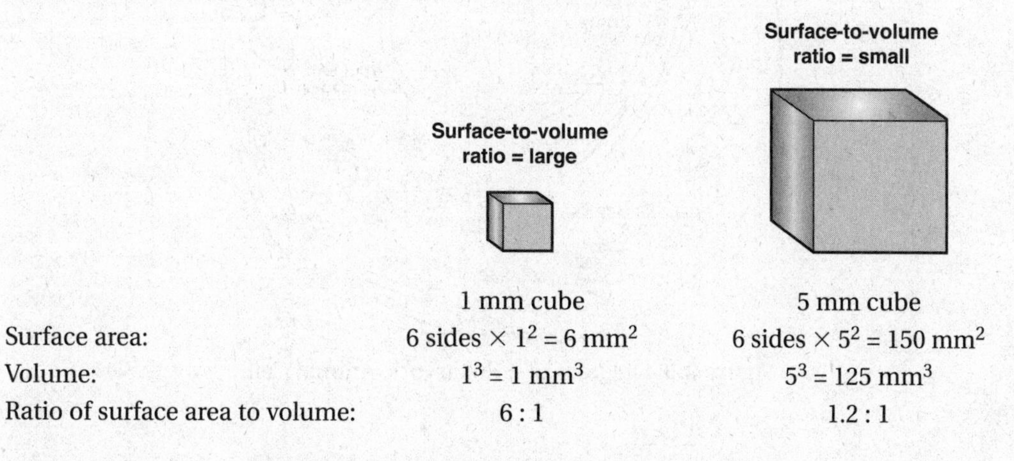

	1 mm cube	5 mm cube
Surface area:	6 sides $\times 1^2 = 6$ mm^2	6 sides $\times 5^2 = 150$ mm^2
Volume:	$1^3 = 1$ mm^3	$5^3 = 125$ mm^3
Ratio of surface area to volume:	6 : 1	1.2 : 1

Figure 3.3 Comparison of Small and Large Cells

Although eukaryotic cells have many organelles in common, they also have organelles that are unique to each cell type, such as cell walls in plant cells. Figure 3.4 shows a typical plant cell.

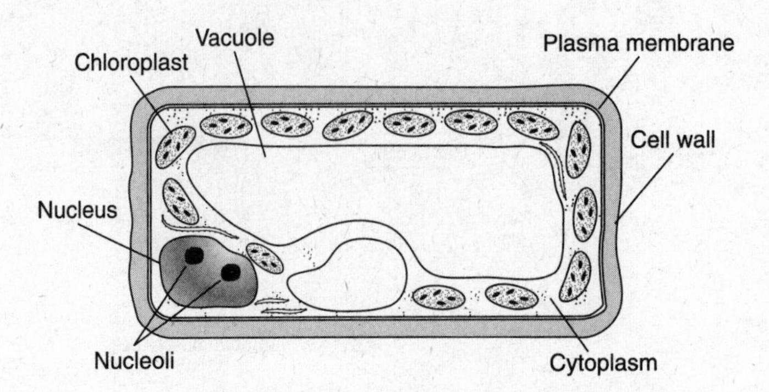

Figure 3.4 Plant Cell

The human body consists of approximately two hundred different cell types, each with a different function and, therefore, a different form. Although different cell types have different appearances, they all contain the same organelles. See Figure 3.5, a sketch of a typical animal cell.

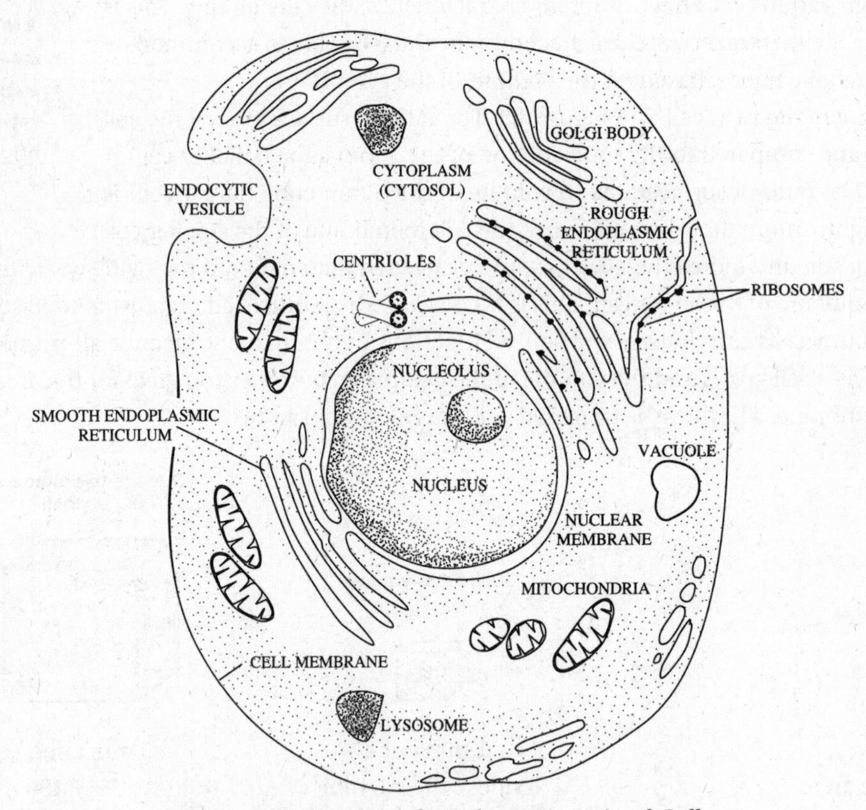

Figure 3.5 Diagram of a Eukaryotic Animal Cell

Nucleolus

The nucleus of a nondividing cell contains one or more prominent nucleoli, where ribosomal RNA (rRNA) is synthesized according to instructions from the DNA. Large and small subunits of ribosomes are also assembled there. A nucleolus combines proteins imported from the cytoplasm with rRNA made in the nucleolus. Nucleoli are not membrane-bound structures but are actually a tangle of chromatin and unfinished ribosomal precursors.

Ribosomes

Ribosomes are protein factories. They are not membrane-bound, and as such, are not considered organelles. Ribosomes can be found freely suspended in the cytosol or bound to the endoplasmic reticulum. Free ribosomes are associated with protein produced for the cell's own use, while ribosomes attached to the endoplasmic reticulum are meant for export out of the cell. Cells, like pancreatic cells, that produce huge amounts of hormones and digestive enzymes contain millions of ribosomes.

Peroxisomes

Peroxisomes are found in both plant and animal cells and perform a specialized function. They contain **catalase**, which converts hydrogen peroxide (H_2O_2), a waste product of respiration in the cell, into water with the release of oxygen atoms. They also detoxify alcohol in liver cells.

Endomembrane System

The endomembrane system regulates protein traffic and performs metabolic functions in cells. It includes the nuclear envelope, endoplasmic reticulum, Golgi apparatus, lysosomes, vesicles, vacuoles, and plasma membrane.

ENE-2 & SYI-1
Internal membranes facilitate cell processes by minimizing competing interactions and by increasing surface area where reactions can occur.

Nucleus

The nucleus contains chromosomes that are wrapped with special proteins into a **chromatin network**. The nucleus is surrounded by a selectively permeable **nuclear envelope** that separates the contents of the nucleus from the cytoplasm. The nuclear envelope contains **pores** to allow for the transport of molecules, like messenger RNA (mRNA), which are too large to diffuse directly through the envelope. (See Figure 9.14.)

Endoplasmic Reticulum

The endoplasmic reticulum (ER) is a membranous system of channels and flattened sacs that traverse the cytoplasm and account for more than half the total membranes in a eukaryotic cell. There are two types:

- **Rough ER** is studded with ribosomes and produces proteins.
- **Smooth ER** has three diverse functions:
  1. Assists in the synthesis of steroid hormones, like sex hormones, and of other lipids
  2. Stores Ca^{++} ions in muscle cells to facilitate normal muscle contractions
  3. Detoxifies drugs and poisons from the body

Golgi Apparatus

The **Golgi apparatus** lies near the nucleus and consists of flattened **membranous sacs**, called **cisternae**, stacked next to one another. The two sides of a Golgi stack are referred to as a ***cis* face** (receiving department—near the ER) and a ***trans* face** (shipping department—away from the ER). Transport **vesicles** from the ER carry material to the *cis* face where they enter the Golgi and are **processed** and **packaged** into **vesicles**. Next, they are shipped out from the *trans* face to various parts of the cell. A cell may have many or even hundreds of these Golgi stacks.

Lysosomes

Lysosomes are sacs of **hydrolytic** (digestive) **enzymes** surrounded by a single membrane. They are the principal site of **intracellular digestion**. With the help of the lysosome, the cell continually renews itself by breaking down and recycling cell parts, a process called **autophagy**. Programmed destruction of cells (**apoptosis**) by their own hydrolytic enzymes is a critical part of the development of multicelled organisms. Lysosomes are not generally found in plant cells.

Mitochondria

The mitochondria are the site of cellular respiration. All cells have many mitochondria; a very active cell could have 2,500 of them. Mitochondria have an outer double membrane and an inner series of membranes called **cristae**. Mitochondria also contain their own DNA. Mitochondria constantly divide and fuse with each other in order to exchange DNA and compensate for one another's defects. This is necessary for normal mitochondrial function, including respiration, cell development, and apoptosis.

The fact that mitochondria contain their own DNA is a major support for the **endosymbiotic theory**. This theory postulates that mitochondria were once free-living prokaryotic cells that took up permanent residence inside larger prokaryotic cells billions of years ago.

Vacuoles

Vacuoles are membrane-bound structures used for storage. They are large vesicles derived from the ER and Golgi apparatus. Mature plant cells generally have a single large **central vacuole**. Many freshwater protists have **contractile vacuoles** to pump out excess water. **Food vacuoles** are formed by the **phagocytosis** of foreign material.

Chloroplast

Chloroplasts contain the green pigment **chlorophyll** that, along with enzymes, absorbs light energy and synthesizes sugar. They are found in plants and algae. In addition to a double outer membrane, they have another inner membrane system called **thylakoids**. (See page 126 for details.)

According to the **theory of endosymbiosis**, chloroplasts were once tiny, free-living prokaryotic cells that were engulfed by a larger prokaryotic cell. Eventually, the engulfed cell became a permanent resident, and the two became one entity. A major piece of evidence for this theory is that chloroplasts have their own DNA that resembles bacterial DNA, not eukaryotic nuclear DNA.

Cytoskeleton

The cytoskeleton of the cell is a complex mesh of protein filaments that extends throughout the cytoplasm. The cytoskeleton has several important roles.

1. It maintains the cell's shape.
2. It controls the position of organelles within the cell by anchoring them to the plasma membrane.
3. It is involved with the flow of the cytoplasm, known as **cytoplasmic streaming**.
4. It anchors the cell in place by interacting with extracellular elements.

The cytoskeleton includes microtubules and microfilaments.

- **Microtubules** are hollow tubes made of the protein tubulin that make up the **cilia**, **flagella**, and **spindle fibers**. Cilia and flagella, which move cells from one place to another, consist of 9 pairs of microtubules organized around 2 singlet microtubules; see Figure 3.6. **Spindle fibers** help separate chromosomes during mitosis and meiosis and consist of microtubules organized into 9 triplets with no microtubules in the center. Flagella, when present in prokaryotes, are not made of microtubules.
- **Microfilaments** are assembled from **actin filaments** and help support the shape of the cell. They enable
  1. Animal cells to form a **cleavage furrow** during cell division
  2. Ameba to move by sending out **pseudopods**
  3. Skeletal muscles to contract as they slide along myosin filaments

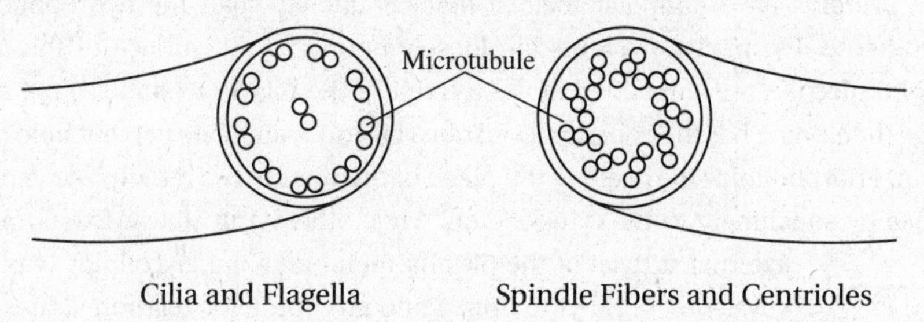

Figure 3.6 Structure of Cilia, Flagella, and Centrioles

STUDY TIP

Cilia and flagella have the 9 + 2 configuration.

Centrioles, Centrosomes, and the Microtubule Organizing Centers

Centrioles, **centrosomes**, or **microtubule organizing centers** (**MTOCs**) are nonmembranous structures that lie outside the nuclear membranes. They organize spindle fibers, and give rise to the spindle apparatus required for cell division (see Figure 3.7). Two centrioles oriented at right angles to each other make up one centrosome and consist of 9 triplets of microtubules (just like spindle fibers) arranged in a circle. Plant cells lack centrosomes, but have MTOCs. In animal cells, the MTOC is synonymous with centrosome.

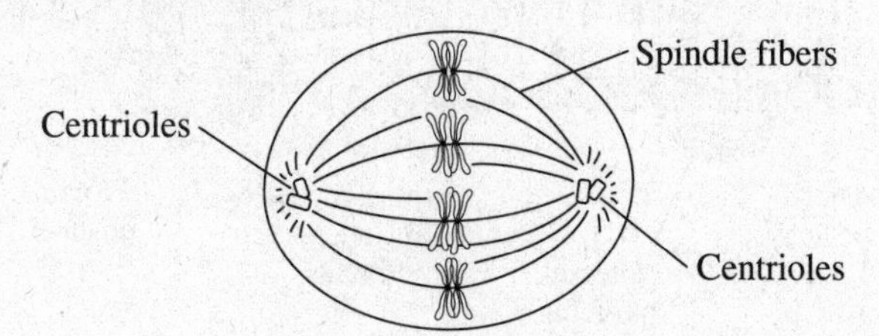

Figure 3.7 Spindle Fibers

Cell Wall

The **cell wall** is one cell structure that is not found in animal cells. Plants and algae have cell walls made of **cellulose**. The cell walls of fungi are usually made of **chitin**. Those of prokaryotes consist of other polysaccharides and complex polymers. The **primary cell wall** is immediately outside the plasma membrane. Some cells produce a second cell wall underneath the primary cell wall, called the **secondary cell wall**. When a plant cell divides, a thin gluey layer is formed between the two new cells, which becomes the **middle lamella**.

Plasma Membrane

The **cell** or **plasma membrane** is a **selectively permeable** membrane that regulates the steady traffic that enters and leaves the cell. S. J. Singer is famous for his description of the cell membrane in 1972, which he called the **fluid mosaic model**. The eukaryotic plasma membrane consists of a **phospholipid bilayer** with proteins dispersed throughout the layers. A phospholipid is amphipathic, meaning it has both a hydrophobic and hydrophilic region (see Figures 2.10 and 3.8).

Nonpolar molecules, such as hydrocarbons, carbon dioxide, and oxygen, are *hydrophobic* and can therefore dissolve in the lipid bilayer and cross it readily without the need of membrane proteins. However, the hydrophobic interior of the membrane impedes the flow of *hydrophilic* substances, such as **ions** and **polar molecules**. Polar molecules, such as glucose, do not pass through the inner phospholipid layer easily. Even water, a very small polar molecule, does not pass through easily.

Integral proteins have nonpolar regions that completely span the hydrophobic interior of the membrane. **Peripheral proteins** are loosely bound to the surface of the membrane. **Cholesterol molecules** are embedded in the interior of the bilayer to stabilize the membrane. The average membrane has the consistency of olive oil and is about 40 percent lipid and 60 percent protein. Phospholipids move along the plane of the membrane rapidly. Some proteins are kept in place by attachment to the cytoskeleton, while others drift slowly. Extending from the external surface of the plasma membrane are glycolipids (carbohydrates covalently bonded to lipids) and glycoproteins (carbohydrates covalently bonded to proteins). Both structures may serve as signaling molecules that distinguish one cell type from another. Glycoproteins on the surface of red blood cells are responsible for ABO and Rh blood types; see Figure 3.8.

> **ENE-2**
> Be able to explain how all membrane structure relates to its function.

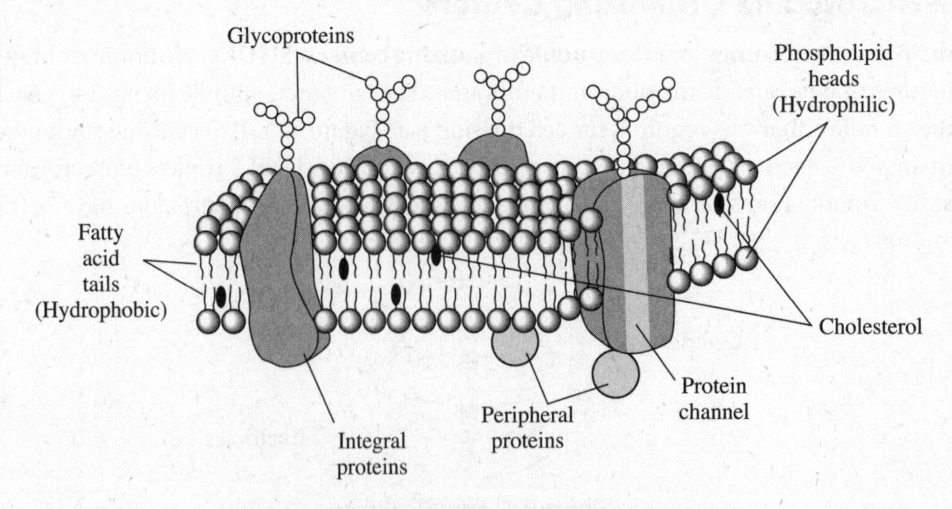

Figure 3.8 Detail of the Plasma Membrane

Proteins in the plasma membrane provide a wide range of functions.

- **TRANSPORT:** Molecules, electrons, and ions are carried through channels, pumps, carriers, and electron transport chains, which manufacture ATP.
- **ENZYMATIC ACTIVITY:** One membrane-bound enzyme is adenylate cyclase, which synthesizes cyclic AMP (cAMP) from ATP.
- **SIGNAL TRANSDUCTION:** Binding sites on protein receptors fit chemical messengers (signaling molecule) like hormones. The protein changes shape and relays the message to the inside of the cell.
- **CELL-TO-CELL RECOGNITION:** Some glycoproteins serve as identification flags that are recognized by other cells.
- **CELL-TO-CELL ATTACHMENTS:** Desmosomes, gap junctions, and tight junctions are examples.
- **ATTACHMENT TO THE CYTOSKELETON AND EXTRACELLULAR MATRIX:** This helps maintain cell shape and stabilizes the location of certain membrane proteins.

TRANSPORT INTO AND OUT OF THE CELL

Transport is the movement of substances into and out of a cell. Transport can be either active or passive. Active transport requires energy (ATP). Passive transport requires no energy. Figure 3.9 shows an overview of passive and active transport.

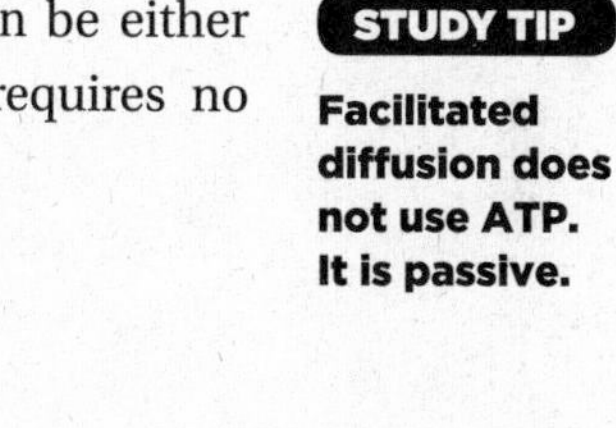
STUDY TIP

Facilitated diffusion does not use ATP. It is passive.

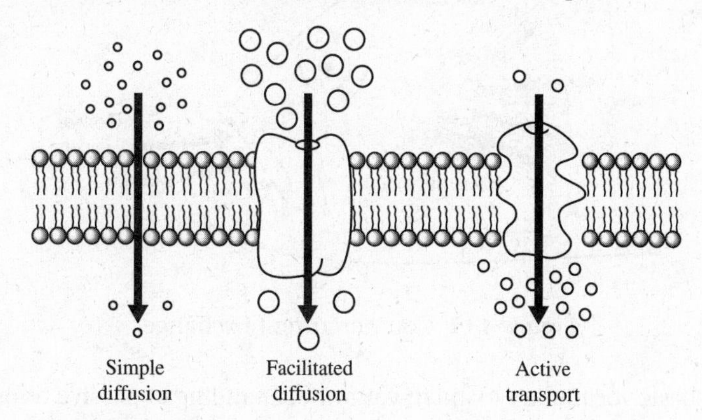

Figure 3.9 Overview: Active and Passive Transport

Passive Transport

Passive transport is the movement of molecules *down a concentration gradient from a region of high concentration to a region of low concentration* until equilibrium is reached. Examples of passive transport are **diffusion** and **osmosis.**

There are two types of **diffusion**: **simple** and **facilitated**. Simple diffusion does not involve protein channels, but facilitated diffusion does. An example of simple diffusion is found in the glomerulus of the human kidney, where solutes dissolved in the blood diffuse into Bowman's capsule of the nephron.

Facilitated diffusion requires a **hydrophilic protein channel** that will speed up the passive transport of specific substances across the membrane; see Figure 3.10. One type of channel transports single ions, such as Na^+, K^+, Ca^{2+}, and Cl^-. Many **ion channels** function as **gated channels** that open and close in response to a stimulus. **Voltage gated ion channels** are important in the passage of an impulse along an axon (see page 354). Neither simple nor facilitated diffusion requires energy.

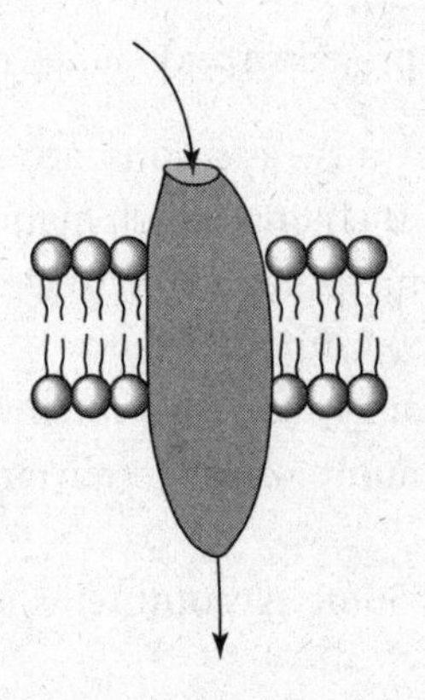

Figure 3.10 Protein Channel

A special case of simple diffusion is called **countercurrent exchange**—the flow of adjacent fluids in opposite directions that maximizes the rate of simple diffusion. One example of countercurrent exchange can be seen in fish gills. Blood flows toward the head in the gills, while water flows over the gills in the opposite direction. This process maximizes the diffusion of respiratory gases and wastes between the water and the fish; see Figure 3.11.

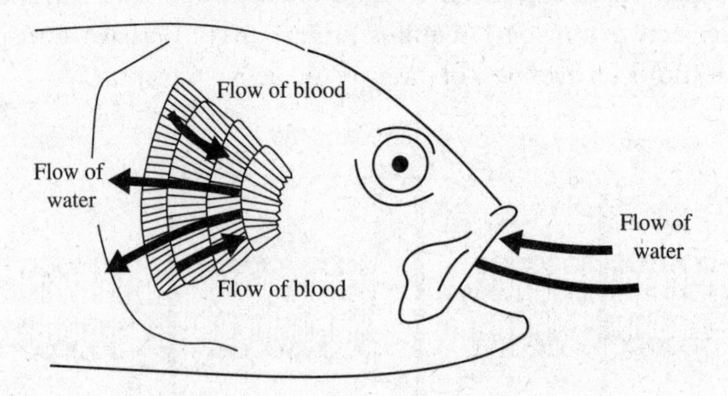

Figure 3.11 Countercurrent Exchange

Here is some basic vocabulary to aid in your understanding of passive transport.

- **DIFFUSION:** random movement of molecules or other particles from a higher concentration to a lower concentration
- **OSMOSIS:** the term used for a specific type of diffusion of *water across a membrane*
- **SOLVENT:** the substance that does the dissolving
- **SOLUTE:** the substance that dissolves
- **HYPERTONIC:** having greater concentration of *solute* than another solution
- **HYPOTONIC:** having lesser concentration of *solute* than another solution
- **ISOTONIC:** two solutions containing equal concentration of *solutes*
- **OSMOTIC POTENTIAL:** the tendency of water to move across a permeable membrane into a solution
- **WATER POTENTIAL:** Scientists look at movement of water in terms of water potential. **Water potential**, symbolized by the Greek letter psi, ψ, results from two factors: solute concentration and pressure. The water potential for *pure water is zero*; the addition of solutes lowers water potential to a value less than zero. Therefore, the water potential inside a cell is a negative value. Water will move across a membrane from the solution with the higher water potential to the solution with the lower water potential.

Figure 3.12 shows a sketch of two containers. Solution A is hypertonic to solution B.

REMEMBER

Water diffuses toward the hypertonic area.

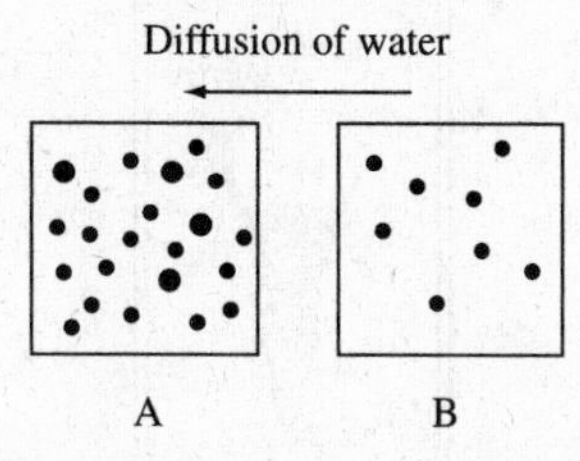

Figure 3.12 Hypertonic and Hypotonic Solutions

Example 1: In Figure 3.13, the cell is in an isotonic solution. Water diffuses in and out, but there is no net change in the size of the cell.

ENE-2 & SP 3
Understand diffusion and osmosis well enough to design an experiment to demonstrate how they affect living cells. See Lab #4 on page 422.

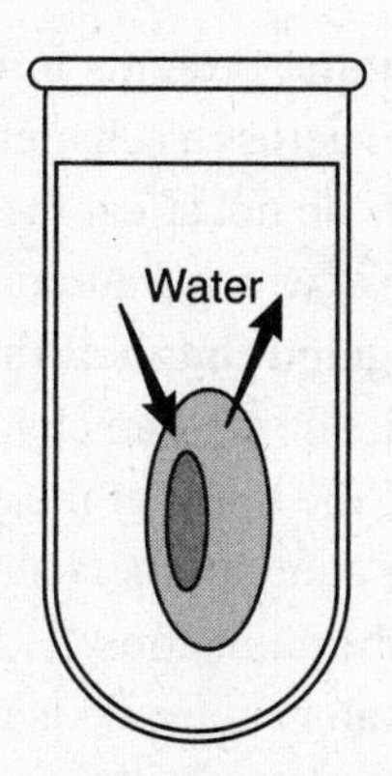

Figure 3.13 Cell in Isotonic Solution

Example 2: The cell in Figure 3.14 is in a **hypotonic** solution. The concentration of solute in the beaker is less than the concentration of solute in the cell. Therefore, *water* will flow into the cell, causing the cell to swell or burst. If the cell is a plant cell, the cell wall will prevent the cell from bursting. The cell will merely swell or become **turgid**. This turgid pressure is what keeps plants like celery crisp. If a plant loses too much water (dehydrates), it loses its **turgor pressure** and wilts.

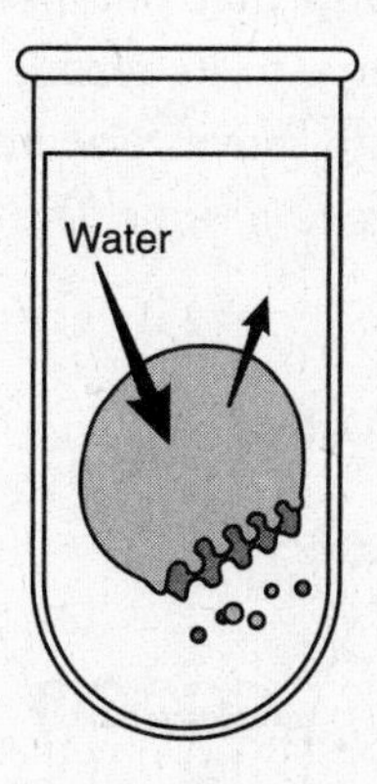

Figure 3.14 Cell in Hypotonic Solution

Example 3: The cell in Figure 3.15 is in a **hypertonic** solution. The concentration of **solute** in the beaker is greater than the concentration of solute in the cell. Therefore, water will flow out of the cell because water flows from high concentration of water to low concentration of water. As a result, the cell shrinks, exhibiting **plasmolysis**.

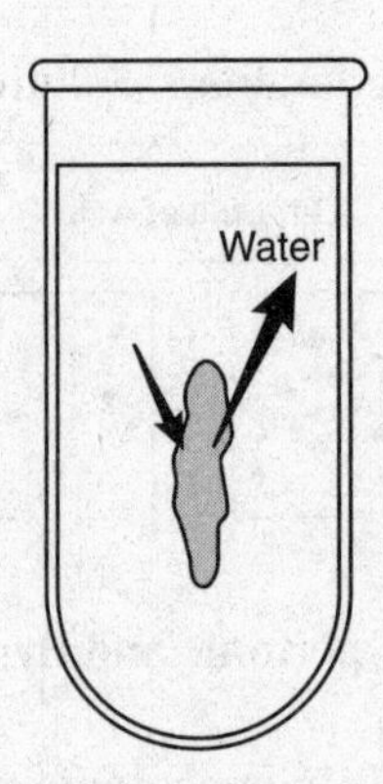

Figure 3.15 Cell in Hypertonic Solution

Aquaporins are special water channel proteins found in certain cells that facilitate the diffusion of massive amounts of water across a cell membrane, up to 3 billion (3×10^9) water molecules per second. These channels do not affect the water potential gradient or the direction of water flow but, rather, the rate at which water diffuses down its gradient. It is possible that aquaporins can also function as **gated channels** that open and close in response to variables such as turgor pressure of a cell. The sudden change in a cell in response to changes in tonicity as seen in Figure 3.15 may be the result of the action of aquaporins.

It is interesting to note that while aquaporins facilitate the rapid passage of H_2O across membranes, they will not carry hydronium ions (H_3O^+). This is most likely because the hydronium ion is charged, while a water molecule is not. Aquaporins, like many channels, are very specific.

Active Transport

Active transport is the movement of molecules *against a gradient,* which requires energy, usually in the form of **ATP**. There are many examples of active transport.

1. **Pumps** or **carriers** carry particles across the membrane by active transport.

   a) The **sodium-potassium pump** that pumps Na^+ and K^+ ions across a nerve cell membrane to return the nerve to its resting state is an example; see Figure 3.16. The sodium-potassium pump moves Na^+ and K^+ ions against a gradient, pumping two K^+ ions for every three Na^+ ions. The Na-K pump is the major **electrogenic pump** in animals.

ENE-2
Homeostasis is maintained by the constant movement of molecules across a membrane.

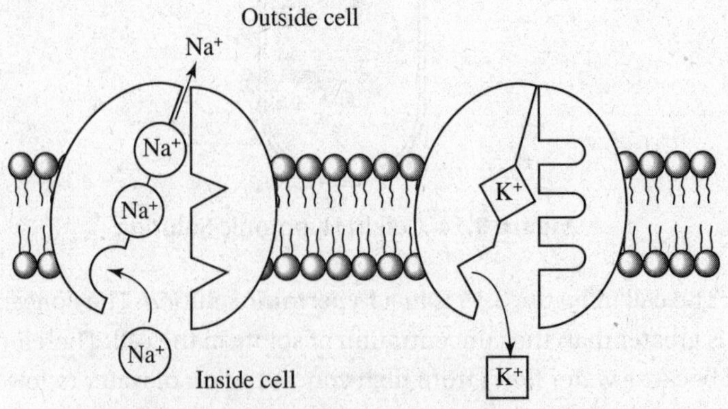

Figure 3.16 Sodium-Potassium Pump

b) **Proton pumps** are electrogenic pumps that actively pump protons (H^+) across a membrane against a gradient. By generating voltage across membranes, they store energy that can be tapped for cellular work. One important example is in the cristae membranes of mitochondria in the production of ATP during cellular respiration.

c) **Cotransport** is another type of membrane pump. In this case, a proton pump is linked to the transport of a second substance. As shown in the diagram below, a cotransporter protein translocates sucrose *in plants* against a gradient and into the cell, but only if a sucrose molecule travels in the company of protons. While protons diffuse down its gradient, sucrose is being carried against its gradient. See Figure 3.17.

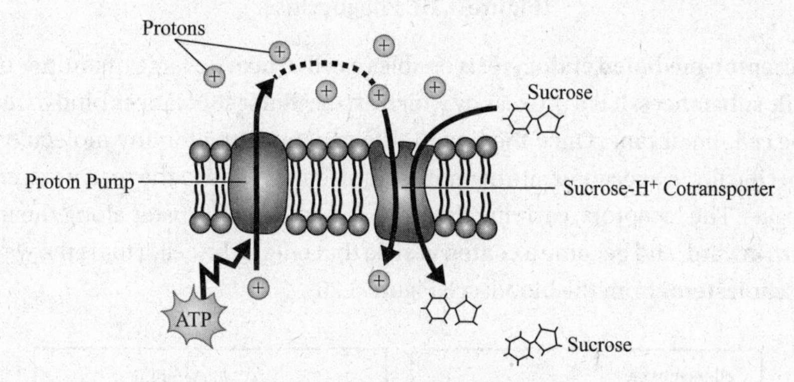

Figure 3.17 Cotransport

2. The **contractile vacuole** in freshwater Protista pumps out excess water that has diffused inward because the cell lives in a hypotonic environment.
3. **Exocytosis**—Internal vesicles fuse with the plasma membrane to secrete macromolecules out of a cell. For example, in nerve cells, vesicles release neurotransmitters into a synapse.
4. **Endocytosis**—A cell takes in macromolecules and particles by forming new vesicles derived from the plasma membrane.

a) **Pinocytosis**, *cell drinking*, is the uptake of large, dissolved particles. The plasma membrane invaginates around the particles and encloses them in a vesicle; see Figure 3.18.

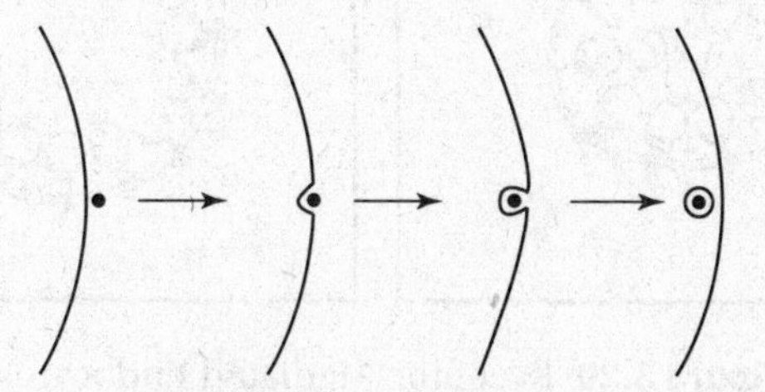

Figure 3.18 Pinocytosis

b) **Phagocytosis** is the engulfing of large particles or small cells by pseudopods. The cell membrane wraps around the particle and encloses it into a vacuole. This is the way human white blood cells engulf bacteria and also the way in which ameba gain nutrition; see Figure 3.19.

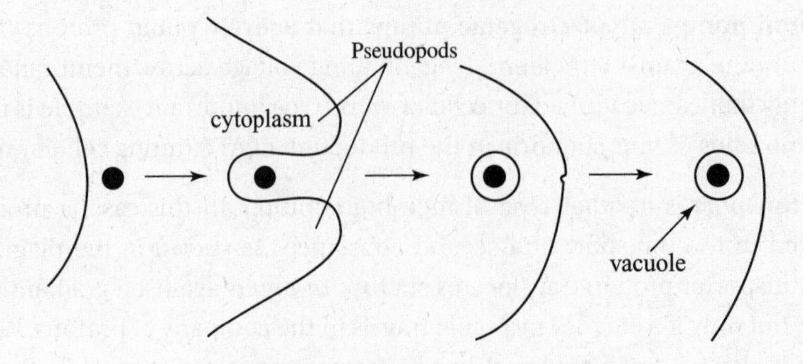

Figure 3.19 Phagocytosis

c) **Receptor-mediated endocytosis** enables a cell to take up large quantities of very **specific substances**. It is a process by which extracellular substances bind to receptors on the cell membrane. Once the **ligand** (the general name for any molecule that binds specifically to a receptor site of another molecule) binds to the receptors, endocytosis begins. The receptors, carrying the ligand, migrate and cluster along the membrane, turn inward, and become a **coated vesicle** that enters the cell. This is the way cells take in **cholesterol** from the blood; see Figure 3.20.

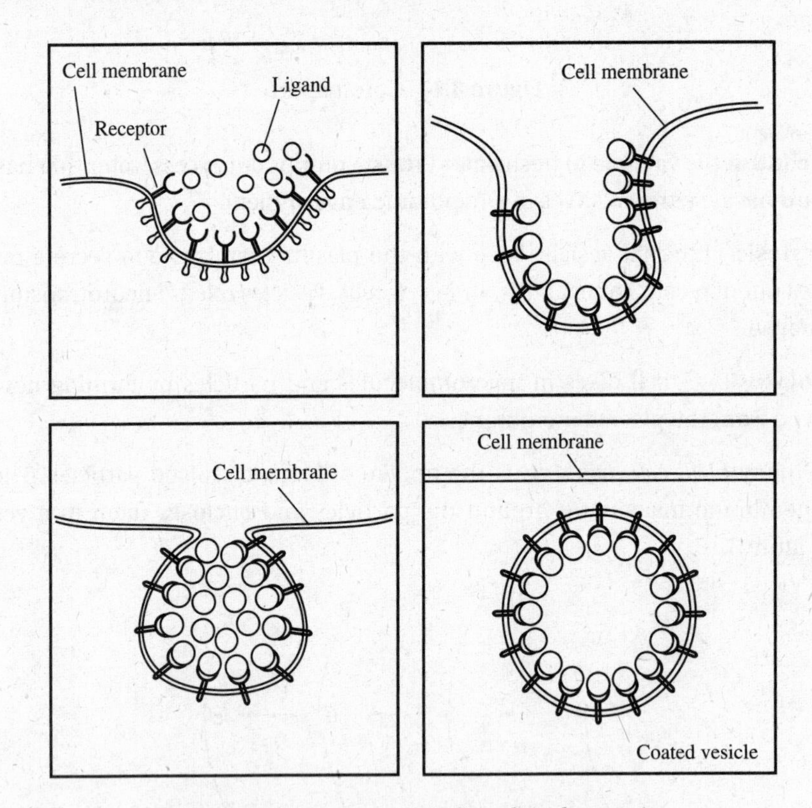

Figure 3.20 Receptor-Mediated Endocytosis

Bulk Flow

Bulk flow is a general term for the overall movement of a fluid in one direction in an organism. In humans, **blood** moves around the body by bulk flow as a result of blood pressure created by the pumping heart. **Sap** in trees moves by bulk flow from the leaves to the roots due to active transport in the phloem. Bulk flow movement is always from **source** (where it originates) **to sink** (where it is used).

CELL COMMUNICATION

Whether they are primitive bacterial cells or cells that make up complex organisms, cells communicate with other cells using **chemical signals**.

STUDY TIP

Cell communication is an important topic. Quorum sensing is a great example.

Quorom Sensing in Bacteria

Bacterial cells are able to communicate through a phenomenon called **quorum sensing**. This process allows bacteria to monitor their population density and use that information to control **gene expression**. Universal quorum sensing languages even enable bacteria of *different* species to "talk" to each other. Many bacterial behaviors are regulated by quorum sensing, including horizontal gene transfers or the process by which a harmless bacterium can transform into a virulent one. Here is an example of quorum sensing.

The marine bacterium *Vibrio fischeri* lives symbiotically inside a variety of Hawaiian squid. The squid provide shelter and a stable source of nutrients for the bacteria, which live in an enclosed light-emitting organ within the squid's body. The bacteria, in return, provide light (**bioluminescence**) that benefits the squid. Through the process of quorum sensing, the bacteria monitor the size of their population and produce light when their population is large enough to be seen and effective. When the *V. fischeri* population lights up the squid, the squid becomes camouflaged and can feed without being attacked by predators.

How do the bacteria "know" when their population is large enough? They release signaling molecules into their environment. As the bacteria population increases, so does the concentration of signaling molecules until there are enough molecules to bind to receptors inside or on the surface of all the bacteria and trigger a cascade of chemical reactions. These reactions alter the expression of certain genes in the bacteria, turning on the ones that code for enzymes, like **luciferase**, which trigger the expression of the protein **luciferin** and produce light.

Direct Contact

Multicellular organisms contain trillions of cells that also must communicate with each other. They do this in a variety of complex ways.

EVO-2
Cell communication processes share common features that reflect a shared evolutionary history.

Signaling substances dissolved in the cytosol can pass freely between adjacent cells. In animals, **gap junctions** permit the passage of materials directly from the cytoplasm of one cell to the cytoplasm of an adjacent cell. In muscle tissue of the **heart**, the flow of ions through gap junctions coordinates the contractions of cardiac cells. See Figure 3.21. **Plasmodesmata** connect one plant cell to the next. They are analogous to gap junctions in animal cells.

Direct contact between membrane-bound molecules is the basis of the human immune system. See Figures 15.1 and 15.3.

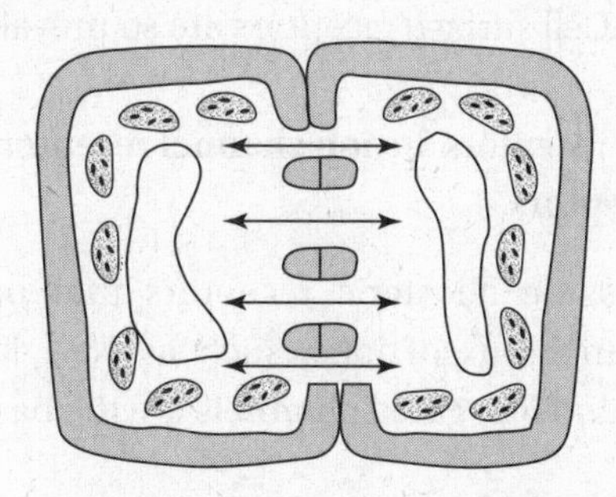

Figure 3.21 Plasmodesmata

Local Signaling

There are three types of local signaling by cells:

1. **PARACRINE SIGNALING**—where chemical signals travel a relatively short distance. See the discussion of **embryonic induction**, whereby one layer of cells induces a nearby layer to transform into another cell type, on page 397.

2. **SYNAPTIC SIGNALING**—where an axon of a nerve cell releases **neurotransmitters**. See the discussion of neurons on page 352.

3. **AUTOCRINE SIGNALING**—where helper T cells release cytokines that signal themselves to release more cytokines. See the discussion of the immune system on page 369.

Long-Distance Signaling

Plants use chemicals called **hormones** for long-distance signaling. In animals, the endocrine system has endocrine glands to release hormones in the blood to reach target organs or structures. Another example of long-distance signaling is how helper T white blood cells send out an alarm to the entire immune system that the body is being invaded.

Three Stages of Cell Signaling

Regardless of the type of cell communication, local or long-distance, cell signaling occurs in three stages: reception, transduction, and response.

IMPORTANT

Reception: A chemical signal is detected when the signaling molecule binds to a receptor, either *on the surface* of a cell or *inside* the cell.

RECEPTION—a signal molecule (**ligand**) from outside the target cell binds to a receptor either on the cell membrane of or inside the target cell.

TRANSDUCTION—the signal is converted to a form that can bring about a specific cellular response.

RESPONSE—a specific cellular response occurs, either by regulation of transcription or by cytoplasmic action.

WHEN THE RECEPTORS ARE ON THE CELL SURFACE

Cell membrane receptors span the entire thickness of the membrane and are therefore in contact with both the extracellular environment and the cytoplasm. Since **water-soluble (hydrophilic) signaling molecules** cannot diffuse through the membrane, they bind to the part of the receptor on the cell surface that changes the shape on the cytoplasmic side of the same receptor. Thus, the signal is transmitted from outside the cell to the cytoplasm. Once inside the cell, the signal is carried by a **second messenger**. The most common second messenger is **cyclic AMP (cAMP)**. Notice that the **first messenger**, the signaling molecule or **ligand**, never enters the cell. Cell surface receptors are so prevalent that they make up about 30% of all human proteins.

Three types of cell surface receptors are **ion channel receptors**, **G protein–coupled receptors**, and **protein kinase receptors**.

1. **Ion-channel receptors** are allosteric receptors that open and shut a gate in a membrane, allowing an influx of ions, such as Na^+, K^+, Ca^{2+}, or Cl^- ions. These ion-channel receptors and **ion-gated channels** are the basis of normal nerve function. See Figure 3.22.

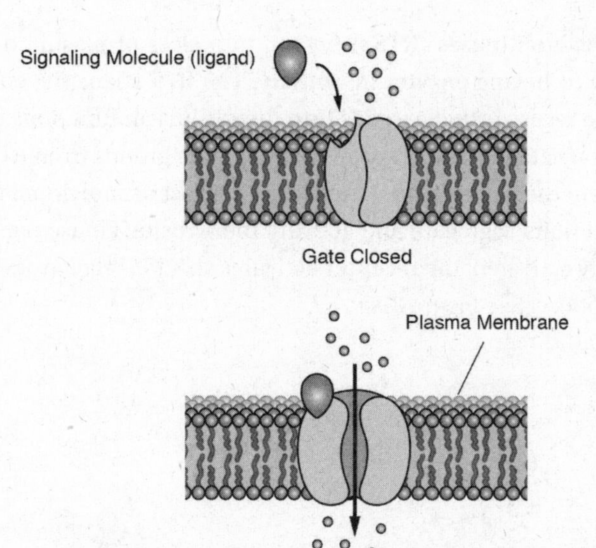

Figure 3.22 Ion-Channel Receptor

STUDY TIP

You need to know the 3 types of cell surface receptors for the AP Biology exam.

2. **G protein–coupled receptors (GPCR)** are cell surface receptors that work with the help of G protein. When a ligand, the **first messenger**, binds to the *extracellular* domain of the receptor, it changes the conformation of the *cytoplasmic side* of the receptor. This, in turn, activates G protein—an on-off molecular switch. When G protein is activated, or "on," it bonds to GTP (guanosine triphosphate, a high-energy nucleotide similar to ATP), which activates the enzyme adenylyl cyclase. This enzyme, in turn, catalyzes the conversion of ATP to **cyclic-AMP (cAMP)**, the **second messenger**. It is this cAMP that activates other molecules inside a cell, leading to a cellular response. See Figure 3.23.

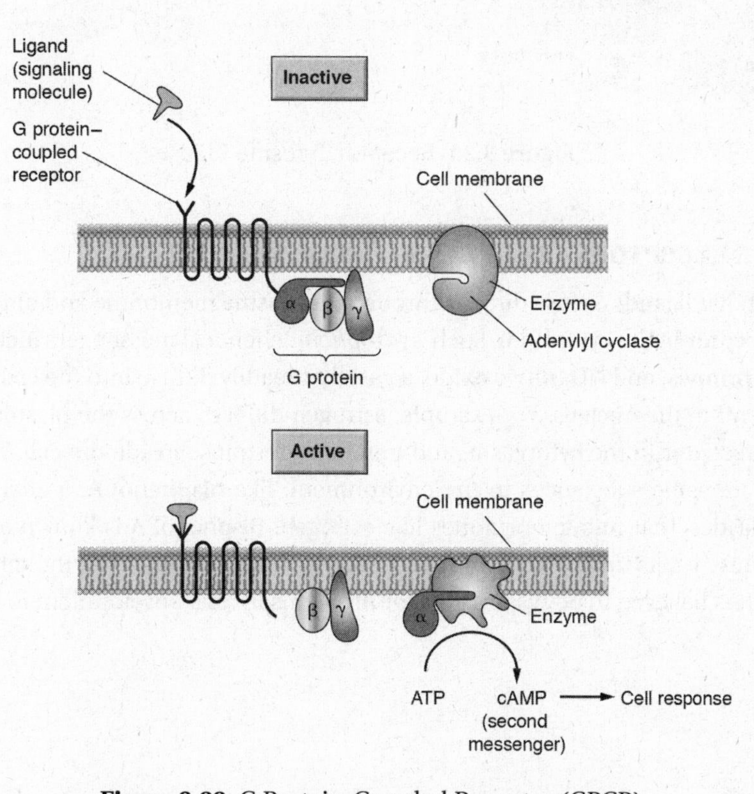

Figure 3.23 G Protein-Coupled Receptor (GPCR)

3. **Receptor Tyrosine Kinases (RTKs)** belong to a class of plasma membrane receptors characterized by having **enzymatic activity**. The RTK spans the entire membrane and the part of the receptor that extends into the cytoplasm functions as a tyrosine kinase, an enzyme that catalyzes the transfer of phosphate groups from ATP to the amino acid tyrosine. Before the ligand binds, the receptors exist as individual units. After binding, the individual units aggregate and activate the tyrosine kinase region, which bonds to ATP. Once fully activated, the receptor activates specific relay proteins that each lead to a cellular response. See Figure 3.24.

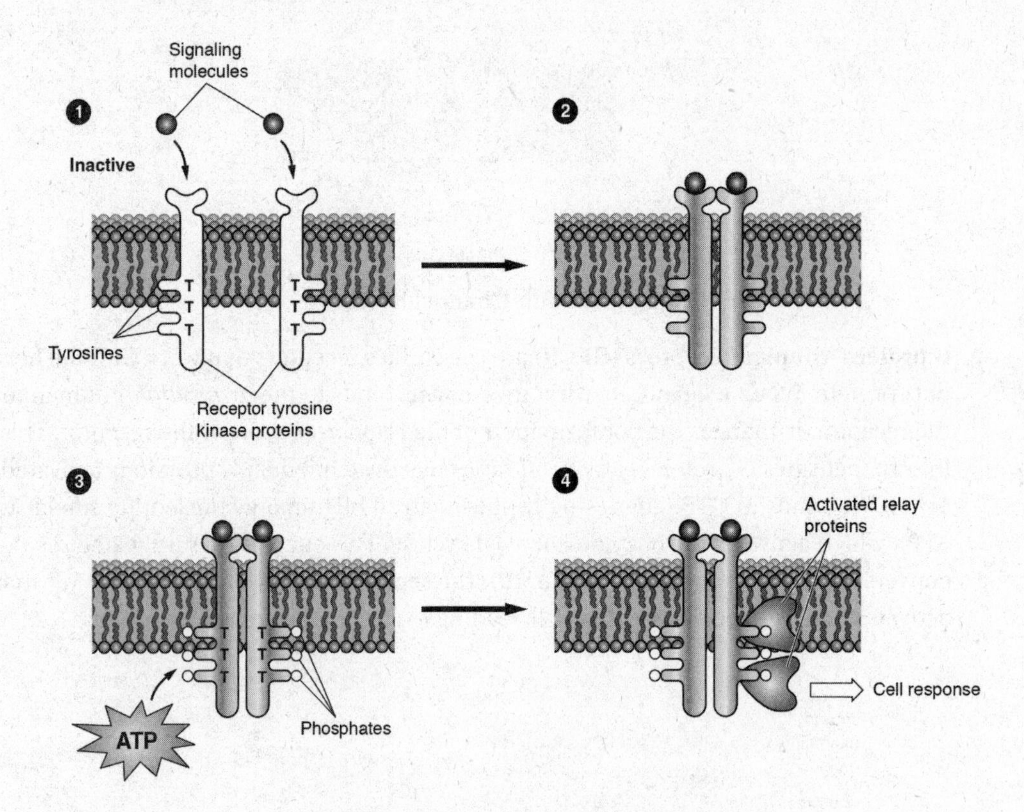

Figure 3.24 Receptor Tyrosine Kinase

WHEN THE RECEPTORS ARE LOCATED INSIDE THE CELL

Small, nonpolar ligands diffuse directly through the plasma membrane and bind to an **intracellular receptor** in the cytoplasm. Such *hydrophobic* chemical messengers include **steroids**, **thyroid hormones**, and **NO**, **nitric oxide**, a gas, that readily diffuse into the cell and switch a gene on or off in the nucleus. For example, estrogen diffuses across the plasma membrane, binds to a receptor in the cytoplasm, and initiates a response inside the cell. We now know that there are some substances in the environment, like **bisphenol A**, a chemical present in many plastics, that mimic hormones like estrogen. Bisphenol A is known as an *estrogen mimic* because it sets up an estrogen signal transduction pathway, even though no estrogen is present. It is believed to be responsible for problems in fetal development in humans.

Transduction

Once activated, the receptor converts (transduces) a molecular signal into a cell response. This transduction sometimes occurs in a single step, but more often it requires a sequence of changes in a series of different molecules—*a signal transduction pathway.*

Signal transduction pathway is a multistep process in which a small number of extracellular signal molecules produce a major cellular response, **a cascade effect**, like falling dominoes. The advantage of this complex set of reactions is that the multistep pathway provides more opportunities to amplify the signal greatly. Signal transduction pathways exist in both yeast and animal cells, and in bacteria and plants. **The striking similarity in these pathways among diverse organisms suggests that the pathways evolved hundreds of millions of years ago in a common ancestor.**

> **ENE-2 & IST-3**
> Be able to describe the key elements of signal transduction pathways by which a signal is converted to a cellular response.

One representative example of a signal transduction pathway is the one that demonstrates how the hormone epinephrine causes a liver cell to release glucose into the bloodstream. Epinephrine binds to a receptor that activates the G protein, followed by the production of an intermediate molecule, **cAMP**, the **second messenger**, which distributes and amplifies the signal throughout the cell. See Figure 3.25 on page 52.

The bold numbers in parentheses in the diagram show a cascade of reactions that begins with one molecule of epinephrine binding to a receptor and ending with thousands of molecules of glucose released into the bloodstream.

Response

Ultimately, signal transduction pathways lead to a multitude of responses, either in the cytoplasm or in the nucleus. Many signaling pathways regulate protein synthesis by turning specific genes on or off. Regardless of the outcome, you need to remember five things about **signal transduction pathways**.

1. They are characterized by a signal, a transduction, and a response.
2. They are highly specific and regulated.
3. One signal molecule can cause a cascade effect, releasing thousands of molecules inside a cell.
4. They regulate cellular activity, altering gene expression, protein activity, or protein synthesis.
5. These pathways evolved millions of years ago in a common ancestor.

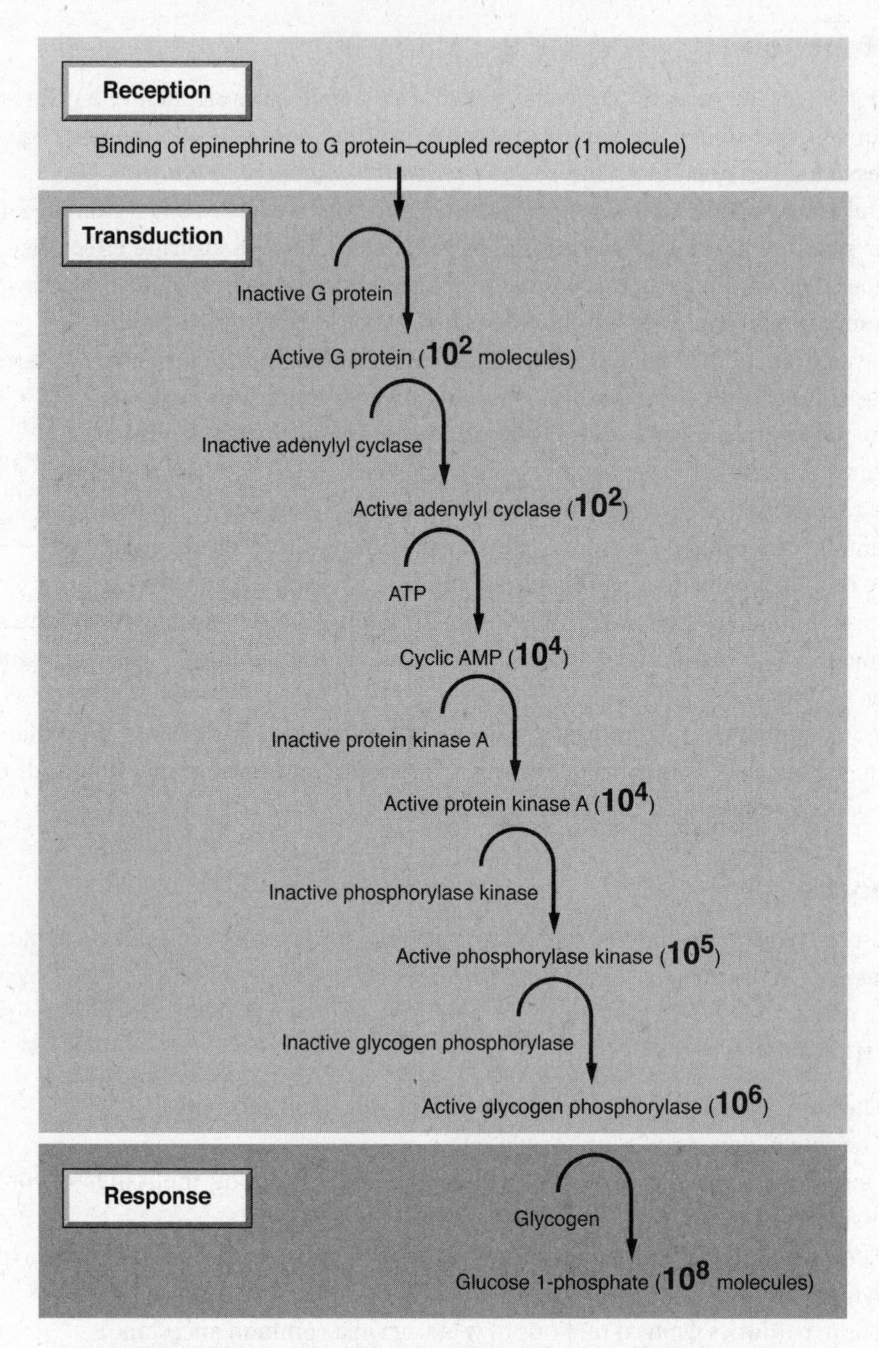

Figure 3.25 Cascade Effect of a Signal Transduction Pathway

APOPTOSIS—PROGRAMMED CELL DEATH

Cells that are infected, damaged, or simply have come to the end of their life span die by **apoptosis**, a genetically programmed series of events that result in cell death. During this process, the DNA, organelles, and other cytoplasmic components are systematically chopped up. The parts are packaged in vesicles that are engulfed by special scavenger cells. Apoptosis relies on several different *cell-signaling pathways* and signals from outside, and sometimes from inside, the cells.

Cells carry out apoptosis for several reasons:

1. During embryonic development, when cells or tissues are no longer needed, they die and are engulfed by neighboring cells. A familiar example is provided by the cells in the tail of the tadpole, which undergo apoptosis during frog metamorphosis.
2. The cell has sustained too much genetic damage that could lead to cancer. This is common for epithelial cells on the surface of the skin, which have been exposed to extensive solar radiation. Skin cells normally die by apoptosis and are replaced every few days.
3. In plant cells, apoptosis is an important defense against infection by fungus and bacterium. By dying, cells at the site of infection leave no living tissue to spread infection inside the plant.
4. In mammals, including humans, several different pathways involving enzymes called **caspases** carry out apoptosis. Signals from different sources—inside or outside the cell—trigger the apoptosis pathway. The trigger can be from a neighboring cell, which makes use of a signal transduction pathway to begin the process. A signal may instead come from inside the cell itself, from irreparably damaged DNA in the nucleus, or from the endoplasmic reticulum when excessive protein misfolding has occurred.

ENE-3
Be able to describe the role of a programmed cell death in the development and maintenance of homeostasis.

The fact that the mechanism of apoptosis is basically similar in single-celled yeast and in mammalian cells indicates that *the mechanism for apoptosis evolved early in the evolution of eukaryotic cells.*

CHAPTER SUMMARY

This chapter supported Big Ideas: EVO, ENE, IST, and SYI.

- All cells share some characteristics because they all arise from a common ancestral cell.
- Eukaryotic cells are particularly efficient because they are compartmentalized by a selectively permeable endomembrane system, which allows for the separation of different chemical processes from each other.
- The fluid-mosaic model describes a selectively permeable plasma membrane that separates the cell from its environment but that also allows for the regulated exchange of materials with the environment.
- A sufficiently high ratio of surface area (cell membrane) to volume (cytosol) is necessary for the exchange of materials into and out of the cell.
- Cells communicate with each other and with the environment through a complex system of reception, transduction, and responses.

PRACTICE QUESTIONS

1. Which of the following is correct about G protein, which is shown in the diagram below?

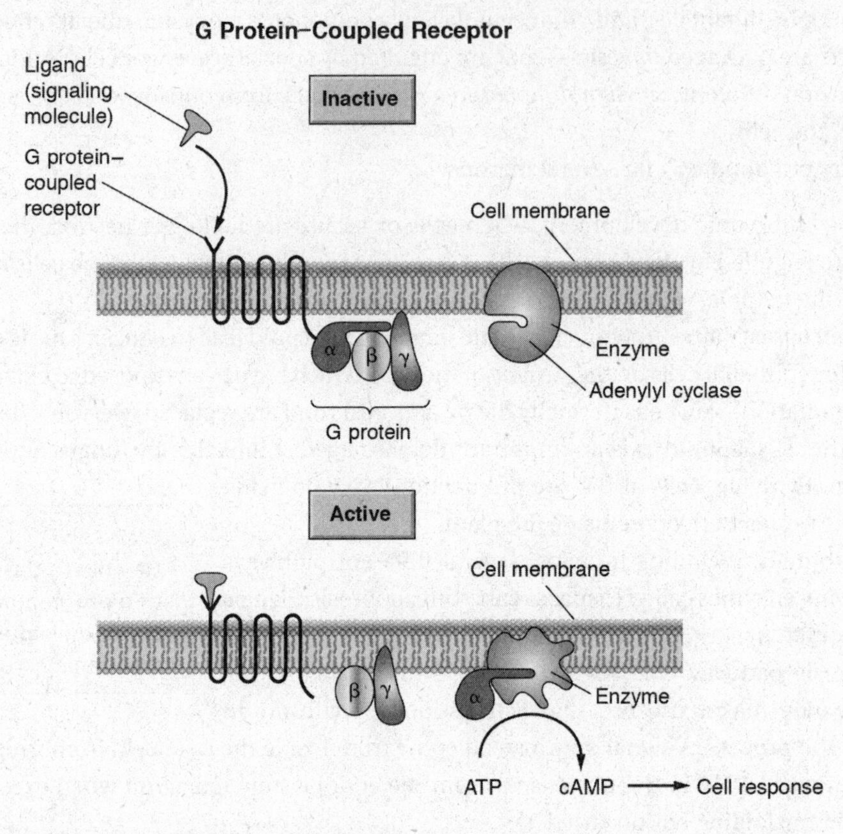

(A) It is a membrane-bound receptor.
(B) It is a cytoplasmic receptor that bonds to hydrophobic substances.
(C) It is an on-off molecular switch that gets activated by the membrane-bound receptor.
(D) It is an ion-channel receptor protein that is the basis of normal neuron function.

2. Which figure depicts an animal cell placed into a solution hypotonic to the cell?

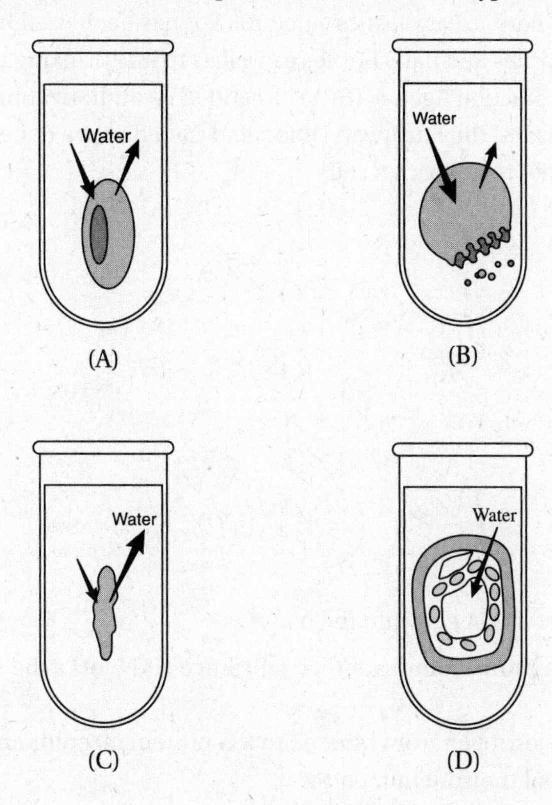

3. Signal transduction pathways are complex, involving many different molecules and many steps. What is the main advantage of sending a signal through a signal transduction pathway as opposed to sending a signal through direct contact between two cells?

   (A) Signal transduction pathways are quicker.
   (B) Signal transduction pathways offer more opportunities to amplify the signal.
   (C) Direct contact messaging is limited to simple cells like bacterial cells.
   (D) There is less chance for error in signal transduction pathways.

4. Bisphenol A (BPA) is an organic, synthetic chemical that has been used in the manufacturing of hard, clear plastics since 1957. It has been used in the manufacturing of plastic water bottles and baby bottles as well as to line drinking water pipes. The Environmental Protection Agency (EPA) has ended its authorization for the use of BPA in many products, and the European Union and Canada have done the same. Below is a diagram of how BPA reacts with a cell.

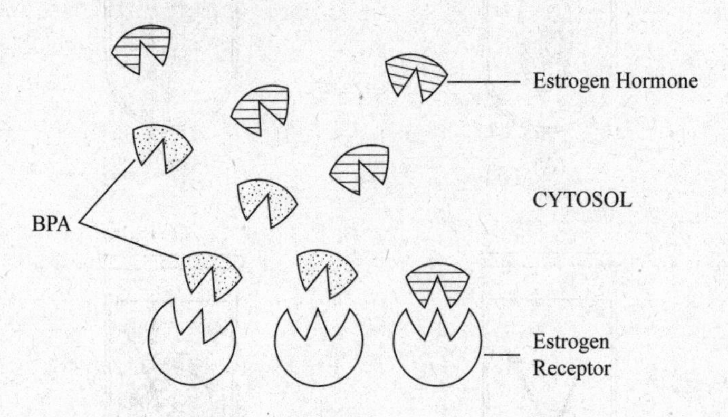

What is the danger if BPA gets into the body?

(A) BPA converts estrogen into another substance and blocks the normal estrogen pathway.
(B) BPA prevents estrogen from binding to a G protein receptor and blocks the normal estrogen signal transduction pathway.
(C) BPA combines with estrogen and diverts estrogen's normal function.
(D) BPA competes with estrogen for its membrane receptors and subverts normal estrogen function.

5. Which of the following statements is correct about apoptosis?

(A) The process of apoptosis has only been identified in advanced animals, such as tadpoles.
(B) It normally triggers an aged cell to undergo mitosis.
(C) It is an important part of normal embryonic development.
(D) The signals that trigger the process always come from outside a cell.

6. A plant cell is placed into a beaker that contains a sucrose solution. The setup is shown in the diagram below.

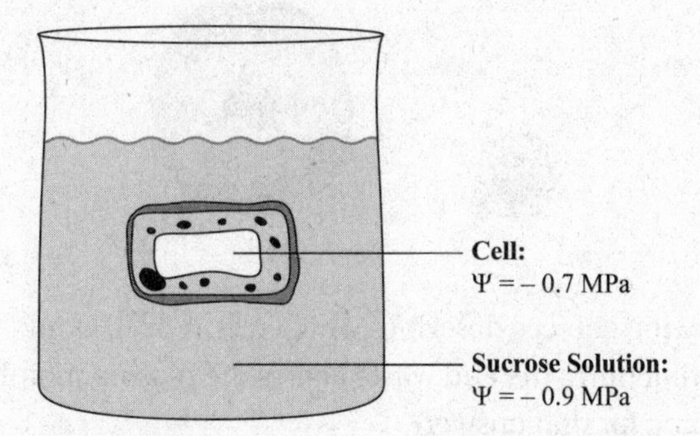

Which of the following is correct about what will occur and gives the most accurate reason for that occurrence?

(A) The cell will shrink because the water potential (Ψ) is greater inside the cell.
(B) The cell will shrink because the water potential (Ψ) is greater outside the cell.
(C) The cell will swell because the water potential (Ψ) is greater inside the cell.
(D) The cell will swell because the water potential (Ψ) is greater outside the cell.

7. You tag amino acids with radioactive sulfur because amino acids are incorporated into proteins in a cell. You monitor the location of these tagged amino acids until they are released from the cell. Which of the following best describes their movement?

(A) Nucleus → Golgi → vacuole
(B) Golgi apparatus → ER → vesicles that fuse with plasma membrane
(C) ER → Golgi apparatus → vesicles that fuse with plasma membrane
(D) Nucleus → ER → vacuole

8. Some pancreatic cells produce insulin, a polypeptide that consists of 51 amino acids, in great abundance. Which organelle would you expect to find in pancreatic cells in the greatest numbers?

(A) vacuoles
(B) smooth endoplasmic reticulum
(C) ribosomes
(D) mitochondria

9. Which of the following is correct about microtubules?

(A) They make up flagella in paramecia.
(B) They assist in the formation of pseudopods in amoeboid cells.
(C) They consist of actin fibers.
(D) They assist in the formation of the cleavage furrow in animal cells during cell division.

10. Dr. Lynn Margulis developed the theory of endosymbiosis. This theory explains how

(A) prokaryotic cells came to exist
(B) eukaryotic cells came to exist
(C) sucrose crosses a membrane against a gradient
(D) ligands enter cells

11. Refer to the diagram below.

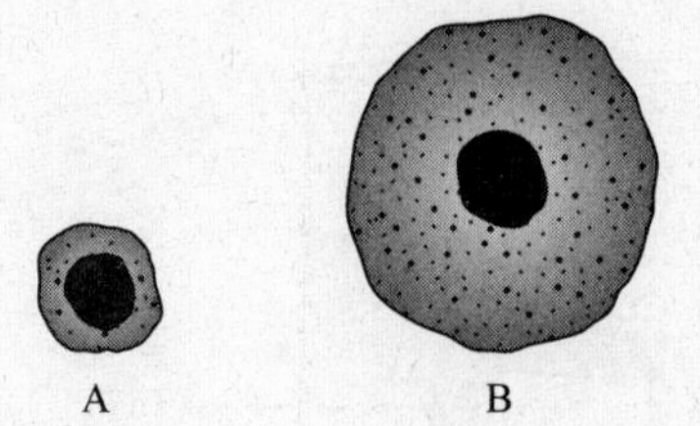

Which of the following choices describes which cell, A or B, would be most efficient in terms of transporting nutrients and waste across the plasma membrane and provides the correct reasoning for that answer?

(A) Cell A, because the ratio of the membrane surface area to the volume of the cytoplasm is much greater in the smaller cell
(B) Cell A, because the ratio of the membrane surface area to the volume of the cytoplasm is much smaller in the smaller cell
(C) Cell B, because the larger the cell, the greater the rate of diffusion
(D) Cell B, because it has the larger nucleus, which exerts greater control on the entire cell, which results in an increased rate of transport across the membrane

12. Below is a diagram of an ameba engulfing a bacterium.

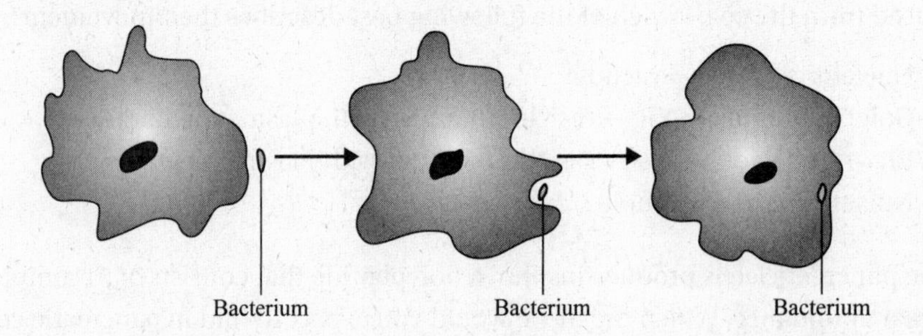

Given the fact that the ameba is so mobile, which of the following is true of an ameba as opposed to a cell that does not move?

(A) An ameba contains a larger nucleus.
(B) An ameba contains more microfilaments.
(C) An ameba contains more microtubules.
(D) An ameba contains more sensory apparatus.

13. Below is an illustration of an important cell organelle.

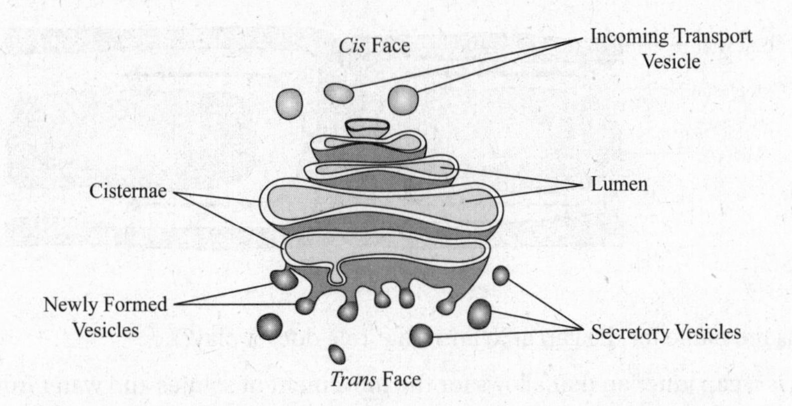

Which of the following statements correctly describes this organelle and what is occurring in the figure?

(A) This is a rough endoplasmic reticulum that is producing proteins in attached ribosomes.
(B) This is a smooth endoplasmic reticulum that is producing proteins in attached ribosomes.
(C) This is a Golgi body that is packaging material, sent from the endoplasmic reticulum, near the *cis* face.
(D) This is a Golgi body that is packaging material, sent from the endoplasmic reticulum, near the *trans* face.

14. Which one of the following would *not* normally diffuse through the lipid bilayer of a plasma membrane and gives the reason for your answer?

(A) CO_2, because it is small and polar
(B) amino acid, because it is acidic and charged
(C) starch, because it is too large
(D) water, because it is strongly polar

15. Which of the following requires ATP?

(A) the uptake of cholesterol by a cell
(B) the facilitated diffusion of glucose into a cell
(C) countercurrent exchange
(D) the diffusion of oxygen into a fish's gills.

16. Which of the following best characterizes the structure of the plasma membrane?

(A) rigid and unchanging
(B) rigid but varying from cell to cell
(C) fluid but unorganized
(D) very active

17. Below is a diagram of two plant cells.

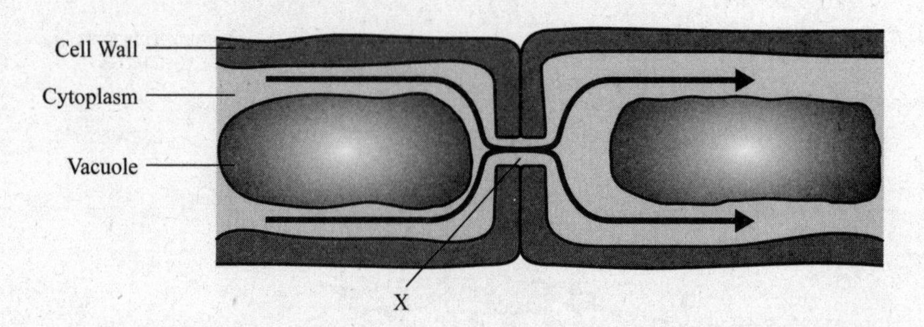

What is the name for the gap at X, and what role does it play?

(A) X is a gap junction that allows for the movement of solutes and water from one cell to another.
(B) X is a gap junction that allows for the passage of sap into the phloem.
(C) X is a plasmodesma that allows for the rapid passage of water and small dissolved solutes from one cell to an adjacent cell.
(D) X is a plasmodesma that allows water to rapidly travel in the xylem.

18. Refer to the figure below, which shows the plasma membrane.

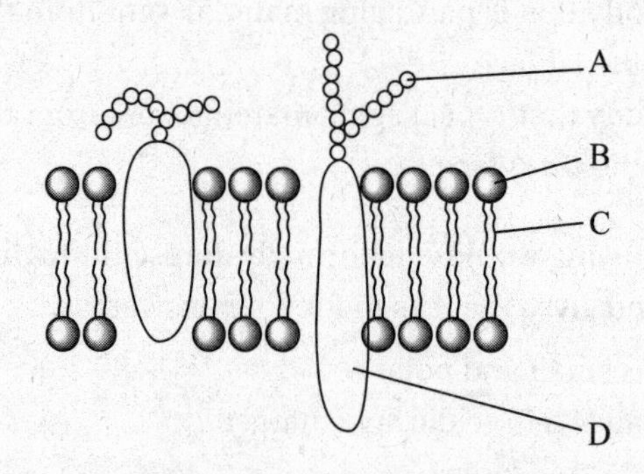

Which of the following correctly matches the part with its function?

(A) Part A is involved with facilitated diffusion.
(B) Part B is cholesterol, which stabilizes the membrane.
(C) Part C allows oil-soluble substances to freely enter a cell.
(D) Part D is an ATP synthase channel.

19. Which of the following is a prokaryotic cell?

(A) a human red blood cell
(B) the bacterium *E. coli* that live in the human intestine
(C) human bone cells
(D) photosynthetic plant cells

20. Below is an illustration of a plasma membrane transporting two substances: sucrose and protons (H^+).

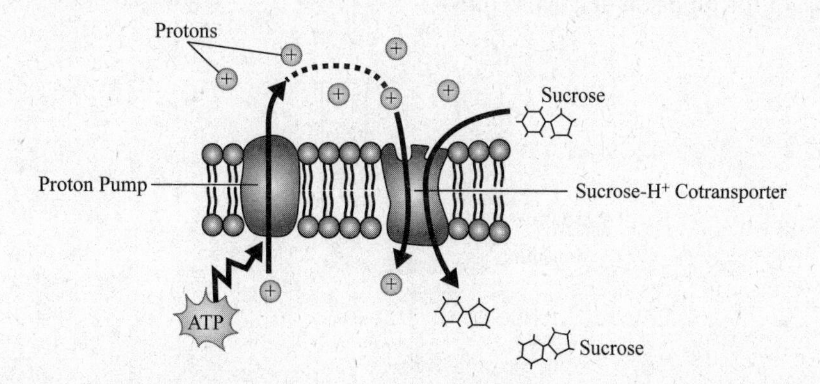

Which of the following statements is supported by what is shown in the illustration?

(A) Sucrose is diffusing down its gradient through a channel, but only in the presence of protons that are also diffusing in the same direction.
(B) Sucrose is diffusing down its gradient through a channel, but only in the presence of protons that are being pumped against a gradient.
(C) Sucrose and protons are simultaneously being pumped against their gradients by a membrane pump.
(D) Sucrose is being carried against a gradient, but only while protons are diffusing down a gradient.

21. Which of the following statements about cell-to-cell communication is correct?

(A) Plant cells cannot communicate with neighboring plant cells because of the presence of a thick cell wall.
(B) Quorum sensing enables bacteria to monitor their population density and to use this information to control gene expression.
(C) Hydrophobic chemical messengers, such as estrogen, rely on a secondary messenger, such as cyclic AMP, to initiate a response from the nucleus.
(D) Regardless of the type of chemical signal, the first thing that happens in a receiving cell is transduction, where the chemical message may be converted into a cascading signal transduction pathway.

22. The illustration below shows a glass tube that contains two different solutions of sugar separated by a selectively permeable membrane (marked with an arrow), which allows for the diffusion of glucose only.

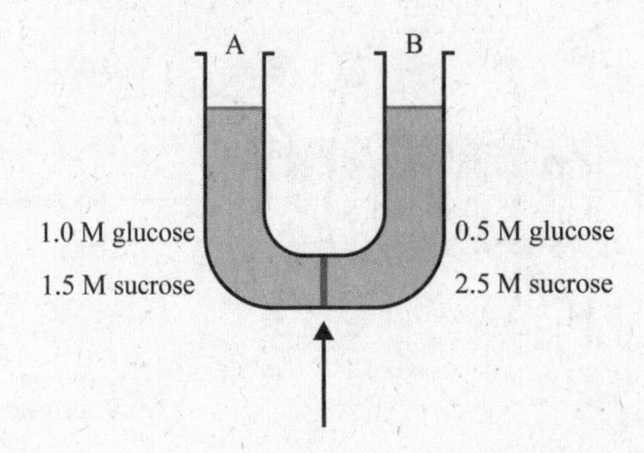

Which of the following choices correctly describes what will occur and gives supporting evidence for that occurrence?

(A) Glucose will diffuse from side B to side A because molecules flow down a gradient.

(B) Water will flow from side A to side B because the water potential (Ψ) is greater on side A.

(C) There will be no net movement of molecules because the concentration of glucose is greater on side A, while the concentration of sucrose is greater on side B.

(D) Sucrose will diffuse from side B to side A because the concentration of sucrose is greater on the right and molecules flow down a gradient from high concentration to low concentration.

23. Which of the following is correct about signal transduction pathways?

(A) Signal transduction pathways are found only in the cells of the most complex animals.

(B) Signal transduction pathways have evolved recently along with the development of enlarged brain size in mammals.

(C) In the signal transduction pathway, a single molecule can stimulate the release of thousands of molecules of product within a cell.

(D) Signal transduction pathways are unique in that receptors that stimulate the pathway are located only on the surface of a plasma membrane.

24. Which of the following initiates a signal transduction pathway that leads to a response in a cell?

(A) Cyclic Amp is formed in the cytoplasm.

(B) A transmembrane receptor changes its conformation.

(C) A kinase transfers a phosphate group from ATP to another molecule inside the cell.

(D) An ion channel opens.

25. Which of the following is correct about chemical signals that bind to receptors located on the cell membrane, rather than receptors in the cytoplasm?
    (A) They are hydrophobic.
    (B) They require a second messenger.
    (C) Nitric oxide is one example.
    (D) They are oil soluble.

26. Which of the following is correct about receptors that are located in the cytoplasm?
    (A) They require a second messenger.
    (B) One example is the ion-channel receptor.
    (C) They bind to hydrophilic substances.
    (D) They bind to a signal molecule in the cytoplasm, which causes a response in the nucleus.

27. Which of the following most accurately describes the advantage of a signal transduction pathway?
    (A) Transduction can occur in a cascade effect.
    (B) The binding of the signaling molecule causes the receptor protein to change its conformation.
    (C) The binding of the signaling molecule with the receptor can happen either on the surface of the cell or in the cytoplasm.
    (D) One binding molecule can trigger the release of thousands of molecules.

28. Which of the following correctly describes the role of cyclic AMP?
    (A) It acts as a second messenger for signaling molecules that cannot pass through the cell membrane.
    (B) It assists in the activation of the tyrosine kinase receptor.
    (C) It activates the G protein–coupled receptor.
    (D) It is an allosteric receptor in an ion-channel receptor.

Answers Explained

1. **(C)** The membrane-bound *receptor* that triggers the G protein is the G protein–coupled receptor (GPCR). G protein itself functions as the switch that transmits the message from outside the cell via the GPCR. (IST-3)

2. **(B)** The solution surrounding the cell is hypotonic to the cell, meaning the solution contains more water than solute or higher water potential than the cell. Water will tend to flow into the cell. If it is an animal cell, the extra water will lyse the cell. Although choice D is swollen or turgid, it is a plant cell. (ENE-2)

3. **(B)** Signal transduction pathways greatly amplify a signal. See Figure 3.25, which shows the binding of a single molecule of epinephrine at the GPCR with a total response of 10^8 molecules of glucose 1-phosphate. (IST-3)

4. **(D)** Normally, estrogen binds with the intracellular estrogen receptor, forming a complex that moves to the nucleus. Once there, the complex causes transcription of genes and a cellular response. The presence of BPA will interfere with this pathway by binding with estrogen receptors. It may either illicit a response without estrogen even being present, or BPA can block receptors and prevent an estrogen response, even though estrogen is present. BPA is in a class of chemicals known as estrogen pathway disruptors. (ENE-2)

5. **(C)** Apoptosis appears across the entire animal kingdom, not only in "advanced" animals, whatever they are. Normally, apoptosis triggers an aged cell to die, not to divide. The signals that trigger a cell to undergo apoptosis come from both outside and inside a cell. (ENE-2)

6. **(A)** The cell will shrink because water potential is greater inside the cell, and water will flow outward. Remember this fact: –0.7 is a higher value than –0.9. Hypotonic solutions have greater water potential than hypertonic ones. Also, adding solutes lowers the water potential. (ENE-2 & SP 2)

7. **(C)** Proteins are produced in the cytoplasm at the ribosomes on the ER and then packaged into vesicles in the Golgi and shipped to the plasma membrane for release. The nucleus is not involved. (SYI-1)

8. **(C)** Insulin is a protein, and proteins are manufactured in the ribosomes. They are located freely in the cytoplasm or as part of the rough endoplasmic reticulum. (SYI-1)

9. **(A)** Microtubules are hollow tubes that consist of the protein tubulin. They are organized in a 9 + 2 configuration and make up flagella, cilia, and spindle fibers. The other three answer choices all refer to microfilaments, which are solid, made of actin, and form a cleavage furrow in a dividing cell. (SYI-1)

10. **(B)** The theory of endosymbiosis explains how eukaryotic cells—those with internal membranes—came about. Small prokaryotic cells took up residence inside larger prokaryotic cells for the mutual benefit of both cells. The mutualistic relationship was so successful that it became permanent. For example, a small photosynthetic alga became engulfed by a larger cell. The larger cell provided all the cell's machinery to both cells, while the smaller cell provided food for both cells. This was the origin of the chloroplast. (EVO-1)

11. **(A)** The surface area of the cell membrane increases more slowly than the volume does. Observe the different formulas: The formula for the surface area of a sphere is $4\pi r^2$ (increases as a square function), while the formula for the volume of a sphere is $\frac{4}{3}\pi r^3$ (increases as a cube function). (ENE-1)

12. **(B)** The ameba has more microfilaments because microfilaments control ameboid movement. While it is often true that a large nucleus is generally characteristic of an active cell, it is not always true. Microtubules make up spindle fibers and are involved with cell division. Amebas have little sensory apparatus. (SYI-1)

13. **(C)** This figure depicts the Golgi body, which packages and secretes material that was received from the endoplasmic reticulum (ER). The side of the Golgi body that faces the ER is the *cis* face. The side that faces away from the ER is the *trans* face. (SYI-1)

14. **(C)** Starch is a large polysaccharide and cannot diffuse through the plasma membrane. CO_2 and O_2 both diffuse through the membrane because they are very small and nonpolar. Some amino acids are small, but some are very large. Amino acids are either polar or nonpolar and either acidic or basic. (ENE-1)

15. **(A)** Cholesterol is a large molecule that requires a special receptor to bring it into a cell. Choices B, C, and D are all about diffusion, which does not require an expenditure of energy. (ENE-2)

16. **(D)** The plasma membrane is organized and made of many small particles that move about readily. Hence the name, fluid mosaic. A membrane is a very active structure. A cell's activity is limited by how fast plasma membranes can take in and get rid of materials. (SYI-1)

17. **(C)** The structure at X is a plasmodesma, an opening in plant cell walls. It allows water and solutes to pass from one cell to another continuously throughout the plant. It is a passive transport system that works by diffusion. (ENE-2 & SYI-1)

18. **(C)** Part C points to the two hydrophobic lipid tails of every phospholipid. Part A is a carbohydrate that is involved in cell-to-cell identification; it is not involved in diffusion. Part B is a water-soluble head of the phospholipids that make up the cell membrane. Cholesterol floats within the membrane. Part D points to a protein channel of some sort. ATP synthase molecules have a very different appearance with a rotor that spins on the cytosol side. (ENE-2)

19. **(B)** Bacteria like *E. coli* are prokaryotes. They do not have any internal membranes. Although human red blood cells do not have a nucleus, they did have one before they matured and left the bone marrow. Human bone cells, osteoblasts, or osteocytes are part of the human body—thus, they are eukaryotic cells. Two domains, Bacteria and Archaea, include all prokaryotes. (EVO-1 & ENE-2)

20. **(D)** Looking at the illustration, you can see that sucrose is going against the gradient. Therefore, you can eliminate choices A and B. Protons are pumped against their gradient in order to create a proton gradient. As shown in the illustration, as protons flow down their gradient, they apparently carry sucrose with them. (ENE-2)

21. **(B)** This is the definition of quorum sensing, a phenomenon in bacteria. The statement in choice A is not correct because chemical messages readily flow from one plant cell to another through plasmodesmata. The statement in choice C is not correct because estrogen is hydrophobic, dissolves through the plasma membrane, and binds with receptors inside the cell. It is hydrophilic ligands that require a secondary messenger, like cyclic AMP. The statement in choice D is not correct, because the first thing that happens in a receiving cell is reception, not transduction. (ENE-2 & IST-3)

22. **(B)** If you add up the total solute on each side, you will see that there is more solute on side B. Therefore, water will tend to flow to side B. There is a higher concentration of glucose on side A, so glucose would not diffuse from side B to side A, making choice A incorrect. Choice C is not correct because there is lots of differential movement in both directions. Choice D is not correct because the information given in the question states that the membrane is not permeable to sucrose. (ENE-2)

23. **(C)** Signal transduction pathways are highly specific and regulated. Their similarity across different kingdoms speaks to a common ancestry. Receptors that stimulate signal transduction pathways are located both on the surface of the membrane, as well as within the cytoplasm. (IST-3)

24. **(B)** The key word is "initiates." The first thing that must happen before transduction can occur or before a response can result is that a receptor must be activated. (IST-3)

25. **(B)** Hydrophobic molecules (acting as the first messenger) bind with receptors on the surface of cell membranes. Once activated, the receptor catalyzes a reaction within the cytoplasm to produce cAMP, a common second messenger. The presence of cAMP in the cytoplasm leads to a cascade effect that ultimately causes a response in the cytoplasm or nucleus. (IST-3)

26. **(D)** Hydrophobic signaling molecules dissolve through the plasma membrane and bind with a receptor in the cytoplasm. In contrast, hydrophilic signaling molecules, those that bond to a receptor on the cell surface, require a second messenger. An ion-channel receptor is an example of a receptor on the surface of the cell. (IST-3)

27. **(D)** Although choices A, B, and C are correct statements, only choice D answers the question. The important thing about signal transduction is that a single signal can be amplified manyfold. (IST-3)

28. **(A)** Tyrosine kinase receptors, G protein–coupled receptors, and ion-channel receptors are all receptors on the surface of the plasma membrane. When the receptor is on the surface of the cell, a second messenger, like cyclic AMP, is required. (IST-3)

Cell Cycle

4

- → THE CELL CYCLE
- → CELL DIVISION AND CANCEROUS CELLS
- → MEIOSIS
- → MEIOSIS AND GENETIC VARIATION
- → REGULATION AND TIMING OF THE CELL CYCLE

Big Ideas: ENE, IST & SYI

Enduring Understandings: ENE-1 & ENE-3; IST-1 & IST-3; SYI-1

Science Practices: 1, 2 & 5

For the complete list of Big Ideas, Enduring Understandings, and Science Practices, refer to the "AP Biology Course and Exam Description" from the College Board: *https://apcentral.collegeboard.org/pdf/ap-biology-course-and-exam-description .pdf?course=ap-biology.*

INTRODUCTION

There are two types of cell division: **mitosis** and **meiosis**. **Mitosis** is responsible for **growth** and **repair**, produces two genetically identical daughter cells referred to as **clones**, and preserves the diploid ($2n$) chromosome number. Mitosis passes a complete genome from parent to the offspring.

BE CAREFUL

Know the difference between mitosis and meiosis.

Meiosis only occurs in sexually reproducing organisms and produces **gametes**. It results in cells that are **haploid**; they have half the chromosome number of the parent cell (n).

Any discussion about cell division must first consider the structure of the chromosome. A chromosome consists of a highly coiled and condensed strand of DNA. A replicated chromosome consists of two **sister chromatids**, where one is an exact copy of the other. The **centromere** is a specialized region that holds the two chromatids together. The **kinetochore** is a disc-shaped protein on the centromere that attaches the chromatid to the **mitotic spindle** during cell division. Figure 4.1 shows a sketch of a replicated chromosome.

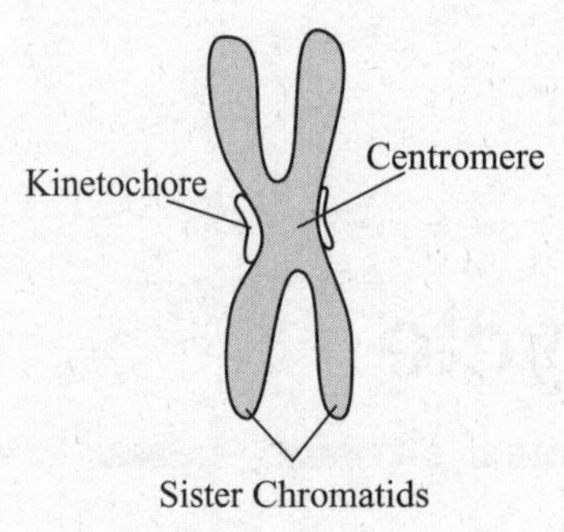

Figure 4.1 Chromosome

THE CELL CYCLE

Living and dividing cells pass through a complex sequence of growth and division called the **cell cycle** as shown in Figure 4.2. The timing and rate of cell division are crucial to normal growth and development. In humans, the frequency of cell division varies with the cell type. Bone marrow cells are always dividing in order to produce a constant supply of red and white blood cells. Liver cells are arrested in G_0 (have stopped dividing) but can be induced to divide or regenerate when liver tissue is damaged. Human intestine cells normally divide about twice per day to renew tissue destroyed during digestion. Specialized cells, like nerve cells, do not divide at all. In every case, however, the entire process is tightly regulated by a complex mechanism involving kinases and allosteric interactions. If something goes awry, the result can be uncontrolled cell division characteristic of cancer.

Two important factors limit cell size and promote cell division: the **ratio of the volume of a cell to its surface area** and the **capacity of the nucleus to control the entire cell**.

Ratio of the Cell Volume to Surface Area

As a cell grows, the surface area of the cell membrane *increases as the square* of the radius, while the volume of the cell *increases as the cube* of the radius. (The formula for the surface area of a sphere is $4\pi r^2$, and the formula for the volume of a sphere is $\frac{4}{3}\pi r^3$.) Therefore, as a cell grows larger, the volume inside the cell increases at a faster rate than the cell membrane increases. Since a cell depends on the cell membrane for an exchange of nutrients and waste products, the ratio of cell volume to membrane size is a major determinant of when the cell divides. See page 35.

Capacity of the Nucleus

The nucleus must be able to provide enough information to produce adequate quantities of all substances to meet the cell's needs. As a result, metabolically active cells are generally small. However, cells that have evolved a strategy to exist as large, active cells do exist in several kingdoms. Large, sophisticated cells like the paramecium have two nuclei that each control different cell functions. Human skeletal muscle cells are giant multinucleate cells. The fungus slime mold consists of one giant cell that contains thousands of nuclei.

Phases of the Cell Cycle

The cell cycle consists of five major phases: G_1, **S**, and G_2 (which together make up **interphase**), **mitosis**, and **cytokinesis**.

IST-1
Be able to describe the events that occur in the cell cycle.

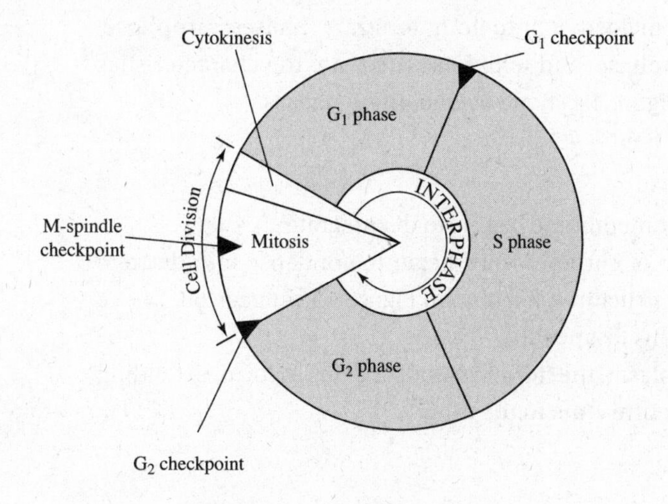

Figure 4.2 The Cell Cycle

INTERPHASE

Interphase consists of G_1 (Gap 1), **S** (Synthesis), and G_2 (Gap 2). The **G_1 phase** is a period of intense growth and biochemical activity. **S** stands for the synthesis or replication of DNA. **G_2** is the phase when the cell continues to grow and to complete preparations for cell division. More than 90 percent of the life of a cell is spent in interphase. When a cell is in interphase, it is not dividing, but it is very active. The chromatin is threadlike, not condensed. Within the nucleus are one or more **nucleoli**. A single **centrosome**, consisting of two **centrioles**, may be seen in the cytoplasm of an animal cell. The centrosome is duplicated during S phase. At the G_2-M transition, the two centrosomes separate from one another and move to opposite poles. Plant cells lack centrosomes but have **microtubule organizing centers**, **MTOC**s, that serve the same function. See Figure 4.3, which shows an animal cell during interphase.

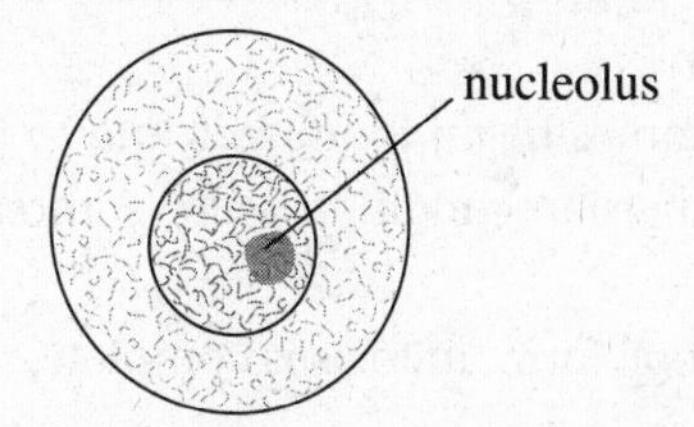

Figure 4.3 Cell During Interphase

STUDY TIP

Although the College Board does not require you to know the names of specific phases, you may find this information useful in understanding the process of mitosis.

MITOSIS

Mitosis consists of the actual dividing of the nucleus. It is a remarkable process because the DNA is passed from one generation to the next with great fidelity. It is a continuous process. However, scientists have divided it into four arbitrary phases: **prophase**, **metaphase**, **anaphase**, and **telophase**. Here are the characteristics of each phase. Figure 4.4 shows each of these phases.

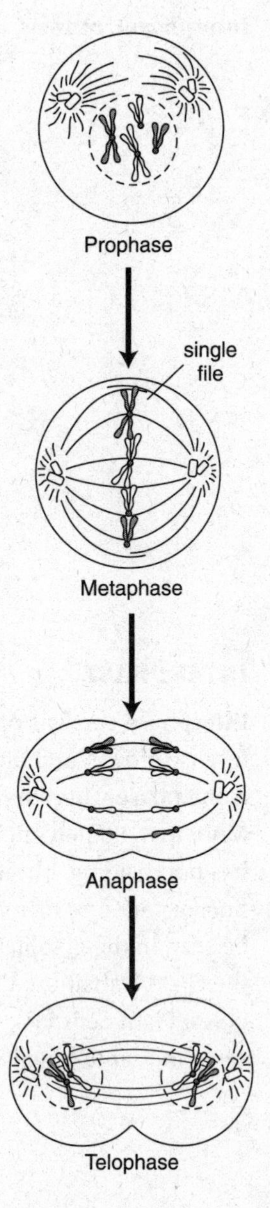

Figure 4.4 Stages of Mitosis

Prophase

- The nuclear membrane begins to disintegrate.
- The strands of chromosomes begin to condense into discrete observable structures, like that in Figure 4.1 on page 68.
- The nucleolus disappears.
- In the cytoplasm, the mitotic spindle begins to form, extending from one centrosome to the other.

Metaphase

- The *chromosomes line up in a single file* located on the equator or metaphase plate.
- Centrosomes are already positioned at opposite poles of the cell.
- Spindle fibers run from the **centrosomes** to the **kinetochores** in the **centromeres**.

Anaphase

- Centromeres of each chromosome separate, as spindle fibers pull apart the sister chromosomes that will move toward the poles.

Telophase

- Chromosomes cluster at opposite ends of the cell, and the nuclear membrane reforms.
- The supercoiled chromosomes begin to unravel and to return to their normal, pre-cell division condition as long, threadlike strands.
- Once two individual nucleoli form, mitosis is complete.

CYTOKINESIS

Cytokinesis consists of the dividing of the cytoplasm. It begins during mitosis, often during anaphase. In animal cells, a **cleavage furrow** forms down the middle of the cell as **actin** and **myosin microfilaments** pinch in the cytoplasm.

In plant cells, a **cell plate** forms during telophase as vesicles from the Golgi coalesce down the middle of the cell. Daughter plant cells do not separate from each other. A new cell wall forms, and a sticky **middle lamella** cements adjacent cells together. See Figure 4.5.

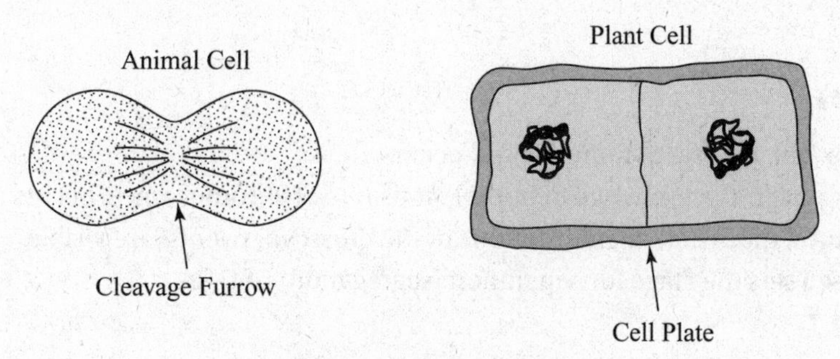

Figure 4.5 Cytokinesis

CELL DIVISION AND CANCEROUS CELLS

Normal cells grow and divide until they become too crowded; then they stop dividing and enter G_0 (G zero). This reaction to overcrowding is called **contact inhibition** or **density-dependent inhibition**. Another characteristic of normal animal cells is **anchorage dependence**. To divide, a cell must be attached or anchored to some surface, such as a Petri dish (in vitro) or an extracellular membrane (in vivo). Cancer cells show neither contact inhibition nor anchorage dependence. They divide uncontrollably and do not have to be anchored to any membrane. That is why cancer cells can migrate or metastasize to other regions of the body.

MEIOSIS

Meiosis generates the genetic diversity that is the raw material for natural selection and evolution.

STUDY TIP

In meiosis I, *homologous pairs* separate.

Meiosis is a form of cell division that produces **gametes** (ova and sperm). These gametes have the **haploid** or **monoploid** chromosome number (***n***), half the genetic material of the parent cell. During meiosis, the nucleus divides twice. Genetic material is randomly separated and recombined so that each gamete differs genetically from every other gamete. Sexual reproduction involves the fusion of two haploid gametes and restores the **diploid** chromosome number to its offspring. The two cell divisions in meiosis are called meiosis I and meiosis II.

Meiosis I, also called the **reduction division**, is the process by which *homologous chromosomes separate.* Each chromosome first pairs up precisely with its homologue into a **synaptonemal complex** by a process called **synapsis** and forms a structure known as a **tetrad** or **bivalent**. This pairing-up process is important. By aligning and binding the homologues together, accurate **crossing-over** is likely to occur. Crossing-over is the process by which *nonsister chromatids exchange genetic material.* It results in the *recombination* of genetic

IST-1
Be able to represent the connection between meiosis and increased genetic diversity necessary for evolution.

material. While crossing-over is occurring, what is visible under a microscope are **chiasmata** (chiasma, singular), Xs on the chromosome where homologous bits of DNA are switching places. *Crossing-over is a highly organized mechanism to ensure greater variation among gametes.*

Meiosis II is like mitosis. In this process, sister chromatids separate into different cells. The two stages of meiosis are further divided into phases. Each meiotic cell division consists of the same four stages as mitosis: prophase, metaphase, anaphase, and telophase.

Meiosis I

STUDY TIP

For details on spermatogenesis and oogenesis see pages 392 and 393.

PROPHASE I

- **Synapsis**, the pairing of homologues, occurs.
- **Crossing-over**, the exchange of homologous bits of chromosomes, occurs.
- **Chiasmata**, the visible manifestations of the crossover events, are visible.
- Prophase I sets the stage for separation (segregation) of DNA.

METAPHASE I

- The homologous pairs of chromosomes are lined up **double file** along the metaphase plate.
- **Spindle fibers** from the poles of the cell are attached to the centromeres of each pair of homologues.

ANAPHASE I

- There is a separation of homologous chromosomes as they are pulled by spindle fibers and migrate to opposite poles.

TELOPHASE I

- Homologous pairs continue to separate until they reach the poles of the cell. Each pole has the haploid number of chromosomes.

CYTOKINESIS I

- Cytokinesis usually occurs simultaneously with telophase I.

In some species, an interphase occurs between meiosis I and meiosis II. In other species, this does not occur. In either case, chromosomes do not replicate between meiosis I and II because chromosomes already exist as double or replicated chromosomes.

Meiosis II

Meiosis II is functionally the same as mitosis and consists of the same phases: prophase, metaphase, anaphase, telophase, and cytokinesis. The chromosome number remains haploid.

Figure 4.6 is a sketch comparing meiosis and mitosis. Each parent cell contains four chromosomes. The cell undergoing meiosis experiences one crossover.

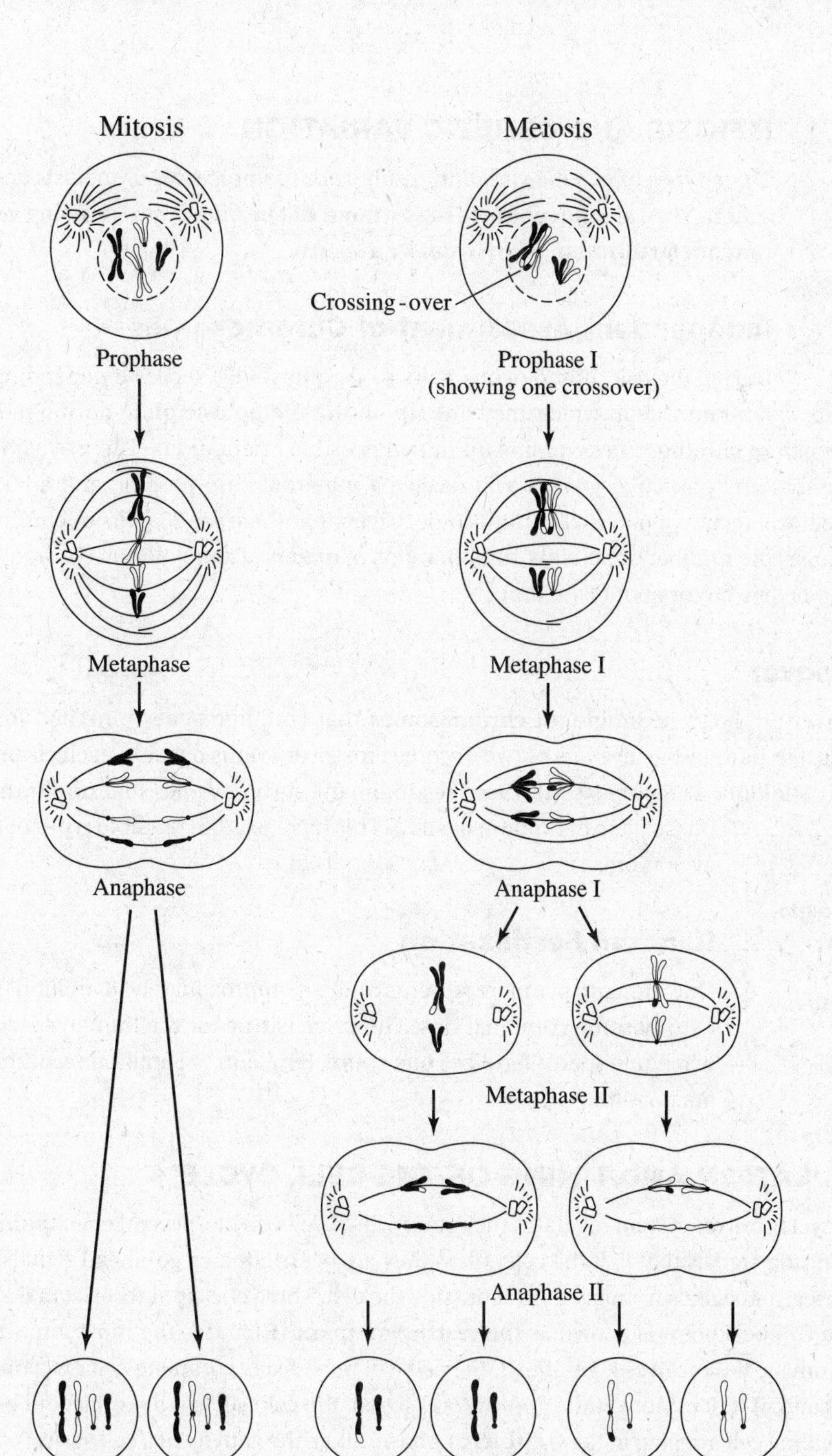

Figure 4.6 Comparison of Mitosis and Meiosis

LOOK CLOSELY

Pay attention to how metaphase differs during mitosis and meiosis I.

> **IST-1**
> In eukaryotes, heritable information is passed to the next generation by the cell cycle and by either mitosis or meiosis plus fertilization.

MEIOSIS AND GENETIC VARIATION

Three types of genetic variation result from the processes of meiosis and fertilization. They are **independent assortment of chromosomes**, **crossing-over**, and **random fertilization of an ovum by a sperm**.

Independent Assortment of Chromosomes

During meiosis, homologous pairs of chromosomes separate depending on the random way in which they line up on the **metaphase plate** during metaphase I. Each pair of chromosomes can line up in two possible orientations. There is a 50 percent chance that any particular gamete will receive a maternal chromosome and a 50 percent chance it will receive a paternal chromosome. Given that there are 23 pairs of chromosomes in humans, the number of possible combinations of maternal and paternal chromosomes in each gamete is 2^{23}, or about 8 million.

Crossover

Crossover produces **recombinant chromosomes** that combine genes inherited from both parents. For humans, an average of two or three crossover events occur in each chromosome pair. In addition, at metaphase II, these recombinant chromosomes line up on the metaphase plate in random fashion. This increases the possible types of gametes even more.

> **IST-1**
> Fertilization increases genetic variation in populations by providing new combinations of genes. It also restores the diploid number.

Random Fertilization

One human ovum represents one of approximately 8 million possible chromosome combinations. The same is true for the human sperm. Thus, when one sperm fertilizes one ovum, 8 million $\times$ 8 million recombinations are possible.

REGULATION AND TIMING OF THE CELL CYCLE

A **cell cycle control system** regulates the rate at which cells divide. Several **checkpoints** act as built-in stop signals that halt the cell unless they are overridden by go-ahead signals. Checkpoints register signals from inside and outside the cell. Three checkpoints exist in $\mathbf{G_1}$, $\mathbf{G_2}$, and **M**. The $\mathbf{G_1}$ checkpoint is known as the **restriction point (R)** and is the most important one in mammals. If it receives a go-ahead, the cell will most likely complete cell division. On the other hand, if it does not get the appropriate signal, the cell will exit the cycle and become a nondividing cell arrested in the **$\mathbf{G_0}$ (G zero) phase**. Since the activity of a cell varies, the rate at which it needs to divide also varies. The timing of the cell cycle is initiated by **growth factors**, like PDGF, and controlled by two kinds of molecules: cyclins and protein kinases.

> **STUDY TIP**
> **This is a great example of cell signaling.**

Cyclins are proteins that get their name because their levels cyclically rise and fall in dividing cells. They are synthesized during every S and G_2 phase, and broken down after every M phase. **Kinases** are a ubiquitous class of proteins that activate other proteins by *phosphorylating* them. The kinases critical to the cell cycle are activated only when they bind to a cyclin. Hence, they are aptly named **cyclin-dependent kinases** or **Cdks**. When a Cdk binds to a cyclin, a **cyclin-Cdk complex** is formed. One particular cyclin-Cdk complex, **MPF**, *triggers the cell's passage from G_2 to mitosis (M)*. MPF stands for **maturation promoting factor**, but you may think of it as *M phase promoting factor.* MPF contributes to molecular events required for chromosome condensation and spindle formation during prophase.

After M phase, during anaphase, MPF switches *off* by initiating a process that leads to the breakdown of cyclin. The noncyclin part of MPF, the Cdk, persists in the cell in an inactive form until it becomes part of MPF again. This can be seen in Figure 4.7.

The details of MPF and PDGF are presented as illustrative examples only (to help you understand how internal and external signals regulate the cell cycle). You are not required to memorize them.

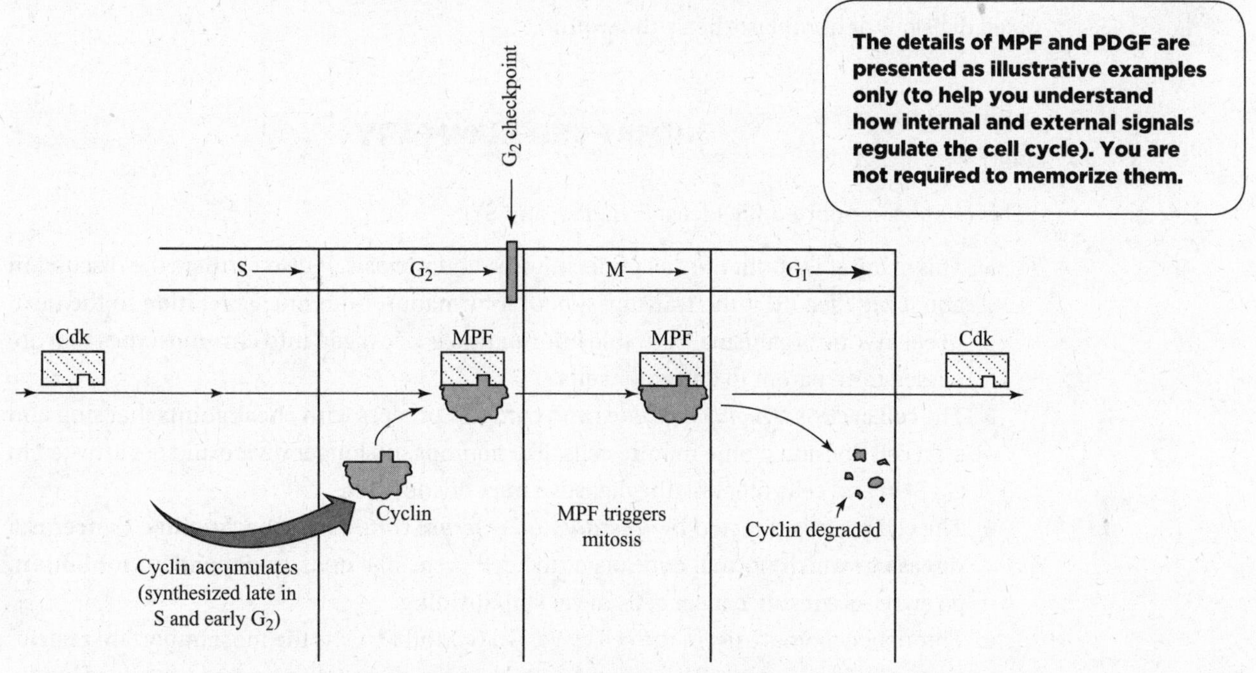

Figure 4.7 Kinase (Cdk) and Cyclin Regulating Cell Cycle
(When MPS passes the G_2 checkpoint, mitosis occurs.)

The activity of a Cdk, cyclin-dependent kinase, rises and falls with changes in the concentration of its cyclin partner. Figure 4.8 shows the fluctuation activity of MPF, the cyclin-Cdk complex. Notice that the peaks of MPF activity correspond to a rise in cyclin concentration. The cyclin levels rise during the S and G_2 phases and then fall abruptly during the M phase. MPF triggers the cell's passage past the G_2 checkpoint into the M phase.

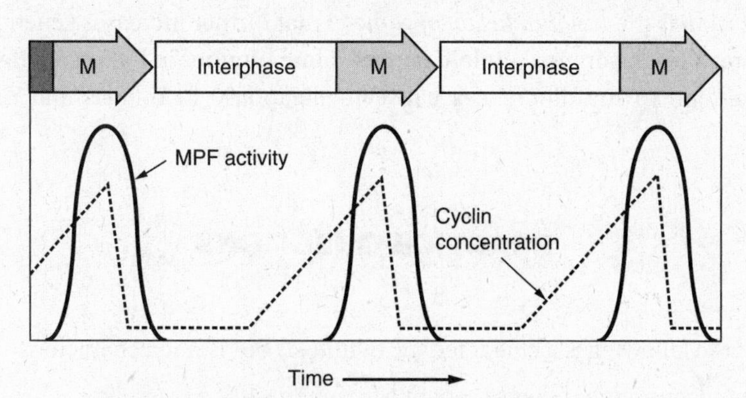

Figure 4.8 Fluctuation of MPF Activity and Cyclin Concentration During the Cell Cycle

Platelet Derived Growth Factor (PDGF)

PDGF is an external signal that drives the cell cycle. Specialized cells called fibroblasts have **PDGF receptors** on their surface membrane. When PDGF binds to a receptor, it triggers a **signal transduction pathway** that allows cells to pass the G_1 checkpoint and divide. When an injury occurs in the body, platelets release PDGF near the injury, and a massive proliferation (cell division) of fibroblasts heals the wound.

CHAPTER SUMMARY

This chapter supported Big Ideas: ENE, IST, and SYI.

- This chapter is about the cell cycle, mitosis, and meiosis. It also furthers the discussion about Big Idea IST—the transmission of information from one generation to the next. In eukaryotic organisms, heritable information is packaged into **chromosomes** that are passed from parent to **daughter cells**.
- The **cell cycle** is a *highly regulated* and complex process with **checkpoints** that stop and start cell division. Some mature cells, like neurons, no longer divide and are arrested in G_0. Skin and cells that line the digestive tract divide often.
- The cell cycle is directed by *internal and external controls* or **checkpoints**. **Cancer** is a disease in which normal controls of the cell cycle, like **density-dependent inhibition,** go awry. As a result, cancer cells never stop dividing.
- Three checkpoints exist in the cell cycle: G_1, G_2, and M. G_1 is the most important **restriction point** in mammals. The cell cycle is initiated by **growth factors** and controlled by two kinds of molecules: **cyclins** and **kinases**. One important **cyclin-kinase complex** is **MPF,** *maturation promoting factor,* which triggers the cell's passage from G_2 to M (metaphase).
- **Mitosis** produces *identical* daughter cells. It allows organisms to grow, replace cells, and reproduce *asexually*. Understand what happens in each stage of mitosis.
- **Meiotic** cell division produces **gametes**, each with the **haploid chromosome number**. During meiosis I, **homologous chromatids** exchange genetic material via **crossing-over** before they separate. Crossing-over increases **genetic diversity**, which is necessary for evolution.
- **Fertilization** is the *random fusion of gametes* that further increases genetic diversity, as well as restores the original **diploid chromosome number** in the **zygote** (fertilized egg).
- Pay attention to the differences between metaphase in mitosis and metaphase in meiosis.

PRACTICE QUESTIONS

1. Which of the following is a characteristic of mitosis but is not a characteristic of meiosis?
   (A) Homologous chromosomes separate in anaphase I.
   (B) Damaged tissue is repaired.
   (C) Chiasmata are visible.
   (D) Prophase is followed by metaphase.

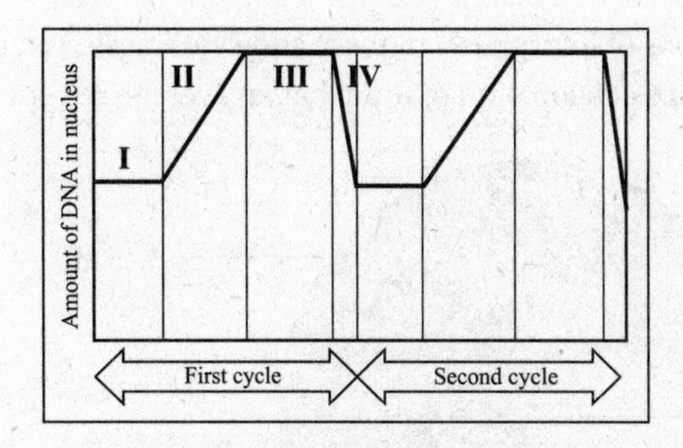

2. The cell cycle consists of interphase (G1, G2, and S), mitosis, and cytokinesis. The graph above shows changing DNA concentrations during the various stages of the cell cycle. Which of the following statements correctly identifies the stage of the cell cycle and also includes the correct support for that claim?
   (A) Stage IV is S because the amount of DNA decreases.
   (B) Stage II is G1 and G2 because the amount of DNA increases.
   (C) Stage II is S because the amount of DNA increases.
   (D) Stage III is interphase because there is no change in the amount of DNA in the nucleus.

3. Researchers are investigating the relationships among different organisms in the animal kingdom based on differences and similarities in their cell cycles. Which of the following scientific questions is most relevant to this investigation?
   (A) How does the production of gametes across different animal kingdoms compare?
   (B) How does the structure of microfilaments that make up spindle fibers in different animal cells compare?
   (C) How do the genes that code for ribosomes compare?
   (D) How do the genes that code for MPF compare?

4. Which of the following is a factor that limits cell size?
   (A) how active a cell is
   (B) what kind of activity a cell is engaged in
   (C) the ratio of the volume to the cell surface area
   (D) whether the cell is a plant cell or animal cell

5. Which of the following is correct about this cell undergoing meiosis?

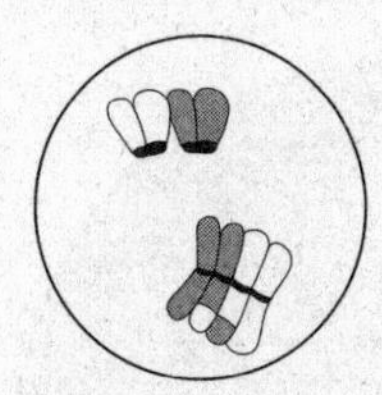

   (A) The cell contains the haploid number of chromosomes.
   (B) The cell is most likely in interphase.
   (C) Homologous chromosomes are undergoing crossing-over.
   (D) Chromosomes are replicating.

6. The cell cycle consists of interphase, mitosis, and cytokinesis. Which of the following pie charts shows the correct allocation of time that an average cell spends in each process?

(A)

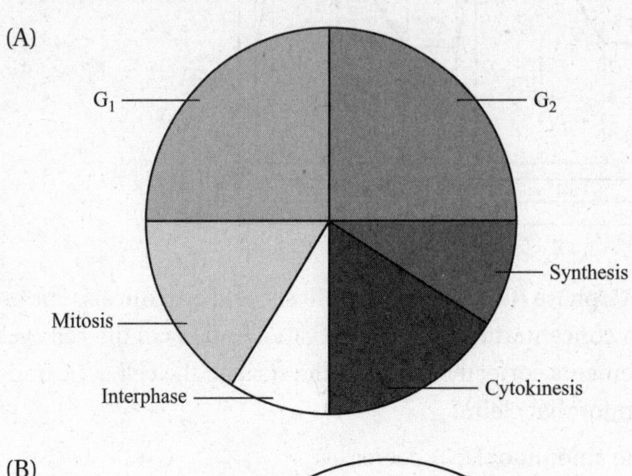

(B)

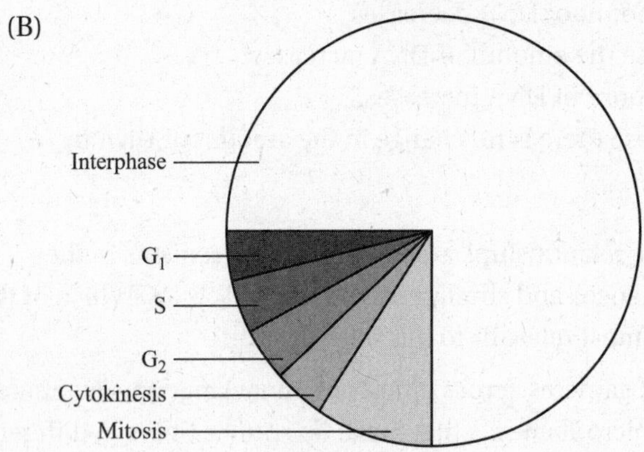

(C)

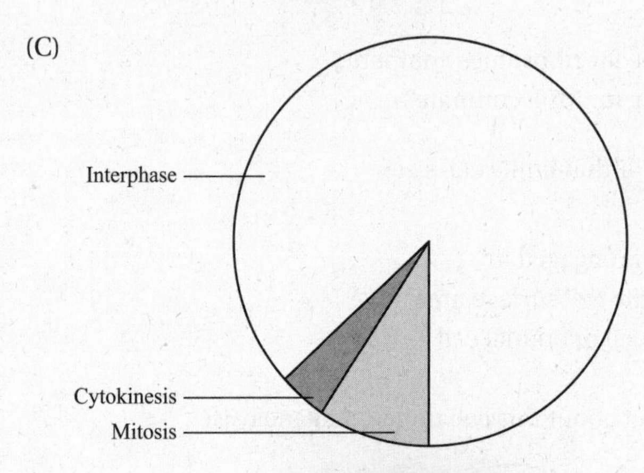

(D)

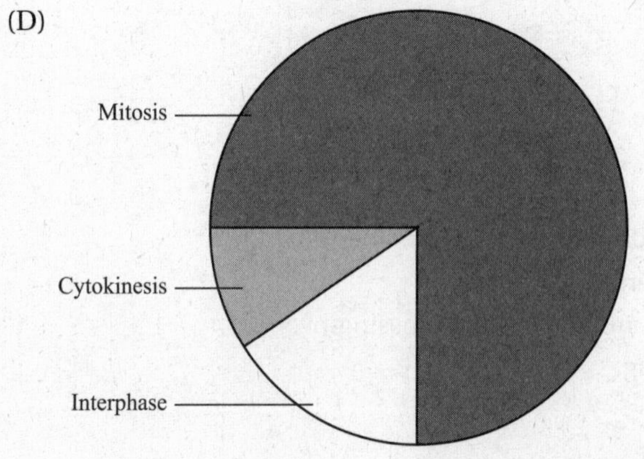

7. Which of the following statements is correct about cell division and the cell cycle?

   (A) MPF triggers a cell's passage from G_2 to metaphase.
   (B) Chiasmata are a significant feature of mitotic cell division.
   (C) If a cell begins meiosis with 12 chromosomes, after meiosis I each daughter cell will still contain 12 chromosomes.
   (D) During metaphase in plant cells, spindle fibers connect each centromere of a chromosome to the centriole.

8. Which of the following statements is correct about cell division and the cell cycle?

   (A) Normal cells growing in culture exhibit density-dependent inhibition.
   (B) In metaphase of meiosis I, chromosomes line up on the metaphase plate single file.
   (C) Chromosomes pair up during crossing-over in preparation for synapsis.
   (D) Cells of the intestine are normally arrested in G_0.

9. Which of the following statements is correct about the cell cycle?

   (A) Normal human body cells spend most of their time and energy dividing.
   (B) Interphase consists of prophase, metaphase, anaphase, and telophase.
   (C) MPF is a cyclin-dependent kinase (Cdk) that triggers the cell to divide.
   (D) Cancer cells must be attached to some surface in order to divide.

Questions 10–11

Assume that each figure below represents a cell that will continue to divide and produce two daughter cells.

A B C D

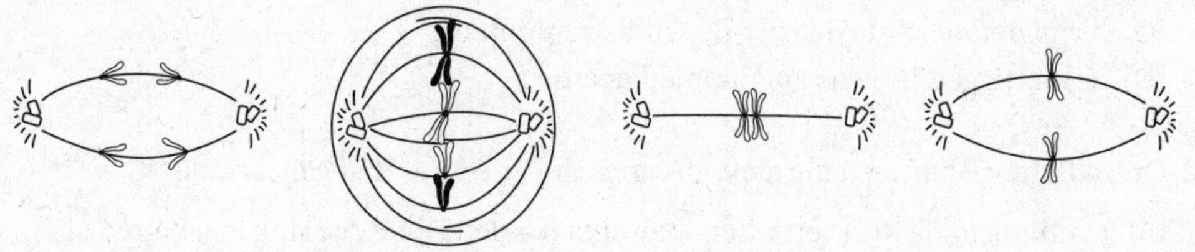

10. Which figure could represent *only* meiotic cell division?

    (A) A
    (B) B
    (C) C
    (D) D

11. In which figure will the daughter cells that result from each cell division show only half the number of chromosomes of the parent cell?

    (A) A
    (B) B
    (C) C
    (D) D

12. During meiosis, homologous pairs of chromosomes line up on the metaphase plate during metaphase I. The illustration below shows one homologous pair that has experienced one crossover event.

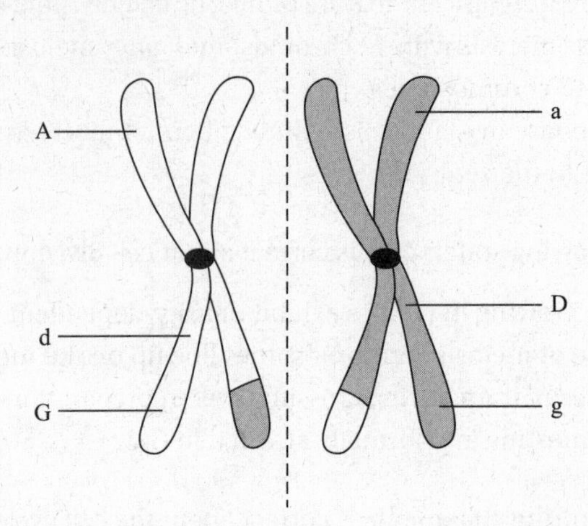

Which of the following choices accurately predicts the gametes that will result from this pair?

(A) AdG, AdG, aDg, ADg
(B) AdG, Adg, aDG, aDg
(C) Aa, dD, Gg
(D) Adg, Adg, aDG, aDG

13. Which of the following is correct about a cell that passes through the G_1 restriction point?

(A) It will enter G_0 and stop dividing.
(B) It will divide.
(C) It will demonstrate density-dependent inhibition.
(D) It will trigger its own programmed death.

14. One difference between normal cells and cancer cells is that cancer cells

(A) continue to divide even when they are crowded while growing in a Petri dish.
(B) usually have more than one nucleus.
(C) are unable to replicate their DNA.
(D) are arrested in G_0.

15. Which of the following is a correct statement about a normal human skin cell?

(A) Meiosis reduces the chromosome number from 46 to 23.
(B) Homologous pairs separate.
(C) Synapsis forms a structure called a synaptonemal complex.
(D) Spindle fibers connect the kinetochore of a chromosome to one of the poles in the cell.

16. In the cells of some organisms, mitosis occurs *without* cytokinesis. What is the result?

    (A) Cells will contain the same number of chromosomes as each other.
    (B) Cells will not survive the process.
    (C) Cells will be abnormally small.
    (D) Cells will contain more than one nucleus.

17. The drug Cytochalasin D inhibits actin formation in a cell while preserving the structure of the rest of the cytoskeleton. Which of the following processes of the cell cycle would be disrupted by exposure to Cytochalasin D?

    (A) chromosomes lining up on the metaphase plate
    (B) DNA synthesis
    (C) formation of the cleavage furrow
    (D) condensing of chromosomes

18. The drug Taxol, which comes from the Pacific yew tree, is a common chemotherapeutic treatment to stop the spread of cancer of the breast. The principal mechanism by which it works is the disruption of the microtubule assembly. As a result of its function, predict the cellular process that Taxol would most directly interfere with.

    (A) degradation of cyclin
    (B) binding of cyclin to Cdks
    (C) mitotic spindle formation
    (D) chromosome replication

Questions 19–20

The following two questions refer to one experiment.

19. Two researchers performed an experiment based on their hypothesis that the progression of the cell cycle is controlled by molecules from within the cytoplasm. To investigate this hypothesis, they chose cultured mammalian cells in different phases of the cell cycle and induced them to fuse. Below is a sketch of two of the cells at the beginning of the experiment.

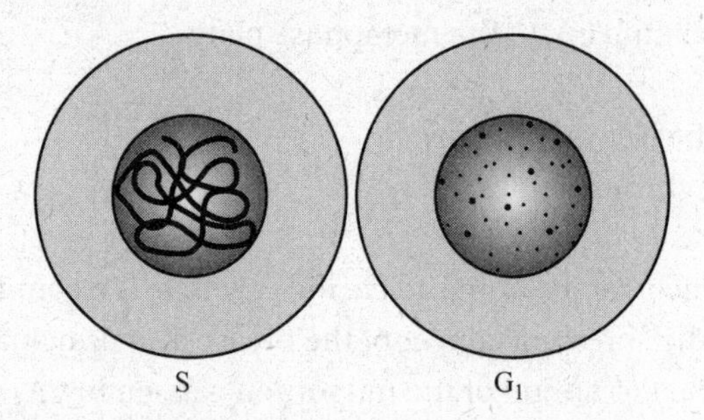

Assuming that the researchers' hypothesis is correct, choose the image that correctly predicts the result of the experiment after fusion.

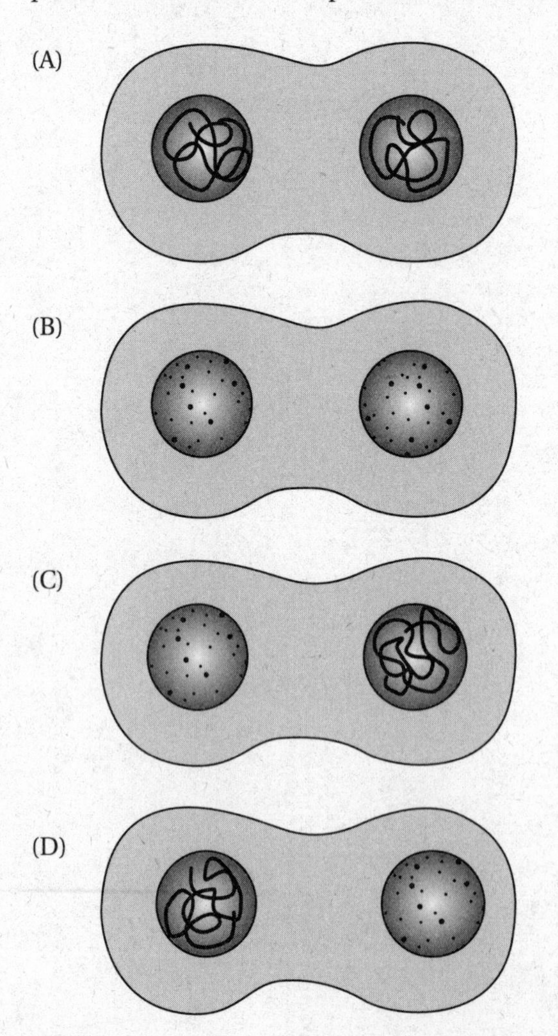

20. Refer to the same two cells sketched in Question 19. If the progression of the phases of the cell cycle was not triggered by molecules from the cytosol, but instead each cell moved into the next phase automatically after completion of its current phase, which of the following correctly predicts what would happen to the original two fused cells?

(A) The cell on the left would enter G_1; the cell on the right would enter G_2.
(B) The cell on the left would enter G_2; the cell on the right would enter G_2.
(C) The cell on the left would enter G_2; the cell on the right would enter S.
(D) The cell on the left would enter G_1; the cell on the right would enter S.

Questions 21–22

The kinetochore is the structural component of the chromosome at which microtubules of the spindle attach. It has been established that during anaphase, microtubules that connect the kinetochore to the poles lose subunits and shorten. Scientists acknowledge two possible theories about the role of the kinetochore in this spindle shortening. The first is that the kinetochore is *pulling* on the spindle fibers (like reeling in a fish), dragging passive chromosomes to the poles. An opposing theory suggests that kinetochores *actively* "walk" on the microtubules of the spindle. To advance scientific knowledge and determine which theory is correct, three researchers carried out the following experiment.

First, they labeled the microtubules of a pig kidney cell in early metaphase with yellow fluorescent dye (YFD). See Figure 1.

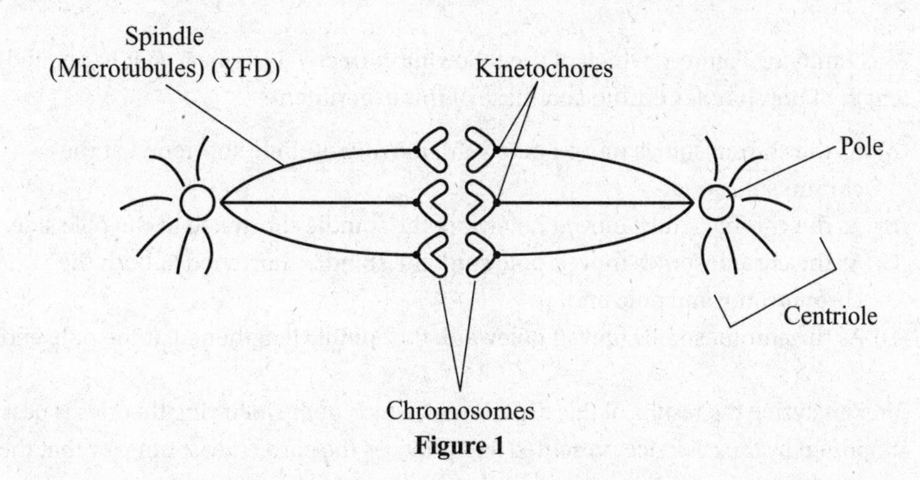

Figure 1

Next, they *marked* (M) an approximately 1.0 micron-wide band across the spindle region of the microtubules between one spindle pole and the chromosomes by using a laser to photobleach and eliminate the YFD from that region. This left the microtubules intact. See Figure 2.

(continues on the next page)

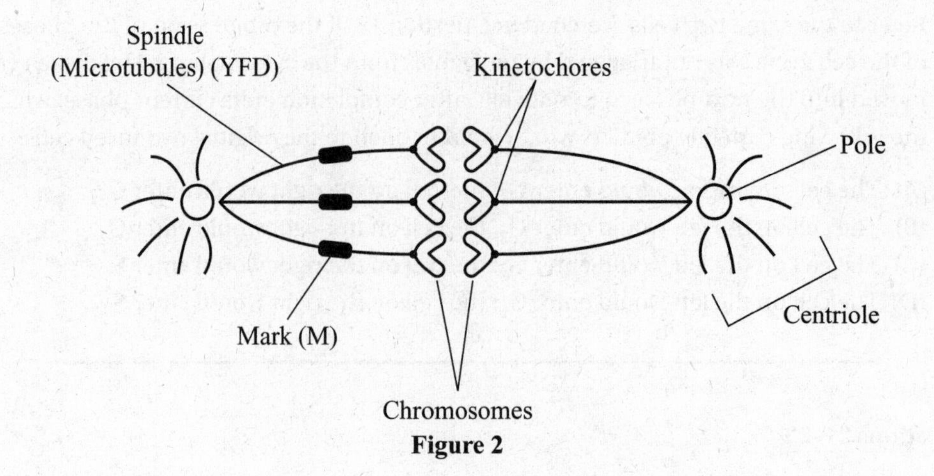

Figure 2

As anaphase continued, the researchers monitored the changes in the microtubule length by measuring the length of the spindle on both sides of the *mark* (M). Figure 3 shows the results of the experiment.

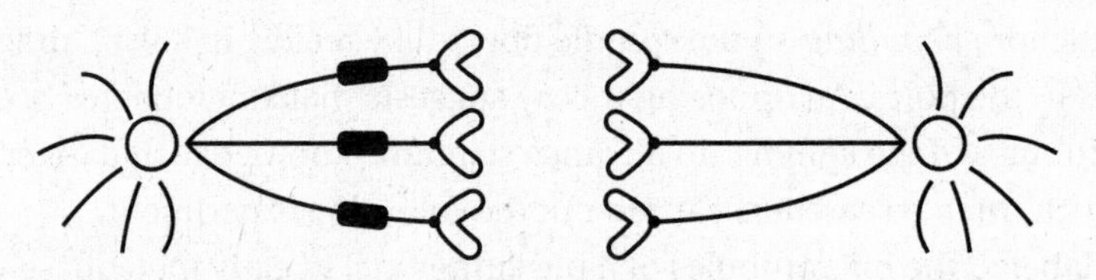
Figure 3

21. After studying Figure 3, which of the following correctly describes what happened to the length of the spindles during anaphase of this experiment?

    (A) As the chromosomes moved poleward, the microtubule shortened at the chromosome end.
    (B) As the chromosomes moved poleward, the spindle shortened at the pole side.
    (C) As the chromosomes moved poleward, the spindle shortened at both the chromosome and pole end.
    (D) As the chromosomes moved poleward, the spindle lengthened at the pole end.

22. After analyzing the results of this experiment, which of the following theories is best supported by the evidence presented and provides the correct reasoning for that theory?

    (A) Chromosomes are passively pulled along the spindle fibers because the chromosome end of the spindle is shortened.
    (B) Chromosomes are passively pulled along the spindle fibers because the pole end of the spindle is shortened.
    (C) Chromosomes "walk" along the spindle fibers as the spindle disassembles at the kinetochore end because the kinetochore end is shortened.
    (D) Chromosomes "walk" along the spindle fibers as the spindle disassembles at the kinetochore end because the pole end is shortened.

Answers Explained

1. **(B)** Mitosis is the type of cell division that is responsible for growth and repair. Choices A and C relate to meiosis, not mitosis, and the production of gametes. Choice D is something that occurs in *both* mitosis and meiosis, but the question asks for the statement that is only a characteristic of mitosis. (IST-1)

2. **(C)** Stage II represents S for synthesis, the only stage in which DNA replicates (doubles). Choice A is not correct because the amount of DNA is decreasing, not doubling, as it does in the S phase. Choice B is not correct because there is no DNA increase in G_1 and G_2. Choice D is not correct because, although stage III shows no change in the amount of DNA, interphase includes the S phase of cell division, where the amount of DNA doubles. (IST-1)

3. **(D)** Only choice D refers to mitosis. Maturation promoting factor (MPF) is an important protein complex that triggers a cell into metaphase of mitosis. Choice A is not the most relevant scientific question because its content relates to meiosis, not the cell cycle. Choice B is not the most relevant scientific question because *microtubules*, not *microfilaments*, make up spindle fibers. If the question had used the term *microtubules*, this would have been a relevant scientific question. Choice C is not the most relevant scientific question because ribosomes do not have anything to do with the cell cycle. (IST-1 & SYI-1)

4. **(C)** Two important factors limit cell size and promote cell division: the ratio of the volume of a cell to the surface area and the capacity of the nucleus. (ENE-3)

5. **(C)** The cell is in prophase, not interphase, because chromosomes have condensed and are visible as discrete structures. During this time, condensed chromosomes pair up and homologous chromosomes exchange DNA. That is the event during meiosis that shuffles genes. During interphase, chromosomes are threadlike and invisible under a light microscope. Chromosomes replicate during the S phase of interphase. The cell contains the diploid chromosome number, not the haploid number. (IST-1)

6. **(C)** Most of the life of a cell is spent in interphase. Therefore, the largest slice of the pie chart should be interphase. It does not matter that G_1, S, and G_2 are not labeled as part of interphase in choice C. Choice B is not correct because, although it allocates a large amount of the pie chart for interphase, it shows G_1, S, and G_2 as separate slices rather than part of the interphase slice. Choices A and D incorrectly show interphase as a small slice of the pie. (IST-1)

7. **(A)** See Figure 4.8. The M phase, which is triggered by MPF, occurs after G_2. Chiasmata are a significant feature of *meiotic* cell division. If a cell begins meiosis with 12 chromosomes, after meiosis I, each daughter cell will contain half that number, 6. Plant cells do not have centrioles. (IST-1 & IST-3)

8. **(A)** Density-dependent inhibition means that when cells growing in culture get crowded, they stop dividing. That is a feature of normal cells growing in culture. Cancer cells do not demonstrate this characteristic. That is why one trait of cancer cells is that they do not stop dividing. In metaphase of meiosis I, chromosomes line up double file, not single file. The process when chromosomes pair up is called synapsis. They do this in preparation for crossing-over, not the other way around. Cells in the intestine are not arrested. They divide every day to replace cells that have been damaged during digestion. (ENE-3)

9. **(C)** The cell cycle is tightly regulated by internal controls or checkpoints. MPF triggers the cell's passage from G_2 into mitosis, the M phase. Choice A is not correct because normal cells spend most of their time and energy in interphase (G_1, S, and G_2), not dividing. Choice B is incorrect because interphase consists of G_1, S, and G_2. Choice D is incorrect because cancer cells do not exhibit anchorage dependence; they can travel throughout the body, which is called metastasis. (IST-1)

10. **(C)** This choice is true because the homologous *pairs* of chromosomes are shown paired up in preparation for separating from each other, as you would expect in meiosis I. The other choices could be either meiosis I or II or mitosis. (IST-1)

11. **(C)** This choice is true for the reason given in the answer explanation to Question 10. Meiosis produces daughter cells with half the number of chromosomes. (IST-1)

12. **(B)** The illustration consists of one homologous pair of chromosomes. Before the crossing-over event, each homologue consisted of two identical sister chromatids. If one chromatid contained genes A, d, and G, its sister chromatid must have had the same A, d, and G genes. The same is true of the other chromosome, in which both sister chromatids contain a, D, and g genes. After crossing-over, the G and g genes on the two touching and adjacent chromatids switch places. The chromatid on the extreme left side remains A, d, and G. After the crossover, the inner chromatid on the left side becomes A, d, and g. The next chromatid (moving right) becomes a, D, and G, while the last chromatid on the right remains unchanged as a, D, and g. This question tests your understanding of the importance of meiosis and crossing-over in producing great genetic variation in gametes and offspring and increased genetic variation for evolution. (IST-1)

13. **(B)** This choice is an important fact about the G_1 restriction point. If a cell passes G_1, it will divide. If it does NOT pass this point, the cell will enter G_0 and stop dividing. (ENE-3)

14. **(A)** Cancer cells do *not* exhibit the trait of density-dependent inhibition, which is described in choice A. Although cancer cells commonly have more than the diploid number of chromosomes, they contain only one nucleus. Cancer cells are not arrested in G_0. On the contrary, they constantly replicate their DNA and divide. (ENE-3)

15. **(D)** Normal human skin cells constantly divide by *mitosis* to repair damage to the skin. Choices A, B, and C are all characteristics of *meiosis* and human cells do not undergo meiosis. Meiotic cell division produces gametes only. (IST-1)

16. **(D)** Choice D correctly describes what happens to a cell that undergoes mitosis without cytokinesis. In fact, this is what happens in human skeletal muscle cells. It explains how the cells end up being multinucleated and very large. (IST 1)

17. **(C)** Disruption of actin, a type of microfilament, would impair or block cytokinesis, the formation of the cleavage furrow or cell plate (in plants) and thereby prevent cell division. Spindle fibers, which connect chromosomes from the metaphase plate to centrioles, consist of microtubules, not microfilaments. DNA synthesis does not involve microfilaments. The condensing of chromosomes does not require microfilaments either. (IST-1)

18. **(C)** Microtubules make up spindle fibers. The configuration is 9 triplets with no microtubules in the center. (Cilia and flagella consist of a 9 + 2 configuration of microtubules.) (SYI-1)

19. **(A)** The image presented in choice A confirms the researchers' hypothesis because, when a cell in the S phase is fused with one in G_1, the G_1 nucleus immediately enters the S phase and DNA is synthesized. This demonstrates that molecules present in the cytoplasm during the S and M phases control the progression of those phases of the cell cycle. Choice B shows the cell in the S phase going backward into G_1, which does not confirm the researchers' hypothesis. Choice C shows both cells changing into the other's phase, which does not make sense. Choice D shows no change from the original sketch and thus does not confirm the researchers' hypothesis. (IST-1 & SP 2)

20. **(C)** The order of phases in the cell cycle are G_1, S, G_2, and M (mitosis). Thus, a cell that is in G_1 would move into S and then G_2. A cell in S phase would move into G_2 and then M. (IST-1 & SP 2)

21. **(A)** Notice the difference between Figure 1 and Figure 2. The spindle is much shorter on the chromosome end in Figure 2. The kinetochore is on the chromosome and is the attachment site for the spindle fiber. (SP 2)

22. **(C)** Since the spindle fiber near the chromosome in Figure 3 is much shorter than it was in Figure 2, and given the two possible theories described in the background information that preceded Questions 21 and 22, the experiment confirms that kinetochores *actively* "walk" on the microtubules of the spindle. (SP 1)

Energy, Metabolism, and Enzymes

5

- → ENERGY
- → METABOLISM
- → ENZYME-CONTROLLED REACTIONS
- → CHARACTERISTICS OF ENZYMES
- → INHIBITION OF ENZYMATIC REACTIONS

Big Ideas: ENE & SYI

Enduring Understandings: ENE-1; SYI-3

Science Practices: 1 & 4

For the complete list of Big Ideas, Enduring Understandings, and Science Practices, refer to the "AP Biology Course and Exam Description" from the College Board: *https://apcentral.collegeboard.org/pdf/ap-biology-course-and-exam-description.pdf?course=ap-biology*

INTRODUCTION

This chapter and the two that follow are all about **energy**. This chapter is about how **enzymes** control reactions governed by the laws of thermodynamics. It lays the groundwork to enable you to understand the next two chapters. Chapter 6, "**Cell Respiration**," is about how cells extract energy from food and convert it to a readily usable form in the cell. Chapter 7, "**Photosynthesis**," explains how certain organisms can convert energy from light and store it in chemical bonds in the cell.

ENE-1
All living systems require a constant input of free energy.

ENERGY

Energy is the ability to do work. **Potential energy** is stored energy that results from an object's position (like a boulder on the edge of a cliff) or structure (like a complex chemical). The more unstable something is, the more potential energy it possesses. The more stable something is, the less potential energy it possesses. **Chemical energy**, a form of potential energy, is energy that is stored in chemical bonds. In general, the more bonds there are in a molecule, the greater its potential energy is. (Fat, a complex molecule, contains more potential and chemical energy than a simple sugar, like glucose.) **Kinetic energy** is related to anything that moves. **Thermal energy** is kinetic energy that is associated with random movement of particles. The higher the temperature of an object, the faster its particles move and the greater its kinetic energy is. When thermal energy is transferred from one object to another, it is called **heat**.

Light is another form of energy. Light consists of the electromagnetic spectrum from gamma radiation (shortest) to radio waves (longest).

Living organisms require a constant input of energy. Biological systems constantly transform this energy from one form to another in order to carry out all life functions. The laws of thermodynamics govern all energy transformations. The **first law of thermodynamics** states that energy cannot be created or destroyed, only transformed from one form to another. For example, plants convert light energy to chemical energy during photosynthesis. This law is also known as the **law of conservation of energy**. The **second law of thermodynamics** states that during energy conversions, the universe becomes more disordered. In other words, entropy increases. We can determine how much free energy is available to do work within a cell by calculating the **Gibb's free energy**, or just free energy, represented by the letter G. During the course of a reaction, if energy is released, the reaction is said to be **exergonic** or exothermic and ΔG is < 0 (negative). If $\Delta G < 0$, the reaction is spontaneous. If in a chemical reaction energy is absorbed, the reaction is **endergonic** or endothermic and ΔG is > 0 (positive). In complex cellular reactions, exergonic and endergonic chemical reactions are coupled. **Exergonic reactions power the endergonic ones.**

METABOLISM

Metabolism is the sum of all the chemical reactions that take place in cells. Some reactions break down molecules (**catabolism**); other reactions build up molecules (**anabolism**). Metabolic reactions take place in a series, or **pathways**, each of which serves a specific function. These multistep pathways are controlled by enzymes and enable cells to carry out their chemical activities with remarkable efficiency.

ENZYME-CONTROLLED REACTIONS

Enzymes do not provide energy for a reaction, and they do not enable a reaction to occur that would not occur on its own. Enzymes orient substrate molecules into positions that favor reactions to occur. Enzymes serve as **catalytic proteins** that speed up reactions by lowering the **energy of activation,** E_A, the amount of energy needed to begin a reaction. In the potential energy diagram in Figure 5.1, the potential energy of the products is less than the potential energy of the reactants, so energy is released and the reaction is exergonic. The dotted line shows the same reaction when an enzyme is introduced. The enzyme serves to lower the energy of activation, and the reaction can proceed more quickly.

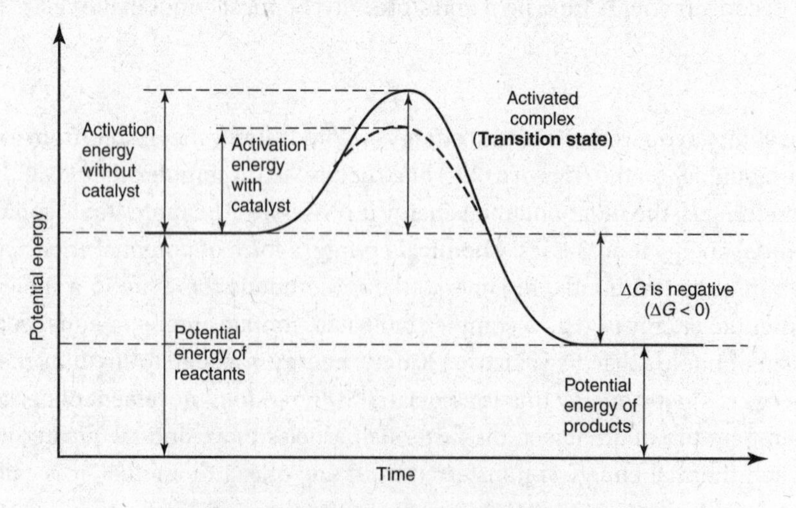

Figure 5.1 Progress of an Exergonic Reaction

In the potential energy diagram in Figure 5.2, the potential energy of the products is greater than the potential energy of the reactants, so energy is absorbed and the reaction is endergonic. The dotted line shows the same reaction when an enzyme is introduced. The enzyme serves to lower the energy of activation, and the reaction can proceed more quickly.

Notice that in both reactions, the only factor altered by the presence of an enzyme is the energy of activation and the energy of the activated complex. The **transition state** is the reactive (unstable) condition of the substrate after sufficient energy has been absorbed to initiate the reaction.

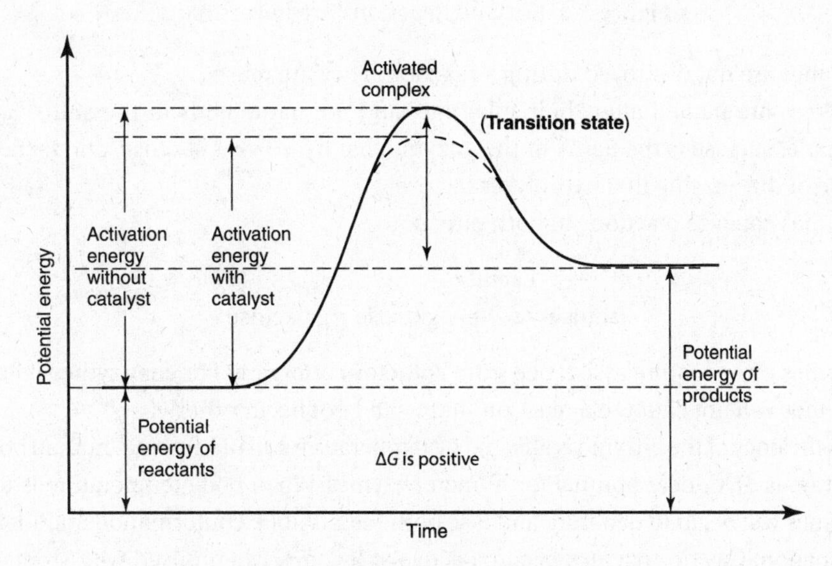

Figure 5.2 Endergonic Reaction

CHARACTERISTICS OF ENZYMES

- Enzymes are **globular proteins** that exhibit **tertiary structure**.
- Enzymes are substrate specific. In Figure 5.3, only substrate A will bind to the enzyme.

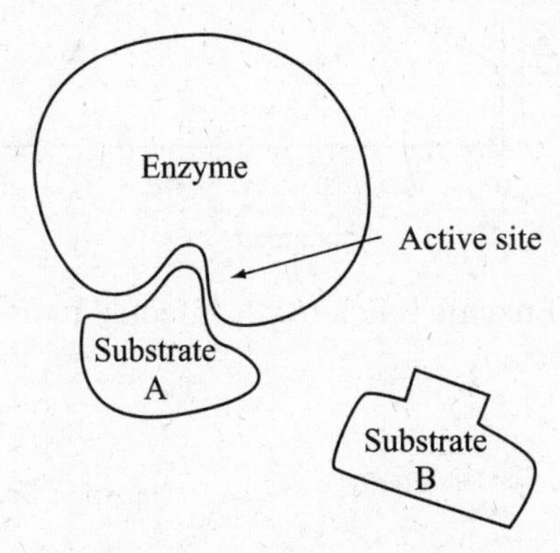

Figure 5.3 Enzymes Are Specific

- The **induced-fit model** describes how enzymes work. As the substrate enters the active site, it induces the enzyme to alter its shape slightly so the substrate fits better. The old lock and key model was abandoned because it falsely implied that the lock and the key were unchanging.
- The enzyme binds to its substrate(s), forming an **enzyme-substrate complex**.

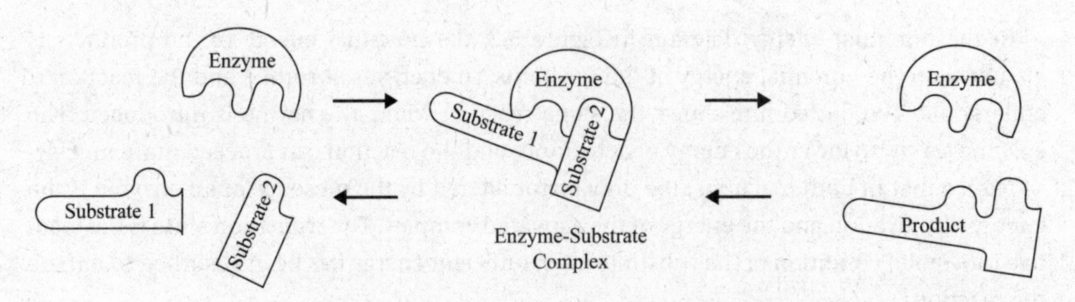

Figure 5.4 Enzyme Reactions Are Reversible

- Enzymes are not destroyed during a reaction. They are reused.
- Enzymes are named after their substrate, and the name ends in the suffix "**ase.**" For example, sucrase is the name of the enzyme that hydrolyzes sucrose, and lactase is the name of the enzyme that hydrolyzes lactose.
- Enzymes catalyze reactions in both directions:

$$\text{lactose} \overset{\text{Lactase}}{\longleftrightarrow} \text{glucose} + \text{galactose}$$

- Enzymes often require assistance from **cofactors** (inorganic) or **coenzymes** (vitamins).
- Enzymes will not catalyze a reaction that would not occur otherwise.
- The efficiency of the enzyme is affected by temperature and pH. Average human body temperature is 37°C, near optimal for human enzymes. When body temperature is too high, enzymes will begin to denature and lose both their unique conformation and their ability to function. Gastric enzymes become active at low pH, when mixed with stomach acid, while intestinal amylase works best in an alkaline environment; see Figures 5.5 and 5.6.

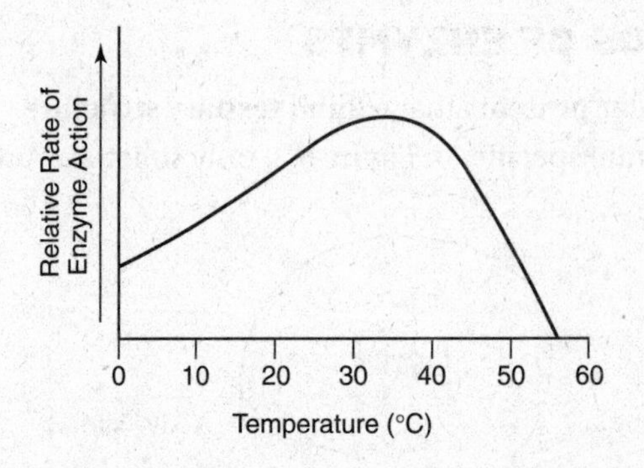

Figure 5.5 Enzyme Efficiency Is Affected by Temperature

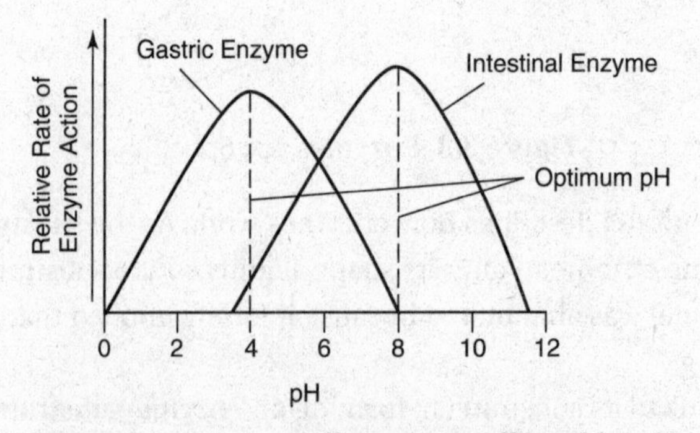

Figure 5.6 Enzyme Efficiency Is Affected by pH

INHIBITION OF ENZYMATIC REACTIONS

Cell metabolism is tightly regulated by controlling when and where different enzymes are active. This is done either by switching on and off the genes that code for enzymes or by regulating the enzymes once they are made. Enzymes that have already been produced are regulated by competitive or noncompetitive inhibition.

Competitive Inhibition

In **competitive inhibition**, some compounds resemble the substrate molecules and compete for the same active site on the enzyme. These mimics or **competitive inhibitors** reduce the amount of product by preventing or limiting the substrate from binding to the enzyme. This kind of inhibition can be overcome by increasing the concentration of substrate. See Figure 5.7. Competitive inhibition can be either reversible or nonreversible.

ENE-1 & SP 1
Understand and be able to sketch how a change in the shape of one enzyme can alter its function.

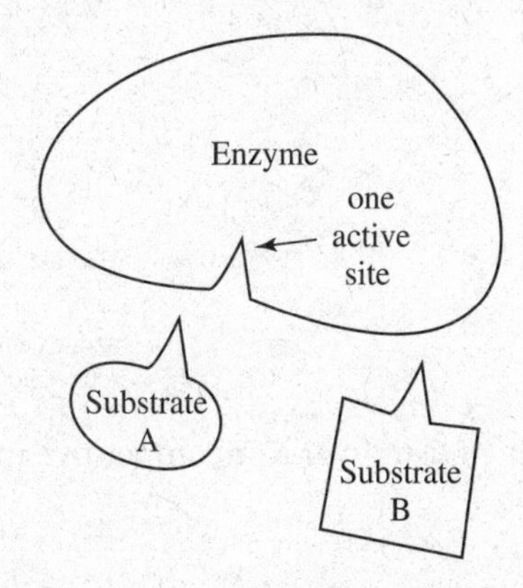

Figure 5.7 Competitive Inhibition

Noncompetitive Inhibition

In contrast with competitive inhibition, in noncompetitive inhibition, some enzymes are **allosteric**, meaning that a change in shape alters their efficiency. Molecules called **noncompetitive inhibitors** or **allosteric regulators** bind to a site distinct and separate from the active site of the enzyme. This binding of the inhibitor to the alternate site causes the enzyme to change shape in a way that inhibits the enzyme from catalyzing substrate into product. Other allosteric enzymes usually toggle between two different conformations (shapes)—one active and the other inactive. The binding of either an activator or an inhibitor locks or stabilizes the allosteric enzyme in either the active or inactive form, respectively. See Figure 5.8. One economical mechanism for regulating a lengthy metabolic pathway is known as **feedback inhibition**, where the *end product* of the pathway is the *allosteric inhibitor* for an enzyme that catalyzes an early step in the pathway: see Figure 5.9.

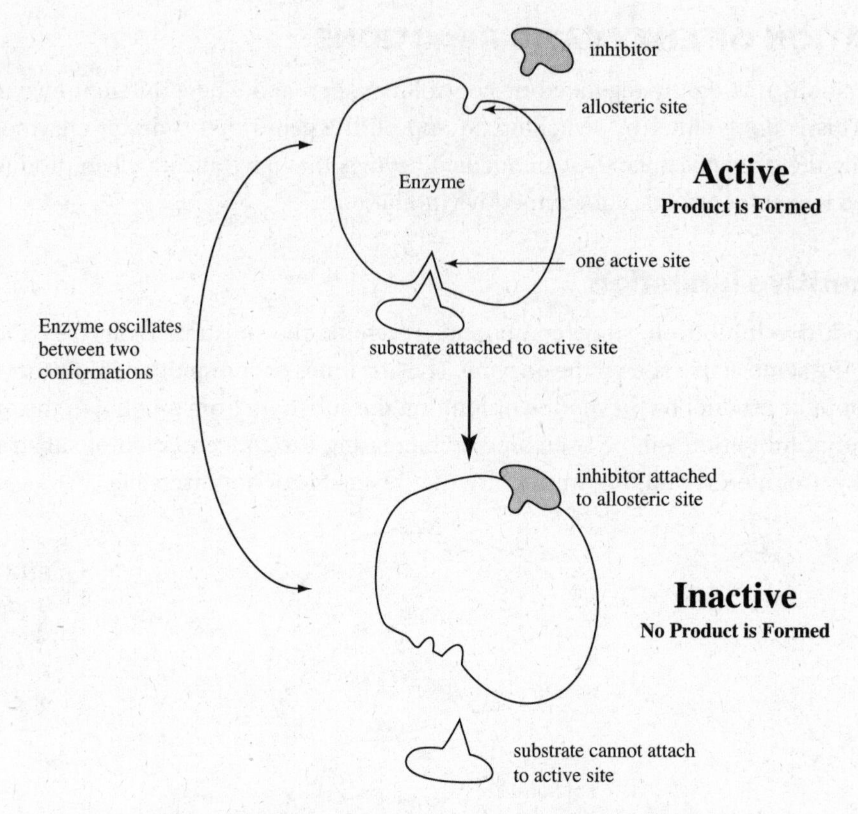

Figure 5.8 Allosteric or Noncompetitive Inhibition

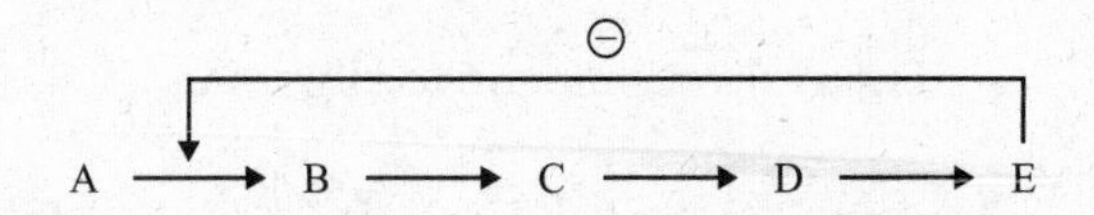

Figure 5.9 Feedback Inhibition

Cooperativity

Cooperativity is a type of *allosteric activation.* The binding of one substrate molecule to one active site of one subunit of the enzyme causes a change in the entire molecule and locks all subunits in an active position. This mechanism amplifies the response of an enzyme to its substrates. See Figure 5.10.

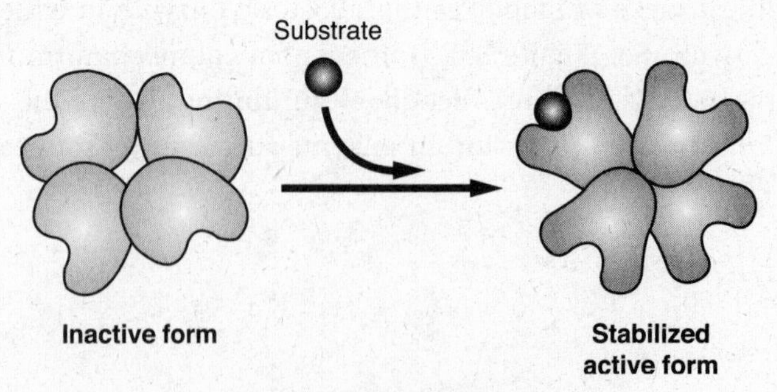

Figure 5.10 Cooperativity

CHAPTER SUMMARY

This chapter supported Big Ideas: ENE and SYI.

- Living organisms require a constant input of energy. They constantly transform energy from one form to another subject to the laws of thermodynamics.
- Free energy (ΔG) changes tell us if a reaction is endergonic or exergonic and spontaneous.
- Enzymes speed up metabolic reactions by making it easier for reactions to occur.
- The shape of an enzyme is critical to how it functions. The shape can be altered (denatured) by extremes in temperature and pH.
- Enzyme function is regulated by inhibitors and feedback mechanisms.

PRACTICE QUESTIONS

1. Which of the following choices best describes the reaction below?

$$A + B \rightarrow AB + \text{energy}$$

(A) The reaction is catalyzed by an enzyme.
(B) The reaction is catabolic.
(C) The potential energy of the reactants is greater than the potential energy of the products.
(D) The reaction takes place in a warm environment.

2. When glucose is broken down in a cell, only about 38% of the energy released is captured by the cell for later use. What happens to the rest of the energy?

(A) It is lost as heat.
(B) It is used to break down other substances besides glucose.
(C) It is stored within the cell to build necessary molecules.
(D) It is used to increase the activity of enzymes.

3. You are performing an enzyme-controlled reaction: A + B + enzyme → AB. You monitor the formation of AB and see that it increases. As instructed, you then add a small amount of substance X to the reaction. To your surprise, the amount of AB being formed decreases. You find that adding more substrate has no effect on the decrease in formation of AB. Which of the following is most likely the explanation for what substance X is?

(A) competitive inhibitor of the enzyme
(B) noncompetitive inhibitor of the enzyme
(C) competitive inhibitor of substance A
(D) competitive inhibitor of substance B

4. You perform an enzyme-controlled reaction: A + B + enzyme → AB + energy. You monitor and record your results on the chart below.

Time (min)	Product Formed	pH
3	0.15	4.0
6	0.25	4.2
9	0.35	4.4
12	0.45	4.6
15	0.35	4.3
18	0.25	4.1

Based on your results, which of the following is true?

(A) The enzyme works best when the pH is alkaline.
(B) The enzyme was most active at pH 4.6.
(C) The temperature must have increased and then decreased during the course of the experiment.
(D) An accelerator must have been added to the system.

5. Below is an equation for a simple enzyme-controlled experiment:

A + B + enzyme → AB

What would be the best way to increase the amount of AB produced?

(A) Introduce a competitive inhibitor.
(B) Raise the activation energy of the reaction.
(C) Add heat to the system.
(D) Increase the free energy (ΔG) of the reaction.

6. Below is a diagram of a branched, enzyme-controlled, cellular pathway. (Minus signs represent inhibition.)

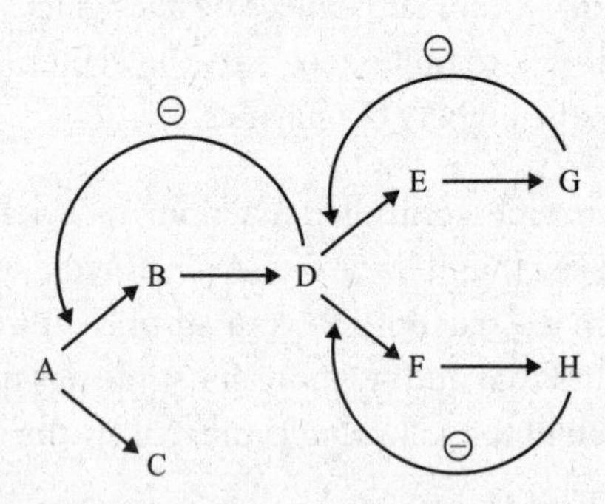

If all inhibitions are in effect, which of the following correctly identifies the product or products formed and the reason for why the product or products formed?

(A) G and H, because they are the end products of the longest pathway
(B) C, because G and H inhibit the production of E and F, and D inhibits the production of B
(C) C, because it is the shortest pathway
(D) B and C, because G and H inhibit the production of E and F, but D does not always inhibit the production of B

7. Which of the following is correct about enzymes?

   (A) Enzymes are proteins that exhibit tertiary structure.
   (B) The bond between an enzyme and its substrate is a strong one.
   (C) The model for enzyme and substrate interactions is like a lock and key.
   (D) During reactions, enzymes increase the number of collisions.

8. Below is a graph that shows the progress of a common cellular reaction carried out in a laboratory under conditions that mimic those of the human body. The only difference is that the reaction vessel does not contain the enzymes that would normally be present in the cell.

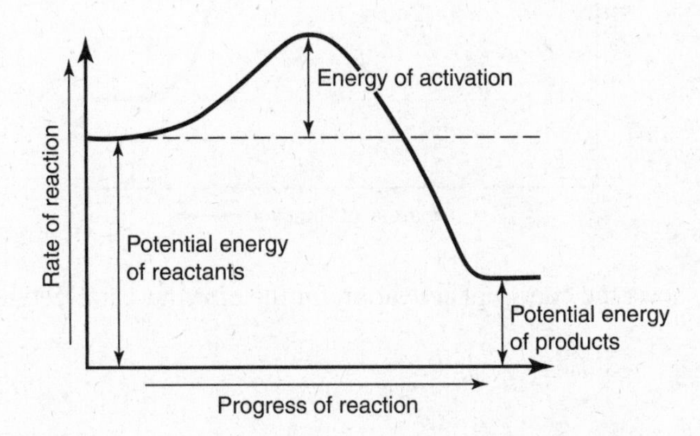

   If the appropriate enzyme were added to the system, which of the following would be true?

   (A) The reaction would not go to completion, and no product would form without an enzyme because there would be no effective collisions.
   (B) The amount of energy stored in the products would be the same with or without enzymes because enzymes only alter the energy needed to start the reaction.
   (C) The energy of activation would be higher in the enzyme-controlled reaction because there would be more effective collisions.
   (D) The potential energy of the reactants would be lower when an enzyme is present because there would be more effective collisions.

9. Which of the following can be used to determine the rate of an enzyme-catalyzed reaction?

   (A) the rate of substrate formed
   (B) the decrease in temperature in the system
   (C) the rate of enzyme used up
   (D) the rate of substrate used up

10. Which of the following best describes the reaction shown below?

$$A + B \rightarrow AB + \text{energy}$$

   (A) hydrolysis
   (B) an exergonic reaction
   (C) an endergonic reaction
   (D) energy production

Questions 11–12

The graph below demonstrates two chemical reactions. One is catalyzed by an enzyme, and one is not.

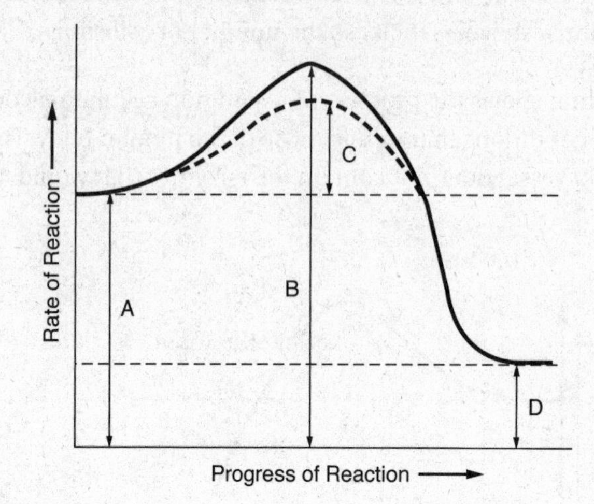

11. Which letter shows the energy of activation for the enzyme-catalyzed reaction?
    (A) A
    (B) B
    (C) C
    (D) D

12. Which letter shows the potential energy of the product?
    (A) A
    (B) B
    (C) C
    (D) D

Questions 13–14

Succinate dehydrogenase (SDH) is an enzyme found in the mitochondrial membrane of eukaryotes and the only enzyme that participates in both the citric acid cycle and the electron transport chain. In its role in cellular respiration, SDH combines with succinate to produce fumarate.

$$\text{Succinate} + \text{SDH} \rightarrow \text{Fumarate}$$

A researcher wanted to explore the role of malonate, another molecule involved in cellular respiration. She added it to the reaction tube that contained SDH and succinate. The result was an accumulation of succinate and a decrease in the production of fumarate.

13. Given the results of this experiment, which of the following statements describes the most likely role of malonate in cellular respiration?

    (A) Malonate is a competitive inhibitor of SDH because it replaced succinate in the reaction and caused a decrease in the formation of fumarate.
    (B) Malonate is a coenzyme for SDH because when it was present, SDH was more efficient and there was an increase in the production of fumarate.
    (C) Malonate is a noncompetitive inhibitor of SDH because it increased the efficiency of SDH and resulted in a decrease in the formation of fumarate.
    (D) Malonate caused a conformational change in succinate that increased its efficiency, resulting in an increase in SDH.

14. Which of the following diagrams most accurately represents the reaction between malonate and SDH according to what the researcher discovered?

    (A)

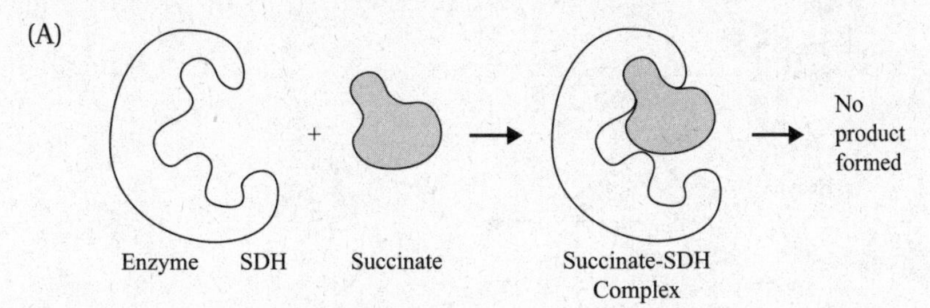

    (B)

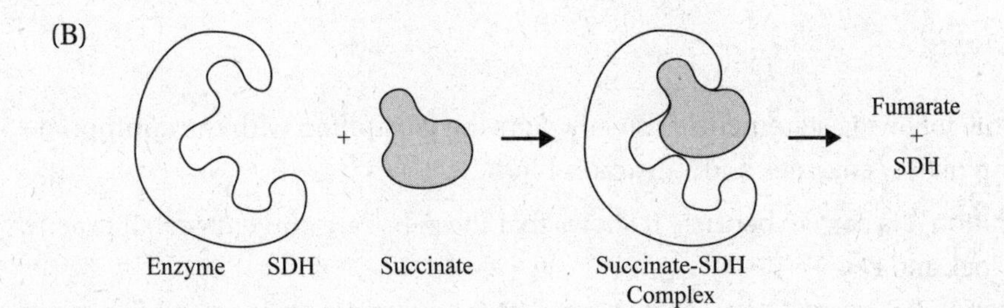

    (C)

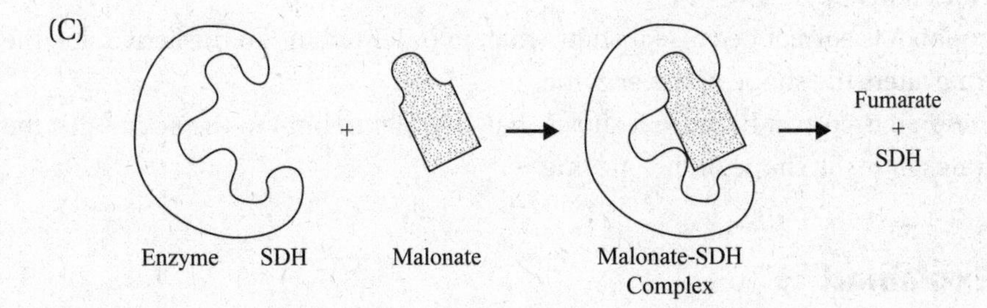

    (D)

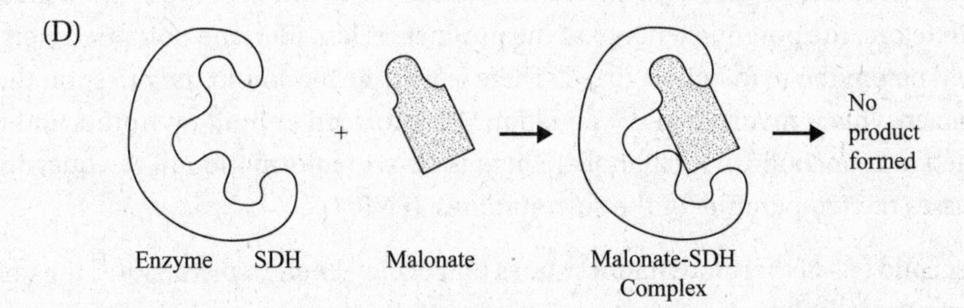

15. Enzymes are specific. They only bind to specific substrates. Below are two illustrations, A and B, that show an enzyme binding to a substrate, forming an enzyme-substrate complex.

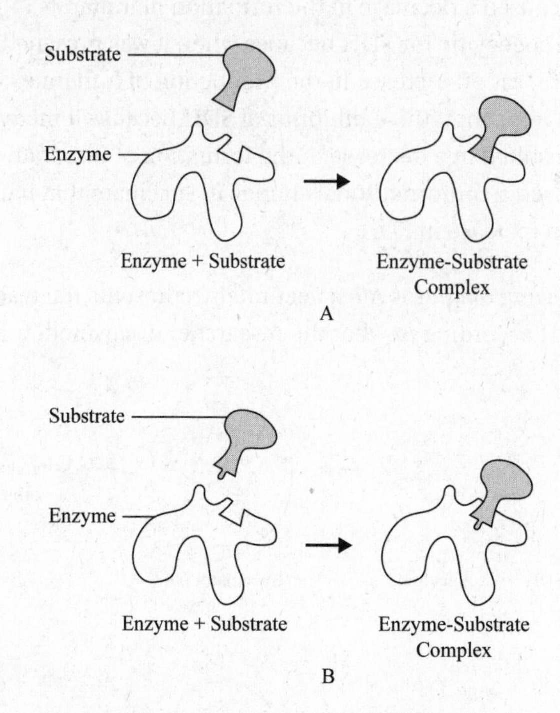

Which of the following statements correctly pairs the illustration with the appropriate explanation of how enzymes and substrates bind?

(A) Illustration A is correct because it shows that the substrate and enzyme fit exactly, like a lock and key.
(B) Illustration B is correct because it shows that the substrate and enzyme fit exactly, like a lock and key.
(C) Illustration A is correct because it shows that, in order to bind to the active site, the substrate alters the shape of the enzyme.
(D) Illustration B is correct because it shows that, in order to bind to the active site, the substrate alters the shape of the enzyme.

Answers Explained

1. **(C)** Since "energy" is on the right side of the equation, we know that energy is given off, and therefore, the potential energy of the products is less than the potential energy that existed before the reaction occurred. There is no information in the question that suggests an enzyme is involved in the equation. AB is formed or built up from A and B, so the reaction is anabolic, not catabolic. There is also no information in the question that discusses the temperature of the surroundings. (ENE-1)

2. **(A)** The second law of thermodynamics states that during energy conversions, the universe becomes more disordered (entropy increases). With each energy transfer, some energy from the system is lost, primarily as heat. (ENE-1)

3. **(B)** Since adding more substrate did not increase production, you can eliminate choice A. Choices C and D make no sense because inhibitors affect enzymes, not substrates. (ENE-1)

4. **(B)** The most product is formed at pH 4.6. The pH increases and then decreases. However, the reaction reaches its maximum at pH 4.6. Alkaline means basic pH, and the whole experiment is carried out in an acid environment. There is no information about temperature. (SP 4)

5. **(C)** Raising the temperature would increase the number of collisions and thus increase the rate of product formation. Introducing a competitive inhibitor would slow the reaction down, decreasing the amount of product, but it would not alter the state of collisions. Raising the activation energy of the reaction would make it more difficult to get the reaction to start and would not alter the number of collisions. Choice D is incorrect because ΔG is the difference between the PE of the reactants and the PE of the products. It is not affected by the absence or presence of enzymes. (ENE-1)

6. **(B)** G and H inhibit the reactions that produce E and F from D, which means that no G and H will be produced. D inhibits the production of B. The only product that is not inhibited when all others are is C. (SP 4)

7. **(A)** Enzymes function as they do because of their unique shape or conformation, which results from their makeup: primary, secondary, and tertiary structure of the polypeptide chain. The bond between an enzyme and its substrate at the active site is a weak one and is only temporary. It consists of Van der Waal and electrostatic forces and hydrogen bonding. The current model for how substrates and enzymes bind is not lock and key; rather, it is "induced-fit" because in order to bind the substrate alters (induces) the enzyme to change its shape. Enzymes do not increase the number of collisions; rather, an increase in heat does. Enzymes make collisions more effective, not faster. (ENE-1)

8. **(B)** The key to understanding graphs that depict enzyme reactions—whether they are endergonic or exergonic—is that the potential energy (PE) of the reactants and of the products is the same before and after the reaction. The only thing that differs between enzyme-catalyzed reactions and reactions without enzymes is the PE of the transition state. It is lower in enzyme-catalyzed reactions. Choice A is not correct because enzymes do not make a reaction occur (that would not occur naturally); they only increase the number of effective collisions. Choice C is not correct because enzymes *decrease* the energy of activation. Choice D is not correct because the potential energy of the reactants remains the same with or without an enzyme present. (SP 4)

9. **(D)** Substrate is used up as the product is formed. Enzymes are never used up; they are reused. So enzyme levels cannot be used to monitor the progress of a reaction. Enzymes lower the energy needed to begin the reaction (the E_A), but they do not affect the temperature of the system. (ENE-1)

10. **(B)** The reaction is exergonic because "energy" is on the right side of the reaction—in other words, the reaction releases energy. Choice A is not correct because hydrolysis is the breaking down of a molecule, but, in this example, AB is being built up. Choice C is not correct because endergonic and endothermic are synonyms for a reaction that requires or absorbs energy. Choice D is not correct because, according to the first law of thermodynamics, energy can never be created or destroyed; it can only be transferred, released, or absorbed. (ENE-1)

11. **(C)** Enzymes lower the energy of activation. (ENE-1)

12. **(D)** The potential energy (PE) of the product is the same for both reactions. (ENE-1)

13. **(A)** The key here is that when malonate was added to the system, succinate appeared in excess and there was a decrease in the product. Therefore, malonate must have blocked the receptor on the SDH enzyme, preventing succinate from doing so. The description of the experiment states that production of fumarate decreased, not increased. (ENE-1)

14. **(D)** The question asks about malonate combining with SDH. Therefore, eliminate choices A and B. When malonate binds to the SDH enzyme, no product is formed. That is what choice D shows. (SP 4)

15. **(D)** The lock and key model is outdated, which eliminates choices A and B. We now know that when the substrate binds to the enzyme, it induces the enzyme to change its shape a bit in order to make a tight fit. That is what is shown in illustration B and described in choice D. (ENE-1 & SP 1)

Cell Respiration

6

- → ATP—ADENOSINE TRIPHOSPHATE
- → GLYCOLYSIS
- → STRUCTURE OF THE MITOCHONDRION
- → AEROBIC RESPIRATION: THE CITRIC ACID CYCLE
- → NAD^+ AND FAD
- → AEROBIC RESPIRATION: THE ELECTRON TRANSPORT CHAIN
- → OXIDATIVE PHOSPHORYLATION AND CHEMIOSMOSIS
- → SUMMARY OF ATP PRODUCTION
- → ANAEROBIC RESPIRATION—FERMENTATION

Big Ideas: EVO, ENE & SYI

Enduring Understandings: EVO-1; ENE-1, ENE-2 & ENE-3; SYI-1

For the complete list of Big Ideas and Enduring Understandings, refer to the "AP Biology Course and Exam Description" from the College Board: *https://apcentral.collegeboard.org/pdf/ap-biology-course-and-exam-description.pdf?course=ap-biology*.

INTRODUCTION

Cell respiration is the means by which cells extract energy stored in food and *transfer* that energy to molecules of **ATP**. Energy that is temporarily stored in molecules of ATP is instantly available for every cellular activity such as passing an electrical impulse, contracting a muscle, moving cilia, or manufacturing a protein. The equation for the complete aerobic respiration of one molecule of glucose (see below) is a highly exergonic process (releases energy and $\Delta G < 0$). See Figure 6.1.

ENE-1
Living things require a constant input of energy.

$$C_6H_{12}O_6 + 6O_2 \rightarrow 6CO_2 + 6H_2O + \text{energy}$$
$$\text{free energy} = \Delta G = -686 \text{ kcal/mole}$$

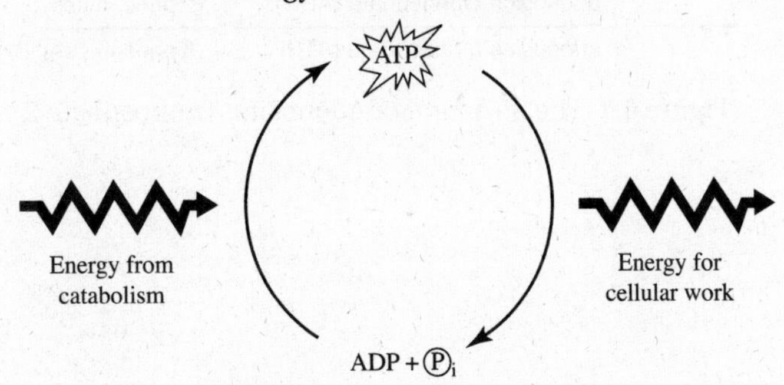

Figure 6.1 The ATP Cycle

> **Be Careful What You Say!**
> Energy cannot be created or destroyed, whereas ATP can be created or destroyed. Energy can be transferred.

There are two types of cell respiration: anaerobic and aerobic. If oxygen is not present (anaerobic respiration), **glycolysis** is followed by either **alcoholic fermentation** or **lactic acid fermentation**. If oxygen is present (aerobic respiration), glycolysis is only the first phase of aerobic respiration. It is followed by the **citric acid cycle (Krebs cycle)**, the **electron transport chain**, and **oxidative phosphorylation.**

Reduction is the gain of electrons (e^-) or hydrogen (H^+), while **oxidation** is the loss of electrons or protons. In the equation showing cellular respiration (see page 103), we see that glucose is oxidized because it loses protons and electrons to oxygen. In the same reaction, oxygen is reduced because it gains protons and electrons from glucose. In any **redox reaction**, one substance is reduced while the other is oxidized.

Beyond the definitions of oxidation and reduction, there is an important concept to understand. As hydrogen (with its electron) is transferred from glucose to oxygen, it is moving from a higher energy level to a lower one, releasing energy in stages. This free energy powers the synthesis of ATP.

ATP—ADENOSINE TRIPHOSPHATE

A molecule of ATP (adenosine triphosphate) consists of **adenosine** (the nucleotide adenine plus ribose) plus three **phosphates.** ATP is an unstable molecule because the three phosphates in ATP are all negatively charged and repel one another. When one phosphate group is removed from ATP by hydrolysis, a more stable molecule, ADP (adenosine diphosphate), results. *The change, from a less stable molecule to a more stable molecule, always releases energy.* ATP provides energy for all cellular activities by transferring phosphates from ATP to another molecule, as seen in Figure 6.1. Figure 6.2 shows the structure of a molecule of ATP.

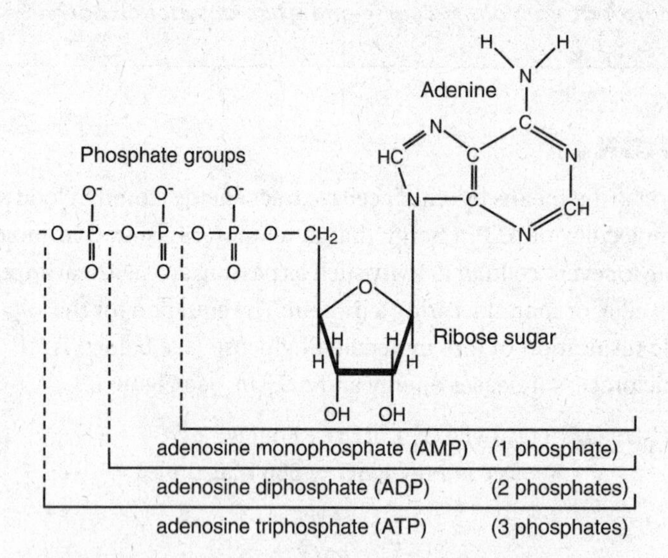

Figure 6.2 The Structure of Adenosine Triphosphate

GLYCOLYSIS

Glycolysis is a ten-step process that breaks down 1 molecule of glucose (a six-carbon molecule) into 2 three-carbon molecules of **pyruvate** or **pyruvic acid** and releases **4** molecules of **ATP**. The energy of activation to begin glycolysis is 2 ATP. After subtracting 2 ATPs from the 4 ATPs released from the reaction, the glycolysis of 1 molecule of glucose results in a *net gain of 2 ATP*. Here is the simplified equation.

$$2\ \text{ATP} + 1\ \text{Glucose} \rightarrow 2\ \text{Pyruvate} + 4\ \text{ATP}$$

Glycolysis occurs in the cytoplasm and releases ATP *without using oxygen*. Each step is catalyzed by a different enzyme. Although this process releases only one-fourth of the energy stored in glucose (most of the energy remains locked in pyruvate), the reaction is critical. The end product, pyruvate, is the *raw material for the Krebs cycle*, which is the next step in aerobic respiration. Without glycolysis to yield pyruvate, aerobic respiration could not occur.

During glycolysis, ATP is produced by **substrate level phosphorylation**—by direct enzymatic transfer of a phosphate to ADP (remember kinases from Chapter 4). Only a small amount of ATP is released this way.

There is one other important thing about glycolysis. The enzyme that catalyzes the third step, **phosphofructokinase (PFK)**, is an **allosteric** enzyme. It inhibits glycolysis when the cell contains enough ATP and does not need to produce any more. If ATP is present in the cell in large quantities, it inhibits PFK by altering the conformation of that enzyme, thus stopping glycolysis.

RELAX

You do not have to memorize the details of glycolysis or the Krebs cycle for the exam.

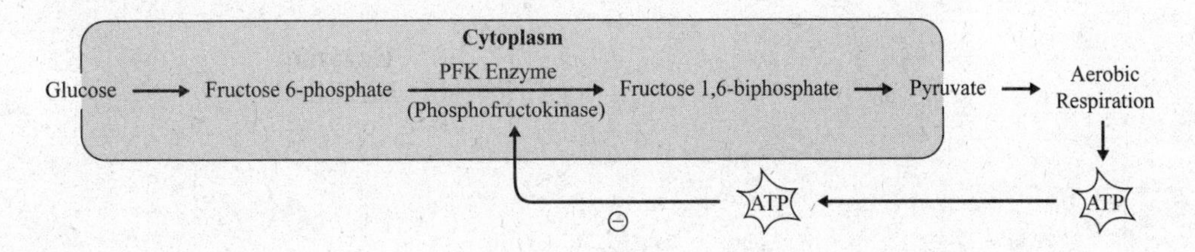

Figure 6.3 Regulation of Cellular Respiration by a Feedback Mechanism

SYI-1
Be able to predict and justify how changes to a protein's shape can affect its function.

This makes sense because as the cell's activities use up ATP, less ATP is available to inhibit PFK and glycolysis continues, ultimately to produce more ATP. *This is an important example of how a cell regulates ATP production through allosteric inhibition;* see Figure 6.4. (Also see page 93.)

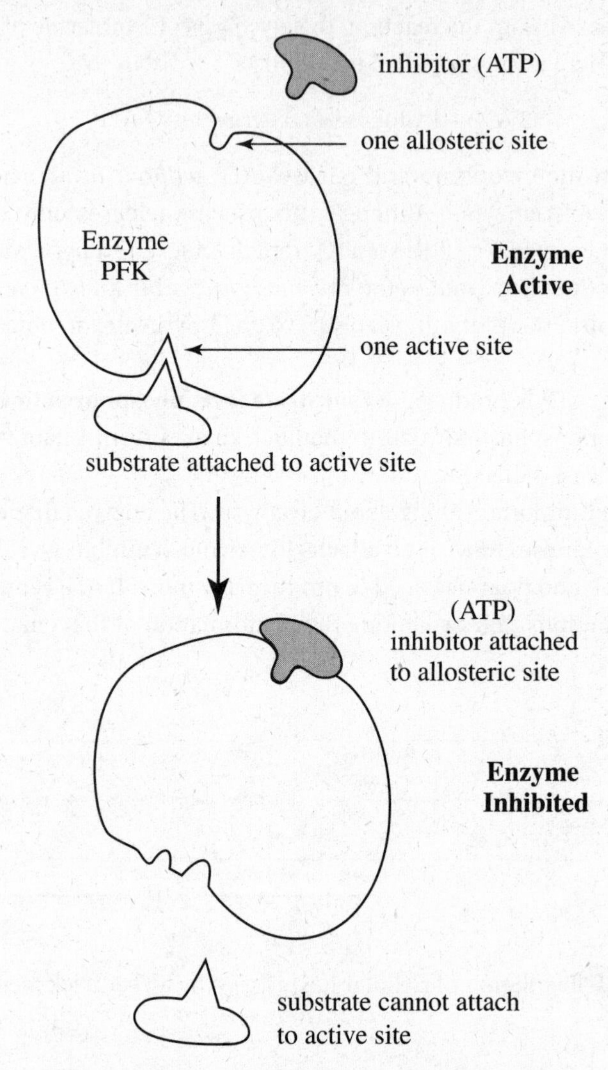

Figure 6.4 Allosteric Inhibition of Enzyme PFK

STUDY TIP

Cytoplasm—Glycolysis

Matrix—Krebs cycle

Cristae membrane—Electron transport chain

Outer compartment—Proton concentration builds up

STRUCTURE OF THE MITOCHONDRION

The mitochondrion is enclosed by a double membrane. The outer membrane is smooth, but the inner or **cristae membrane** is folded. This inner membrane divides the mitochondrion into two internal compartments, the **outer compartment** and the **matrix**. The Krebs, or citric acid, cycle takes place in the matrix; the electron transport chain and oxidative phosphorylation take place in the cristae membrane. Figure 6.5 shows a diagram of the mitochondrion.

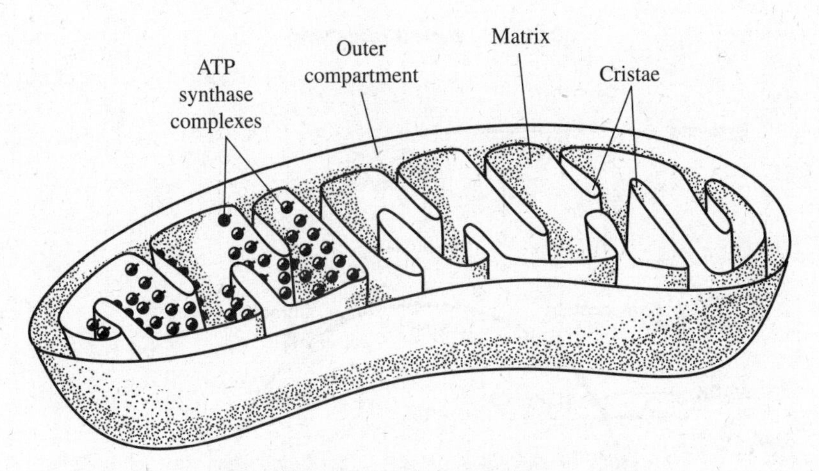

Figure 6.5 Mitochondrion

AEROBIC RESPIRATION: THE CITRIC ACID CYCLE

TIP

The citric acid cycle is also referred to as the Krebs cycle. Don't let the different names confuse you.

When oxygen is present, cells carry out aerobic respiration. It is highly efficient and produces a lot of ATP. This process consists of an anaerobic phase—glycolysis—plus an aerobic phase. The aerobic phase consists of two parts: the citric acid cycle and oxidative phosphorylation.

The **citric acid cycle** is a cyclical series of enzyme-catalyzed reactions also known as the **Krebs cycle**. It takes place in the **matrix of mitochondria** and requires pyruvate, the product of glycolysis. The citric acid cycle completes the oxidation of glucose to CO_2. It turns *twice* for each glucose molecule that enters glycolysis (once for each pyruvate molecule that enters the mitochondrion). The cycle generates 1 ATP per turn by **substrate level phosphorylation**, the direct enzymatic transfer of phosphate to ADP. The remainder of the chemical energy is transferred to NAD^+ and FAD ($NAD^2 + H \rightarrow NADH$; $FAD^+ + 2H \rightarrow FADH_2$). The reduced coenzymes, NADH and $FADH_2$, shuttle high-energy electrons into the electron transport chain in the cristae membrane. Figure 6.6 illustrates a simplified version of the citric acid cycle. Here are some other important points:

ENE-2
Membranes and membrane-bound organelles localize metabolic processes and specific reactions.

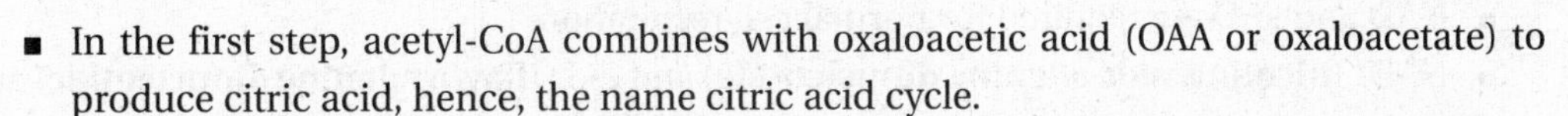

- In the first step, acetyl-CoA combines with oxaloacetic acid (OAA or oxaloacetate) to produce citric acid, hence, the name citric acid cycle.
- Remember that each molecule of glucose is broken down to 2 molecules of pyruvate during glycolysis. Therefore, the respiration of each molecule of glucose causes the Krebs cycle to turn two times.
- Before it enters the Krebs cycle, pyruvate must first combine with coenzyme A (a vitamin) to form **acetyl-CoA**, which does enter the Krebs cycle. The conversion of pyruvate to acetyl-CoA produces **2 molecules of NADH**, 1 NADH for each pyruvate.
- Each turn of the Krebs cycle releases **3 NADH, 1 ATP, 1 FADH**, and the waste product $\mathbf{CO_2}$, which is exhaled. (Remember, two turns of the Krebs cycle occur per glucose molecule.)
- During the Krebs cycle, ATP is produced by **substrate level phosphorylation**—direct enzymatic transfer of a phosphate to ADP. Very little energy is produced this way compared with the amount produced by oxidative phosphorylation.

STUDY TIP

Your teacher may want you to memorize this, but you won't need to know the specifics for the AP exam.

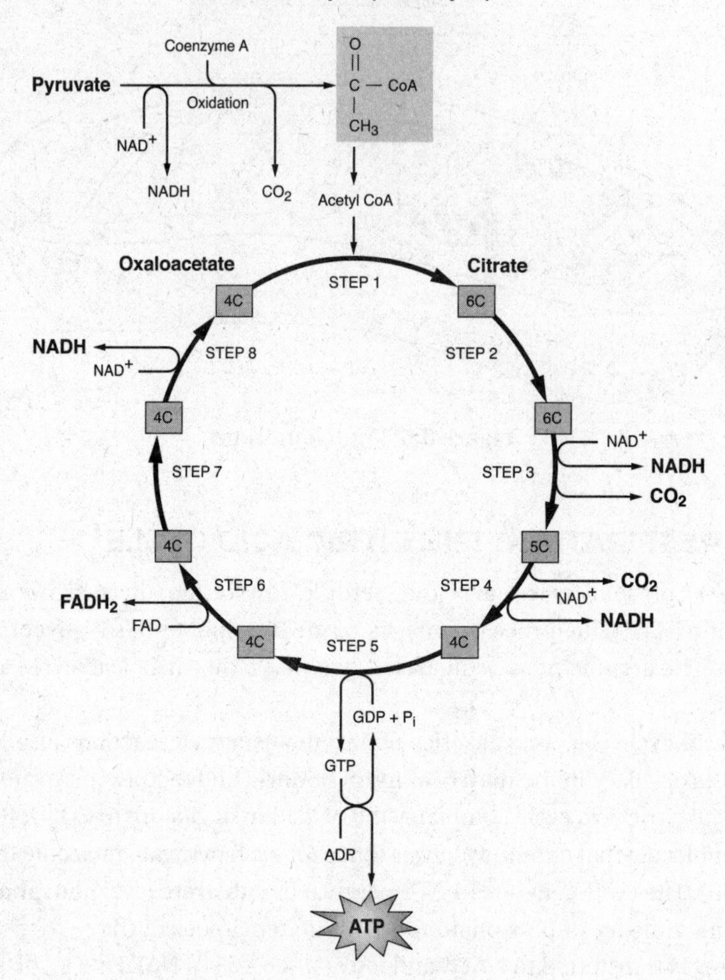

Figure 6.6 The Citric Acid Cycle (Krebs Cycle)

NAD⁺ AND FAD

- NAD and FAD are required for normal cell respiration.
- NAD^+ (**nicotinamide adenine dinucleotide**) and FAD (**flavin adenine dinucleotide**) are **coenzymes** that carry protons or electrons from glycolysis and the citric acid cycle to the electron transport chain.
- The enzyme NAD dehydrogenase or FAD dehydrogenase facilitates the transfer of hydrogen atoms from a substrate, such as glucose, to its coenzyme NAD^+.
- Without NAD^+ to accept protons and electrons from glycolysis and the Krebs cycle, both processes would cease and the cell would die.
- NAD and FAD are vitamin derivatives.
- NAD^+ is the oxidized form. NAD_{re} or NADH is the reduced form. NADH carries 1 proton and 2 electrons.
- FAD is the oxidized form. FAD_{re} or $FADH_2$ is the reduced form.

AEROBIC RESPIRATION: THE ELECTRON TRANSPORT CHAIN

The **electron transport chain (ETC)** is a **proton pump** in the mitochondria that couples two reactions, one exergonic and one endergonic. It uses the energy released from the exergonic flow of electrons to pump protons against a gradient from the matrix to the outer compartment. This results in the establishment of a **proton gradient** inside the mitochondrion. The electron transport chain makes no ATP directly but sets the stage for ATP production during **chemiosmosis**; see Figure 6.7.

> **ENE-2 & SYI-1**
> The structure and function of cell organelles provide essential processes—like energy capture and transformation.

Important Points about the Electron Transport Chain

- The ETC is a collection of molecules embedded in the **cristae membrane** of the mitochondrion.
- There are thousands of copies of the ETC in every mitochondrion due to the extensive folding of the cristae membrane.
- The ETC carries electrons delivered by NADH and $FADH_2$ from glycolysis and the Krebs cycle to oxygen, the **final electron acceptor**, through a series of **redox reactions**. In a redox reaction, one atom gains electrons or hydrogen (**reduction**), and one atom loses electrons or hydrogen (**oxidation**).
- The highly **electronegative** oxygen pulls electrons through the electron transport chain. Electrons lose PE as they "fall down" the ETC toward oxygen.
- **NADH** delivers its electrons to a higher energy level in the chain than does **$FADH_2$**. As a result, NADH provides more energy for ATP synthesis than does $FADH_2$. Theoretically, each NADH produces 3 ATP molecules, while each $FADH_2$ produces 2 ATP molecules.
- The ETC consists mostly of **cytochromes**. These are proteins structurally similar to hemoglobin. Cytochromes are present in all aerobes and are used to trace evolutionary relationships.
- One component of the ETC is labeled Q, which stands for **ubiquinone**, or alternately **coenzyme Q**. The important thing about Q is that it is a **mobile electron carrier**. It diffuses within and along the membrane. If the cristae membrane were not fluid, Q could not move through it, and the ETC could not operate. This phenomenon is a great example of how *the structure of a fluid membrane relates to its function.*
- **Exergonic reactions are coupled with endergonic ones**. The **exergonic flow** of electrons toward the highly electronegative oxygen provides the energy for the **endergonic** pumping of protons from the inner matrix to the outer compartment to create a gradient.

OXIDATIVE PHOSPHORYLATION AND CHEMIOSMOSIS

Most of the energy released during cell respiration occurs in the mitochondria by a process known by the general name of **oxidative phosphorylation**. This term means the phosphorylation of ADP into ATP by the oxidation of the carrier molecules NADH and $FADH_2$. This **energy-coupling mechanism** was elucidated in 1961 by Peter Mitchell, who named it **chemiosmosis**. According to the Mitchell hypothesis, chemiosmosis uses potential energy stored in the form of a **proton (H^+) gradient** to phosphorylate ADP and produce ATP ($ADP + P \rightarrow ATP$).

Important Points about Oxidative Phosphorylation

- It is powered by the *redox reactions* of the electron transport chain.
- Protons are pumped from the **matrix** to the **outer compartment**, against a gradient, by the electron transport chain.
- A proton gradient is created by a **proton pump** between the outer compartment and the inner matrix.

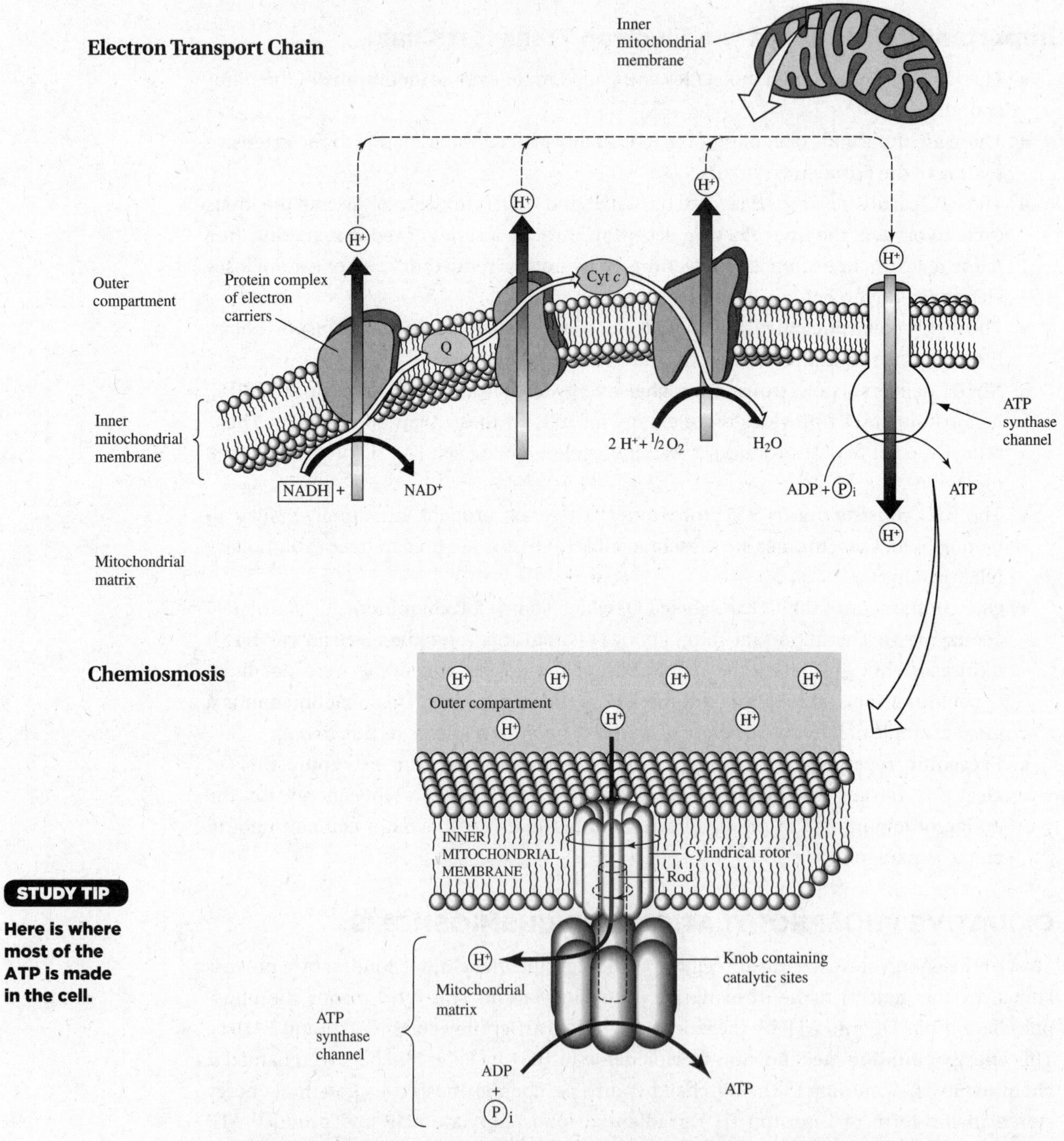

Figure 6.7 The Electron Transport Chain, Proton Gradient, and ATP Synthase

STUDY TIP

Here is where most of the ATP is made in the cell.

- Protons cannot diffuse through the cristae membrane; they can flow *only* down the gradient into the matrix through **ATP synthase channels**. This process is **chemiosmosis**, the key to the production of ATP. *As protons flow through the ATP synthase channels, they generate energy to phosphorylate ADP into ATP.* This process is similar to how a hydroelectric plant converts the enormous potential energy of water flowing through a dam to turn turbines and generate electricity.

SYI-1
Explain how internal membranes and organelles contribute to cell functions.

- **Oxygen** is the **final hydrogen acceptor**, combining half an oxygen molecule with 2 electrons and 2 protons, thus forming **water**. This water is a waste product of cell respiration and is excreted.

SUMMARY OF ATP PRODUCTION

ATP is produced in two ways.

- **Substrate level phosphorylation** occurs when an enzyme, a **kinase**, transfers a phosphate from a substrate directly to ADP. Only a small amount of ATP is produced this way. This is the way energy is produced during glycolysis and the Krebs cycle. (See Figure 6.1.)
- **Oxidative phosphorylation** depends on chemiosmosis. This is the way 90 percent of all ATP is produced from cell respiration. During oxidative phosphorylation, NAD and FAD lose protons (become oxidized) to the electron transport chain, which pumps them to the outer compartment of the mitochondrion, creating a steep proton gradient. This electrochemical or proton gradient powers the phosphorylation of ADP into ATP during chemiosmosis.

During respiration, most energy flows in this sequence:

$$\textbf{Glucose} \rightarrow \textbf{NAD}_{\textbf{re}} \text{ and } \textbf{FAD}_{\textbf{re}} \rightarrow \textbf{electron transport chain} \rightarrow \textbf{chemiosmosis} \rightarrow \textbf{ATP}$$

Theoretically, about 36–38 ATP can be released from the aerobic respiration of 1 molecule of glucose. This is hypothetical because some cells are more efficient than others and cells vary in their efficiency at different times. Remember that during glycolysis, 2 pyruvates are formed and each enters the Krebs cycle separately. So following glycolysis, all numbers for $FADH_2$, NADH, and ATP are doubled.

The catabolism (breakdown) of glucose under aerobic conditions occurs in three sequential pathways: glycolysis, pyruvate oxidation, and the citric acid cycle, which produce the reduced coenzymes NADH and $FADH_2$. These coenzymes are then oxidized by the electron transport chain and ATP is produced by oxidative phosphorylation; see Figure 6.8.

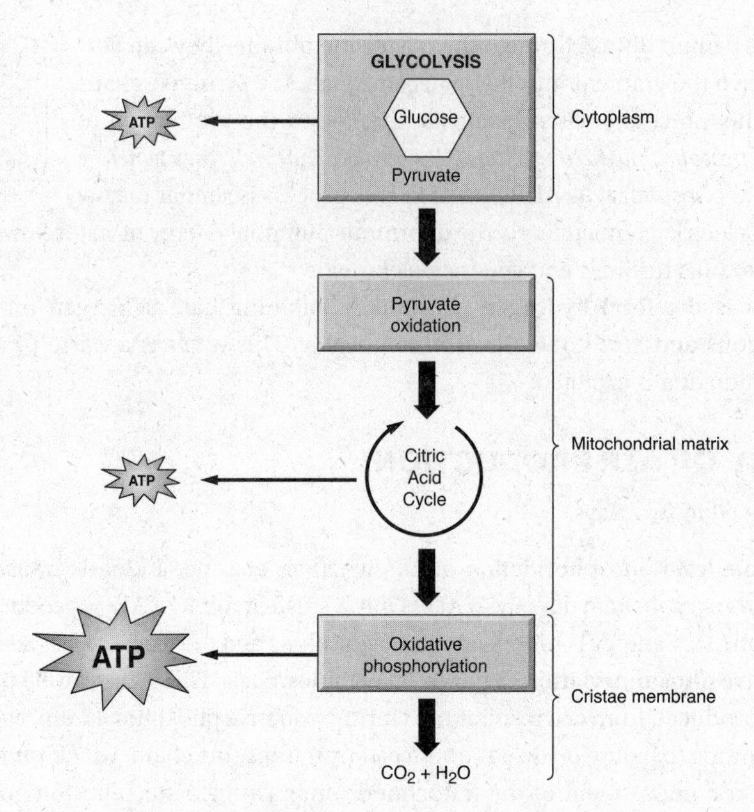

Figure 6.8 The Aerobic Breakdown of Glucose

Figure 6.9 shows an overview of ATP production from cell respiration.

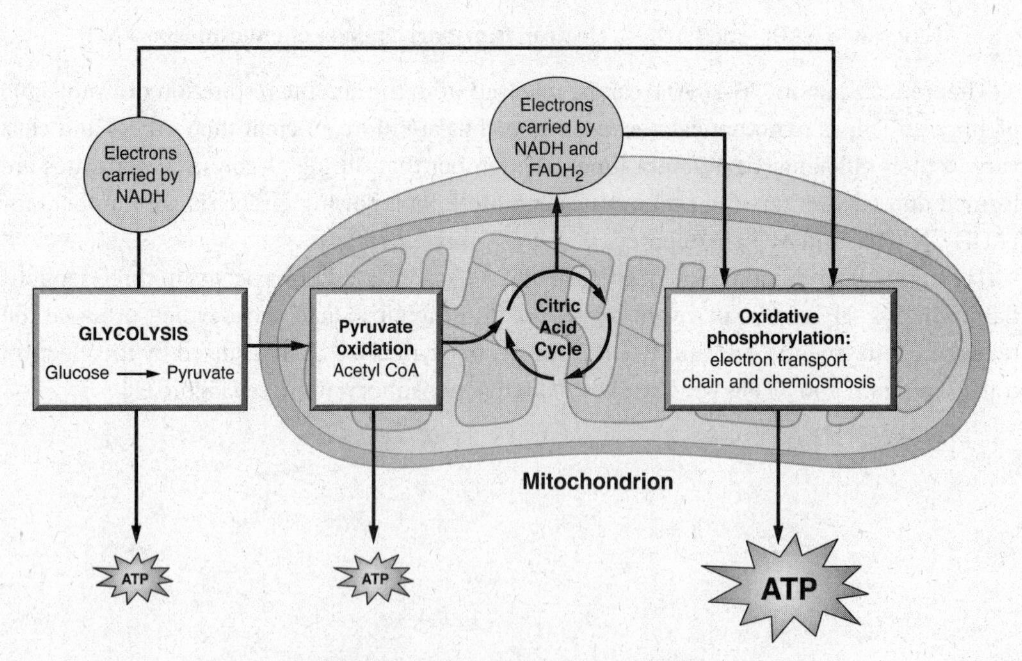

Figure 6.9 Overview of ATP Production

ANAEROBIC RESPIRATION—FERMENTATION

Anaerobic respiration or **fermentation** is not a synonym for glycolysis. It is an anaerobic, catabolic process that consists of glycolysis *plus* alcohol or lactic acid fermentation. Anaerobic respiration originated billions of years ago when there was no free oxygen in Earth's atmosphere. Even today it is the sole means by which anaerobic bacteria such as *botulinum* (the bacterium that causes a form of food poisoning, botulism) release energy from food. There are two types of anaerobes: facultative and obligate. *Facultative anaerobes* can tolerate the presence of oxygen; they simply do not use it. *Obligate anaerobes* cannot live in an environment containing oxygen.

Fermentation can generate ATP during anaerobic respiration *only as long as there is an adequate supply of* NAD^+ *to accept electrons* during glycolysis. Without some mechanism to convert NADH back to NAD^+, glycolysis would shut down. Fermentation consists of glycolysis plus the reactions that regenerate NAD^+. Two types of fermentation are **alcohol fermentation** and **lactic acid fermentation**.

Alcohol Fermentation

Alcohol fermentation or simply **fermentation** is the process by which certain cells convert pyruvate from glycolysis into **ethyl alcohol** and **carbon dioxide** in the **absence of oxygen** and, *in the process, oxidize NADH back to* NAD^+. The bread-baking industry depends on the ability of yeast to carry out fermentation and produce carbon dioxide, which causes bread to rise. The beer, liquor, and wine industry depends on yeast to ferment sugar into ethyl alcohol.

Lactic Acid Fermentation

During **lactic acid fermentation**, pyruvate from glycolysis is reduced to form **lactic acid** or **lactate**. This is the process that the dairy industry uses to produce yogurt and cheese. Also in the process, *NADH gets oxidized back to* NAD^+. **Human skeletal muscles** also carry out lactic acid fermentation when the blood cannot supply adequate oxygen to muscles during strenuous exercise. Lactic acid in the muscle causes fatigue and burning. The lactic acid continues to build up until the blood can supply the muscles with adequate oxygen to repay the oxygen debt. With normal oxygen levels, the muscle cells will revert to the more efficient aerobic respiration and the lactic acid is then converted back to pyruvate in the liver.

CHAPTER SUMMARY

This chapter supported Big Ideas: EVO, ENE, and SYI.

- All living things require a constant input of energy.
- There are a lot of interconnected pathways in this chapter. As you learn them, keep your thoughts on the big picture—**energy capture and transfer**. In Chapter 12, "**Ecology**," we will tie concepts from this chapter with energy flowing through an entire ecosystem.
- **ATP** is important because it stores energy for immediate use.
- **Exergonic** means release of energy: $A + B \rightarrow AB + \text{energy}$.
- **Endergonic** means absorbing energy: $A + B + \text{energy} \rightarrow AB$.

- **Glycolysis** is the first phase of cellular respiration. It does *not* use oxygen to break down 1 glucose molecule into 2 molecules of **pyruvate** with the release of 4 ATP. This ATP is produced by **substrate level phosphorylation,** which is carried out with the help of an **allosteric kinase, phosphofructokinase.**
- The **citric acid cycle** utilizes the pyruvate from glycolysis to produce ATP by **substrate level phosphorylation**. It releases a small amount of ATP and the waste product CO_2. The important products are **NADH** and **$FADH_2$**, which carry protons and electrons to the electron transport chain (ETC) in the **cristae membrane** where oxidative phosphorylation occurs.
- **Oxidative phosphorylation** includes the ETC and chemiosmosis. The **ETC** pumps protons and electrons through the cristae membrane against a gradient through a series of **REDOX reactions** to create a proton gradient.
- **Chemiosmosis** is the process by which a proton gradient powers the production of ATP as protons flow down the gradient through the ATP synthase channel.
- Oxidative phosphorylation, more specifically the ETC, **couples two reactions** within the cristae membrane—an exergonic one (electrons are strongly pulled toward oxygen) to an endergonic one (the pumping of protons against a gradient to create a proton gradient).

PRACTICE QUESTIONS

Questions 1–2

The following questions refer to the equation below for the complete aerobic respiration of one molecule of glucose.

$$C_6H_{12}O_6 + 6O_2 \rightarrow 6CO_2 + 6H_2O + ATP$$

1. Which of the following is correct about this reaction and provides the correct supporting evidence based on this equation?
   (A) Glucose is reduced because hydrogen is transferred to carbon.
   (B) Glucose is reduced because hydrogen is transferred to oxygen.
   (C) Glucose is oxidized because oxygen is transferred to carbon.
   (D) Glucose is oxidized because hydrogen is transferred to oxygen.

2. Which of the following is correct about this reaction and provides the correct supporting evidence for your answer?
   (A) The overall reaction is endergonic because energy is absorbed.
   (B) The overall reaction is endergonic because energy is on the right side of the equation.
   (C) ATP is shown on the right side of the equation because energy is created.
   (D) This reaction shows aerobic respiration because oxygen is one of the raw materials.

3. Which of the following statements is correct regarding cellular respiration?
   (A) The cristae membrane of mitochondria allows protons to freely pass anywhere along its entire length as long as the protons are flowing down a gradient.
   (B) Most energy in aerobic respiration is produced by substrate level phosphorylation.
   (C) The electron transport chain is located in the cytoplasm.
   (D) NAD^+ and FAD carry protons or electrons from glycolysis and the citric acid cycle to the electron transport chain.

4. Which of the following harvests the most energy during cell respiration?
   (A) the Krebs cycle
   (B) chemiosmosis
   (C) 2 ATP + 1 Glucose → 2 Pyruvate + 4 ATP
   (D) electron transport chain

5. Which of the following is correct about cellular respiration?
   (A) Aerobic respiration couples the exergonic flow of electrons flowing through the electron transport chain with the endergonic pumping of protons across the cristae membrane.
   (B) Glycolysis only occurs in anaerobic organisms.
   (C) The citric acid cycle produces pyruvate as a waste product.
   (D) The electron transport chain directly produces ATP.

6. Which of the following is correct about ATP production?
   (A) Oxygen is required in cell respiration to carry protons to the citric acid cycle.
   (B) Substrate level phosphorylation is the mechanism by which ATP is produced in oxidative phosphorylation.
   (C) ATP synthase uses the energy of an ion gradient to power ATP synthesis.
   (D) ATP carries protons across the cristae membrane to create a proton gradient in the cristae membrane.

7. Which of the following is correct about the processes shown?

   The three circles represent three major processes in aerobic respiration.

   glucose → (process *A*) → (process *B*) → (process *C*) → $CO_2 + H_2O$

   (A) Process *A* takes place in mitochondria.
   (B) Process *B* produces ATP by substrate level phosphorylation.
   (C) Process *B* releases NAD^+.
   (D) Process *B* produces the most ATP of all the processes.

Questions 8–9

The following questions refer to the sketch of a mitochondrion shown below.

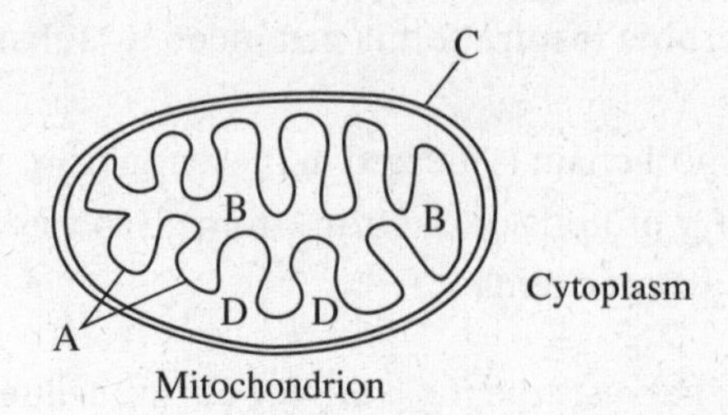

8. Identify the site of the ATP synthase.
   (A) A
   (B) B
   (C) C
   (D) D

9. Identify the site of the citric acid cycle.
   (A) A
   (B) B
   (C) C
   (D) D

---

10. Glycolysis is the most widespread metabolic pathway among all of Earth's organisms. This fact suggests that it evolved very early in the history of life. Which of the following statements best supports this claim?
    (A) Certain primitive bacteria, like *C. botulinum,* can only live in places where oxygen is not present.
    (B) Yeast are simple cells that carry out glycolysis and fermentation.
    (C) The fossil record of the earliest organisms shows that they did not have internal membranes or organelles.
    (D) Mitochondria were once free-living prokaryotes that were engulfed by larger cells and provided these larger cells with the means to carry out aerobic respiration.

11. Which of the following is the most important thing that happens during aerobic respiration?
    (A) Electrons move down the electron transport chain in a series of redox reactions.
    (B) Acetyl-CoA enters the Krebs cycle.
    (C) NAD carries hydrogen to the electron transport chain.
    (D) ATP is produced.

12. What process(es) in your cells produce CO_2 that you exhale?
    (A) oxidation of pyruvate at the end of glycolysis and the citric acid cycle
    (B) glycolysis and oxidative phosphorylation
    (C) the citric acid cycle only
    (D) the citric acid cycle and oxidative phosphorylation

13. Which of the following probably evolved first?

(A) the Krebs cycle
(B) oxidative phosphorylation
(C) glycolysis
(D) the electron transport chain

14. Which is an example of a feedback mechanism?

(A) Phosphofructokinase, an allosteric enzyme in glycolysis, is inhibited by ATP.
(B) Lactic acid gets converted back to pyruvic acid in the human liver.
(C) ATP is produced in mitochondria as protons flow through the ATP synthase channel.
(D) Energy is released from glucose as it decomposes into CO_2 and H_2O.

15. During cell respiration, most ATP is formed as a direct result of the net movement of

(A) electrons flowing against a gradient
(B) electrons flowing through a channel
(C) protons flowing through a channel
(D) protons flowing against a gradient

16. Below is a diagram of a part of cellular respiration.

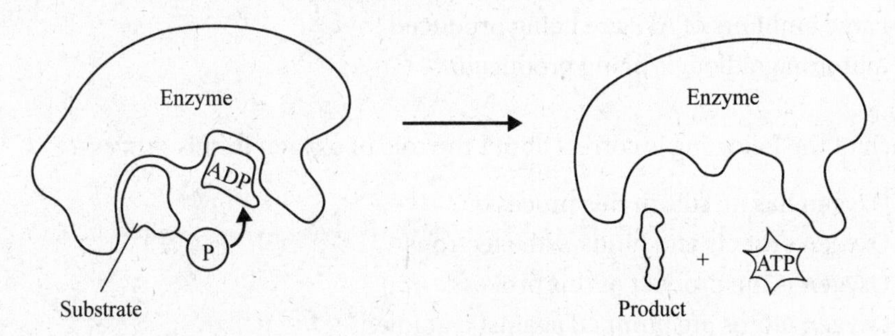

What is the process shown in the diagram, and where does it occur?

(A) The process is oxidative phosphorylation, and it occurs in the cristae membrane.
(B) The process is substrate level phosphorylation, and it occurs during oxidative phosphorylation.
(C) The process is substrate level phosphorylation, and it occurs during the citric acid cycle and glycolysis.
(D) The process is alcoholic fermentation, and it occurs during anaerobic respiration.

Questions 17–18

The following questions refer to the figure below, which depicts a part of cellular respiration.

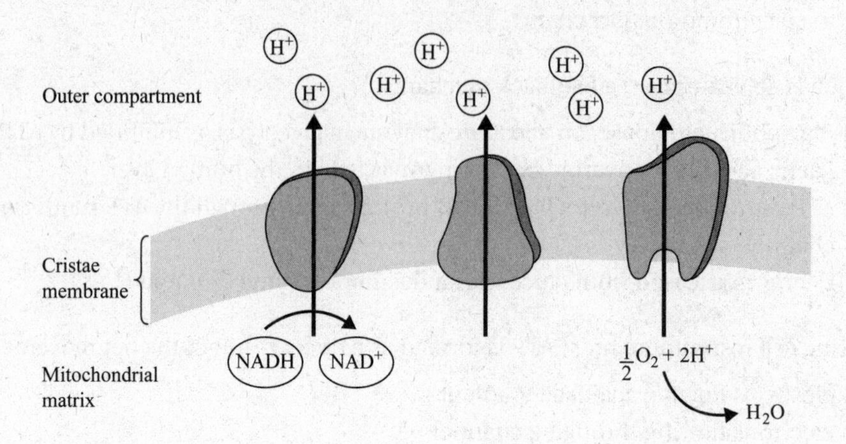

17. Which of the following best describes what is occurring in the figure?

    (A) ATP is inhibiting an enzyme-controlled reaction.
    (B) Glucose is being broken down into pyruvate.
    (C) Large amounts of ATP are being produced.
    (D) A proton gradient is being produced.

18. Which of the following is correct about the role of oxygen in this process?

    (A) Oxygen has no role in this process.
    (B) Oxygen attracts and binds with electrons.
    (C) Oxygen is the product of this process.
    (D) Oxygen atoms are pumped against a gradient

19. The enzyme that catalyzes the third step in glycolysis, PFK, is an allosteric enzyme. When functioning, it catalyzes the reaction of fructose 6-phosphate to fructose 1,6-biphosphate. PFK is inhibited by the presence of high levels of ATP in the cytoplasm of the cell. Which of the following graphs most accurately represents the relationship between PFK activity and ATP?

(A)

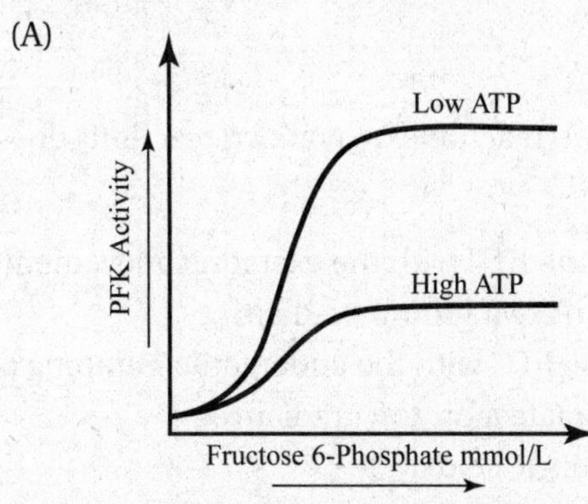

(B)

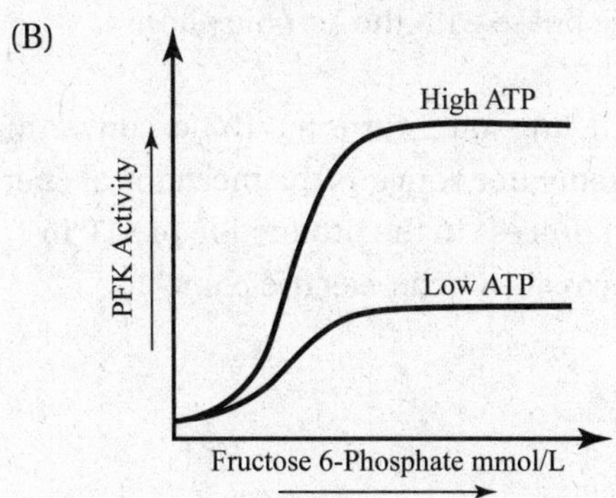

(C)

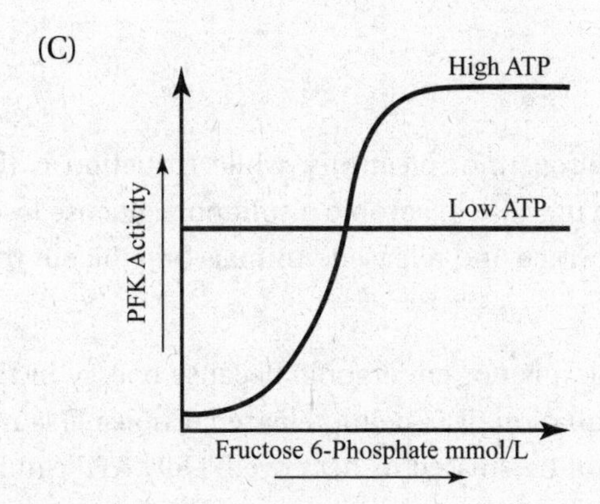

(D)

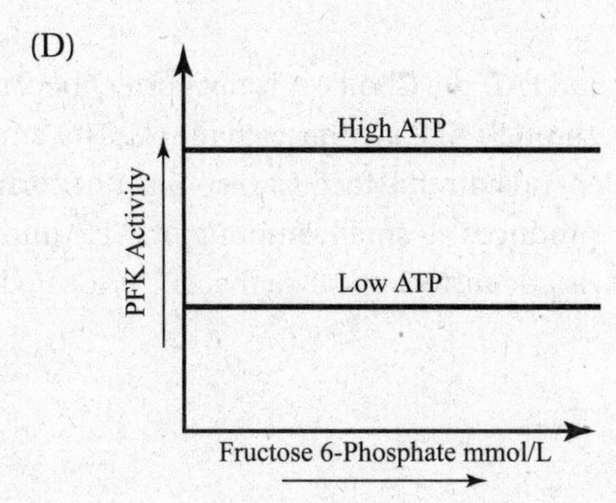

20. Which of the following correctly describes the *immediate* source of energy that drives ATP synthesis in the mitochondrion?

    (A) the flow of electrons down the electron transport chain
    (B) the H^+ concentration gradient across the cristae membrane
    (C) the affinity of oxygen for electrons flowing through the electron transport chain
    (D) ATP → ADP

21. Oxidative phosphorylation is said to couple two reactions. Which two reactions does it bring together?

    (A) the endergonic flow of electrons through the ETC with the exergonic movement of protons across the cristae membrane to create a proton gradient
    (B) the exergonic flow of electrons through the ETC with the endergonic pumping of protons across the cristae membrane to create a proton gradient
    (C) the flow of protons from glycolysis to the citric acid cycle
    (D) the production of ATP from the anaerobic phase with the aerobic phase

22. In a hydroelectric power plant, the energy of falling water turns a turbine, converting kinetic energy to mechanical energy. Next a generator converts the mechanical energy from the turbine into electrical energy. Which process in the production of ATP in cellular respiration is most similar to this process in a hydroelectric plant?

    (A) citric acid cycle
    (B) glycolysis
    (C) oxidative phosphorylation
    (D) fermentation

Answers Explained

1. **(D)** Oxidation is the loss of protons, hydrogen, or electrons, while reduction is the gain of protons, hydrogen, or electrons. In the case of aerobic respiration, glucose loses hydrogen to oxygen, and water vapor is formed and released. Animals breathe out this waste water vapor. (ENE-1)

2. **(D)** Oxygen is a raw material. This reaction is not endergonic because energy in the form of ATP is on the right side of the equation; it is being released. Choice C is not a correct statement because energy cannot be created or destroyed. Only ATP can be created or destroyed. (ENE-1)

3. **(D)** This description is exactly what NAD^+ and FAD do. Choice A is not correct because protons can only pass down their gradient through ATP synthase channels. The entire production of ATP in aerobic respiration depends on this fact. Choice B is not correct because substrate level phosphorylation produces a small amount of ATP during glycolysis and the citric acid cycle. Choice C is not correct because the ETC is located in the matrix of mitochondria. (ENE-1)

4. **(B)** Choice C represents glycolysis. Only a small amount of ATP is produced by substrate level phosphorylation during glycolysis and the Krebs cycle. The electron transport chain creates a proton gradient by pumping protons across the cristae membrane, but no ATP is created. Most ATP is produced by chemiosmosis as protons flow through the ATP synthase channels. (ENE-1)

5. **(A)** Choice A accurately describes the coupled reactions in the ETC. Choice B is not correct because while anaerobic organisms carry out glycolysis, glycolysis is the first stage of aerobic respiration in *all* aerobic organisms. Choice C is not correct because the citric acid cycle uses pyruvate as a raw material; it does not produce it. Choice D is not correct because the ETC does not produce any ATP. It sets the stage for the production of ATP by oxidative phosphorylation. (ENE-1)

6. **(C)** "Ion gradient" could also be written as "proton gradient." Choice A is not correct because oxygen is the final electron acceptor in the ETC. It does not carry protons. Choice B is not correct because substrate level phosphorylation is the mechanism by which ATP is produced during *glycolysis and the citric acid cycle.* Choice D is not correct because it is the ETC that pumps protons across the cristae membrane, creating a gradient. (ENE-1)

7. **(B)** The three processes result in the complete breakdown of glucose. Process *A* represents glycolysis. Process *B* represents the citric acid cycle. Process *C* represents oxidative phosphorylation. Process *A* takes place in the cytoplasm. Process *B* releases NADH, not NAD^+. Process *C* consists of the electron transport chain plus chemiosmosis. It produces the most ATP from cellular respiration. (ENE-1)

8. **(A)** The ATP synthase molecules lie within the cristae membrane. (ENE-1 & ENE-2)

9. **(B)** B represents the citric acid cycle. It is the inner matrix. (ENE-1 & SYI-1)

10. **(C)** Glycolysis is carried out in the cytoplasm and does not rely on internal membranes. It must have been the earliest mechanism by which ATP was produced in the first organisms on Earth, which the fossil record shows had no internal membranes. Choices A and B refer to primitive cells, but they do not explain anything about glycolysis. Also, yeast is aerobic as well as anaerobic. Choice D is not correct because, although it refers to endosymbiosis (the theory that mitochondria were once free-living cells), it does not explain anything about glycolysis. In fact, choice D explains the process by which aerobic respiration evolved in early life, but it does not explain when that process occurred. (ENE-1)

11. **(D)** Choices A, B, and C all describe events that occur during cell respiration and lead up to the production of ATP. However, the most important event in respiration is the production of ATP. (ENE-1)

12. **(A)** This is a fact. No CO_2 is produced during oxidative phosphorylation. (ENE-1)

13. **(C)** Glycolysis occurs in the first phase of cell respiration and does not require oxygen. The first organisms on Earth were probably anaerobes because no free oxygen was available in the atmosphere. Choices A, B, and D all require free oxygen. (ENE-1)

14. **(A)** Choices B–D are all correct statements about respiration, but they are not examples of a feedback mechanism. Choice A is an example of negative feedback because when plenty of ATP is available in the cell to meet demand, respiration slows down, conserving valuable molecules and energy for other functions. (ENE-3)

15. **(C)** Most of the ATP produced during cell respiration comes from oxidative phosphorylation as protons flow through the ATP synthase channel in the cristae membrane in mitochondria. (ENE-1)

16. **(C)** The process is substrate level phosphorylation: the release of energy by the transfer of a phosphate group from any molecule to ADP, forming ATP. This process releases a small amount of energy and occurs during the citric acid cycle and glycolysis. (ENE-1)

17. **(D)** This figure shows the electron transport chain (ETC), where no ATP is produced. The purpose of the ETC is to build a proton or ion gradient or differential between the inner matrix and outer compartment of mitochondria. This proton gradient represents great potential energy. When protons flow down the gradient through ATP synthase channels (the only place they can flow across the cristae membrane), energy is released. This energy phosphorylates ATP. ATP made this way is called oxidative phosphorylation. Choices A and B describe glycolysis. Choice C describes oxidative phosphorylation at the ATP synthase channel. However, the ATP synthase channel is not shown in this figure. (ENE-1)

18. **(B)** Oxygen, which is very strongly electronegative, pulls the electrons through the ETC. This flow of electrons is exergonic and powers the pumping of protons (ions) against a gradient. (ENE-1)

19. **(A)** When ATP levels in the cytoplasm are high, PFK is inhibited. The reverse is also true. The only graph that shows the correct relationship is the graph for choice A. (ENE-3)

20. **(B)** This choice is the essence of how ATP is produced during oxidative phosphorylation. Protons flow down a gradient through the ATP synthase channel. The force of that flow powers the phosphorylation of ADP into ATP. (ENE-1)

21. **(B)** Oxygen pulls electrons down a gradient in the ETC. This process is exergonic and therefore releases energy. The *pumping* of protons against a gradient across the cristae membrane to the outer compartment is endergonic. The first reaction powers the second reaction. (ENE-1)

22. **(C)** The issue is one of energy conversion. In a hydroelectric power plant, energy is converted from potential energy to mechanical energy to electrical energy. In a mitochondrion, electrons flow down the ETC, releasing energy along the way. This released energy is converted to create a proton (H^+) gradient across the cristae membrane. As protons flow down the proton gradient through the ATP synthase channel, the kinetic energy released is converted to chemical bond energy. (ENE-1)

Photosynthesis 7

Big Ideas: EVO, ENE & SYI
Enduring Understandings: EVO-1; ENE-1; SYI-1
Science Practices: 1, 2 & 3

For the complete list of Big Ideas, Enduring Understandings, and Science Practices, refer to the "AP Biology Course and Exam Description" from the College Board: *https://apcentral.collegeboard.org/pdf/ap-biology-course-and-exam-description.pdf?course=ap-biology*.

INTRODUCTION

Photosynthesis is the process by which light energy is converted to chemical bond energy, and carbon is fixed into organic compounds. It provides all the oxygen we breathe, the food we eat, and the fossil fuel we consume. The general formula is:

$$6CO_2 + 12H_2O \xrightarrow{\text{light}} C_6H_{12}O_6 + 6H_2O + 6O_2$$

There are two main processes of photosynthesis: the **light-dependent** and the **light-independent reactions**. The light-dependent reactions use light energy directly to produce ATP. The light-independent reactions consist of the Calvin cycle, which produces sugar. To power the production of sugar, the Calvin cycle uses ATP formed during the light reactions. Both reactions occur only when light is present. See Figure 7.6 for an overview of photosynthesis.

STUDY TIP

Photosynthesis is a reduction reaction because CO_2 gains hydrogen from water.

STUDY TIP

Photo—Light

Synthesis—Sugar production: think of the Calvin cycle

PHOTOSYNTHETIC PIGMENTS

Photosynthetic pigments absorb light energy and use it to provide energy to carry out photosynthesis. Plants contain two major groups of pigments: the chlorophylls and carotenoids. **Chlorophyll *a*** and **chlorophyll *b*** are green and absorb all wavelengths of light in the red, blue, and violet ranges. The **carotenoids** are yellow, orange, and red. They absorb light in the

INTERPRETING THIS GRAPH

Chlorophyll *a* absorbs light in the violet, blue, and red ranges. It reflects green, yellow, and orange light.

blue, green, and violet ranges. **Xanthophyll**, another photosynthetic pigment, is a carotenoid with a slight chemical variation. Pigments found in red algae, the **phycobilins**, are reddish and absorb light in the blue and green range. Chlorophyll *b*, the carotenoids, and the phycobilins are known as **antenna pigments** because they capture light in wavelengths other than those captured by chlorophyll *a*. Antenna pigments absorb photons of light and pass the energy along to chlorophyll *a*, which is directly involved in the transformation of light energy to sugars. Figure 7.1 is a graph that shows the absorption spectrum for photosynthetic pigments that were extracted from a leaf.

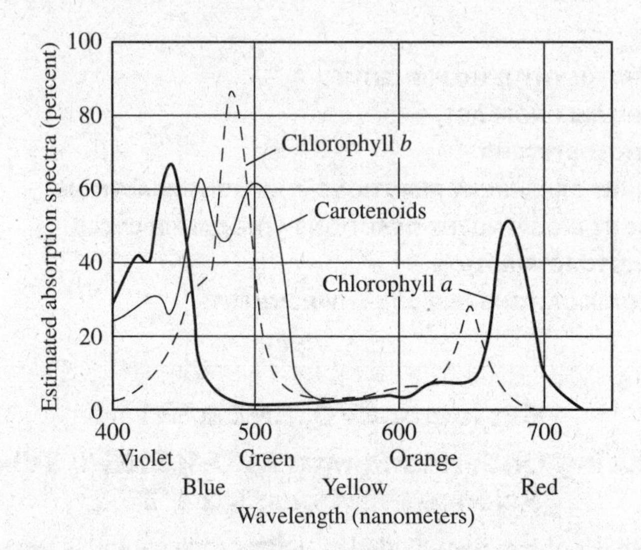

Figure 7.1 Absorption Spectrum for Photosynthetic Pigments Extracted from a Leaf

Figure 7.2 is an **action spectrum** with wavelengths of light plotted against rate of photosynthesis. The data are obtained from a living plant—not from isolated photosynthetic pigment as in the absorption spectrum above. This figure is different from Figure 7.1 because Figure 7.2 includes data from all photosynthetic pigments in the plants: chlorophyll *a*, chlorophyll *b*, and carotenoids.

SYI-3
Different types of chlorophyll give a plant greater flexibility to exploit light as an energy source.

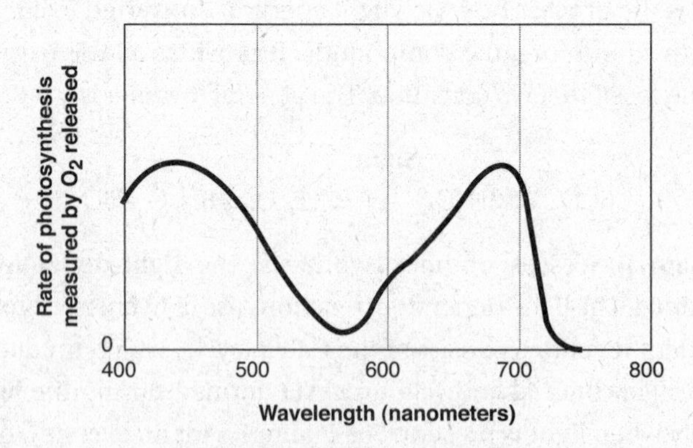

Figure 7.2 Action Spectrum

Chlorophyll *a* is the pigment that participates directly in the light reactions of photosynthesis. It is a large molecule with a single **magnesium** atom in the head surrounded by alternating double and single bonds. The head, called the porphyrin ring, is attached to a long hydrocarbon tail. The double bonds within the head of a molecule play a critical role in the light reactions. They are the source of the electrons that flow through the electron transport chains during photosynthesis. Figure 7.3 is a drawing of chlorophyll *a*.

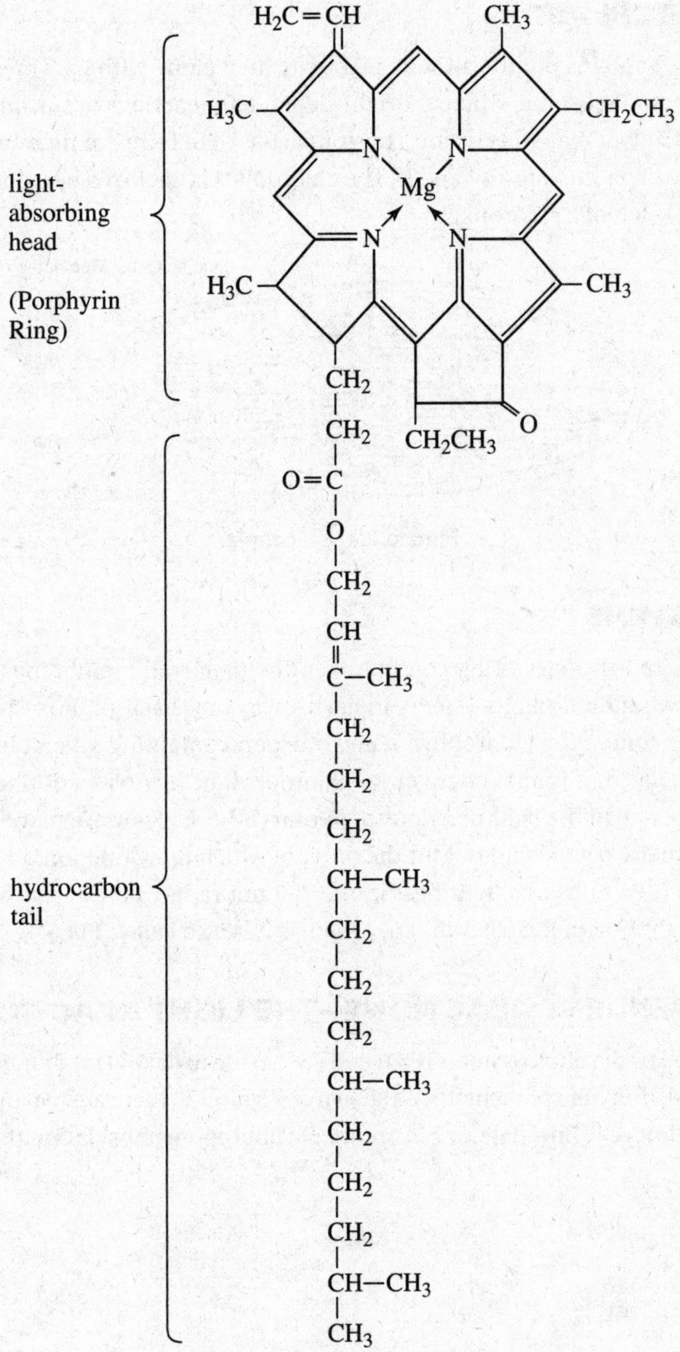

$H_2C{=}CH$

CH_3

H_3C

CH_2CH_3

N N Mg N N

H_3C

CH_3

light-
absorbing
head

(Porphyrin
Ring)

CH_2

CH_2

CH_2CH_3

O

$O{=}C$

O

CH_2

CH

C—CH_3

CH_2

CH_2

CH_2

CH—CH_3

CH_2

CH_2

CH_2

CH—CH_3

CH_2

CH_2

CH_2

CH—CH_3

CH_3

hydrocarbon
tail

Figure 7.3 Chlorophyll *a*

NOTE

This figure is included for reference.

THE CHLOROPLAST

The chloroplast contains photosynthetic pigments that, along with enzymes, carry out photosynthesis. It contains **grana**, where the light-dependent reactions occur, and **stroma**, where the light-independent reactions occur. The grana consist of layers of membranes called **thylakoids**, the site of photosystems I and II. The chloroplast is enclosed by a double membrane. Figure 7.4 is a sketch of a chloroplast.

> **EVO-1**
> Photosynthesis first evolved in prokaryotes. Scientific evidence supports the idea that this early photosynthesis produced our oxygenated atmosphere.

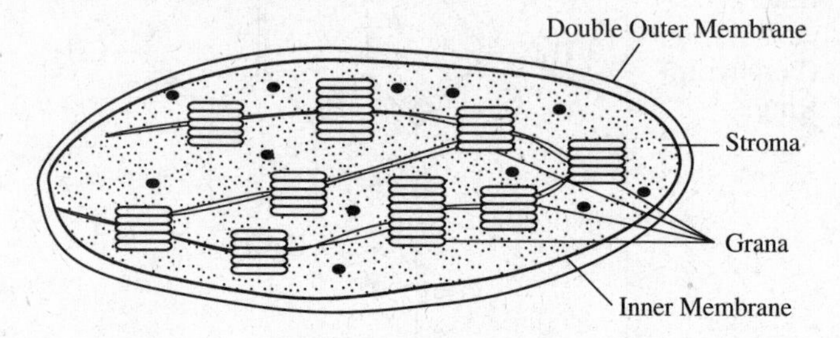

Figure 7.4 Chloroplast

PHOTOSYSTEMS

Photosystems are light-harvesting complexes in the thylakoid membranes of chloroplasts. There are a few hundred photosystems in each thylakoid. Each photosystem consists of a **reaction center** containing chlorophyll *a* and a region containing several hundred antenna pigment molecules that funnel energy into chlorophyll *a*. Two types of photosystems, **PS I** and **PS II**, cooperate in the light reactions of photosynthesis. They are named in the order in which they were discovered and not for the order in which they function. PS II operates first, followed by PS I. PS I absorbs light best in the 700 nm range; hence, it is also called **P700**. PS II absorbs light best in the 680 nm range; hence, it is also called **P680**.

LIGHT-DEPENDENT REACTIONS—THE LIGHT REACTIONS

Light is absorbed by the photosystems (PS II and PS I) in the thylakoid membranes and electrons flow through electron transport chains. As shown in Figure 7.5, there are two possible routes for electron flow: **noncyclic flow** (Figure 7.5) and **cyclic photophosphorylation** (Figure 7.7).

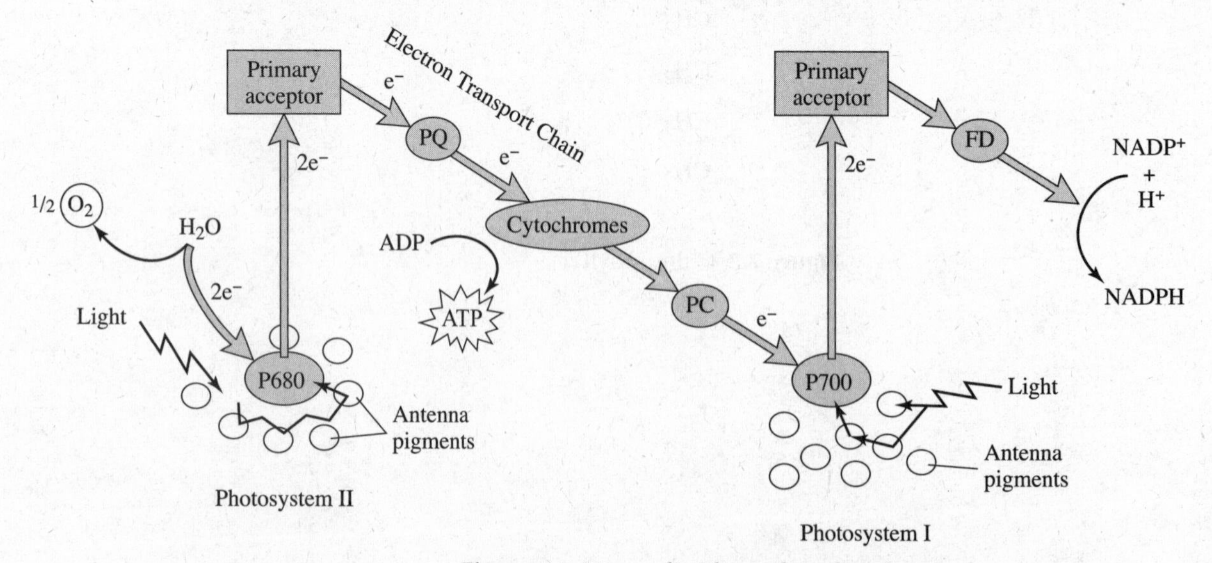

Figure 7.5 Noncyclic Photophosphorylation

Noncyclic Photophosphorylation

During **noncyclic photophosphorylation**, electrons enter two electron transport chains. The products are *ATP* and *NADPH (nicotinamide dinucleotide phosphate)*. The process begins in PS II and proceeds through the following steps.

REMEMBER

Electrons flow from water to P680 to P700 to NADP, which carries them to the Calvin cycle.

- **PHOTOSYSTEM II—P680.** Energy is absorbed by P680. Electrons from the double bonds in the head of chlorophyll *a* become energized and move to a higher energy level. They are captured by a **primary electron acceptor**.
- **PHOTOLYSIS.** Photolysis is the splitting of water. It provides electrons to replace those lost from chlorophyll *a* in P680. Photolysis splits water into two electrons, two protons (H^+), and one oxygen atom. Two oxygen atoms combine to form one O_2 molecule, which is released into the air as a waste product of photosynthesis.

$$H_2O \rightarrow 2H^+ + 2e^- + O_2 \uparrow$$

- **ELECTRON TRANSPORT CHAIN.** Electrons from P680 pass along an electron transport chain consisting of several molecules including **cytochromes** and several other proteins, and ultimately end up in P700 (PS I). This flow of electrons is exergonic and provides energy to produce ATP by **chemiosmosis**, the same way ATP is produced in mitochondria. Because this ATP synthesis is powered by light, it is called **photophosphorylation**. See Figure 7.6.
- **CHEMIOSMOSIS.** This is the process by which ATP is formed during the light reactions of photosynthesis. Protons that were released from water during photolysis are pumped by the thylakoid membrane from the stroma into the **thylakoid space (lumen)**. ATP is formed as these protons diffuse down the gradient from the thylakoid space, through the **ATP synthase channels**, and into the stroma. The ATP produced here provides the energy that powers the Calvin cycle.
- **NADP** becomes reduced when it picks up the two protons that were released from water in P680. Newly formed **NADPH** carries hydrogen to the Calvin cycle to make sugar in the light-independent reactions.
- **PHOTOSYSTEM I—P700.** Energy is absorbed by P700. Electrons from the head of chlorophyll *a* become energized and are captured by a primary electron receptor. This process is similar to the way it happens in P680. One difference is that the electrons that escape from chlorophyll *a* are replaced with electrons from photosystem II, P680, instead of from water. Another difference is that this electron transport chain produces NADPH, not ATP.

Here is an overview of noncyclic photophosphorylation.

light	→	P680 oxygen released	→	ATP produced	→	P700	→	NADPH produced (NADPH carries H^+ to the Calvin cycle)

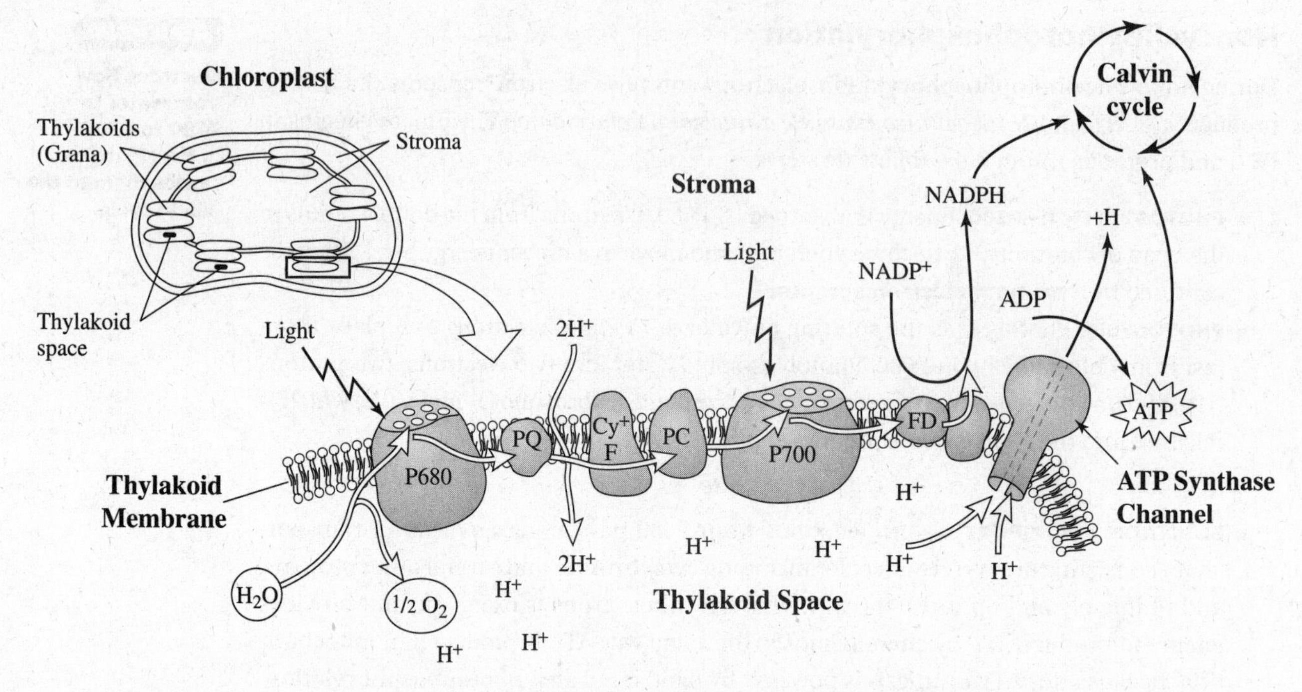

Figure 7.6 Overview of Photosynthesis

Cyclic Photophosphorylation

The sole purpose of cyclic photophosphorylation is to produce ATP. No NADPH is produced, and no oxygen is released.

The production of sugar that occurs during the Calvin cycle consumes enormous amounts of ATP, so periodically, the chloroplast runs low on ATP. When it does, the chloroplast carries out **cyclic photophosphorylation** to replenish the ATP levels. Cyclic electron flow takes photoexcited electrons on a short-circuit pathway. Electrons travel from the P680 electron transport chain to P700, to a primary electron acceptor, and then back to the cytochrome complex in the P680 electron transport chain. Cyclic photophosphorylation is shown in Figure 7.7.

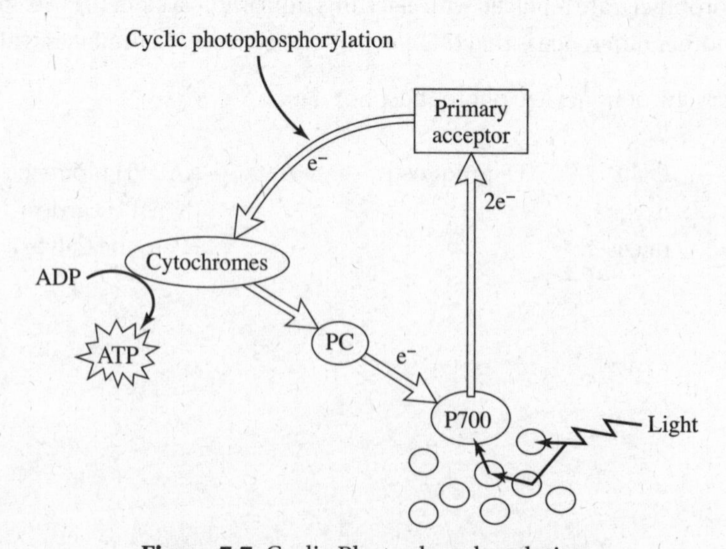

Figure 7.7 Cyclic Photophosphorylation

LIGHT-INDEPENDENT REACTIONS—THE CALVIN CYCLE

The **Calvin cycle** is the main business of the **light-independent reactions**. It is a cyclical process that produces the 3-carbon sugar **PGAL (phosphoglyceraldehyde)**. Carbon enters the stomates of a leaf in the form of CO_2 and becomes *fixed* or incorporated into PGAL.

Here are the important aspects of the Calvin cycle, as shown in Figure 7.8.

- The process that occurs during the Calvin cycle is **carbon fixation**.
- It is a **reduction reaction** since carbon is gaining protons and electrons.
- CO_2 enters the Calvin cycle and becomes attached to a 5-carbon sugar, **ribulose biphosphate (RuBP)**, forming a 6-carbon molecule. The 6-carbon molecule is unstable and immediately breaks down into two 3-carbon molecules of **3-phosphoglycerate (3-PGA G3P)**. The enzyme that catalyzes this first step is ribulose biphosphate carboxylase **(rubisco)**.
- The Calvin cycle does not directly depend on light. Instead, it uses the products of the light reactions: ATP and NADPH.
- The Calvin cycle, like the light-dependent reactions, **occurs only in the light**.

STUDY TIP

Your teacher may want you to know this for class, but you will not see all these details on the AP exam. You do, however, need to understand how the Calvin cycle works.

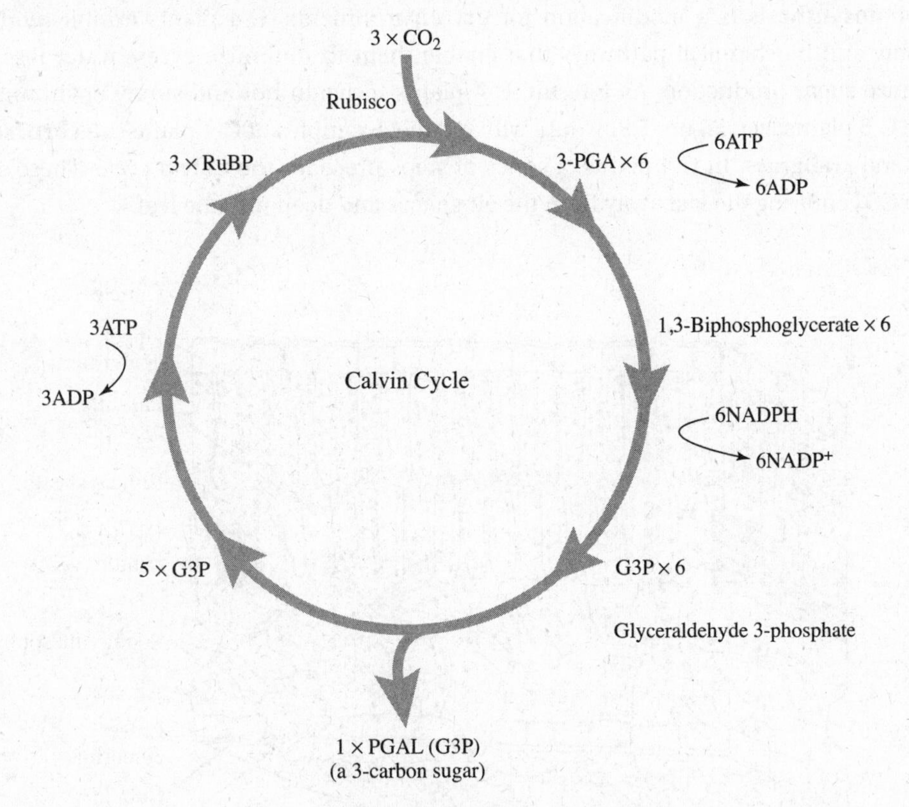

Figure 7.8 The Calvin Cycle

PHOTORESPIRATION

In most plants, CO_2 enters the Calvin cycle and is fixed into **3-phosphoglycerate (3-PGA)** by the enzyme **rubisco**. These plants are called **C-3 plants** because the first step produces the compound 3-PGA, which contains three carbons. This first step is not very efficient because rubisco can also bind with O_2 as well as with CO_2. When rubisco binds with O_2 instead of CO_2, this process, **photorespiration**, diverts the process of photosynthesis in two ways.

1. Unlike normal respiration, no ATP is produced.
2. Unlike normal photosynthesis, no sugar is formed.

Instead, **peroxisomes** break down the products of photorespiration. If photorespiration does not produce any useful products, why do plants carry it out at all? This process is probably a vestige from ancient Earth billions of years ago when the atmosphere had little or no free oxygen to divert rubisco and sugar production.

MODIFICATIONS FOR DRY ENVIRONMENTS

C-4 photosynthesis is a modification for dry environments. C-4 plants exhibit modified anatomy and biochemical pathways that enable them to minimize excess water loss and maximize sugar production. As a result, C-4 plants thrive in hot and sunny environments where C-3 plants (see Figure 7.9) would wilt and die. Examples of C-4 plants are **corn**, **sugar cane**, and **crabgrass**. In C-4 plants, a series of steps precedes the Calvin cycle. These steps pump CO_2 entering the leaf away from the air spaces and deep into the leaf.

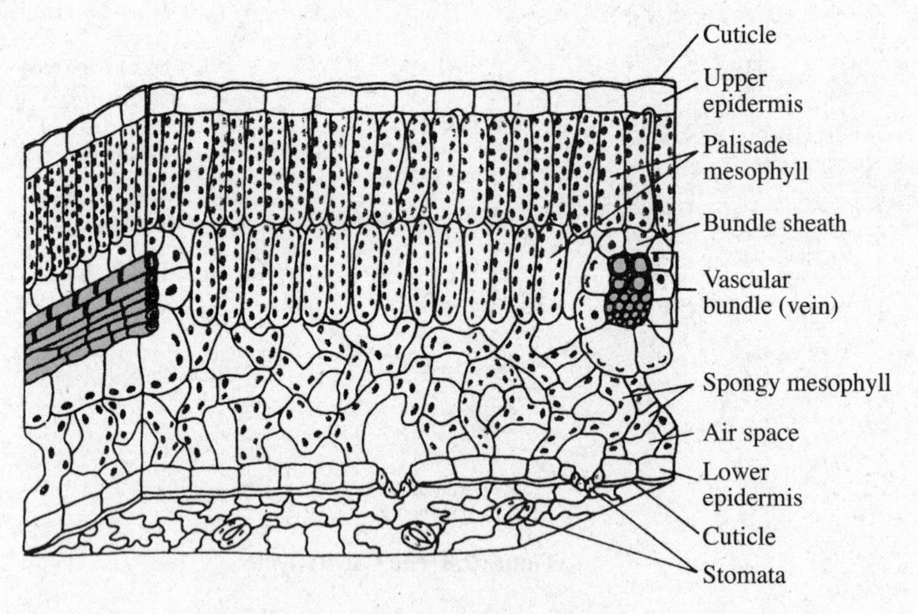

Figure 7.9 Cross Section of a C-3 Leaf

CHAPTER SUMMARY

This chapter supported Big Ideas: EVO, ENE, and SYI.

- Living things require a constant supply of energy.
- Photosynthesis converts light energy to chemical energy. In eukaryotic autotrophs, photosynthesis occurs in **thylakoid membranes** and **stroma** of **chloroplasts**. Here is the general equation: $6CO_2 + 12H_2O \rightarrow C_6H_{12}O_6 + 6O_2$.
- Photosynthesis is a **reduction reaction** because in order for glucose to form, hydrogen binds to carbon in CO_2. The production of sugar occurs by a process called **carbon fixation**.
- The **light-dependent reactions** begin when light is absorbed by **photosynthetic pigments** in light-harvesting complexes called **photosystems**, located within the **grana** (stacks of **thylakoids**). Different photosynthetic pigments absorb light of differing wavelengths, expanding the range of light that can power photosynthesis. Electrons become excited and enter an **electron transport chain** (linear flow of electrons) that results in the production of **ATP** by **chemiosmosis** (similar to cell respiration) and **NADPH**. **Oxygen** is a waste product of the light reactions.
- Water is split, and electrons (e^-) are transferred along with protons (H^+) from water to CO_2, reducing the CO_2 to sugar.
- The **light-independent reactions** only occur when there is light. Although they do not use light directly, these reactions use the *products* of the light-dependent reactions: ATP and protons carried by NADPH from the light-dependent reactions. These substances reduce CO_2 to **PGAL** or **G3P** (3-carbon sugar). **Carbon fixation**, which is a reduction of CO_2, occurs during the **Calvin cycle**.
- Plants absorb CO_2 from the atmosphere and give off water vapor and O_2 all through the same openings—stomates in leaves. This presents a problem. Plants must keep their stomates open long enough to take in enough CO_2 to make sugar but not so long that they lose too much water vapor and dry out. Plants have evolved different strategies to minimize water loss. One example is **C-4 photosynthesis.**
- You should learn about the anatomy and leaf structure in plants. (Be sure to also review Chapter 14, "**Plants.**")
- You do **not** have to memorize the details of the Calvin cycle.

PRACTICE QUESTIONS

1. The graph below shows an absorption spectrum for an unknown pigment molecule. Which answer choice identifies the color of this pigment and gives the correct reason for your answer?

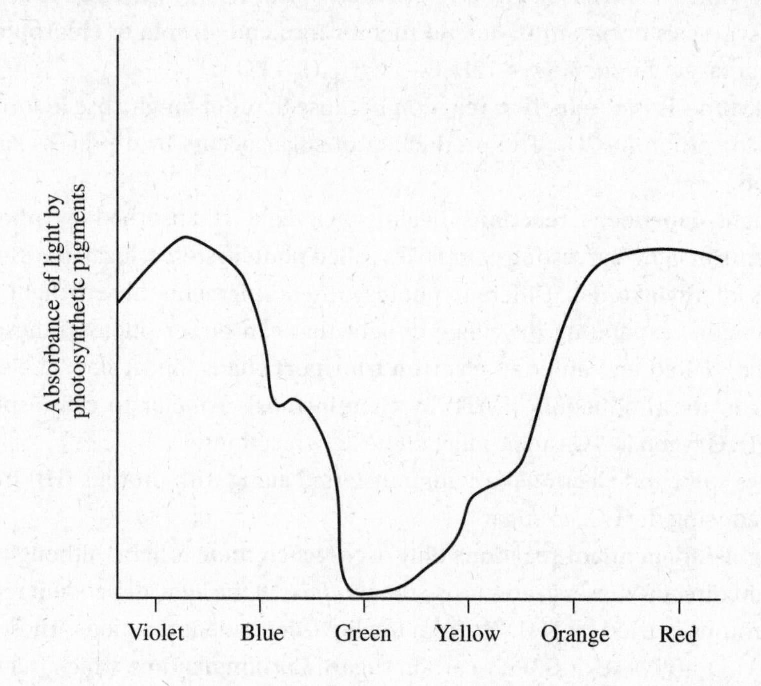

   (A) The pigment is red-violet because it reflects red-violet light and absorbs all others.
   (B) The pigment is red-violet because it reflects red-violet light and is seen as red-violet.
   (C) The pigment is green because it reflects green light and absorbs all other wavelengths.
   (D) The pigment is green because it absorbs green light and is seen as green.

2. Which of the following statements is correct about cyclic photophosphorylation?

   (A) ATP is produced by light energy.
   (B) This process occurs when a chloroplast runs low on ATP.
   (C) Both NADPH and ATP are produced during this process.
   (D) Only P680 is directly involved, not P700.

3. Which of the following statements is correct about photosynthesis?

   (A) PGAL (G3P) is produced in the grana.
   (B) Carbon dioxide is released from the Calvin cycle.
   (C) Oxygen is released from the light-dependent reactions.
   (D) Photorespiration increases ATP levels when a plant is in distress.

4. Which of the following is correct about the Calvin cycle?
   (A) The Calvin cycle occurs in the thylakoid membrane.
   (B) The Calvin cycle mostly occurs in the absence of light.
   (C) Six turns of the Calvin cycle are required to produce glucose.
   (D) The Calvin cycle is a series of reactions that result in the oxidation of glucose.

Questions 5–6

Questions 5 and 6 refer to the following diagram of a chloroplast.

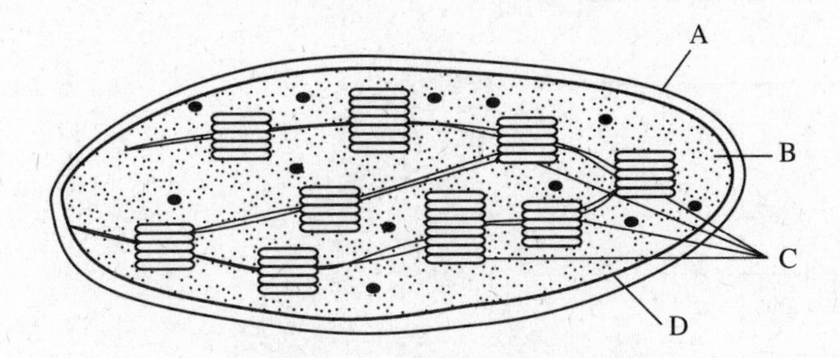

5. Which of the following correctly identifies the process with the structure?
   (A) Photolysis occurs at A.
   (B) An electron transport chain is located at C.
   (C) ATP is produced on the inner membrane at D.
   (D) Cyclic photophosphorylation occurs at B.

6. Which of the following correctly identifies the process with the structure?
   (A) P680 can be found in large quantities at B.
   (B) G3P can be found in large quantities at B.
   (C) Photosynthetic pigments can be found in large quantities at B.
   (D) Thylakoid membranes make up the membranes at A.

---

7. Which of the following is correctly matched with its location in a leaf?
   (A) ATP synthase molecules—grana
   (B) Photosystems—double outer membrane of the chloroplast
   (C) Calvin cycle—thylakoid membrane
   (D) Electron transport chain—stroma

8. Theodor Engelmann carried out a now famous experiment in 1883 to determine which wavelengths of light are most important for photosynthesis to occur in green plants. He placed a single strand of the green algae *Spirogyra* on a microscope slide along with a drop of water containing aerobic (oxygen-requiring) bacteria. At first, the bacteria distributed themselves randomly on the slide. Engelmann then placed the slide on the stage of a microscope and illuminated it with light that had been passed through a prism. The prism separated the light into its component wavelengths. See the figure below.

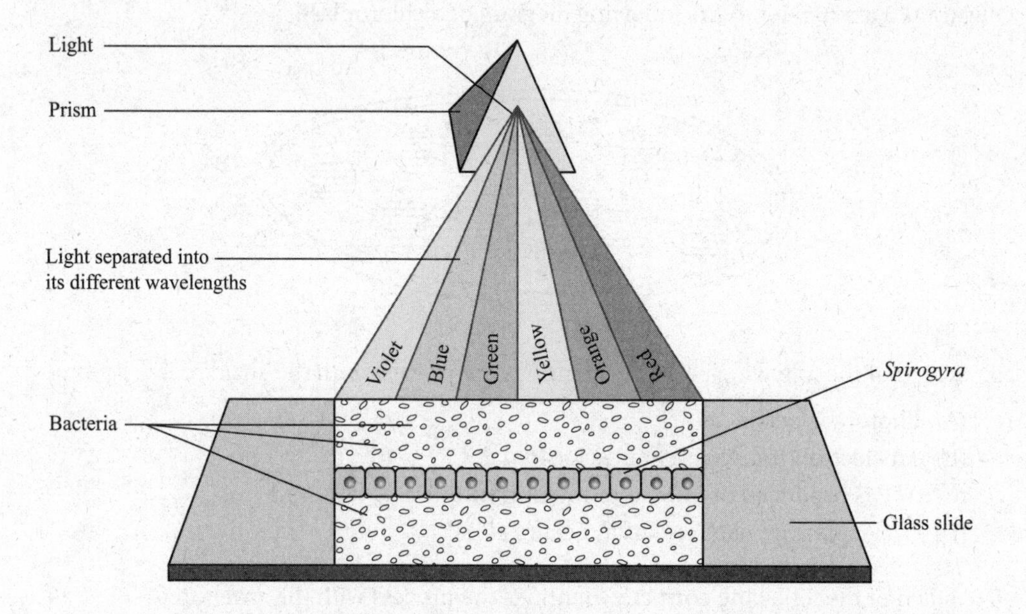

After several minutes, the bacteria became attracted to regions along the *Spirogyra* strand where oxygen had been released by photosynthesis.

Given what you know about photosynthesis and light, from the following four images, choose the one that most accurately shows what the slide would look like after one hour when the bacteria had been allowed to distribute themselves.

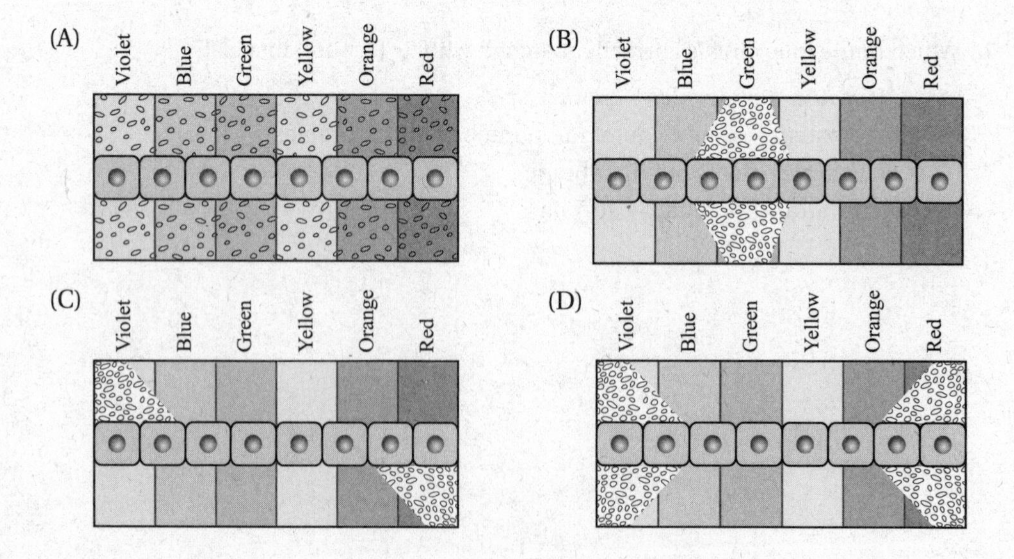

9. Which of the following is correct about light-independent reactions?
   (A) Carbon is fixed into molecules of G3P.
   (B) ATP is produced.
   (C) They occur in the thylakoid membranes.
   (D) They rely on ETCs.

10. Which of the following is associated with photosystem I?
   (A) It is the first in a series of electron transport chains.
   (B) It replaces its lost electrons with electrons from photosystem II.
   (C) It is active only at night.
   (D) It depends on antenna pigments, not on chlorophyll *a*.

11. Which of the following is correct regarding the role of antenna pigments in photosynthesis?
   (A) They expand the wavelengths of light that can be used for photosynthesis.
   (B) They cannot absorb light.
   (C) They shuttle hydrogen from the light-dependent reactions to the light-independent reactions.
   (D) They store light for those plants that need to carry out photosynthesis in the dark.

12. Refer to the diagram below. On the left, there is a sketch of chromatography paper that your teacher set up. She squashed some spinach onto the bottom of the paper by rolling the edge of a coin over a spinach leaf, making a green line on the paper. Then, she placed the bottom edge of the paper into a solvent. On the right, there is a sketch of how the paper looked like after 20 minutes in the solvent.

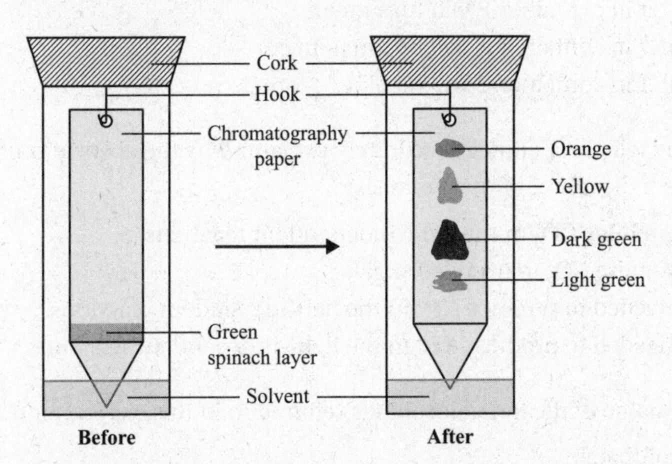

   Which of the following is correct regarding the different layers of color that appeared on the paper after 20 minutes in the solvent?
   (A) The process broke apart the bonds that were holding each chemical together.
   (B) The yellow pigment was denser than the green pigment.
   (C) They were all different versions of chlorophyll *a*.
   (D) They were all photosynthetic pigments.

Questions 13–15

You are conducting an experiment to determine the effect of different wavelengths of light on the rate of photosynthesis in aquatic plants. You use the absorption of CO_2 as an indicator of photosynthesis. You place elodea, a small aquatic plant, into each of three identical freshwater-filled glass containers. You blow equal amounts of CO_2 into each container using a straw, and you also add a chemical indicator to each container. At the beginning of the experiment, with high CO_2 levels in each container, the indicator is yellow. Refer to the diagram below.

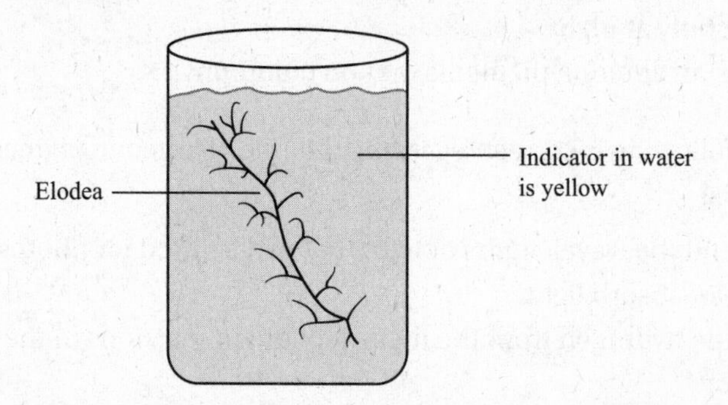

If the level of CO_2 decreases, the indicator turns green. With all CO_2 removed, the indicator turns blue. You place container A in normal sunlight, container B under green light only, and container C under red light only. You then observe the three vessels after 24 hours.

13. Predict what you will observe after 24 hours.
    (A) The water in container B will turn blue.
    (B) The water in container B will turn green.
    (C) The water in container C will be unchanged.
    (D) The water in container C will be blue.

14. Which of the following choices accurately explains why the absorption of CO_2 is an appropriate indicator of photosynthesis?
    (A) Plants produce CO_2 in the light-independent reactions.
    (B) Plants require CO_2 to produce sugars.
    (C) CO_2 is needed to produce ATP in the light-dependent reactions.
    (D) CO_2 is needed to produce ATP in the light-independent reactions.

15. What is the name of the indicator that is referred to in this experiment?
    (A) phenolphthalein
    (B) red cabbage juice
    (C) Bromothymol blue
    (D) methyl orange

16. When light energy boosts electrons from photosystem P680 to the electron acceptor, what replaces the missing electrons?

    (A) photosystem I
    (B) photosystem II
    (C) water
    (D) NADPH

17. In an experiment, chloroplasts are isolated from cells and placed in an Erlenmeyer flask with all the chemicals necessary for the chloroplasts to carry out ATP synthesis. Next, you place the flask under strong white light. Predict what will happen to the rate of photosynthesis if a compound that makes membranes permeable to protons (H^+) is added to the flask, and give the correct reason for your prediction.

    (A) ATP production will decrease because the compound introduced will shut down the electron transport chain in the thylakoid membrane.
    (B) ATP production will decrease because no proton gradient can be maintained across the thylakoid membrane.
    (C) ATP production will increase because a stronger proton gradient will be established across the thylakoid membrane.
    (D) ATP production will increase because the Calvin cycle will be unable to produce ATP.

Answers Explained

1. **(C)** If light is absorbed, it is not reflected. Only reflected colors are visible. The graph shows that red and violet are most absorbed and that green is most reflected. Therefore, the color of the pigment is green. (ENE-1)

2. **(A)** Electrons undergoing cyclic photophosphorylation move from P680 to P700 and then cycle back to P680. The sole purpose of cyclic photophosphorylation is the production of ATP. No NADPH is produced, and no oxygen is released. This process is necessary when the cell needs more ATP because ATP has been used up by the Calvin cycle. Sugar is produced during the light-independent reactions only, not during the light-dependent ones. (ENE-1)

3. **(C)** Oxygen is released, via the process of photolysis, during the light-dependent reactions. Choice A is not correct because PGAL or G3P is produced in the stroma from the Calvin cycle. Choice B is not correct because carbon dioxide is not released during photosynthesis; rather, carbon dioxide is a raw material for the Calvin cycle. Choice D is not correct because photorespiration does *not* produce ATP, nor does it produce sugar. Rather, photorespiration occurs when rubisco binds with oxygen instead of CO_2. (ENE-1)

4. **(C)** Three turns of the Calvin cycle produce the 3-carbon PGAL (G3P), while six turns of the Calvin cycle produce glucose. Choice A is incorrect because the Calvin cycle occurs in the stroma. Choice B is incorrect because the Calvin cycle (which is part of the light-independent reactions) occurs when there *is* light because this process depends on the products of the light-dependent reactions, which only occur when there is light. Choice D is incorrect because the Calvin cycle results in the reduction of CO_2, not the oxidation of glucose. (ENE-1)

5. **(B)** A is the double outer membrane of the chloroplast. B is the stroma, which is the site of the light-independent reactions. C is the grana, which contain thylakoid membranes and carry out the light-dependent reactions, including photolysis. D is the inner membrane, but no relevant reactions occur there. Cyclic photophosphorylation occurs in the light-dependent reaction—in the grana at C. (ENE-1 & SYI-1)

6. **(B)** B is the stroma, which is where the light-independent reactions occur. This is also where the Calvin cycle occurs and where PGAL or G3P is produced. Choices A, C, and D are incorrect because they all relate to the light-dependent reactions, which occur in the grana at C. (ENE-1 & SYI-1)

7. **(A)** ATP is produced in the light-dependent reactions of the grana, which is where you find photosystems and electron transport chains. The matches in choices B, C, and D are all incorrect. (ENE-1 & SYI-1)

8. **(D)** This experiment was a brilliant one. As cells carry out photosynthesis, oxygen is released. In this experiment, aerobic bacteria organize themselves close to an oxygen source—those cells that carry out photosynthesis and are therefore releasing oxygen. That leaves one question—which lights provide the highest rate of photosynthesis? The answer is violet light and red light. Choice D has more bacteria clustered in violet light and red light. Choices A and B do not show that distribution and are thus incorrect. Choice C is not correct because it shows the bacteria in a distribution that does not make sense. (ENE-1 & SP 3)

9. **(A)** Carbon fixation produces sugars or G3P in the Calvin cycle of light-independent reactions. Choices B, C, and D relate to light-dependent reactions. (ENE-1)

10. **(B)** Photosystem II occurs first; PS I occurs second. PS II loses electrons into the ETC and replaces them from water, which plants take in. When PS I loses electrons into the ETC, it replaces them with electrons from PS II. Choice A is not correct because photosystem II occurs first. Choice C is not correct because both PS I and PS II only occur in the light. Choice D is not correct because both photosystems depend on chlorophyll *a* and the antenna pigments. (ENE-1)

11. **(A)** The role of antenna pigments is to absorb light and pass it on to chlorophyll *a.* Choice B is not correct because antenna pigments *do* absorb light. Choice C is not correct because it is in fact NADP that shuttles hydrogen from the light-dependent reactions to the light-independent reactions. Choice D is not correct because the antenna pigments do not store light; they pass it on to chlorophyll *a.* (ENE-1)

12. **(D)** You have likely completed a chromatography lab in your AP Biology class. The lab described in this question demonstrated that the green from the spinach actually contained a mixture of different photosynthetic pigments that all separated during the process. Choice A is not correct because the separation was physical; there were no bonds being broken. Choice B is not correct because the carotenoids run the fastest and farthest (moving upward away from the solvent) because they are the least dense. Choice C is not correct because the sample of spinach is a mixture of four different photosynthetic pigments: chlorophyll *a* and *b* and two carotenoids. Chromatography is a technique to separate mixtures into their components. (SP 1)

13. **(D)** The solution is yellow when it contains a lot of CO_2. As photosynthesis uses up the CO_2 in the solution, the solution and indicator will turn from yellow to green and then to blue. The colors of light that provide the most energy for photosynthesis are violet and red. Green light is reflected and thus provides no energy for photosynthesis. Sunlight contains all the colors of light; some provide a lot of energy for plants, and others do not. However, a beam of focused red light contains the most energy, even more so than sunlight, and would be the most efficient at turning the indicator from yellow to blue. (SP 1)

14. **(B)** Plants use the CO_2 they take in to make sugars. Therefore, as photosynthesis continues, and sugar is manufactured, CO_2 is used up. You could also measure the amount of sugar that appears inside a plant cell to monitor photosynthesis. However, watching an indicator change color is much easier. (ENE-1)

15. **(C)** Bromothymol blue turns from yellow (in the presence of an acid) to blue as a solution becomes basic. When CO_2 dissolves in water, it becomes carbonic acid, which makes the water acidic and turns Bromothymol blue to yellow. Phenolphthalein is another pH indicator. It turns from clear (in the presence of an acid) to strong pink in the presence of a base. (SP 1)

16. **(C)** When electrons in PS II (P680) get excited by light, they are captured by an electron acceptor and shunted into an electron transport chain that produces ATP. These excited electrons end up in PS I (P700). The process of photolysis removes electrons from water to replace those electrons that left PS II. (ENE-1)

17. **(B)** ATP synthesis during the light-dependent reactions of photosynthesis depends on the maintenance of a proton gradient forming across the thylakoid membrane. Just as in ATP production in cellular respiration, both processes depend on the fact that protons can only flow down the gradient across a membrane at ATP synthase channels. If a chemical were introduced that would make the thylakoid membrane (or cristae membrane in mitochondria) permeable to protons, there would be no ATP produced. (ENE-1)

8 Heredity

→ BASICS OF PROBABILITY
→ LAW OF DOMINANCE
→ LAW OF SEGREGATION
→ MONOHYBRID CROSS
→ BACKCROSS OR TESTCROSS
→ LAW OF INDEPENDENT ASSORTMENT
→ BEYOND MENDELIAN INHERITANCE
→ POLYGENIC INHERITANCE
→ MORE EXCEPTIONS TO MENDELIAN INHERITANCE
→ GENES AND THE ENVIRONMENT
→ LINKED GENES
→ CROSSOVER AND LINKAGE MAPPING
→ THE PEDIGREE
→ X INACTIVATION—THE BARR BODY
→ MUTATIONS
→ NONDISJUNCTION

Big Ideas: EVO, IST & SYI
Enduring Understandings: EVO-2; IST-1; SYI-3
Science Practices: 3 & 5

For the complete list of Big Ideas, Enduring Understandings, and Science Practices, refer to the "AP Biology Course and Exam Description" from the College Board: *https://apcentral.collegeboard.org/pdf/ap-biology-course-and-exam-description.pdf?course=ap-biology*.

INTRODUCTION

The father of modern genetics is **Gregor Mendel**, an Austrian monk who, in the 1850s, bred garden peas in order to study patterns of inheritance. Mendel was successful because he brought an experimental and mathematical approach to the study of inheritance. First, he studied traits that were clear-cut, with no intermediates between varieties. Second, he collected data from a large sample, hundreds of plants from each of several generations. Mendel collected ten thousand plants in all. Third, he applied statistical analysis to his carefully collected data.

Until the nineteenth century, people thought that inheritance was blended, a mixture of fluids that passed from parents to children. In contrast, Mendel's theory of genetics is one of

particulate inheritance in which inherited characteristics are carried by discrete units that he called *elementes*. These *elementes* eventually became known as genes.

BASICS OF PROBABILITY

Probability is the likelihood that a particular event will happen. If an event is an absolute certainty, its probability is 1. If the event cannot happen, its probability is 0. The probability of anything else happening is between 0 and 1. Probability cannot predict whether a particular event will actually occur. However, if the sample is large enough, probability can predict an average outcome.

Understanding probability is important to the study of genetics because predicting outcomes is what Punnett squares enable us to do. What is the chance that two brown-eyed people can give birth to a child with blue eyes? That is probability, and that is what this chapter is all about.

When Do You Multiply? Multiplication Rule

To find the probability of **two independent events** happening, **multiply** the chance of one happening by the chance that the other will happen. For example, the chance of a couple having two boys depends on two independent events. The chance of the first child being a boy is $\frac{1}{2}$; the chance of the next child being a boy is $\frac{1}{2}$. Therefore, the chance that the couple will have two boys is $\frac{1}{2} \times \frac{1}{2} = \frac{1}{4}$. The chance of having three boys is $\frac{1}{2} \times \frac{1}{2} \times \frac{1}{2} = \frac{1}{8}$.

When Do You Add? Addition Rule

When more than one arrangement of events producing the specified outcome is possible, the probabilities for each outcome are added together. For example, if a couple is planning on having two children, what is the chance that they will have one boy and one girl (in either order)? Here is how you solve this problem. The probability of having a boy and then a girl is $\frac{1}{2} \times \frac{1}{2} = \frac{1}{4}$. The probability of having a girl and then a boy is $\frac{1}{2} \times \frac{1}{2} = \frac{1}{4}$. Therefore, the probability of having one boy and one girl (in either order) is $\frac{1}{4} + \frac{1}{4} = \frac{1}{2}$.

LAW OF DOMINANCE

Mendel's first law is the **law of dominance**, which states that when two organisms, each **homozygous** (pure) for two opposing traits are crossed, the offspring will be **hybrid** (carry two different alleles) but will exhibit only the **dominant trait**. The trait that remains hidden is known as the **recessive trait**.

Parent (P):	*TT*	×	*tt*
	Pure tall		Pure dwarf
Offspring (F_1):		*Tt*	
		All hybrid tall	

	T	*T*
t	*Tt*	*Tt*
t	*Tt*	*Tt*

Law of dominance
All offspring are tall

LAW OF SEGREGATION

The **law of segregation** states that during the formation of gametes, the two traits carried by each parent separate. See Figure 8.1.

SYI-3
Segregation and independent assortment of chromosomes result in genetic variation.

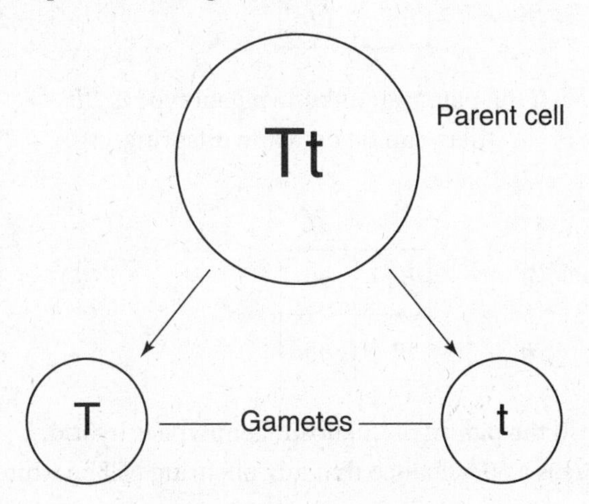

Figure 8.1 Law of Segregation

The cross that best exemplifies this law is the **monohybrid cross**, *Tt* × *Tt*. In the monohybrid cross, a trait that was not evident in either parent appears in the F_1 generation.

MONOHYBRID CROSS

The **monohybrid cross** (*Tt* × *Tt*) is a cross between two organisms that are each hybrid for one trait. The **phenotype** (appearance) ratio from this cross is 3 tall (shaded) to 1 dwarf plant (unshaded). The **genotype** (type of genes) **ratio**, 1:2:1, is given as percentages: 25 percent homozygous dominant, 50 percent heterozygous, and 25 percent homozygous recessive. These results are always the same for any monohybrid cross.

F_1: *Tt* × *Tt*

F_2: *TT*, *Tt*, or *tt*

	T	*t*
T	*TT*	*Tt*
t	*Tt*	*tt*

Monohybrid cross

BACKCROSS OR TESTCROSS

The **testcross** or **backcross** is a way to determine the genotype of an individual plant or animal showing only the dominant trait. It involves a cross between real organisms. The individual in question (*B*___) is crossed with a homozygous recessive individual (*bb*). If the individual being tested is in fact homozygous dominant, all offspring of the testcross will be *Bb* and will show the dominant trait. There can be no offspring showing the recessive trait. If the individual being tested is hybrid (*Bb*), one-half of the offspring can be expected to show the recessive trait. Therefore, if any offspring show the recessive trait, the parent of unknown genotype must be hybrid.

	B	*B*	
b	*Bb*	*Bb*	*B* = black
b	*Bb*	*Bb*	*b* = white

If the parent of unknown genotype is *BB*,
there can be no white offspring.

	B	*b*	
b	*Bb*	*bb*	*B* = black
b	*Bb*	*bb*	*b* = white

If the parent of unknown genotype is hybrid,
there is a 50% chance that any offspring will be white.

LAW OF INDEPENDENT ASSORTMENT

The **law of independent assortment** applies when a cross is carried out between two individuals hybrid for two or more traits that are **not on the same chromosome**. This type of cross is called a **dihybrid cross**. This law states that during gamete formation, the alleles of a gene for one trait, such as height (*Tt*), segregate independently from the alleles of a gene for another trait, such as seed color (*Yy*). Figure 8.2 represents a dihybrid individual (*TtYy*) where the traits will assort independently. In it, *T* = tall, *t* = short, *Y* = yellow seed, and *y* = green seed.

EVO-2 & SYI-3
Segregation and independent assortment apply to genes on different chromosomes.

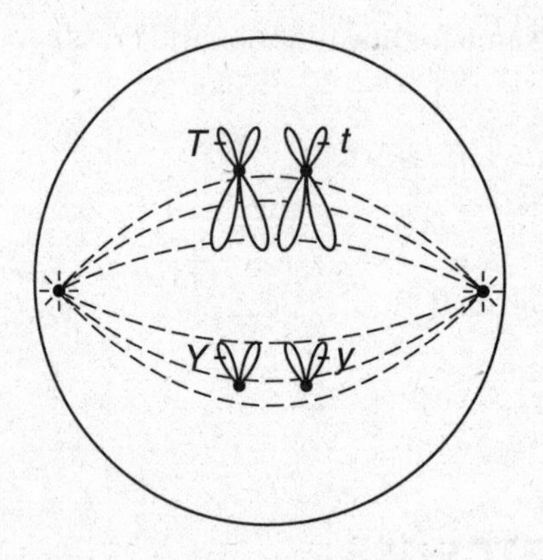

Figure 8.2 The Dihybrid Cross

The genes for height and seed color are not on the same chromosome and will assort independently. The only factor that determines how these alleles **segregate** or assort is how the homologous pairs line up in metaphase of meiosis I, which is random.

During metaphase I, if the homologous pairs happen to line up like this:

T | t

Y | y

they will produce these gametes: *TY* and *ty*

If the homologous pairs happen to line up like this:

T | t

y | Y

they will produce these gametes: *Ty* and *tY*

In contrast, if the gene for tall is **linked** to the gene for yellow seed color and the gene for short is linked to the gene for green seed color, the genes will **not assort independently**. If a plant is tall, it will have yellow seeds. If a plant is short, it will have green seeds. See Figure 8.3.

> **IST-1**
> Genes that are adjacent and *close* to each other on the same chromosome tend to move as a unit and not segregate.

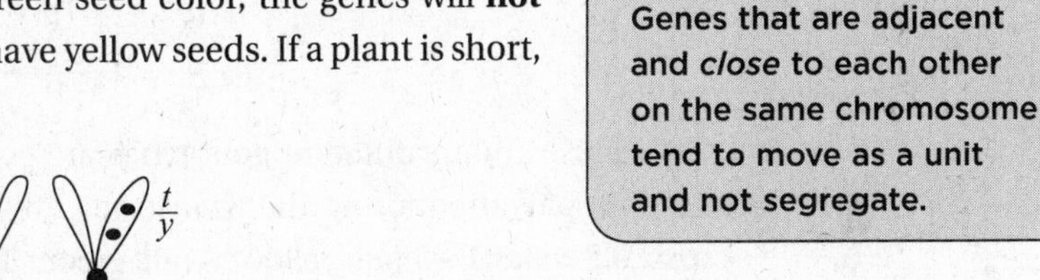

Figure 8.3 Linked Genes

The following describes a cross that adheres to the law of independent assortment. Two flowers are crossed that are homozygous for different traits and have the genes for height and seed color on different homologous chromosomes.

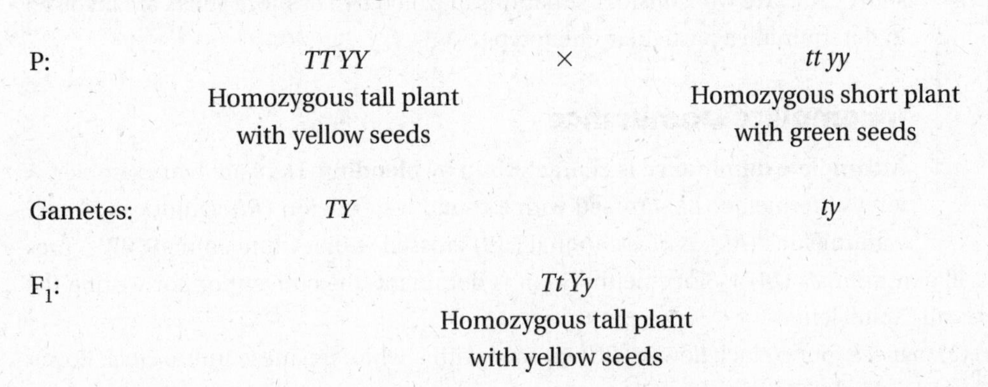

P: *TT YY* (Homozygous tall plant with yellow seeds) × *tt yy* (Homozygous short plant with green seeds)

Gametes: *TY* and *ty*

F_1: *Tt Yy*
Homozygous tall plant
with yellow seeds

The *TT YY* parent can produce gametes carrying only *TY* genes. The *tt yy* parent can produce gametes carrying only *ty* genes. A Punnett square is not needed because only one outcome is possible in the F_1: *Tt Yy*. The phenotype of all members of the F_1 generation is tall plants with yellow seeds. Their genotype is known as **dihybrid**.

The Dihybrid Cross

A cross between two F_1 plants is called a **dihybrid cross** because it is a cross between individuals that are hybrid for two different traits, such as height and seed color. This cross can produce four different types of gametes: *TY*, *Ty*, *tY*, and *ty*. The figure below shows how to set up the Punnett square for this cross. There are 16 squares; therefore, all the ratios are out of 16.

Dihybrid cross	*TY*	*Ty*	*tY*	*ty*
TY	*TTYY*	*TTYy*	*TtYY*	*TtYy*
Ty	*TTYy*	*TTyy*	*TtYy*	*Ttyy*
tY	*TtYY*	*TtYy*	*ttYY*	*ttYy*
ty	*TtYy*	*Ttyy*	*ttYy*	*ttyy*

Obviously, many different genotypes are possible in the resulting F_2 generation, but you need not pay attention to them. Just pay attention to the **phenotype ratio** of the dihybrid cross. It is **9:3:3:1**—9 tall, yellow; 3 tall, green; 3 short, yellow; and 1 short, green. To show this as a probability:

F_2: 9/16 tall, yellow 3/16 tall, green 3/16 short, yellow 1/16 short, green

BEYOND MENDELIAN INHERITANCE

Mendelian principles apply to traits determined by a single gene for which there are only two alleles. An example is height in pea plants (one gene), which has only two alleles (tall and short). Now we will consider situations in which two or more genes are involved in determining a particular phenotype.

IST-1
The inheritance pattern of many traits cannot be explained by simple Mendelian inheritance.

Incomplete Dominance

Incomplete dominance is characterized by **blending**. Here are two examples. A long watermelon (*LL*) crossed with a round watermelon (*RR*) produces all oval watermelons (*RL*). A black animal (*BB*) crossed with a white animal (*WW*) produces all gray animals (*BW*). Since neither trait is dominant, the convention for writing the genes is all capital letters.

A red Japanese four o'clock flower (*RR*) crossed with a white Japanese four o'clock flower (*WW*) produces all pink offspring (*RW*):

	R	*R*
W	*RW*	*RW*
W	*RW*	*RW*

If two pink four o'clocks are crossed, there is a 25 percent chance that the offspring will be red, a 25 percent chance the offspring will be white, and a 50 percent chance the offspring will be pink.

	R	*W*
R	*RR*	*RW*
W	*RW*	*WW*

REMEMBER

Incomplete dominance is a blending. In codominance, both traits show.

Codominance

In **codominance**, *both traits show.* A good example is the *MN* blood groups in humans. (These are not related to *ABO* blood groups.) There are three different blood groups: *M*, *N*, and *MN*. These groups are based on two distinct molecules located on the surface of the red blood cells. There is a **single gene locus** at which **two allelic variants** are possible. A person can be homozygous for one type of molecule (*MM*), be homozygous for the other molecule (*NN*), or be hybrid and have both molecules (*MN*) on their red blood cells. The *MN* genotype is not intermediate between *M* and *N* phenotypes. Both *M* and *N* traits are expressed because both molecules are present on the surface of the red blood cells; see Figure 8.4.

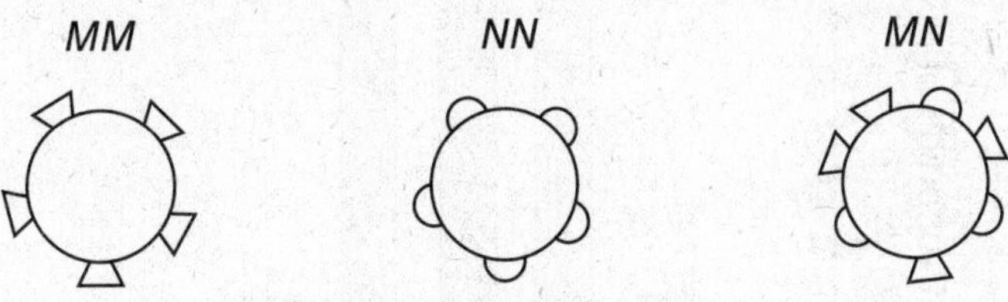

Figure 8.4 Codominance

Multiple Alleles

Many genes in a population exist in only two allelic forms. For example, pea plants can be either tall (*T*) or short (*t*). When *there are more than two allelic forms of a gene*, that is referred to as **multiple alleles**. In humans there are four different blood types: **A**, **B**, **AB**, and **O** determined by the presence of specific molecules on the surface of the red blood cells. There are three alleles, *A*, *B*, and *O*, which determine the four different blood types. A and B are codominant and are often written as $\mathbf{I^A}$ and $\mathbf{I^B}$. (*I* stands for immunoglobulin.) When both alleles are present, they both manifest themselves, and the person has AB blood type. In addition, O is a recessive trait and is often written as **i**. A person can have any one of the six blood genotypes shown in Table 8.1.

Table 8.1

Human Blood Types and Genotypes

Blood Type	Genotype	
A	Homozygous *A*:	*AA* I^AI^A
A	Hybrid *A*:	*Ai* I^Ai
B	Homozygous *B*:	*BB* I^BI^B
B	Hybrid *B*:	*Bi* I^Bi
AB	*AB*	I^AI^B
O	Homozygous recessive	*ii*

POLYGENIC INHERITANCE

> **SYI-3**
> Many traits are the product of multiple genes.

Many characteristics such as skin color, hair color, and height result from a blending of several separate genes that vary along a continuum. These traits are known as **polygenic**. Two parents who are short carry more genes for shortness than for tallness. However, they can have a child who inherits mostly genes for tallness from both parents and who will be taller than the parents. This wide variation in genotypes always results in a bell-shaped curve in an entire population. See Figure 8.5, which shows the distribution of skin pigmentation across a population.

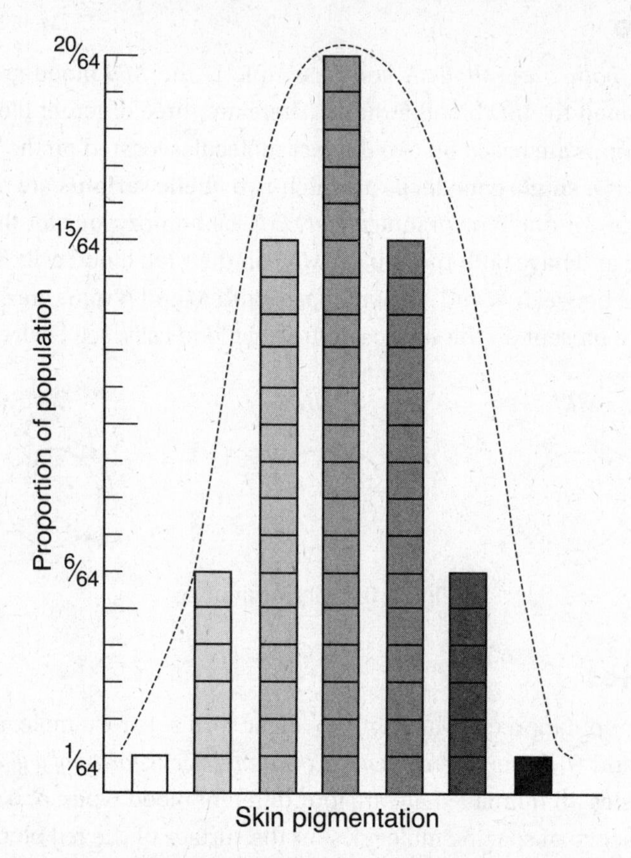

Figure 8.5 Polygenic Inheritance

MORE EXCEPTIONS TO MENDELIAN INHERITANCE

Two other inheritance patterns are exceptions to Mendelian inheritance: genomic imprinting and extranuclear genes. **Genomic imprinting** is a variation in phenotype depending on whether a trait is inherited from the mother or from the father. It occurs during gamete formation and is caused by the silencing of a particular allele by methylation of DNA. Therefore, a zygote expresses only one allele of the imprinted gene. The imprint is carried to all body cells and passed down from generation to generation. Unlike sex-linked genes, which are located on the X chromosome, imprinted genes are located on autosomes.

Extranuclear genes are genes located in **mitochondria** and **chloroplasts**. The DNA in these organelles is small, circular, and carries only a small number of genes. Extranuclear genes have been linked to several rare and severe inherited diseases in humans. Since the products of most mitochondrial genes are involved with energy production, defects (mutations) in these genes cause weakness and deterioration in muscles. Mitochondrial DNA is inherited only from the mother because the father's mitochondria do not enter the egg during fertilization.

GENES AND THE ENVIRONMENT

The environment can alter the expression of genes. In fruit flies, the expression of the mutation for vestigial wings (short, shriveled wings) can be altered by the temperature of the environment. When raised in a hot environment, fruit flies that are homozygous recessive for vestigial wings can grow wings almost as long as normal wild-type wings.

Many human diseases have a **multifactorial basis**. There is an underlying genetic component with a significant environmental influence. Examples are heart disease, diabetes, cancer, alcoholism, schizophrenia, and bipolar disorder. Similarly, the development of intelligence is the result of an interaction of genetic predisposition and the environment, or **nurture and nature**.

SYI-3
Environmental factors influence the expression of the genotype in an organism.

Penetrance

Penetrance is the proportion or percentage of individuals in a group with a given genotype that actually shows the expected phenotype. For example, many people who inherit a mutant allele for BRCA1 develop breast cancer in their lifetimes. For reasons that are unknown, some people with the allele do not develop breast cancer. Perhaps this is caused by an interaction with some other gene or some factor from the environment.

LINKED GENES

Genes on the same chromosome are called **linked genes**. Since there are many more genes than chromosomes, thousands of genes are linked. Humans have 46 chromosomes in every cell. Therefore, humans have 46 linkage groups. Linked genes tend to be inherited together and do not assort independently (unless they are separated by a crossover event during meiosis and gamete formation).

Sex-linkage

REMEMBER
Sex-linked traits are located on the X chromosome.

Of the 46 human chromosomes, 44 (22 pairs) are **autosomes** and 2 are **sex chromosomes**, X and Y. Traits *carried on the X chromosome* are called **sex-linked**. Few genes are carried on the Y chromosome. Females (XX) inherit two copies of the sex-linked genes. If a sex-linked trait is due to a **recessive mutation**, a female will express the phenotype only if she carries two mutated genes (X–X–). If she carries only one mutated X-linked gene, she will be a **carrier** (X–X). If a sex-linked trait is due to a **dominant mutation**, a female will express the phenotype with only one mutated gene (X–X). Males (XY) inherit only one X-linked gene. As a result, if the male inherits a mutated X-linked gene (X–Y), he will express the gene. Recessive sex-linked traits are much more common than dominant sex-linked traits; so males suffer with sex-linked conditions more often than females do.

SYI-3
Some traits are not sex-linked but are instead sex-limited, where expression depends on the sex of the individual. Human males cannot produce milk even though they have mammary glands.

Here are some important facts about sex-linked traits.

- Common examples of recessive sex-linked traits are **color blindness**, **hemophilia**, and Duchenne muscular dystrophy.
- *All daughters of affected fathers are carriers* (shaded squares).

Punnett square	X–	Y
X	X–X	XY
X	X–X	XY

- Sons cannot inherit a sex-linked trait from the father because the son inherits the Y chromosome from the father.
- A son has a 50 percent chance of inheriting a sex-linked trait from a carrier mother (shaded square).

Punnett square	X	Y
X–	X–X	X–Y
X	XX	XY

- There is no carrier state for X-linked traits in males. If a male has the gene, he will express it.
- It is uncommon for a female to have a recessive sex-linked condition. In order to be affected, she must inherit a mutant gene from *both* parents.

CROSSOVER AND LINKAGE MAPPING

The farther apart two genes are on one chromosome, the more likely they will be separated from each other during meiosis because a crossover event will occur between them. At the site at which a crossover and recombination occur, one can see a **chiasma**, a physical bridge built around the point of exchange. The result of a cross-over is a **recombination**. *Crossover and recombination are major sources of variation in sexually reproducing organisms.*

> **SYI-3**
> The probability that genes on the same chromosome will segregate as a unit is a function of the distance between the genes.

Figure 8.6 shows one crossover between homologous chromosomes. Without the crossover, the resulting four gametes would contain the following genes: *AB, AB, ab,* and *ab.* There would be only two different types of gametes, *AB* and *ab.* With one crossover (as shown), the four resulting gametes contain the following genes: *AB, Ab, aB,* and *ab.* There would be four different types of gametes. That one crossover results in twice the variation in the type of gametes possible.

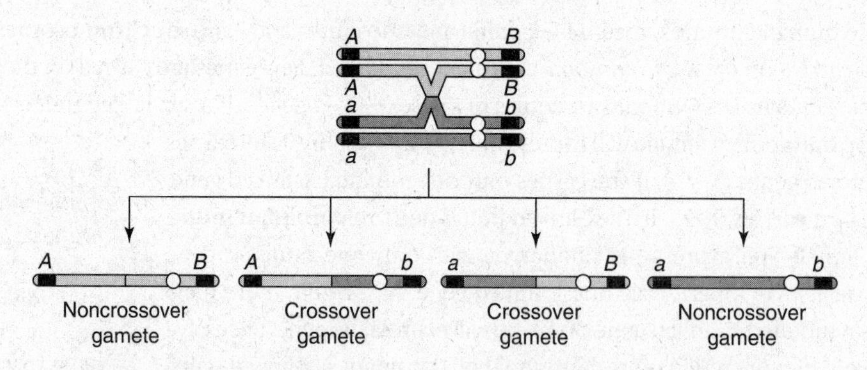

Figure 8.6 One Crossover Between Homologous Chromosomes

By convention, one **map unit** distance on a chromosome is *the distance within which recombination occurs 1 percent of the time.* The rate of crossover gives no information about the actual distance between genes, but it tells us the order of the linked genes on the chromosome.

Here is one example. Genes *A, B,* and *D* are linked. The crossover or **recombination frequencies** for *B* and *D* is 5 percent, for *B* and *A* is 30 percent, and for *D* and *A* is 25 percent.

The **linkage map** that can be constructed from this data is *BDA* or *ADB*. Whether you read it forward or backward does not matter.

A ← 25 → D ← 5 → B

A ← 30 → B

Recombination Frequencies

Here is a *testcross* between a dihybrid and a homozygous recessive for both traits

F_1 dihybrid	×	Pure black, vestigial-winged fly
Gg Nn	×	*gg nn*

Here are the ratios that were expected and the results:

Expected ratios from the testcross if all traits are located on different chromosomes, and therefore assort independently.	**Wild-type, gray normal**	**Black vestigial**	**Gray vestigial**	**Black normal**
	Gg Nn (parental phenotype) 1	*gg nn* (parental phenotype) 1	*Gg nn* (nonparental phenotype) 1	*gg Nn* (nonparental phenotype) 1
Expected ratios if genes are linked and located on the same chromosome. Alleles will always be inherited together.	*Gg Nn* (parental phenotype) 1	*gg nn* (parental phenotype) 1	0	0
Actual results:	965	944	206	185

In this experiment, the actual results do not conform to either prediction. The reason is that the genes for body color and wing size are located on the same chromosome. They are linked, although not permanently. The existence of small numbers of *nonparental phenotypes* (shaded in) can be explained only by an occasional break in the linkage—where **crossover** occurred. Remember that *crossing-over accounts for the recombination of linked genes.* Here is how to determine recombination frequency given data such as this.

Determining Recombination Frequencies

Use the following formula to calculate recombination frequencies

$$\frac{\text{Number of recombinants}}{\text{Total number of offspring}} \times 100$$

Use the results from the cross above.

Number of recombinants: 206 + 185 = 391

Total number of offspring: 965 + 944 + 206 + 185 = 2300

$$\frac{391}{2300} \times 100 = 17\%$$

The recombination frequency for these linked genes, *Gg* and *Nn*, is 17%.

THE PEDIGREE

A **pedigree** is a family tree that indicates the phenotype of one trait being studied for every member of a family. *Geneticists use the pedigree to determine how a particular trait is inherited.* By convention, females are represented by a circle and males by a square. The carrier state is not always shown. If it is, though, it is sometimes represented by a half-shaded-in shape. A shape is completely shaded in if a person exhibits the trait.

The pedigree in Figure 8.7 shows three generations of deafness. Try to determine the pattern of inheritance. First, eliminate all possibilities. Dominance can be ruled out (either sex-linked or autosomal) because in order for a child to have the condition, she or he would have had to receive one mutant gene from one afflicted parent, and nowhere is that the case. (All afflicted children have unaffected parents.) Also, you can rule out sex-linked recessive, because in order for F_3 generation daughter #1 to have the condition, she would have had to inherit two mutant traits (X–X–), one from each parent. However, her father does not have the condition. Therefore, the trait must be autosomal recessive.

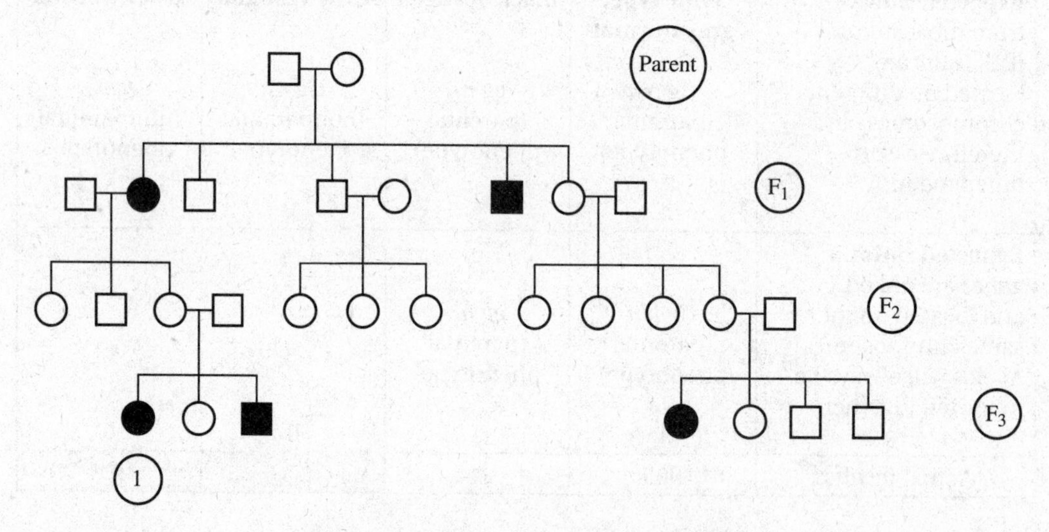

Figure 8.7 Three Generations of Deafness

X INACTIVATION—THE BARR BODY

STUDY TIP

All female body cells have one Barr body. Normal male cells have none.

Early in the development of the embryo of a female mammal, one of the X chromosomes is inactivated in every **somatic** (body) **cell**. This inactivation occurs randomly. The process results in an embryo that is a **genetic mosaic**; some cells have one X inactivated, some cells have the other X inactivated. Therefore, all the cells of female mammals are not identical. The inactivated chromosome condenses into a dark spot of chromatin and can be seen at the outer edge of the nucleus of all somatic cells in the female. This dark spot is called a **Barr body**.

Proof of X chromosome inactivation can be seen in the genetics of the female calico cat where the alleles for black and yellow fur are carried on the X chromosome. Male cats, having only a single X chromosome, can be either yellow (X^YY) **or** black (X^BY). Calico cats, which are almost always female, have coats with patches of both yellow and black (X^BX^Y). These patches of fur developed from embryonic cells with different deactivated X chromosomes. Some fur-producing cells contained the X^B active chromosome and produce black fur. Other fur-producing cells contain the X^Y active chromosome and produce yellow fur. The result is a cat with yellow and black patches of fur, the characteristic calico appearance. See Figure 8.8.

Black Female

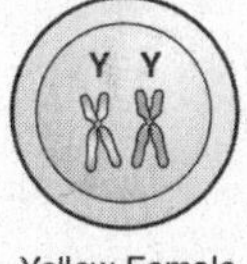

Yellow Female

Calico Female

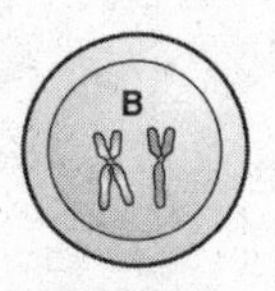

Black Male

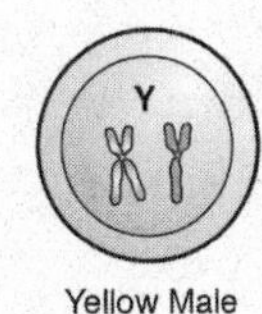

Yellow Male

Figure 8.8 X Inactivation in Calico Cats

Another example of X chromosome inactivation is evident in humans. A certain X-linked recessive mutation prevents the development of sweat glands. A woman who is heterozygous for this trait is not merely a carrier. Because of X inactivation, she has patches of normal skin and patches of skin that lack sweat glands.

MUTATIONS

Mutations are any changes in the genome. They can occur in the somatic (body) cells and be responsible for the spontaneous development of cancer, or they can occur during gametogenesis and affect future offspring. Even though radiation and certain chemicals cause mutations, when and where mutations occur is random.

There are two types of mutations: **gene mutations** and **chromosome mutations**. Gene mutations are caused by a change in the DNA sequence. Some human genetic disorders caused by gene and chromosome mutations are listed and described further in Table 8.2. The nature of gene mutations at the DNA level is discussed in the next chapter.

IST-1
Certain human genetic diseases can be caused by the inheritance of a single gene or by a chromosomal abnormality.

Although gene mutations cannot be seen under a microscope, **chromosome mutations** can. A technique called a **karyotype** shows the size, number, and shape of chromosomes and can reveal the presence of certain abnormalities. Karyotypes can be used to scan for chromosomal abnormalities in developing fetuses. Figure 8.9 shows a karyotype of a male with Down syndrome due to the presence of an extra chromosome 21. Figure 8.10 shows a karyotype of a normal human female with 46 pairs (the normal number) of chromosomes, including two X chromosomes (sex chromosomes).

IST-1
Changes in chromosome number often result in humans with developmental limitations.

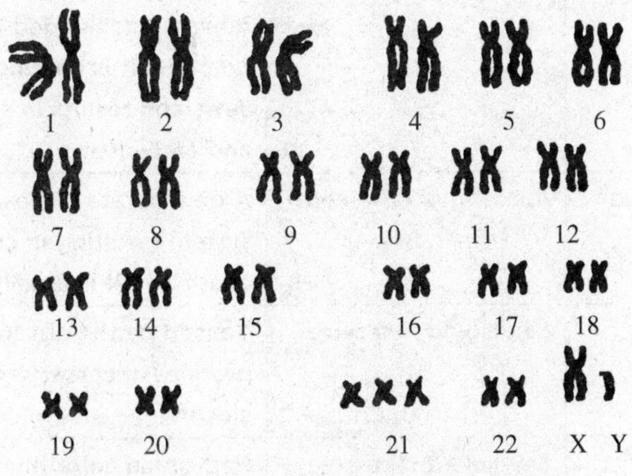

Figure 8.9 Karyotype of a Male with Trisomy 21

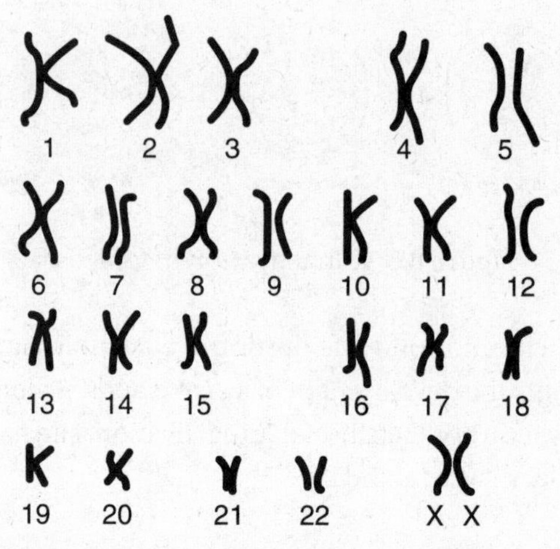

Figure 8.10 Normal Karyotype of a Female

The bottom of Table 8.2, "Chromosomal Disorders," describes three conditions that result from nondisjunction in the formation of the ovum or the sperm.

Table 8.2 Chromosomal Disorders

Gene and Chromosome Mutations		
Genetic Disorder	**Pattern of Inheritance**	**Description**
Phenylketonuria (PKU)	Autosomal recessive	Inability to break down the amino acid phenylalanine. Requires elimination of phenylalanine from diet, otherwise serious mental retardation will result.
Cystic fibrosis	Autosomal recessive	The most common lethal genetic disease in the U.S. 1 out of 25 Caucasians is a carrier. Characterized by buildup of extracellular fluid in the lungs, digestive tract, etc.
Tay-Sachs disease	Autosomal recessive	Onset is early in life and is caused by a lack of the enzyme necessary to break down lipids needed for normal brain function. It is common in Ashkenazi Jews and results in seizures, blindness, and early death.
Huntington's disease	Autosomal dominant	A degenerate disease of the nervous system resulting in certain and early death. Onset is usually in middle age.
Hemophilia	Sex-linked recessive	Caused by the absence of one or more proteins necessary for normal blood clotting.
Color blindness	Sex-linked recessive	Red-green color blindness is rarely more than an inconvenience.
Duchenne muscular dystrophy	Sex-linked recessive	Progressive weakening of muscle control and loss of coordination.

Table 8.2 Chromosomal Disorders (continued)

Gene and Chromosome Mutations		
Sickle cell disease	Autosomal recessive	A mutation in the gene for hemoglobin results in deformed red blood cells. Carriers of the sickle cell trait are resistant to malaria.
Chromosomal Disorder	**Pattern of Inheritance**	**Description**
Down syndrome	47 chromosomes due to trisomy 21	Characteristic facial features, mental retardation, prone to developing Alzheimer's and leukemia
Turner syndrome	XO 45 chromosomes due to a missing sex chromosome	Small stature, female
Klinefelter syndrome	XXY 47 chromosomes due to an extra X chromosome	Have male genitals, but the testes are abnormally small and the men are sterile

Chromosomal aberrations include:

- **Deletion**—when a fragment lacking a centromere is lost during cell division
- **Inversion**—when a chromosomal fragment reattaches to its original chromosome but in the reverse orientation
- **Translocation**—when a fragment of a chromosome becomes attached to a nonhomologous chromosome
- **Polyploidy**—when a cell or organism has extra sets of chromosomes

NONDISJUNCTION

Nondisjunction is an error that sometimes occurs during meiosis in which homologous chromosomes fail to separate as they should. See Figure 8.11. When this happens, one gamete receives two of the same type of chromosome and another gamete receives no copy. The remaining chromosomes may be unaffected and normal. If either aberrant gamete unites with a normal gamete during fertilization, the resulting zygote will have an abnormal number of chromosomes. Any abnormal number of chromosomes is known as **aneuploidy**. If a chromosome is present in triplicate, the condition is known as **trisomy**. People with Down syndrome have an extra chromosome 21. The condition is referred to as **trisomy 21**. Cancer cells grown in culture almost always have extra chromosomes. An organism in which the cells have an extra set of chromosomes is referred to as **triploid** ($3n$). An organism with the $4n$ chromosome number is known as **tetraploid**. Strawberries are **octoploid**. An organism with extra sets of chromosomes is referred to as **polyploid**. Hugo de Vries, the scientist who coined the term **mutation**, was studying plants that were polyploidy. Polyploidy is common in plants and results in plants of abnormally large size. In some cases, it is responsible for the evolution of new species. As the word is used today, mutation refers to any **genetic** or **chromosomal abnormality**.

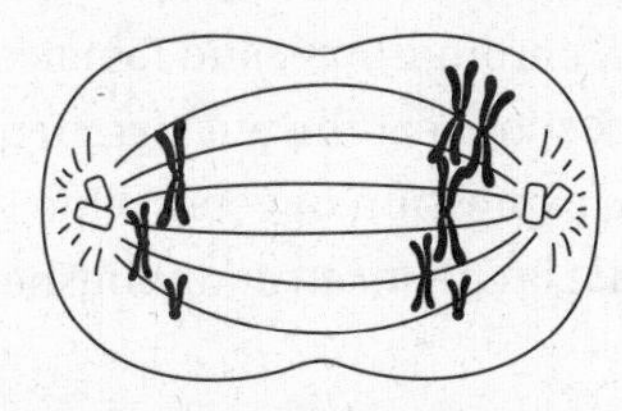

Figure 8.11 Nondisjunction

CHAPTER SUMMARY

This chapter furthered the discussion presented in Big Ideas: EVO, IST, and SYI.

- Be able to apply *simple mathematical operations* using **Punnett squares** and the **rules of probability** to determine how specific traits are inherited. Be able to analyze and interpret information about inheritance of a particular trait shown on a **pedigree**. Understand that many traits are controlled by *single genes* and can be accounted for by Mendel's laws: **Laws of Dominance**, **Segregation**, and **Independent Assortment. Segregation** and **independent assortment** of genes apply only to genes on different chromosomes and result in genetic variation in offspring.
- Be aware that the inheritance pattern of many traits **cannot** be predicted by Mendel's laws. Examples include **incomplete dominance**, **codominance**, **multiple alleles**, **genomic imprinting**, and **genes in mitochondria**, which are inherited from only the mother. In addition, **polygenic inheritance**—which are traits that span a range of characteristics along a continuum like height, hair color, and skin color in humans—result from the influence of *multiple genes.*
- **Mutations** are random changes in the genome. Some occur in **genes** and cause human diseases like **PKU**, **cystic fibrosis**, **Tay-Sachs disease**, **Huntington's disease**, **hemophilia**, and **sickle cell disease**. Some mutations, like **nondisjunction**, relate not to genes but to **chromosomal mutations**. They are caused by errors during meiosis and gamete formation. Examples are **Down syndrome**, **Turner syndrome**, and **Klinefelter syndrome.** Chromosomal mutations are identified on a **karyotype** because the number and placement of chromosomes is other than 46.
- A **backcross** or **testcross** is an actual mating carried out between an animal of unknown genotype that shows the dominant trait (*Bb* or *BB*) with one that exhibits the recessive trait (*bb*). The purpose is to determine the genotype of the animal that expresses the dominant trait.
- **Linked genes**, which are genes on the same chromosome, are often inherited together, especially if they are close together. However, if two linked genes are far apart on a chromosome, a **crossover** event might separate them during meiosis. So they will *not* be inherited together. The results of a crossover are a **recombination** of alleles and an increase in **genetic diversity**, which are necessary for evolution. **Linkage maps** can be constructed from data of **recombination frequencies**.
- **Sex-linked genes** are genes located on the X chromosome. Examples are **hemophilia**, **Duchenne muscular dystrophy**, and **color blindness**. Females can be **carriers** (X–X). However, they must have two affected X chromosomes to express the condition (X–X–). Males, on the other hand, express the condition if they inherit only one X-linked mutation (X–Y). A father *cannot* pass an X-linked condition to his sons because they inherit their Y chromosome from him. However, all his daughters will inherit his affected X chromosome and be carriers. Men who express an X-linked condition inherit it from their mothers.
- During development of the mammalian female, one of the X chromosomes normally becomes inactivated. This phenomenon equalizes the dosage of X chromosomes in males, who have only one X chromosome, and females, who have two. The inactivated X chromosome forms a dark spot near the nuclear membrane called a **Barr body**.
- The environment alters the expression of genes. One example is IQ in humans, which has a genetic component that is enhanced or diminished by stimulation, or lack thereof, from the environment.

PRACTICE QUESTIONS

1. Which is TRUE about a testcross?

   (A) It is a mating between two hybrid individuals.
   (B) It is a mating between a hybrid individual and a homozygous recessive individual.
   (C) It is a mating between an individual of unknown genotype and a homozygous recessive individual.
   (D) It is a mating to determine which individual is homozygous recessive.

2. Two genes, *A* and *B*, are on two different chromosomes. The probability of allele *A* segregating into a gamete is $\frac{1}{2}$, while the probability of allele *B* segregating into a gamete is $\frac{1}{4}$. What is the probability that both alleles will segregate into the same gamete?

   (A) $\frac{1}{4} + \frac{1}{2}$
   (B) $\frac{1}{4}$ divided by $\frac{1}{2}$
   (C) $\frac{1}{2}$ divided by $\frac{1}{4}$
   (D) $\frac{1}{2} \times \frac{1}{4}$

3. A round watermelon is crossed with a long watermelon and all the offspring are oval. If two oval watermelons are crossed, what is the percent of watermelons that will be round?

   (A) 0
   (B) 25%
   (C) 50%
   (D) 75%

4. In a certain species of plant, the trait for tall is dominant (*T*) and the trait for short is recessive (*t*). The trait for yellow seeds is dominant (*Y*) and the trait for green seeds is recessive (*y*). A cross between two plants results in 292 tall yellow plants and 103 short green plants. Which of the following are most likely to be the genotypes of the parents?

   (A) *TtYY* × *Ttyy*
   (B) *TTYy* × *TTYy*
   (C) *TTyy* × *TTYy*
   (D) *TtYy* × *TtYy*

5. A child is born with blood type O. Which of the following is correct about her parents?

   (A) Both parents must have blood type O.
   (B) One parent must have blood type O; the other parent might have any blood type.
   (C) One parent must have blood type O; the other parent must have blood type A.
   (D) Both parents must have at least one recessive gene (i).

6. *ABCDEF → ABEDCF*

A rearrangement in the linear sequence of genes as shown above is known as a/an

(A) translocation
(B) deletion
(C) addition
(D) inversion

7. A botanist crossed two flowers. One flower was tall with a smooth seed coat; the other was short with a wrinkled seed coat. The following distribution of traits was observed in the offspring.

Phenotype	Number of Offspring
Tall with a smooth seed coat	83
Short with a wrinkled seed coat	86
Short with a smooth seed coat	8
Tall with a wrinkled seed coat	9

Which of the following statements is best supported by the data above?

(A) Genes for the traits for height and seed coat are located far apart on the same chromosome, and a crossover occurred.
(B) Genes for the traits for height and seed coat are located close together on the same chromosome, and a crossover occurred.
(C) Genes for the traits for height and seed coat are on different chromosomes and assorted independently.
(D) These are the expected results from a cross between one parent plant that is hybrid (tall with a smooth seed coat), while the other is homozygous recessive (short with a wrinkled seed coat).

8. Chipmunk eye color is controlled by two genes. An autosomal gene controls the color of the pigments in the eye and has two alleles: a dominant allele (P), which results in pink eyes, and a recessive allele (p), which results in yellow eye color. A sex-linked gene controls the expression of the colored pigments and also has two alleles: a dominant allele (C), which allows for the expression of the colored pigments, and a recessive allele (c), which does not allow for the expression of the colored pigments. Individuals without a C allele have white eyes regardless of the alleles of other eye-color genes.

Which of the following represents a cross between a white-eyed female and a pink-eyed male?

(A) Pp $X^cX^c \times$ Pp X^CY
(B) Cc $X^PX^P \times$ cc X^pY
(C) Pp $X^CX^c \times$ Pp X^CY
(D) Cc $X^pX^p \times$ Cc X^PY

9. Below is an illustration of a karotype of a human female. How many autosomes does a human female normally have?

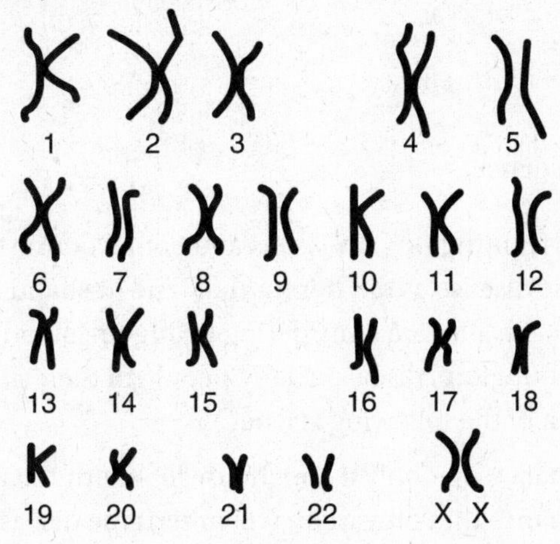

(A) 2
(B) 22
(C) 23
(D) 44

10. A couple has 6 children, all girls. If the mother gives birth to a seventh child, what is the probability that the seventh child will be a girl?

(A) $\frac{6}{7}$

(B) $\frac{1}{128}$

(C) $\frac{1}{2}$

(D) 1

11. Assume that two genes, *A* and *B*, are not linked. If the probability of allele *A* being in a gamete is $\frac{1}{2}$ and the probability of allele *B* being in a gamete is $\frac{1}{2}$, then the probability of BOTH *A* and *B* being in the same gamete is

(A) $\frac{1}{2}$

(B) $\frac{1}{4}$

(C) 1

(D) $\frac{1}{8}$

12. In guinea pigs, the gene for short hair is dominant, and the gene for long hair is recessive. The gene for black hair color is dominant over the gene for white hair color, which is recessive. A cross between two guinea pigs produces a litter of 9 short-haired black guinea pigs and 3 short-haired white guinea pigs. What is the most likely genotype of the parent guinea pigs in this cross?

(A) *SS BB* × *SS BB*
(B) *SS BB* × *SS Bb*
(C) *Ss Bb* × *Ss Bb*
(D) *SS Bb* × *Ss Bb*

13. In one strain of mice, fur color ranges from white to darkest brown with every shade of brown in between. This pattern of inheritance for fur color is most likely controlled by

(A) multiple genes
(B) a single gene with many alleles
(C) pleiotropy
(D) incomplete dominance

14. The gene that causes Huntington's disease is autosomal dominant. Consider one couple where the wife has the disease as did her mother. The husband does not. The children, now that they are 18 years old, are considering taking the blood test to determine if they have the gene for Huntington's disease. Many people in their situation have chosen not to take the test. Which of the following is true?

(A) There is a 50% chance that only a daughter will inherit the disease.
(B) There is a 50% chance that only a son will inherit the disease.
(C) It is certain that one of the children has the gene.
(D) There is a 50% chance that any child of these parents would have the disease.

15. Two traits, *A* and *B,* are linked, but they are usually not inherited together. The most likely reason is

(A) they are not on the same chromosome
(B) they are not sex-linked
(C) they are on the same chromosome but are far apart
(D) they are close together on the same chromosome

16. A cross was made between two fruit flies: a white-eyed female and a wild male (red-eyed). One hundred F_1 offspring were produced. All the males were white-eyed and all the females were wild. When these F_1 flies were allowed to mate, the F_2 flies were observed and the following data was collected.

	Females		Males
P:	White-eyed	×	Wild (red-eyed)
F_1:	59 wild		51 white-eyed
F_2:	24 wild 26 white-eyed		23 wild 27 white-eyed

What is the most likely pattern of inheritance for the white-eyed trait?

(A) autosomal dominant
(B) autosomal recessive
(C) sex-linked dominant
(D) sex-linked recessive

17. Which is true of a man who has a sex-linked condition?

(A) There is a 100% chance that each daughter will inherit the condition from the father.
(B) There is a 100% chance that each daughter will inherit the trait from the father.
(C) There is a 50% chance that each daughter will inherit the condition from the father.
(D) There is a 50% chance that each daughter will inherit the trait from the father.

18. The figure below shows a pedigree for a family that carries the gene for Huntington's disease. Individuals who express a particular trait are shown shaded in.

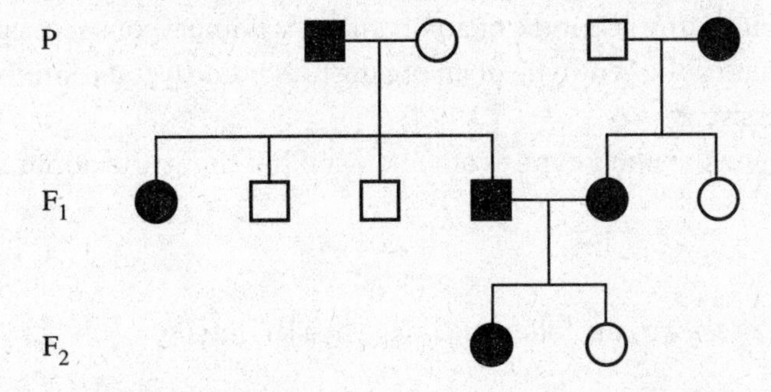

What is the genotype of the daughter in the F_2 generation who does not have the disease?

(A) *HH*
(B) *Hh*
(C) *hh*
(D) *X–X*

19. This figure shows a pedigree of the blood types for a family. What is the genotype for person number 14?

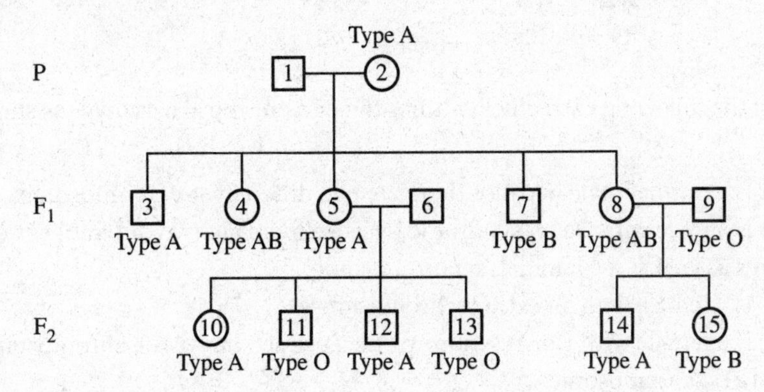

(A) I^AI^A
(B) I^Ai
(C) ii
(D) I^AI^B

20. Male black guinea pig #1 is crossed with albino guinea pig #2. All 14 offspring are black. In a second mating, a different male black guinea pig #3 is crossed with the same albino guinea pig #2 as in the first mating. This time, the mating produces 8 black guinea pigs and 6 albinos. Which statement offers the best explanation for what genetics are at work?

(A) Guinea pig #1 is homozygous dominant; guinea pig #3 is hybrid.
(B) Guinea pig #1 is hybrid; guinea pig #3 is homozygous dominant.
(C) Guinea pig #1 is homozygous dominant; guinea pig #3 is homozygous recessive.
(D) Guinea pig #1 is hybrid; guinea pig #3 is homozygous recessive.

21. Which of the following accurately describes the reason for completing a testcross?

(A) to determine the genotype of an organism that shows the recessive trait
(B) to determine if the genotype of an organism is homozygous recessive
(C) to determine if the genotype of an organism is homozygous dominant or homozygous recessive
(D) to determine the phenotype of an organism that shows the dominant trait

<u>**Questions 22–23**</u>

Questions 22 and 23 refer to the following image of a karyotype.

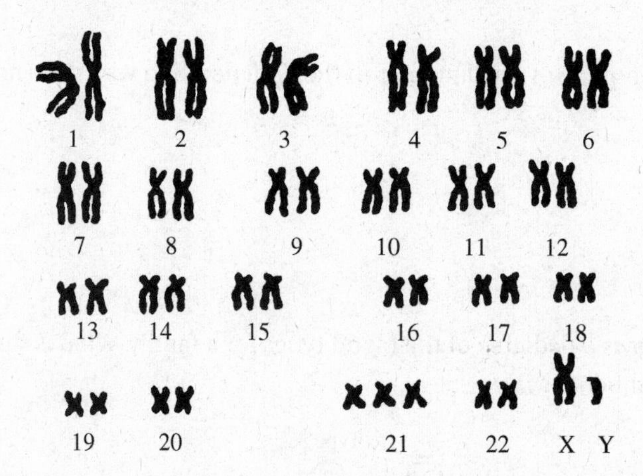

22. Which of the following correctly identifies the person whose karyotype is shown above?

(A) This is a normal male because there are two different sex chromosomes.
(B) This is a male with Down syndrome because there are two different sex chromosomes as well as a triploid 21st chromosome.
(C) This is a female with an extra X chromosome.
(D) This is a female with Down syndrome because she has two X chromosomes and an extra 21st chromosome.

23. Which of the following best explains how this person came to have the chromosomes shown in this karyotype?

(A) Twenty-three chromosomes were inherited from the mother, and twenty-three chromosomes were inherited from the father.
(B) There was an error in mitosis in the growing embryo.
(C) This was the result of an error in the formation of the sperm or ova.
(D) This situation resulted from a single crossover event.

Answers Explained

1. **(C)** A testcross is an actual mating between a pure recessive animal and an animal that shows the dominant phenotype but whose genotype is unknown. (IST-1)

2. **(D)** Since both events are independent of each other, you multiply the probability of one by the probability of the other. (SYI-3 & SP 5)

3. **(B)** Here is the cross.

	R	*L*
R	*RR*	*RL*
L	*RL*	*LL*

RR is round, *RL* is oval, and *LL* is long. The inheritance is an example of incomplete dominance. (SYI-3)

4. **(D)** Since there are four genes (*T, t, Y,* and *y*) but only two phenotypes in the offspring, the traits for height and seed color must be linked, that is, on the same chromosome. So solve the problem this way. Consider the traits separately at first. The phenotype ratio in the offspring for height is 3 : 1, tall to short. Therefore, the parents must be *Tt* and *Tt*. The phenotype ratio in the offspring for seed color is 3 : 1, yellow to green. Therefore, the parents must be *Yy* and *Yy*. Now, put both genotypes together. The parents must be *TtYy* and *TtYy*. (IST-1 & SP 5)

5. **(D)** In blood type, both A and B are dominant over type O. A person with blood type O must have two recessive genes (ii); one must come from each parent. The parents, however, could be hybrid: I^Ai or I^Bi. They could also be type O (ii), but they do not have to be. (EVO-2 & IST-1)

6. **(D)** The section of the strand shows genes *CDE* inverted. (IST-1)

7. **(A)** Notice that, except for the odd results of 8 and 9, the ratio in the offspring is close to 1 to 1 (83:86). That means that the parents had to have been hybrid (*TtSs*) and homozygous recessive (*ttss*), regardless of which traits are dominant and which are recessive. Where do the unexpected offspring (8 short/smooth and 9 tall/wrinkled) come from? They result from a crossover in which two traits, smooth and wrinkled, exchanged genes.

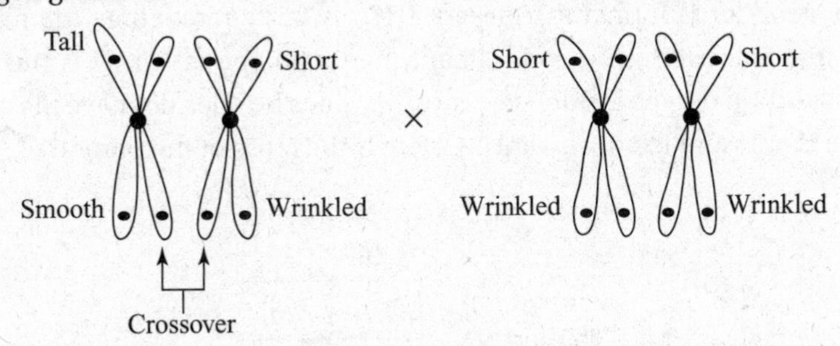

In order for linked genes to crossover, they must be located far apart on one chromosome. Choice B cannot be correct because traits located close together would be inherited together and not undergo crossover. Choice C cannot be correct because, if these four traits had assorted independently, the ratios of the offspring would have been 9:3:3:1. Choice D cannot be correct because it doesn't account for the odd results of the 8 and 9 offspring. (IST-1)

8. **(A)** For the autosomal gene: P = pink pigments; p = yellow pigments. For the sex-linked gene (on the X-chromosome): C, which allows for colored pigments, is dominant and = X^C; c, which prevents colored pigments regardless of the presence of P or p, = X^c. In choice A (Pp X^cX^c × Pp X^CY), the female lacks the sex-linked gene for the deposition of pigment. Therefore, she has white eyes. The male has dominant C, so he has pigmented eyes. Since he has at least one dominant P, he has pink eyes. That satisfies what is asked in the question. (IST-1)

9. **(D)** Autosomes are the chromosomes **other than** sex chromosomes (X and Y). An adult female has 44 autosomes plus two sex chromosomes (XX). (IST-1)

10. **(C)** No matter how many children a couple has, the chance that the child will be a boy *or* a girl is always $\frac{1}{2}$. Although it is true that whether the sperm carries an X or a Y sex chromosome determines the sex of the child, that it is irrelevant to the question here. (EVO-2 & IST-1)

11. **(B)** Since *A* and *B* are not linked, they assort independently. To find the probability of two independent events happening, multiply the chance of one happening by the chance of the other happening. (SYI-3 & SP 5)

12. **(D)** You can eliminate choice A because all the offspring would show only the dominant traits. You can eliminate choice B because the parents could not produce any offspring showing the recessive coat color. You can eliminate choice C because that would produce offspring in a 9:3:3:1 ratio—a dihybrid cross, which is not a choice. That leaves choice D. Of the choices, choice D is the best one to produce the offspring described. The parents could also be *SS Bb* × *SS Bb*. (IST-1 & SP 5)

13. **(A)** Examples of polygenic inheritance in humans are genes for skin color and height. (SYI-3)

14. **(D)** The gene for Huntington's disease (HD) is fairly rare. Since the mother has the disease and since the gene is dominant, her genotype is *Hh*. *H* is the gene for HD. The husband's genotype is pure recessive, *hh*, since he does not have the condition. The HD gene is autosomal, not sex-linked. Here is the cross of the parents:

		Mother	
		H	*h*
Father	*h*–	*Hh*	*hh*
	h–	*Hh*	*hh*

(IST-1 & SP 5)

15. **(C)** If genes are on the same chromosome but far apart, they will often be inherited separately because they will often be separated by crossover. (IST-1)

16. **(D)** Here is the first cross.

	X	Y
X–	X–X	X–Y
X–	X–X	X–Y

All the female offspring are carriers (X–X), and all the male offspring have white eyes (X–Y).

Here is the second cross.

	X–	Y
X–	X–X–	X–Y
X	X–X	XY

There is a 50 percent chance that a male will be white-eyed and a 50 percent chance he will be red-eyed. There is a 50 percent chance a female will be white-eyed and a 50 percent chance she will be red-eyed (a carrier). (IST-1)

17. **(B)** Here is the cross.

Punnett square	X–	Y
X	X–X	XY
X	X–X	XY

X– shows the presence of the sex-linked trait. The father has the condition, and he passes it to all of his daughters. However, the daughters do not have the condition because they have a normal X to provide enzymes that are missing from the impaired X–. The daughters are merely carriers. (IST-1)

18. **(C)** Females are represented as circles and males as squares unless otherwise stated. You should know that Huntington's disease is inherited as autosomal dominant. The F_2 daughter who does not have the condition must have inherited one healthy gene from each parent. She must be *hh*. (EVO-2, IST-1 & SP 3)

19. **(B)** Person 14 has blood type A. It can be either I^AI^A or I^Ai. Since his father has type O blood, ii, person 14 must have inherited the A from his mother and the i from his father. His genotype therefore is I^Ai. (EVO-2 & IST-1)

20. **(A)** The albino guinea pig is pure recessive. The crossings between guinea pigs #1 and #2, and between #3 and #2 are backcrosses or testcrosses. The matings are carried out to reveal the genotypes of the black individuals. With the results of the first mating, it is most likely that black guinea pig #1 is homozygous dominant. If it had been hybrid, there probably would have been at least one white offspring. The fact that there are several white offspring from the second mating indicates that the second black male, #3, is hybrid for certain. (IST-1)

21. **(D)** A testcross is carried out on an organism that shows the dominant trait to determine if it is homozygous dominant or hybrid (*BB* or *Bb*). You cross the organism in question with an organism that shows the recessive trait (*bb*). For example, if white fur is a recessive trait (*b*), and an organism has white fur, you know the organism's genotype by looking at it (*bb*). Below are the Punnett squares and results from a testcross:

Punnett Square A

	B	*B*
b	*Bb*	*Bb*
b	*Bb*	*Bb*

Punnett Square B

	B	*b*
b	*Bb*	*bb*
b	*Bb*	*bb*

If the parent with the unknown genotype is homozygous dominant (*BB*), there will never be any offspring that show the recessive trait in a testcross (see Punnett Square A). However, if the parent with the unknown genotype is hybrid (*Bb*), the chance of getting an offspring with white fur is 50% (see Punnett Square B). (IST-1)

22. **(B)** The karyotype shows an X and a Y chromosome. Therefore, this is a male. There is an extra chromosome at the 21st position, a situation called triploid. This is a diagnostic for Down syndrome. (IST-1)

23. **(C)** There are 47 chromosomes in this karyotype. One gamete donated the normal 23 chromosomes; the other gamete donated an abnormal 24 chromosomes. The extra chromosomes resulted from an error called nondisjunction, which occurred during meiosis, in which one pair of chromosomes failed to separate. Choice A is not correct because it describes normal meiotic division, which would produce normal gametes, normal offspring, and a normal karyotype. Choice B is not correct because the error occurred during meiosis and the formation of gametes, not during mitosis in the baby's cells. Choice D is not correct because crossover involves an exchange of genes; this question involved an entire chromosome. (IST-1)

The Molecular Basis of Inheritance

9

- → THE SEARCH FOR HERITABLE MATERIAL
- → STRUCTURE OF NUCLEIC ACIDS
- → DNA REPLICATION IN EUKARYOTES
- → FROM DNA TO PROTEIN
- → GENE MUTATION
- → THE GENETICS OF VIRUSES AND BACTERIA
- → PRIONS
- → THE HUMAN GENOME
- → REGULATION OF GENE EXPRESSION
- → RECOMBINANT DNA; CLONING GENES
- → TOOLS AND TECHNIQUES OF BIOTECHNOLOGY
- → ETHICAL CONSIDERATIONS

Big Idea: IST

Enduring Understandings: IST-1, IST-2 & IST-4

Science Practices: 1, 2 & 5

For the complete list of Big Ideas, Enduring Understandings, and Science Practices, refer to the "AP Biology Course and Exam Description" from the College Board: *https://apcentral.collegeboard.org/pdf/ap-biology-course-and-exam-description.pdf?course=ap-biology.*

INTRODUCTION

Today, everyone knows that DNA is the molecule of heredity. We know that DNA makes up chromosomes and that genes are located on the chromosomes. Today, we can even see the location of particular genes by tagging them with fluorescent dye.

However, until the 1940s, many scientists believed that proteins, not DNA, were the molecules that make up genes and constitute inherited material. Several factors contributed to that belief. First, proteins are a major component of all cells. Second, they are complex macromolecules that exist in seemingly limitless variety and have great specificity of function. Third, a great deal was known about the structure of proteins and very little was known about DNA. The work of many brilliant scientists has transformed our knowledge of the structure and function of the DNA molecule and led to the acceptance of DNA as the molecule responsible for heredity.

This chapter covers the history of the search for the heritable material, the structure of nucleic acids, and how DNA makes proteins. It also includes an extensive review of genetic engineering and recombinant DNA techniques.

THE SEARCH FOR HERITABLE MATERIAL

STUDY TIP

While you may see questions that are based on these classic experiments on the AP exam, you likely won't be tested on the names of the scientists.

Frederick Griffith (1928) performed experiments with several different strains of the bacterium *Diplococcus pneumoniae.* Some strains are virulent and cause pneumonia in humans and mice, and some strains are harmless. Griffith discovered that *bacteria have the ability to transform harmless cells into virulent ones by transferring some genetic factor from one bacteria cell to another.* This phenomenon is known as **bacterial transformation**, and the experiment is known as the **transformation experiment**. See the information about bacterial transformation later in this chapter.

Avery, MacLeod, and McCarty (1944) published their classic findings that Griffith's **transformation factor** is, in fact, DNA. This research proved that DNA was the agent that carried the genetic characteristics from the virulent dead bacteria to the living nonvirulent bacteria. *This provided direct experimental evidence that DNA, not protein, was the genetic material.*

Hershey and Chase (1952) carried out experiments that lent strong support to *the theory that DNA is the genetic material.* They tagged bacteriophages with the radioactive isotopes ^{32}P and ^{35}S. Since proteins contain sulfur but not phosphorus and DNA contains phosphorus but not sulfur, the radioactive ^{32}P labeled the DNA of the phage viruses, while ^{35}S labeled the protein coat of the phage viruses. Hershey and Chase found that when bacteria were infected with phage viruses, the radioactive phosphorus in the phage always entered the bacterium, while the radioactive sulfur remained outside the cells. This proved that *DNA from the viral nucleus, not protein from the viral coat, was infecting bacteria and producing thousands of progeny.*

Rosalind Franklin (1950–53), while working in the lab of Maurice Wilkins, carried out the X-ray crystallography analysis of DNA that showed DNA to be a helix. (Dr. Franklin's Photo 51 is famous.) Her work was critical to Watson and Crick. Although Maurice Wilkins shared the Nobel prize with Watson and Crick, Rosalind Franklin did not. She had died by the time the prize was awarded, and the prize is not awarded posthumously.

IST-1
The proof that DNA is the carrier of genetic information involved a number of important historic experiments.

Watson and Crick (1953), while working at Cambridge University, *proposed the double helix structure of DNA* in a one-page paper in the British journal *Nature.* Throughout the 1940s, until 1953, many scientists worked to understand the structure of DNA. All the data that Watson and Crick used to build their model of DNA derived from other scientists who published earlier. Two major pieces of information they used were the biochemical analysis of DNA (from Erwin Chargaff) and the X-ray diffraction analysis of DNA (from Rosalind Franklin). However, the fact that much of the components of DNA were known before Watson and Crick began their model building does not detract from the brilliance of their achievement. Understanding the structure of DNA gives a foundation to understand how DNA could replicate itself. Watson and Crick received the Nobel prize in 1962 for correctly describing the structure of DNA.

Meselson and Stahl (1958) *proved that DNA replicates in a semiconservative fashion,* as Francis Crick predicted. They cultured bacteria in a medium containing heavy nitrogen (^{15}N), allowing the bacteria to incorporate this heavy nitrogen into their DNA as they replicated and divided. These bacteria were then transferred to a medium containing light nitrogen (^{14}N) and allowed to replicate and divide only once. The bacteria that resulted from this final replication were spun in a centrifuge and found to be midway in density between the bacteria grown in heavy nitrogen and those grown in light nitrogen. This demonstrated that the new bacteria contained DNA consisting of one heavy strand and one light strand. See Figure 9.1 of semiconservative replication.

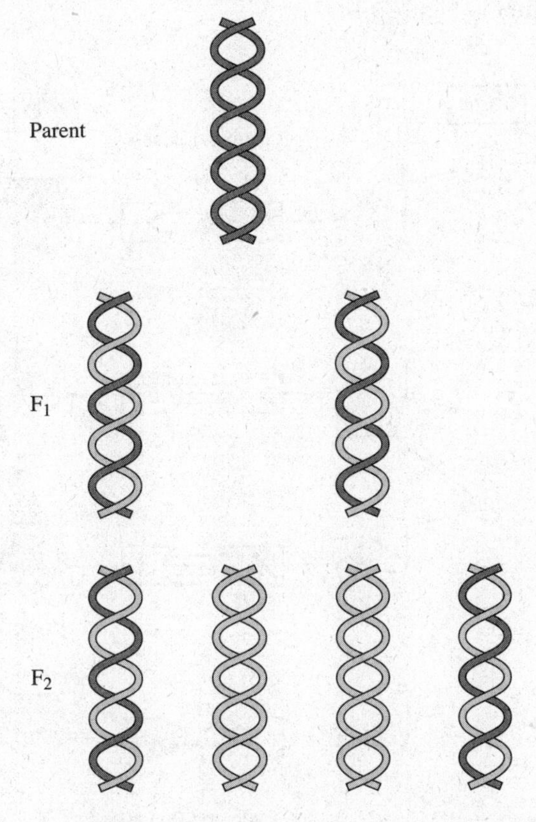

Figure 9.1 Semiconservative Replication

STRUCTURE OF NUCLEIC ACIDS

Deoxyribonucleic Acid (DNA)

REMEMBER

The two strands of DNA run in opposite directions.

The DNA molecule is a **double helix**, shaped like a twisted ladder, consisting of two strands running in opposite directions (antiparallel); see Figure 9.2. One strand runs **5′ to 3′** (right side up), the other **3′ to 5′** (upside down). DNA is a polymer consisting of repeating units of **nucleotides**. In DNA, these consist of a **5-carbon sugar** (**deoxyribose**), a **phosphate**, and a **nitrogen base**. The carbon atoms in deoxyribose are numbered 1 to 5. There are four nitrogenous bases in DNA: **adenine (A)**, **thymine (T)**, **cytosine (C)**, and **guanine (G)**. Of the four nitrogenous bases, adenine and guanine are **purines**, and thymine and cytosine are **pyrimidines**. The nitrogenous bases of opposite chains are paired to one another by **hydrogen bonds**: the adenine nucleotide bonds by a **double hydrogen bond** to the thymine nucleotide, and the cytosine nucleotide bonds by a **triple hydrogen bond** to the guanine nucleotide; see Figure 9.3.

IST-1
Both DNA and RNA exhibit base pairing rules that are conserved (unchanged) through evolutionary history.

DNA gets packed and unpacked in the nucleus as needed. Eukaryotic DNA combines with a large amount of proteins called **histones** from which it separates only briefly during replication. This complex of DNA plus histones is called by the general name **chromatin**. The double helix of DNA wraps twice around a core of histones, forming structures called **nucleosomes** that look like beads on a string, as depicted later in this chapter in Figure 9.14.

REMEMBER

A bonds with T.

A = T

C bonds with G.

C ≡ G

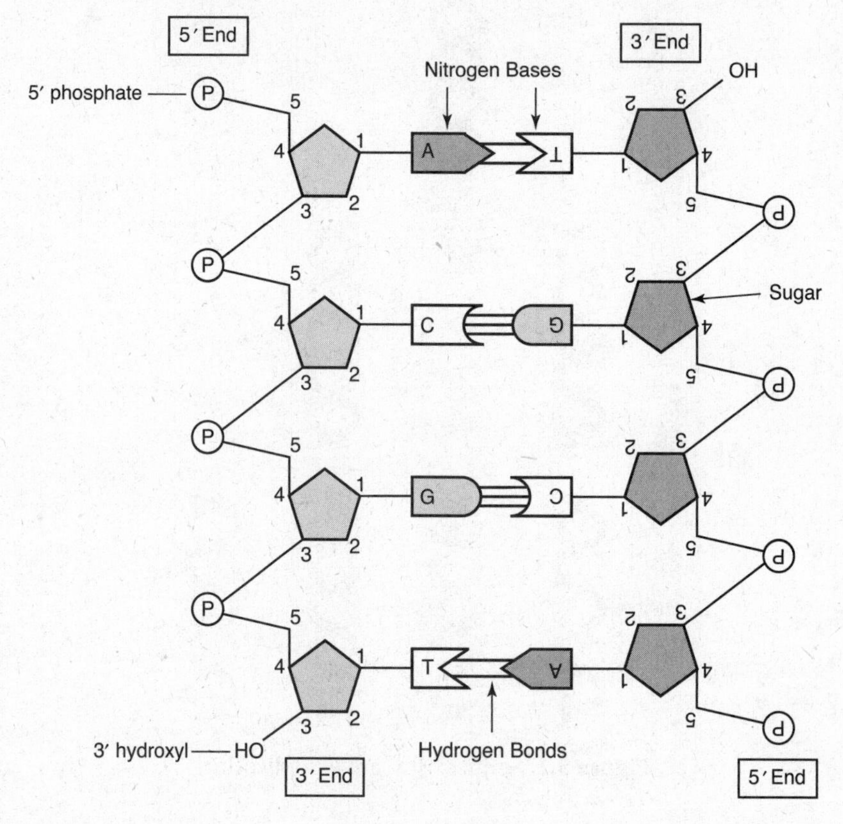

Figure 9.2 DNA

Ribonucleic Acid (RNA)

RNA is a single-stranded helix consisting of repeating nucleotides: adenine, cytosine, guanine, and **uracil (U)**, which replaces thymine. The 5-carbon sugar in RNA is **ribose**.

Figure 9.3 shows structural formulas for the purines—adenine and guanine—and for the pyrimidines—cytosine, thymine, and uracil. Notice that purines have a double-ring structure, while pyrimidines have a single-ring structure.

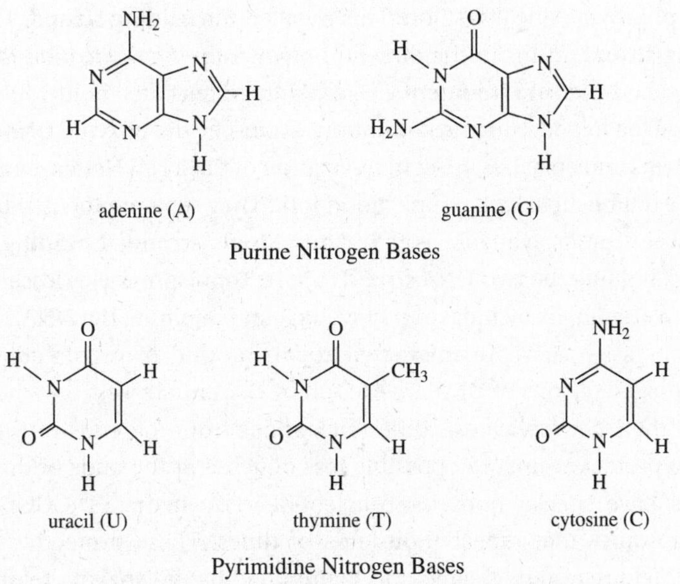

adenine (A)　　guanine (G)

Purine Nitrogen Bases

uracil (U)　　thymine (T)　　cytosine (C)

Pyrimidine Nitrogen Bases

Figure 9.3 Structure of the 5 Nitrogenous Bases

DNA REPLICATION IN EUKARYOTES

DNA replication, the making of an exact replica of the DNA molecule by **semiconservative replication**, was predicted by Watson and Crick and proven by Meselson and Stahl. The DNA double helix unzips, and each strand serves as a **template** for the formation of a new strand composed of complementary nucleotides: A with T and C with G. The two new molecules each consist of one old strand and one new strand. The following describes DNA replication in eukaryotes. Be sure to also review Figures 9.1 and 9.4.

- Replication begins at special sites called **origins of replication**, where the two strands of DNA separate to form **replication bubbles**. Thousands of these bubbles can be seen along the DNA molecule by using electron microscopy. Replication bubbles speed up the process of replication along the giant DNA molecule that consists of *6 billion nucleotides.* A replication bubble expands as replication proceeds in *both directions at once.*
- At each end of the replication bubble is a **replication fork**, a Y-shaped region where the new strands of DNA are elongating. Eventually, all the replication bubbles fuse.
- The enzyme **DNA polymerase** catalyzes the antiparallel elongation of the new DNA strands. (At least 15 different types of DNA polymerase have been identified, but only one is involved in the elongation of the DNA strand.)
- DNA polymerase builds a new strand from the 5′ to the 3′ direction by moving along the template strand and pushing the replication fork ahead of it. In humans, the rate of elongation is about 50 nucleotides per second.
- *DNA polymerase cannot initiate synthesis*; it can only add nucleotides to the 3′ end of a preexisting chain. This preexisting chain actually consists of RNA and is called **RNA primer**. An enzyme called **primase** makes the primer by joining together RNA nucleotides.
- One of your cells can replicate its entire DNA in a few hours.
- DNA polymerase replicates the two original strands of DNA differently. Although it builds both new strands in the 5′ to 3′ direction, one strand is formed *toward the replication*

fork in an unbroken, linear fashion. This is called the **leading strand**. The other strand, the **lagging strand**, forms in the direction *away from the replication fork* in a series of segments called **Okazaki fragments**. Okazaki fragments are about 100–200 nucleotides long and will be joined into one continuous strand by the enzyme **DNA ligase**.

- Other proteins and enzymes assist in replication of the DNA. **Helicases** are enzymes that untwist the double helix at the replication fork. They separate the two parental strands, making these strands available as templates. **Single-stranded binding proteins** act as scaffolding, holding the two DNA strands apart. **Topoisomerases** lessen the tension on the tightly wound helix by breaking, swiveling, and rejoining the DNA strands.
- DNA polymerases carry out **mismatch repair**, a kind of proofreading that corrects errors. Damaged regions of DNA are excised by **DNA nuclease**.
- Each time the DNA replicates, some nucleotides from the ends of the chromosomes are lost. To protect against the possible loss of genes at the ends of the chromosomes, eukaryotes have special nonsense nucleotide sequences (TTAGGG) at the ends of the chromosomes that repeat thousands of times. These protective ends are called **telomeres**. Telomeres are created and maintained by the enzyme **telomerase**. Normal body cells contain little telomerase, so every time the DNA replicates, the telomeres get shorter. This may serve as a clock that counts cell divisions and causes the cell to stop dividing as the cell ages.

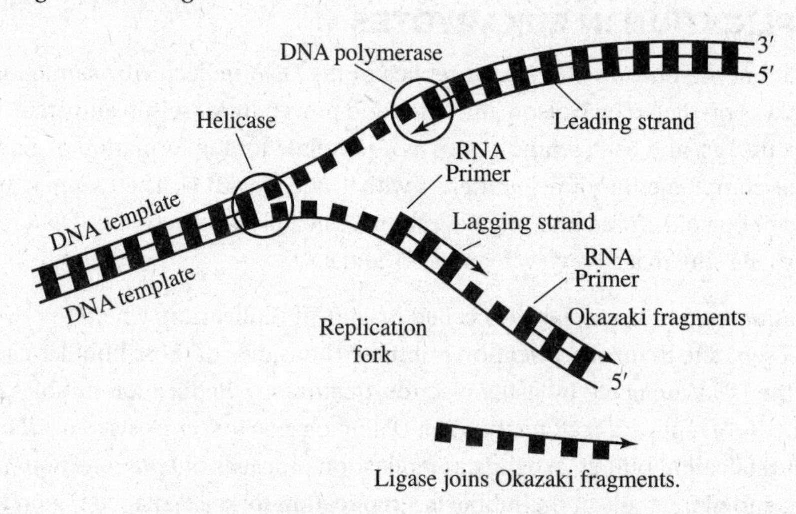

Figure 9.4 DNA Replication at Replication Fork

> **IST-1**
> Genetic information flows from a sequence of nucleotides in a gene to a sequence of amino acids in a protein.

FROM DNA TO PROTEIN

The process whereby DNA makes proteins has been worked out in great detail. To summarize, the **triplet code** in DNA is **transcribed** into a **codon sequence** in messenger-RNA (mRNA) inside the nucleus. Next, this newly formed strand of **RNA**, known as pre-RNA, is **processed** or modified in the nucleus. Then the codon sequence leaves the nucleus and is **translated** into an amino acid sequence (a polypeptide) in the cytoplasm at the ribosome.

If the strand of DNA triplets to be transcribed is	5′-AAA TAA CCG GAC-3′
Then the strand of mRNA **codons** that forms is	3′-UUU AUU GGC CUG-5′
The transfer RNA (tRNA) **anticodon** strand complementary to the mRNA strand is	AAA UAA CCG GAC

Figure 9.5 shows an overview of transcription, RNA processing, and translation.

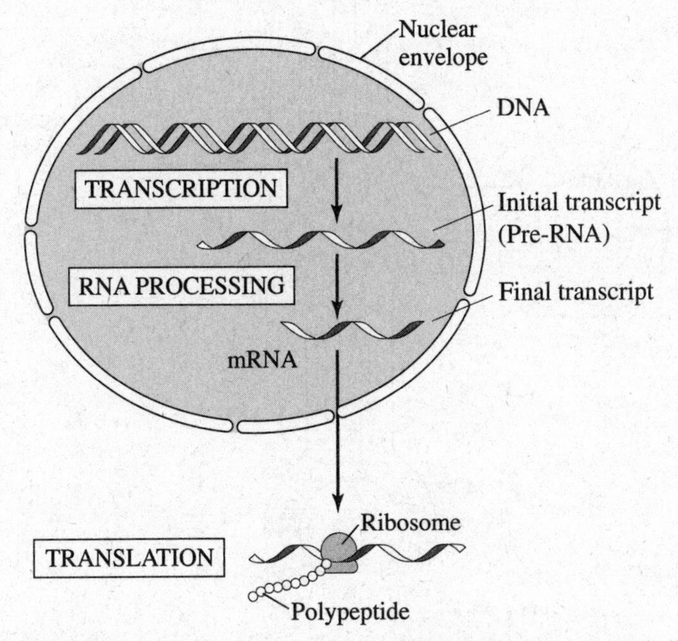

Figure 9.5 Transcription and Translation in a Eukaryotic Cell

Transcription

Transcription is the process by which the information in a DNA sequence is copied (transcribed) into a complementary RNA sequence. Although there are many kinds of RNA, three types are directly involved in protein synthesis.

1. **MESSENGER RNA (mRNA) IS INVOLVED IN TRANSCRIPTION:** When a sequence of DNA is expressed, one of two strands of DNA is copied into mRNA according to the base-pairing rules: C with G and A with U (in RNA, uracil replaces the thymine in DNA).
2. **RIBOSOMAL RNA (rRNA) IS INVOLVED IN TRANSLATION:** rRNA is structural. Along with proteins, it makes up the ribosome, which consists of two subunits, one large and one small. The ribosome has one mRNA binding site and three tRNA binding sites, known as A, P, and E sites. A ribosome is a protein synthesis factory. See Figure 9.8.
3. **TRANSFER RNA (tRNA) CARRIES AMINO ACIDS FROM THE CYTOPLASMIC POOL OF AMINO ACIDS TO mRNA AT THE RIBOSOME:** tRNA is shaped like a cloverleaf and has a binding site for an amino acid at one end and another binding site for an **anticodon** sequence that binds to mRNA at the other. See Figure 9.6.

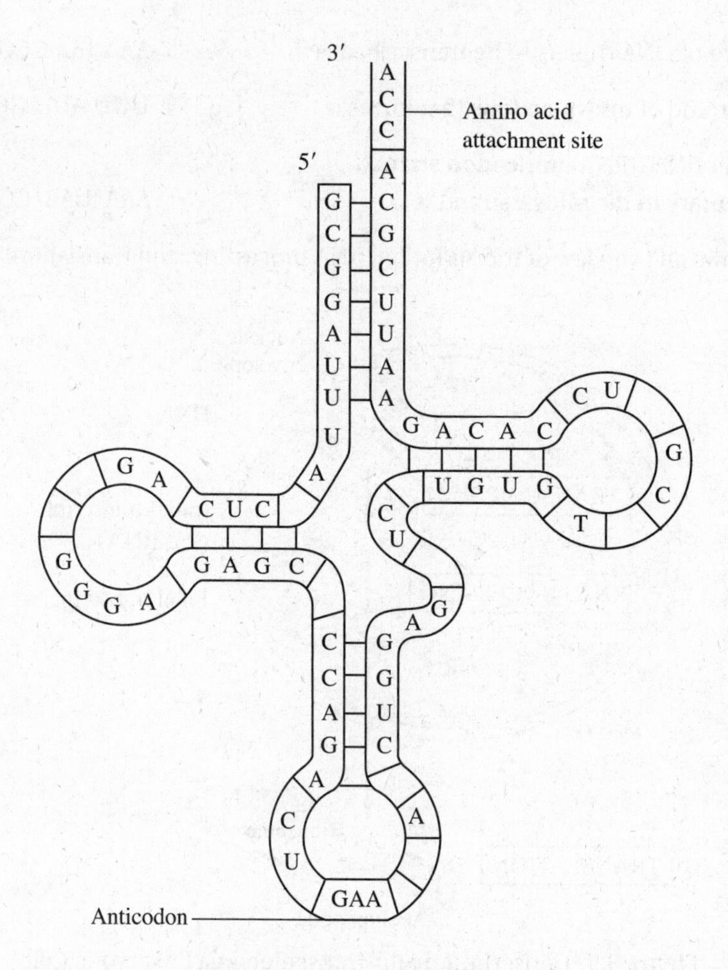

Figure 9.6 Transfer RNA (tRNA)

- Transcription consists of three stages: **initiation**, **elongation**, and **termination**.
- **Initiation** begins when an enzyme, ***RNA polymerase**, recognizes and binds to DNA at the* ***promoter region***. The promoter "tells" RNA polymerase where to begin transcription and which of the two strands to transcribe. A collection of proteins called **transcription factors** recognize a key area within the promoter, the **TATA box** (because of its many repeating thymine and adenine nucleotides), and mediate the binding of RNA polymerase to the DNA. The completed assembly of transcription factors and RNA polymerase bound to the promoter is called a **transcription initiation complex**. Once RNA polymerase is attached to the promoter, DNA transcription of the DNA **template** begins.
- **Elongation** of the strand continues as *RNA polymerase adds nucleotides to the 3′ end of a growing chain*. RNA polymerase pries the two strands of DNA apart and attaches RNA nucleotides according to the base pairing rules: C with G and A with U. The stretch of DNA that is transcribed into an mRNA molecule is called a **transcription unit**. Each unit consists of triplets of bases called **codons** (for example, AAU, CGA) that code for specific amino acids. A single gene can be transcribed into mRNA simultaneously by several molecules of RNA polymerase following each other in a caravan fashion. Like DNA polymerases, RNA polymerase has mechanisms for proofreading during transcription. Because mRNA is usually short-lived, any errors in mRNA are not as potentially harmful as errors in the DNA sequence.

- **Termination** is the final stage in transcription. Elongation continues for a short distance after the RNA polymerase transcribes the **termination sequence** (AAUAAA). At this point, mRNA is cut free from the DNA template.

RNA Processing

Before the newly formed pre-RNA strand is shipped out of the nucleus to the ribosome in the cytoplasm, it is altered or **processed** by a series of enzymes. Here are the details.

- A **5′ cap** consisting of a modified guanine nucleotide is added to the 5′ end. This cap helps the RNA strand bind to the ribosome in the cytoplasm during translation.
- A **poly (A) tail**, consisting of a string of adenine nucleotides, is added to the 3′ end. This tail protects the RNA strand from degradation by hydrolytic enzymes, and facilitates the release of mRNA from the nucleus into the cytoplasm.
- Noncoding regions of the mRNA called **introns** or **intervening sequences** are removed or **spliced** out by **snRNPs**, small nuclear ribonucleoproteins, within **spliceosomes**. This removal allows only **exons**, which are expressed sequences, to leave the nucleus. As a result of this processing, the mRNA that leaves the nucleus is a great deal shorter than the original transcription unit. If the average length of a human DNA molecule that is transcribed is about 27,000 nucleotides long, the *primary transcript* is also the same length. However, an average-size protein of 200 amino acids requires only 1,200 nucleotides of RNA to code for it. Therefore, out of the original 27,000 nucleotides, 15,800 noncoding nucleotides must be *removed* or *spliced out* during RNA processing. See Figure 9.5.

Alternative Splicing

Before the human genome was sequenced by the Human Genome Project, scientists expected that they would find about 100,000 genes. In fact, they discovered that humans have only about 22,000 genes. This surprised everyone but can be explained when you recognize the mechanism of **alternative RNA splicing**. In this process, different RNA molecules are produced from the same primary transcript, depending on which RNA segments are treated as **exons** and which are treated as **introns**. (Exons are expressed sequences. Introns are intervening, noncoding sequences.) *Regulatory proteins* specific to a cell type control intron-exon choices by binding to regulatory sequences within the primary transcript. See Figure 9.7.

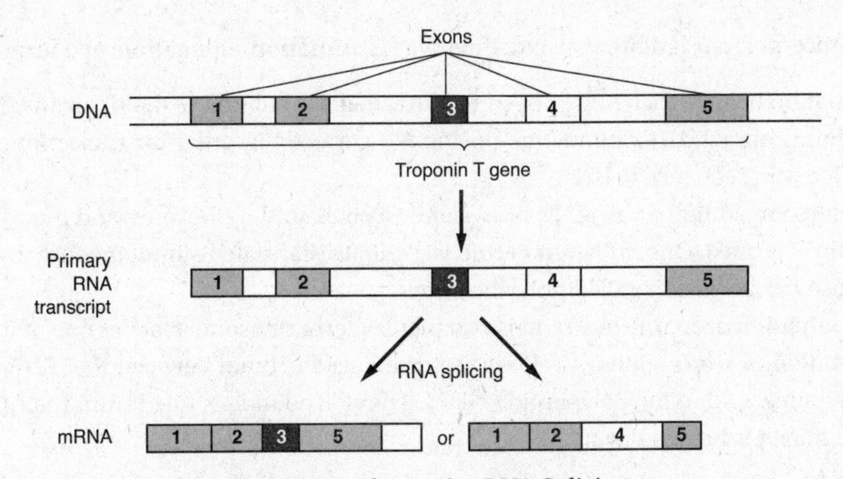

Figure 9.7 Alternative RNA Splicing

Translation of mRNA—Synthesis of a Polypeptide

Translation *is the process by which the codons of an mRNA sequence are changed into an amino acid sequence*; see Figure 9.8. Amino acids present in the cytoplasm are carried by tRNA molecules to the codons of the mRNA strand at the ribosome according to the base pairing rules (A with U and C with G). One end of the tRNA molecule bears a specific amino acid, and the other end bears a nucleotide triplet called an **anticodon**. Unlike mRNA, which is broken down immediately after it is used, tRNA is used repeatedly. The energy for this process is provided by **GTP (guanosine triphosphate)**, a molecule closely related to ATP. Each amino acid is joined to the correct tRNA by a specific enzyme called **aminoacyl-tRNA synthetase**. There are only 20 different aminoacyl-tRNA synthetases, one for each amino acid. There are 64 codons; 61 of them code for amino acids. One codon, **AUG**, has two functions; it codes for the amino acid methionine and is also a **start codon**. Three codons, **UAA**, **UGA**, and **UAG**, are **stop codons** and terminate translation. Some tRNA molecules have anticodons that can recognize two or more different codons. This occurs because the pairing rules for the third base of a codon are not as strict as they are for the first two bases. This relaxation of base pairing rules is known as **wobble**. For example, the codons UCU, UCC, UCA, and UCG all code for the amino acid serine.

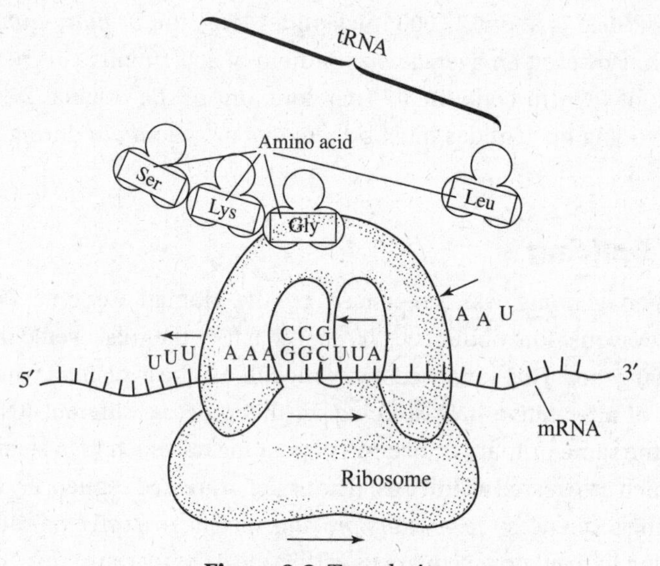

Figure 9.8 Translation

The process of translation consists of three stages: **initiation**, **elongation**, and **termination**.

- **Initiation** begins when mRNA becomes attached to a subunit of the ribosome. This first codon is always **AUG**. It must be positioned correctly in order for transcription of an amino acid sequence to begin.
- **Elongation** continues as tRNA brings amino acids to the ribosome and a polypeptide chain is formed. One mRNA molecule is generally translated simultaneously by several ribosomes in clusters called **polyribosomes**.
- **Termination** of an mRNA strand is complete when a ribosome reaches one of three **termination** or **stop codons**. A **release factor** breaks the bond between the tRNA and the last amino acid of the polypeptide chain. The polypeptide is freed from the ribosome, and mRNA is broken down.

The Genetic Code

The complete genetic code is shown in Figure 9.9. There are 64 (4^3) possible combinations of the four nitrogenous bases. Notice that AUG, which codes for methionine, also codes for **start**, the initiation signal for translation. Three of the codons are **stop codons**, termination signals of translation.

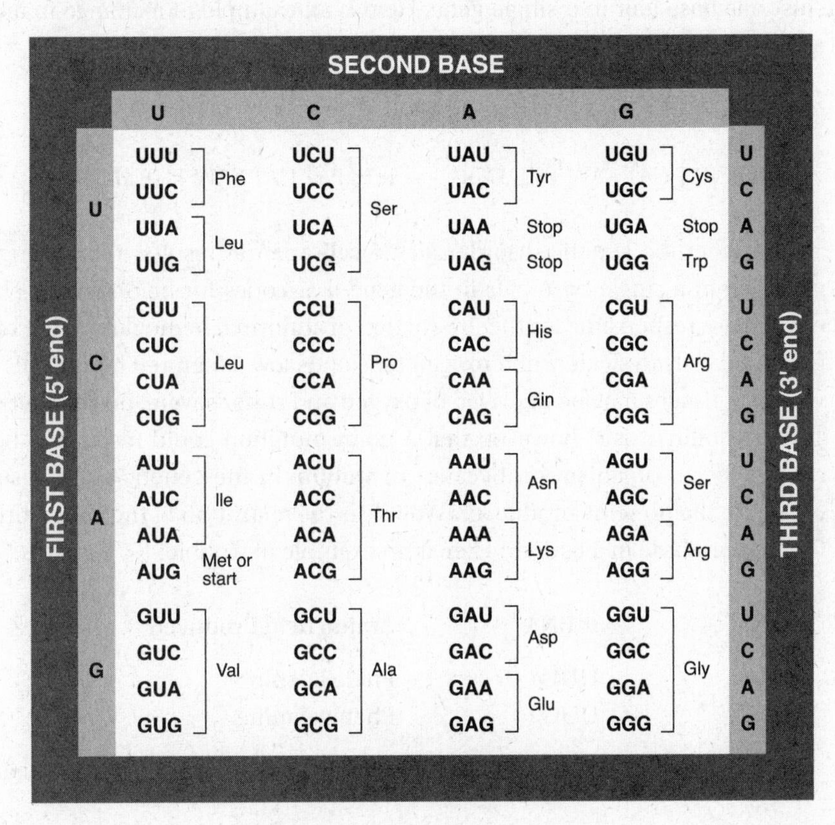

Figure 9.9 The Genetic Code

Remember this important statement about the genetic code: *There are redundancies in the code, but there is no ambiguity.* Examine the chart. The fact that there are four codons for leucine means there is redundancy. In every case, one codon codes for one particular amino acid. So there is no ambiguity.

With very few exceptions within bacteria, mitochondria, and chloroplasts, this code is universal and unifies all life. It indicates that the code originated early in the evolution of life on Earth and that all living things descended from those first ancestral cells.

GENE MUTATION

Mutations are permanent changes in genetic material. They occur *spontaneously* and *randomly.* They can be caused by **mutagenic agents**, including toxic chemicals and radiation. A mutation in **somatic** (body) cells disrupts normal cell functions. Mutations that occur in gametes are transmitted to offspring and can change the **gene pool** of a population. *Mutations are the raw material for natural selection.*

Some regions of DNA are more vulnerable to mutations than others. For example, regions of As and Ts are subject to more breakages than regions of Cs and Gs because A and T are connected by a double hydrogen bond whereas Cs and Gs are more strongly connected by a triple hydrogen bond.

Point Mutation

The simplest mutation is a **point mutation**. This is a **base-pair substitution**, a chemical change in just one base pair in a single gene. Here is an example of a change in an English sentence analogous to a point mutation in DNA:

Point Mutation

THE FAT CAT SAW THE **D**OG → THE FAT CAT SAW THE **H**OG.

IST-2 & IST-4
The processing of genetic information is imperfect and causes mutations that are a source of genetic variation.

The inherited genetic disorder **sickle cell anemia** results from a single point mutation in a single base pair in the gene that codes for hemoglobin. This point mutation is responsible for the production of abnormal hemoglobin that can cause red blood cells to sickle when oxygen tension is low. When red blood cells sickle, a variety of tissues may be deprived of oxygen and suffer severe, permanent damage. The possibility exists, however, that a point mutation could result in a beneficial change for an organism or, because of **wobble** in the genetic code, result in no change in the proteins produced. (Wobble is the relaxation of the base pairing rules for the third base in a codon.) Here is an example of wobble:

DNA	mRNA	Amino Acid Produced
AAA	UUU	Phenylalanine
AAG	UUC	Phenylalanine
↑ mutation		No change occurs in the amino acid.

The mutation shown above is called a **silent mutation** because it does not result in any change in the amino acid sequence.

Insertion or Deletion

A second type of gene mutation results from a single nucleotide **insertion** or **deletion.** To continue the three-letter word analogy, a deletion is the loss of one letter, and an insertion is the addition of a letter into the DNA sentence. Both mutations result in a **frameshift**, because the entire **reading frame** is altered.

IST-2 & IST-4
Changes in genotype can result in changes in phenotype.

Deletion of the letter E shifts the reading frame:

↓
THE FAT CAT SAW THE DOG → THF ATC ATS AWT HED OG

Insertion of the letter T shifts the reading frame:

↓
THE FAT CAT SAW THE DOG → THE FTA TCA TSA WTH EDO G

As a result of the frameshift, one of two things can happen. Either a mutated polypeptide is formed or no polypeptide is formed.

THE GENETICS OF VIRUSES AND BACTERIA

Since the early part of the twentieth century when Griffith discovered the transformation factor, knowledge of genetics has been based on work with the simplest biological systems—viruses and bacteria. Scientists' understanding of replication, transcription, and translation of DNA was worked out using bacteria as a model. Their understanding of how viruses and bacteria infect cells is the basis for how diseases are treated and how vaccines are developed. A worldwide industry of genetic engineering and recombinant DNA relies on bacteria like *Escherichia coli* and viruses like the phage viruses for research and therapeutic endeavors. Whereas Gregor Mendel depended on the garden pea and Thomas Hunt Morgan depended on the fruit fly, researchers now depend on bacteria and viruses.

The Genetics of Viruses

A virus is a parasite that can live only inside another cell. It commandeers the host cell machinery to transcribe and translate all the proteins it needs to fashion new viruses. In the process, thousands of new viruses are formed and the host cell is often destroyed. A virus consists of DNA or RNA enclosed in a protein coat called a **capsid**. Some viruses also have a viral **envelope** that is derived from membranes of host cells, cloaks the capsid, and aids the virus in infecting the host. Each type of virus can infect only one specific cell type because it gains entrance into a cell by binding to *specific receptors* on the cell surface. For example, the virus that causes colds in humans infects only the membranes of the respiratory system, and the virus that causes AIDS infects only one type of white blood cell. In addition, one virus can usually only infect one species. The range of organisms that a virus can attack is referred to as the **host range** of the virus. A sudden emergence of a new viral disease that affects humans, such as AIDS or H1N1, may result from a mutation in the virus that expands its host range.

- **BACTERIOPHAGES**—Also known as **phage** viruses, **bacteriophages** are viruses that infect bacteria. They are the most complex and best understood viruses. The bacteriophage can reproduce in two ways.

  1. In the **lytic cycle**, the phage enters a host cell, takes control of the cell machinery, replicates itself, and then causes the cell to burst, releasing a new generation of infectious phage viruses. These new viruses infect and kill thousands of cells in the same manner. A phage that replicates only by a lytic cycle is a virulent phage.
  2. In the **lysogenic cycle**, viruses replicate without destroying the host cell. The phage virus becomes incorporated into a specific site in the host's DNA. It remains dormant within the host genome and is called a **prophage**. As the host cell divides, the phage is replicated along with it and a single infected cell gives rise to a population of infected cells. At some point, an environmental trigger causes the prophage to switch to the **lytic phase**. Viruses capable of both modes of reproducing, lytic and lysogenic, within a bacterium are called **temperate viruses**.

IST-2
Viral replication can introduce genetic variation into its hosts.

- **RETROVIRUSES**—These are viruses that contain RNA instead of DNA and replicate in an unusual way. Following infection of the host cell, the retrovirus RNA serves as a template for the synthesis of complementary DNA (cDNA) because it is complementary to the RNA from which it was copied. *Thus, these retroviruses reverse the usual flow of information from DNA to RNA.* This reverse transcription occurs under the direction of an enzyme called **reverse transcriptase.** A retrovirus usually inserts itself into the host genome, becomes a permanent resident, called a **prophage**, and is capable of making multiple copies of the viral genome for years. An example of a retrovirus is HIV (human immunodeficiency virus), which causes AIDS.
- **TRANSDUCTION**—Phage viruses acquire bits of bacterial DNA as they infect one cell after another. This process, which leads to genetic recombination, is called **transduction.** Two types of transduction occur: **generalized** and **restricted (specialized). Generalized transduction** moves random pieces of bacterial DNA as the phage lyses one cell and infects another during the lytic cycle. **Restricted transduction** involves the transfer of specific pieces of DNA. During the lysogenic cycle, a phage integrates into the host cell at a specific site. At a later time, when the phage ruptures out of the host DNA, it sometimes carries a piece of adjacent host DNA with it and inserts this host DNA into the next host it infects.

> **IST-2**
> The reproductive cycles of viruses transfer genetic information via transposons and transduction.

The Genetics of Bacteria

The bacterial chromosome is a circular, double-stranded DNA molecule, tightly condensed into a structure with a small amount of protein. It is located in a **nucleoid** region that has no nuclear membrane. Bacteria replicate their DNA in **both directions** from a **single point of origin.**

> **IST-2**
> In general, prokaryotes have circular chromosomes while eukaryotes have multiple, linear chromosomes.

Although bacteria can reproduce by a primitive sexual method called **conjugation**, the main mode of reproduction is asexual, by **binary fission.** Binary fission results in a population with all identical genes, but mutations do occur spontaneously. Although mutations are rare, bacteria reproduce by the millions, and even one mutation in every 1,000 replications can amount to significant variation in the population as a whole.

- **Bacterial transformation** was discovered by **Frederick Griffith** in 1928 when he performed experiments with several different strains of the bacterium *Diplococcus pneumoniae.*

  Transformation is either a natural or an artificial process that provides a mechanism for the recombination of genetic information in some bacteria. Small pieces of extracellular DNA are taken up by a living bacterium, ultimately leading to a stable genetic change in the recipient cell. Bacterial transformation is very easy to carry out today.
- A **plasmid** is a foreign, small, circular, self-replicating DNA molecule that inhabits a bacterium. A bacterium can harbor many plasmids and will express the genes carried by the plasmid. These genes may impart an advantage to the host bacterium. The first plasmid discovered was the **F plasmid**. F stands for fertility. Bacteria that contain the F plasmid are called F^+; those that do not carry the plasmid are called F^-. The F plasmid contains genes for the production of **pili**, cytoplasmic bridges that connect to an adjacent cell and that allow DNA to move from one cell to another in a form of primitive sexual reproduction called **conjugation**. Another plasmid, the

R plasmid, makes the cell in which it is carried resistant to specific antibiotics, such as ampicillin or tetracycline. In addition, the R plasmid can be transferred to other bacteria by conjugation. Bacteria that carry the R plasmid have a distinct evolutionary advantage over bacteria that are not resistant to antibiotics. Resistant bacteria will be selected for (survive) and their populations will increase while nonresistant bacteria die out. This is exactly what is happening today as an increasing number of populations of pathogenic bacteria, such as the one that causes tuberculosis, are becoming resistant to antibiotics. This is cause for serious concern in the health community.

The Operon

The **operon** was discovered in the bacterium *E. coli* by **Jacob and Monod** in the 1940s. Although it is found only in bacteria, it is an important model of **gene regulation**. An operon is essentially a set of genes and the switches that control the expression of those genes. There are two types of operons: the **repressible** (**tryptophan**) operon and the **inducible** (**lac**) operon.

REMEMBER

If a free-response question on the AP exam is about regulation, the operon is a perfect example.

THE TRYPTOPHAN OPERON

The **tryptophan operon** consists of a **promoter** and five adjacent structural genes (A, B, C, D and E) that code for the five separate enzymes necessary to synthesize the amino acid tryptophan; see Figures 9.10 and 9.11. As long as **RNA polymerase** binds to the promoter, one long strand of mRNA containing start and stop codons is transcribed. If adequate tryptophan is present, tryptophan itself acts as a **corepressor** activating the **repressor**. The activated repressor binds to the **operator**, preventing RNA polymerase from binding to the promoter. Without RNA polymerase attached to DNA at the promoter, transcription ceases. The tryptophan operon is known as a repressible operon, meaning it is always switched on unless the repressor is activated.

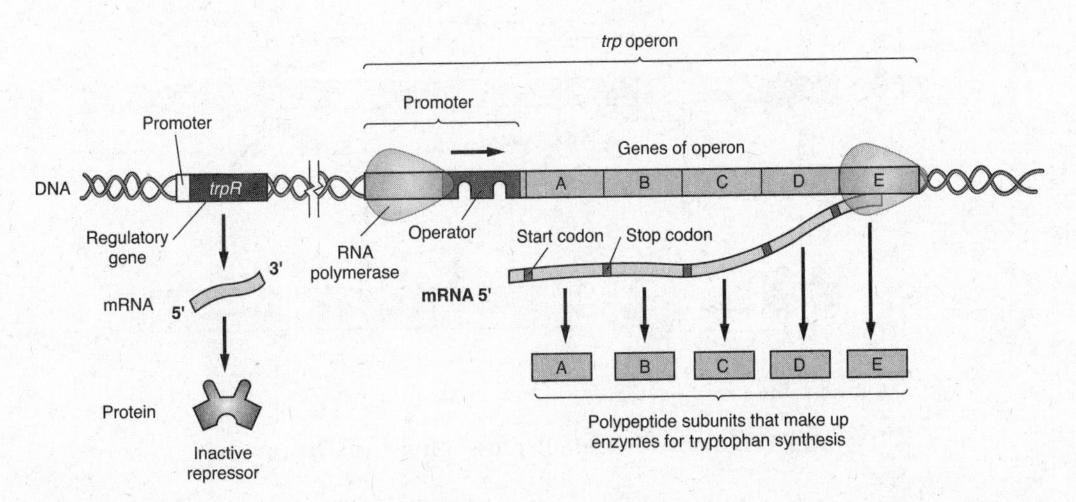

Figure 9.10 Tryptophan Absent, Repressor Inactive, Operon On → Tryptophan Produced

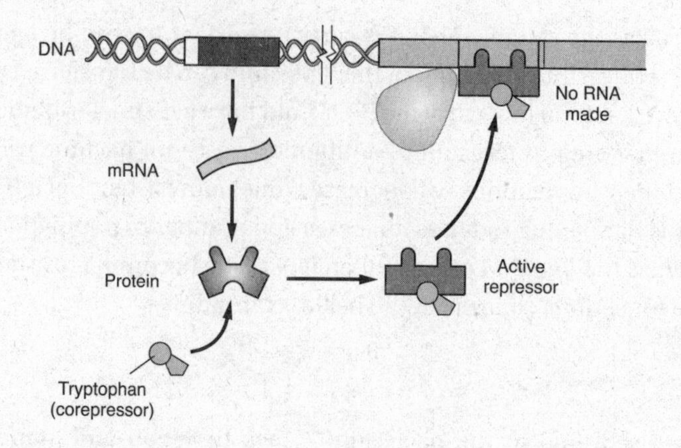

Figure 9.11 Tryptophan Present, Repressor Active, Operon Off

THE LAC OPERON

In order for the *E. coli* in our intestines to utilize lactose as an energy source, three enzymes must be synthesized to break down lactose into glucose and galactose. These enzymes, β-galactosidase, permease, and transacetylase, are coded for by three genes in the *lac* operon (Z, Y, and A); see Figures 9.12 and 9.13. In order for these three genes to be transcribed, the **repressor** must be prevented from binding to the operator and RNA polymerase must bind to the promoter region. Allolactose, an isomer of lactose, is the **inducer** that facilitates this process by binding to the **active repressor** and inactivating it. When a person drinks milk, they ingest allolactose, the inducer, which deactivates the repressor, allowing RNA polymerase to bind to DNA. When RNA polymerase binds to DNA, transcription of the *lac* genes occurs and lactose can be utilized as an energy source.

> **IST-2**
> Both positive and negative control mechanisms regulate gene expression in bacteria and viruses.

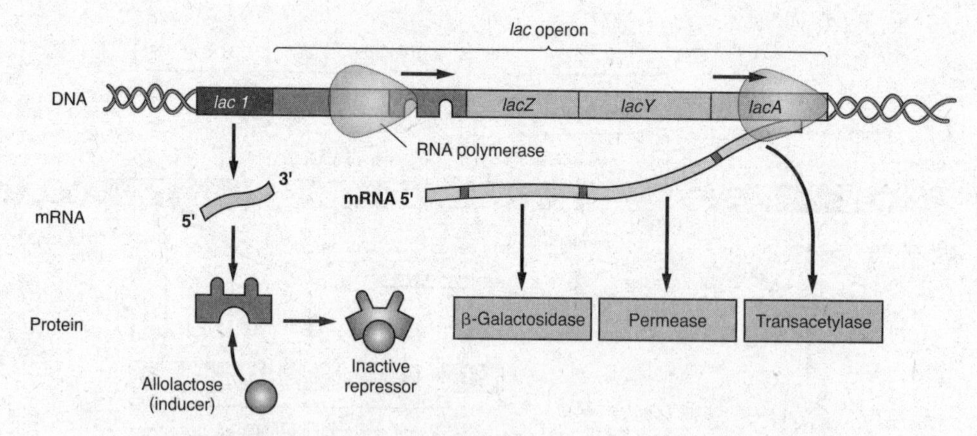

Figure 9.12 Lactose Present, Repressor Inactive, Operon On

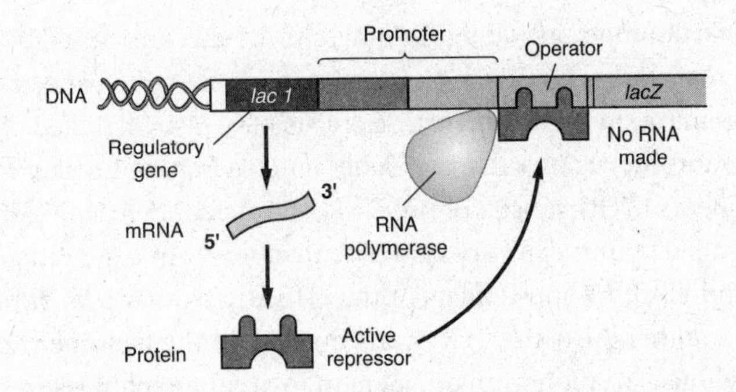

Figure 9.13 Lactose Absent, Repressor Active, Operon Off

CAP AND cAMP—POSITIVE GENE REGULATION

When glucose and lactose are both present in the intestine, *E. coli* preferentially metabolize glucose and the enzymes for breaking down glucose are always present. However, when lactose is present and glucose is in short supply, *E. coli* switch to lactose as an energy source. This ability depends on the interaction of an allosteric regulatory protein, CAP (catabolite activator protein), and cAMP (cyclic-AMP). Since the attachment of CAP to the promoter directly stimulates gene expression, this mechanism is an example of **positive gene regulation**.

VOCABULARY FOR THE OPERON

- **RNA polymerase**: Enzyme that transcribes a new RNA chain by linking ribonucleotides to nucleotides on a DNA template.
- **Operator**: Sequence of nucleotides near the start of an operon to which the active repressor can attach. The binding of the repressor prevents RNA polymerase from attaching to the promoter and transcribing the operon's genes.
- **Promoter**: Nucleotide sequence in the DNA of a gene that is the binding site of RNA polymerase, positioning the RNA polymerase to begin to transcribe RNA at the appropriate position.
- **Repressor**: Protein that inhibits gene transcription. In the operon of prokaryotes, repressors bind to the operator.
- **Regulator gene**: Gene that codes for a repressor. It is located some distance from its operon and has its own promoter.

PRIONS

Prions are not cells nor are they viruses. They are misfolded versions of a protein normally found in the brain. If prions get into a normal brain, they cause all the normal versions of the protein to misfold in the same way. Prions are infectious and cause several brain diseases: **scrapie** in sheep, **mad cow disease** in cattle, and **Creutzfeldt-Jakob disease** in humans. All known prion diseases are fatal.

THE HUMAN GENOME

The human genome consists of 3 billion base pairs of DNA and includes about 22,000 genes. Surprisingly, only a tiny fraction (1%) of our DNA actually gets translated into proteins. The newest research reveals that much of this nongene DNA gets *transcribed* into RNA and consists of **regulatory** and **repetitive sequences** that alter the expression of genes.

A number of genetic disorders, including Huntington's disease, are caused by abnormally long stretches of **tandem repeats** (back-to-back repetitive sequences) *within* affected genes. Many of these tandem repeats make up the **telomeres**. Scientists have also identified certain noncoding regions of DNA, **polymorphic regions**, that are highly variable from one region to the next.

Short tandem repeats (STR) consist of units of 2–5 nucleotides. One example is GTTAC. The number of repeats of a unit can vary from site to site within a genome. There can be as many as several hundred thousand repeats of the GTTAC or other STR at one site but only half that number in another. Also, the STR at any one site in the genome varies from person to person. This individual variation from one person to another enables forensic scientists to create a person's **genetic** or **DNA profile**, which has been used to prosecute many individuals and is the basis of several television shows. On the other hand, more than 350 wrongly convicted people have been exonerated and freed from prison by these same genetic profiles. A group called "The Innocence Project," founded in 1992 by Peter Neufeld and Barry Scheck at Cardozo School of Law, works to exonerate the wrongly convicted through DNA testing.

REGULATION OF GENE EXPRESSION

> **IST-2**
> Be able to describe the connection between the regulation of gene expression and observable differences within cells and between individuals in a population.

An excellent model for regulation of gene expression is the **operon** in bacteria. (See page 181.) However, regulation of genes in humans is far more complex. Although every cell in your body contains the same 3 billion base pairs of DNA, a typical cell only expresses a small percentage of its genes at any one time. The expression of genes is tightly regulated by different mechanisms. These mechanisms are described below. See Figure 9.15 for a pictorial overview of all the places where the expression of genes can be altered.

Regulation at the Chromatin Structure Level

Eukaryotic DNA is packaged with proteins called **histones** into an elaborate complex known as **chromatin**, the basic unit of which is the **nucleosome**. Changes to the histone structure alter chromatin configuration, binding it more tightly or more loosely, thus making DNA less or more accessible for transcription and expression. The inhibition is reversible.

Acetylation of histone tails, adding of acetyl groups (–$COCH_3$), promotes the loosening of that chromatin structure and permits transcription. Removing acetyl groups blocks transcription.

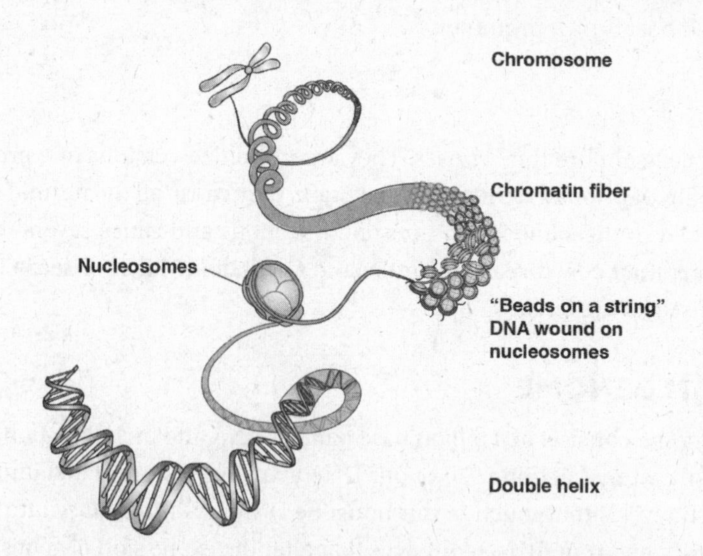

Figure 9.14 A DNA Molecule Condensing into a Chromosome

Regulation by Methylation of DNA

Methylation of certain bases—adding methyl groups (CH_3) to **DNA**—*silences* the DNA temporarily or for long periods of time. The reverse—removing methyl groups (CH_3) from highly methylated regions—can turn genes on. The switching off of genes by methylation is probably responsible for the long-term X-chromosome deactivation in females, as well as long-term deactivation of genes necessary for normal cell differentiation in embryonic development.

Epigenetic Inheritance

Epigenetic inheritance refers to alterations to the genome that do *not* directly involve the nucleotide sequence. Unlike mutations, which are permanent, these changes are reversible. The mechanism behind epigenetics is not well understood, but we do know that environmental factors like diet, stress, and prenatal nutrition can alter the expression of genes. Epigenetics may explain why one identical twin can develop schizophrenia, while the other one, who has the identical genome, does not develop the disorder.

Regulation at the Transcription Level

Transcription is a highly regulated process. To initiate transcription, **RNA polymerase** must bind to the promoter. This process requires the assistance of **transcription factors**, whose presence or absence enables or inhibits transcription.

Regulation at the Post-Transcriptional Regulation Level

Alternative RNA splicing is an important means of regulating gene expression in which different mRNA molecules are produced from the same **primary transcript**, depending on which RNA segments are treated as **introns** (intervening sequences) and which are treated as **exons** (expressed sequences). Regulatory proteins specific to a cell type control intron-exon choices by binding to RNA sequences within the primary transcript. It has been estimated that 90% of human-protein coding genes are subject to alternative splicing. (See Figure 9.7.)

Degradation of mRNA

The span of time after gene transcription before the **degradation of mRNA** is another opportunity to regulate gene expression. Bacterial mRNA molecules are degraded within minutes of their synthesis. This rapid degradation of mRNA may be the reason that bacteria change their patterns of protein synthesis and are so adaptable to changes in the environment. In contrast, human mRNA may continually translate protein for hours or weeks. Molecules of mRNA in developing red blood cells are stable and may translate hemoglobin molecules repeatedly for an extended time.

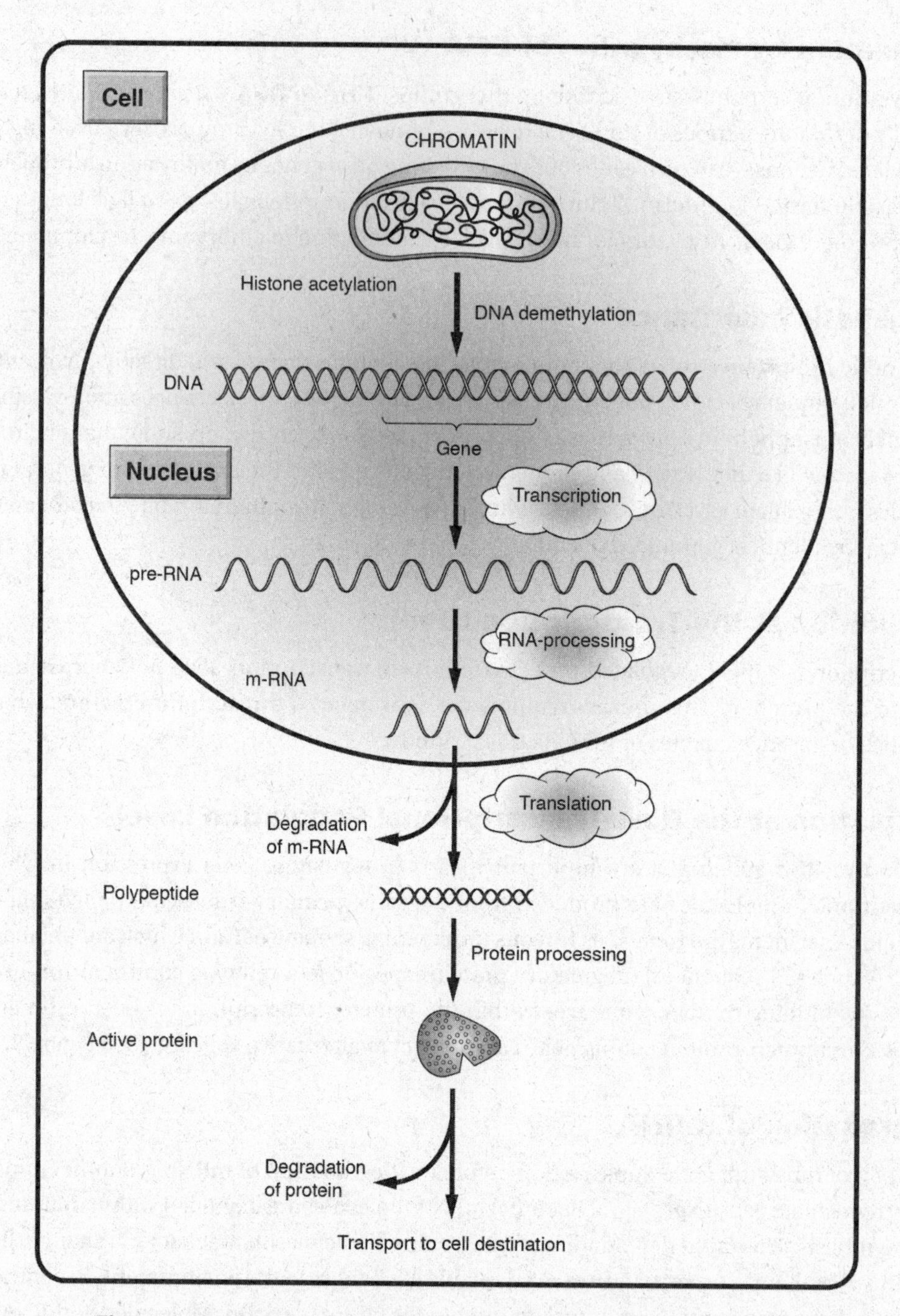

Figure 9.15 Overview of Regulation of Gene Expression Within a Cell

New Role for RNA—Controlling Gene Expression

Recent data show that as much as 90% of non-protein-coding DNA is transcribed into various kinds of **noncoding RNA (ncRNA)**. These ncRNAs bind to and are assisted by specialized binding proteins called Argonaute proteins.

While scientists are still learning about these noncoding regions, we now know that they are not simply "junk," but they regulate much of our DNA. Three types of ncRNA have been extensively studied in recent years and new information is published almost weekly.

MicroRNA (miRNA)—is a small, single-stranded RNA that is about 22 nucleotides long. It does not code for protein. Instead, it targets specific mRNA molecules—to either *degrade* them or *block their translation.* It has been estimated that the expression of at least one-half of all human genes may be regulated by miRNA.

Small interfering RNA (siRNA)—are similar to miRNA in size and function. The blocking of gene expression by siRNA is called **RNA interference (RNAi)**. The Nobel Prize in Physiology or Medicine for 2006 was awarded to Andrew Fire and Craig Mello for their discovery of how **interfering RNA**, or **RNAi**, can play a role in silencing genes. It does this by binding to and destroying mRNA before it can be translated.

Piwi-associated RNA (piRNA)—recently discovered—is a large class of ncRNAs that guide PIWI proteins to complementary RNAs which are derived from transposable elements (elements that can move within the genome, also known as "jumping genes"). Similar to RNAi, they protect germ line cells from attacks by transposons.

Proteins Are Modified at the Post-Translation Level

The final opportunity for controlling gene expression occurs following translation. After emerging from a ribosome, a newly made protein may spontaneously fold into its correct shape and begin "working" immediately. On the other hand, some newly made proteins must be activated before they can begin to function. One example is insulin, which is released from a ribosome in an inactive form, and only becomes an active hormone after being cleaved by an enzyme.

RECOMBINANT DNA; CLONING GENES

Recombinant DNA means taking DNA from two or more sources and combining them into one molecule. This occurs in nature through **viral transduction**, **bacterial transformation**, and **conjugation** and when **transposons** or "jumping genes" move around the genome. Scientists can also manipulate and engineer genes in vitro (in the laboratory). The branch of science that uses **recombinant DNA techniques** for practical purposes is called **biotechnology** or **genetic engineering**.

Many tools and techniques have been developed to manipulate and engineer genes. The following sections contain a discussion of the uses for genetic engineering, an explanation of some techniques that are used, and a brief discussion of some ethical issues within the field.

The potential uses of **recombinant DNA** or **gene cloning** are many:

- To produce a **protein product**, such as human insulin, in large quantities as an inexpensive pharmaceutical.
- To **replace** a **nonfunctioning gene** in a person's cells with a functioning gene by **gene therapy**. Scientists are currently conducting clinical trials in this area with disappointing results. Sometimes the human subjects become ill from the viral **vector** used to carry the gene. Other times, the gene is inserted successfully and begins to produce the necessary protein but stops working in a short time. If scientists can master this technique, many lives will be improved.
- To prepare **multiple copies of a gene** itself **for analysis**. Since most genes exist in only one copy on a chromosome, the ability to make multiple copies is of great value as a research tool.
- To **engineer bacteria** to clean up the environment. Scientists have engineered many bacteria; one modified species can even eat **toxic waste**.

The Technique of Gene Cloning

- Isolate a gene of interest, for example, the gene for human insulin.
- Insert the gene into a **plasmid**.
- Insert the plasmid into a **vector**, a cell that will carry the plasmid, such as a bacterium. To accomplish this, a bacterium must be made **competent**, which means it must be able to take up a plasmid.
- **Clone** the gene. As the bacteria reproduce themselves by **fission**, the plasmid and the selected gene are also being **cloned**. Millions of copies of the gene are produced.
- Identify the bacteria that contain the selected gene and harvest them from the culture.

TOOLS AND TECHNIQUES OF BIOTECHNOLOGY

Restriction Enzymes

Restriction enzymes were discovered in the late 1960s and are a basic biotechnology tool. They are extracted from bacteria, which use them to fend off attacks by invading bacteriophages. Restriction enzymes cut DNA at specific **recognition sequences** or **sites**, such as GAATTC. Often these cuts are staggered, leaving single-stranded **sticky ends** to form a temporary union with other sticky ends. The fragments that result from the cuts made by restriction enzymes are called **restriction fragments.**

IST-1
Be able to explain how heritable information can be manipulated using technologies.

Scientists have now isolated hundreds of different restriction enzymes. They are named for the bacteria in which they were found. Common examples are ***Eco*RI** (which was discovered in *E. coli*), ***Bam*HI**, and ***HIND*III**. Restriction enzymes have many uses, including **gene cloning.**

Gel Electrophoresis

Gel electrophoresis separates large molecules of DNA on the basis of their rate of movement through an **agarose gel** in an electric field. The smaller the molecule, the faster it runs. DNA, which is negative (due to the presence of phosphate groups, PO_4^{3-}), flows from **cathode** (–) to **anode** (+). The concentration of the gel can be altered to provide a greater impediment to the DNA, allowing for finer separation of smaller pieces.

Electrophoresis is also commonly used to separate proteins and amino acids. If DNA is going to be run through a gel, it must first be cut up by **restriction enzymes** into pieces small enough to migrate through the gel. Once separated on a gel, the DNA can be analyzed in many ways. The DNA strands can be **sequenced** to determine the sequence of bases A, C, T, and G. The gel can also be used in a comparison with other DNA samples. A **DNA probe** can also identify the location of a specific sequence within the DNA.

Figure 9.16 shows gel electrophoresis of DNA that was cut with restriction enzymes. Lane 1 has four bands of DNA: three larger pieces and one short piece. Lane 2 contains two pieces of DNA: one large one and one tiny one. Lane 3 contains one very large, uncut piece of DNA. Lane 4 contains two pieces of DNA. *The smaller the piece of DNA, the farther it has traveled from the well.*

DNA Probe

A **DNA probe** is a **radioactively labeled single strand** of nucleic acid molecule used to tag a specific sequence in a DNA sample. The probe bonds to the complementary sequence wherever it occurs, and the radioactivity enables scientists to detect its location. The DNA probe is used to identify a person who carries an inherited genetic defect, such as sickle cell anemia, Tay-Sachs disease, Huntington's disease, and hundreds of others.

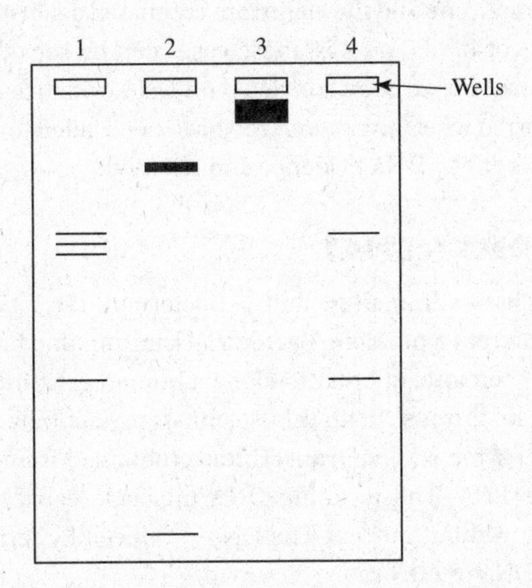

Figure 9.16 Gel Electrophoresis of DNA Fragments

Polymerase Chain Reaction (PCR)

Devised in 1985, **polymerase chain reaction** is a cell-free, automated technique by which a piece of DNA can be rapidly copied or amplified. Billions of copies of a fragment of DNA can be produced in a few hours. The DNA piece that is to be amplified is placed into a test tube with ***Taq* polymerase** (a heat-stable form of DNA polymerase extracted from extremophile bacteria), along with a supply of **nucleotides** (**A**, **C**, **T**, and **G**) and **primers** necessary for DNA synthesis. Once the DNA is amplified, these copies can be studied or used in a comparison with other DNA samples.

The PCR technique has limitations:

- Some information about the nucleotide sequence of the target DNA must be known in advance in order to make the necessary primers.
- The size of the piece that can be amplified must be very short.
- Contamination is a major problem. If a few skin cells from the technician who is working with the sample accidently contaminate the sample, that could make obtaining accurate results difficult or impossible. Such an error could have dire consequences with a crime scene sample.

Restriction Fragment Length Polymorphisms (RFLPs)

A **restriction fragment** is a segment of DNA that results when DNA is treated with restriction enzymes. When scientists compared **noncoding regions** (junk DNA) of human DNA across a population, they discovered that the **restriction fragment pattern** is different in every individual. These differences have been named **restriction fragment length polymorphisms** or **RFLPs**, pronounced "riflips." A RFLP analysis of someone's DNA gives a human **DNA fingerprint** that looks like a barcode.

Each person's RFLPs are unique, except in identical twins, and are inherited in a Mendelian fashion. Because they are inherited in this way, they can be used very accurately in **paternity**

suits to determine, with absolute certainty, if a particular man is the father of a particular child. In addition, RFLPs are routinely used to identify the perpetrator in rape and murder cases. DNA from the crime scene and the victim are compared against DNA from the suspect. Because of the accuracy of RFLP analysis, these cases can be solved with a high degree of certainty, and some suspects have been convicted on DNA evidence alone. By contrast, several incidents have occurred where prisoners who have been jailed for many years for violent crimes were proven innocent by DNA evidence and released.

Complementary DNA (cDNA)

When scientists try to clone a human gene in a bacterium, the introns (long intervening, noncoding sequences) present a problem. Bacteria lack introns and have no way to edit them out after transcription. Therefore, in order to clone a human gene in a bacterium, scientists must insert a gene with no introns. To do this, scientists extract fully processed mRNA from cells and then use the enzyme **reverse transcriptase** (obtained from **retroviruses**) to make DNA transcripts of this RNA. The resulting DNA molecule carries the complete coding sequence of interest but without introns. The DNA produced by retroviruses in this way is called **complementary DNA** or **cDNA**.

CRISPR

CRISPR stands for Clustered Regularly Interspersed Short Palindromic Repeats. It is a gene editing tool that gives scientists the ability to change an organism's DNA. CRISPR is guided by the enzyme Cas9 to modify a stretch of DNA. The whole gene editing unit is referred to as CRISPR-Cas9.

CRISPR-Cas9 was adapted from a naturally occurring genome editing system in bacteria used for protection against attacking viruses. The bacteria capture snippets of DNA from invading viruses and use them to create DNA segments known as CRISPR arrays. These arrays allow the bacteria to "remember" the viruses (or closely related ones). If the viruses attack again, the bacteria produce RNA segments from the CRISPR arrays to target the viruses' DNA. The bacteria then use Cas9 or a similar enzyme to cut the DNA apart, disabling the virus.

Jennifer Doudna and Emmanuelle Charpentier were the first to propose that CRISPR-Cas9 could be used for programmable gene editing. Their work has since been further developed by many research groups for the treatment of diseases, including sickle cell anemia, cystic fibrosis, Huntington's disease, and HIV. In November 2017, the journal *Science* reported the first time CRISPR was used to alter a person's genome: in an instance where a person was suffering from a rare metabolic disorder called Hunter syndrome.

ETHICAL CONSIDERATIONS

Many people are concerned about potential problems arising from genetic engineering. This section discusses some of those concerns.

Safety

Much of the milk available in stores comes from cows that have been given a genetically engineered bovine growth hormone (BGH) to increase the quantity of milk they produce. Many people are concerned that this hormone will find its way into the milk and cause problems for the people who drink it.

Vegetable seeds have been genetically engineered to produce special characteristics in the vegetables people eat. Again, individuals are concerned that the genes that have been inserted into the vegetables may be dangerous to those who eat them.

Privacy

DNA probes are being coupled with the technology of the semiconductor industry to produce **DNA chips** that are about $\frac{1}{2}$ inch square and can hold personal information about someone's genetic makeup. The chips scan a person for mutations in over 7,000 genes, including mutations in the immune system or the breast cancer genes (BRCA I and II) and for a predisposition to other cancers or heart attacks.

Although many people might like to know if they carry these mutations, the possibility that the personal information on a DNA chip might not remain private has caused much controversy. For example, if your health insurance company learned that you carry a harmful gene, it might not insure you or might charge you a higher premium. Similarly, if a company where you have applied for employment learns about a defect in your personal genetic makeup, based on the possibility that you might be disabled with a serious illness in the future, it might refuse to hire you.

CHAPTER SUMMARY

- This chapter advanced the discussion presented in Big Idea: IST. It details how genetic information gets from one generation to the next via DNA. It includes the historic search for heritable material, DNA structure, replication, transcription, translation, and regulation of genes.
- The work of many scientists carrying out a myriad of experiments proved that DNA, not protein, is the heritable material. You must understand and be able to analyze the work or experiments of Griffith, Avery, Franklin, Watson and Crick, and Meselson and Stahl.
- The DNA molecule consists of a **double helix**, a polymer of repeating units of **nucleotides**. Each nucleotide consists of a **sugar** (**deoxyribose**), a **phosphate**, and a **nitrogen base**: **adenine** (A), **guanine** (G), **cytosine** (C), or **thymine** (T). Inside a nucleus, DNA is packaged with a large amount of proteins called **histones** and then supercoiled into the familiar X- or Y-shaped structure we recognize as a chromosome.
- DNA replicates in a **semiconservative pattern**. Each strand serves as a template for the formation of a new strand composed of complementary nucleotides: A with T and C with G. The process requires many enzymes, including **primase** and **DNA polymerase**. The enzyme DNA polymerase also carries out proofreading by the process of **mismatch repair**.
- The sequence of nucleotides in DNA is also the template for **protein synthesis**. Through the processes of **transcription** in the nucleus and **translation** at ribosomes, genetic information is transmitted with a high degree of fidelity. Errors, called **mutations**, randomly occur. Remember that changes in an organism's genotype may alter its phenotype and make the organism more or less adapted to its environment. Understand how to use the genetic code shown on page 177 to determine what polypeptide would form from a particular mRNA sequence.

- Focus on **regulation** of **gene expression**. Gene expression can be altered *before transcription* (by altering **chromatin structure** or by **methylating DNA**), *during transcription* (how, in an **operon**, RNA polymerase binds to or is blocked from binding to a promoter), or *after transcription* (by **alternate RNA slicing** and **RNA processing**).
- Understand tools and techniques of **biotechnology** and how you would employ them to do an experiment or solve a problem. For example, you are shown an image of an electrophoresis gel containing bands of DNA from a child and from a man whom the child's mother claims is the child's biological father. Can you determine if the man is the father by comparing the two sets of bands? Can you also explain why your answer is correct?

PRACTICE QUESTIONS

1. Which of the following enhances the expression of a gene?

   (A) adding methyl groups to DNA
   (B) packaging chromatin more tightly
   (C) the attachment of microRNA to a sequence of DNA
   (D) the attachment of RNA polymerase to DNA

2. Which statement describes when the lactose operon is turned on and transcribing?

   (A) Active repressor binds to the operator.
   (B) RNA polymerase binds to the operator.
   (C) Allolactose or lactose binds to and changes the shape of the repressor.
   (D) A regulator gene produces the repressor.

3. A researcher was studying genetic traits in bacteria. He used two strains of the bacterium *S. pneumoniae.* X-strain is a pathogen that is known to cause pneumonia in mice because it has an outer capsule that protects it from the mouse's immune system. Y-strain lacks the gene for the protective capsule and does not cause pneumonia in mice. To confirm this, the researcher injected two different mice with the two different strains of bacteria. The mouse that was injected with X-strain grew sick and died. The other mouse, who was injected with Y-strain, remained healthy.

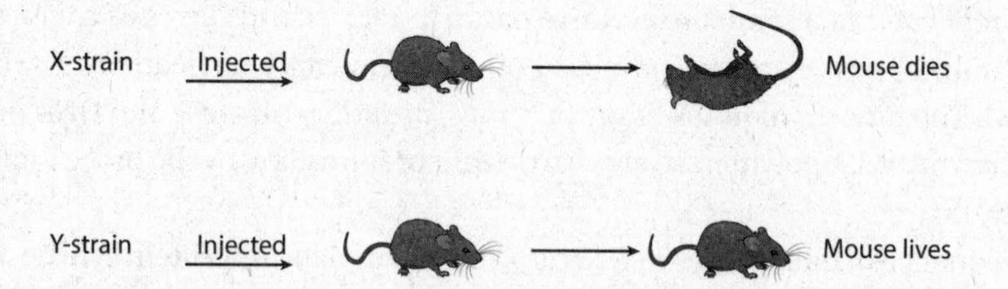

   Based on his initial findings, the researcher carried out another experiment. His hypothesis was that there can be a horizontal transfer of genes between and among bacteria. He injected an X-strain of *S. pneumoniae* that had been heat-killed into a mouse. This time, the mouse remained healthy. Next, he injected a mouse with the heat-killed X-strain along with the living Y-strain.

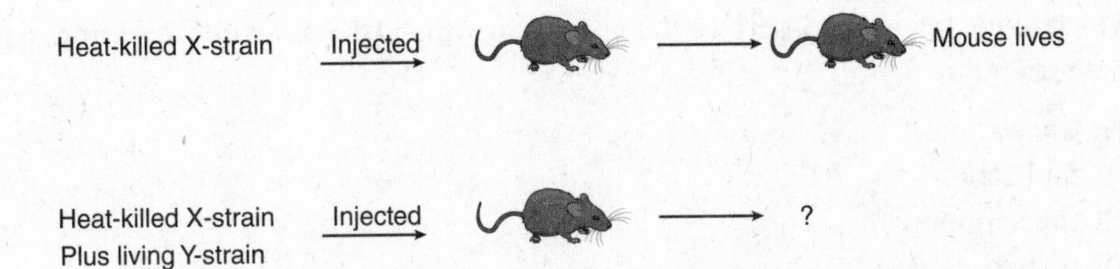

Assume that the researcher's hypothesis was correct and that bacteria can transfer genetic material from one bacterium to another. Predict what happened when the heat-killed X-strain plus the living Y-strain was injected into the mouse.

(A) The mouse lived, and a biopsy revealed there was live Y-strain bacteria in the mouse's bloodstream.
(B) The mouse lived, and a biopsy revealed there was live X-strain bacteria in the mouse's bloodstream.
(C) The mouse died, and an autopsy revealed there was live Y-strain bacteria in the mouse.
(D) The mouse died, and an autopsy revealed there was live X-strain bacteria in the mouse.

4. The number of active copies of a gene is important for proper development. Which of the following describes a normal process to establish proper dosage in a cell?

(A) methylation of the X chromosome
(B) alternative splicing of an RNA sequence
(C) transduction as described by Griffith
(D) the presence of reverse transcriptase in a eukaryotic cell

5. Predict which of the following mutations would likely be most harmful to an organism.

(A) a deletion of three nucleotides within the first intron
(B) a single point substitution
(C) a single nucleotide deletion close to the start of the coding sequence
(D) a single nucleotide deletion close to the end of the coding sequence

6. The findings from the Human Genome Project reveal that humans have only about 22,000 genes, the same as a roundworm. Which of the following best explains how the more complex humans have so few genes?

(A) More than one polypeptide can be produced from one gene.
(B) The human genome consists of a high proportion of exons.
(C) The human genome has a large number of spliceosomes.
(D) In humans, the mRNA produced by transcription does not get broken down but translates into polypeptides without ceasing.

7. Which of the following describes the process by which DNA makes messenger RNA?

(A) translation
(B) replication
(C) transcription
(D) transformation

8. Which of the following describes the process by which DNA is synthesized from a template strand?

   (A) translation
   (B) replication
   (C) transcription
   (D) transformation

9. Which of the following describes the process by which foreign DNA is taken up by a bacterial cell?

   (A) translation
   (B) replication
   (C) transcription
   (D) transformation

10. Which of the following describes the process by which a polypeptide strand is synthesized using mRNA as a template?

   (A) translation
   (B) replication
   (C) transcription
   (D) transformation

Questions 11–13

The following three questions are based on your knowledge of gel electrophoresis and this image of a gel:

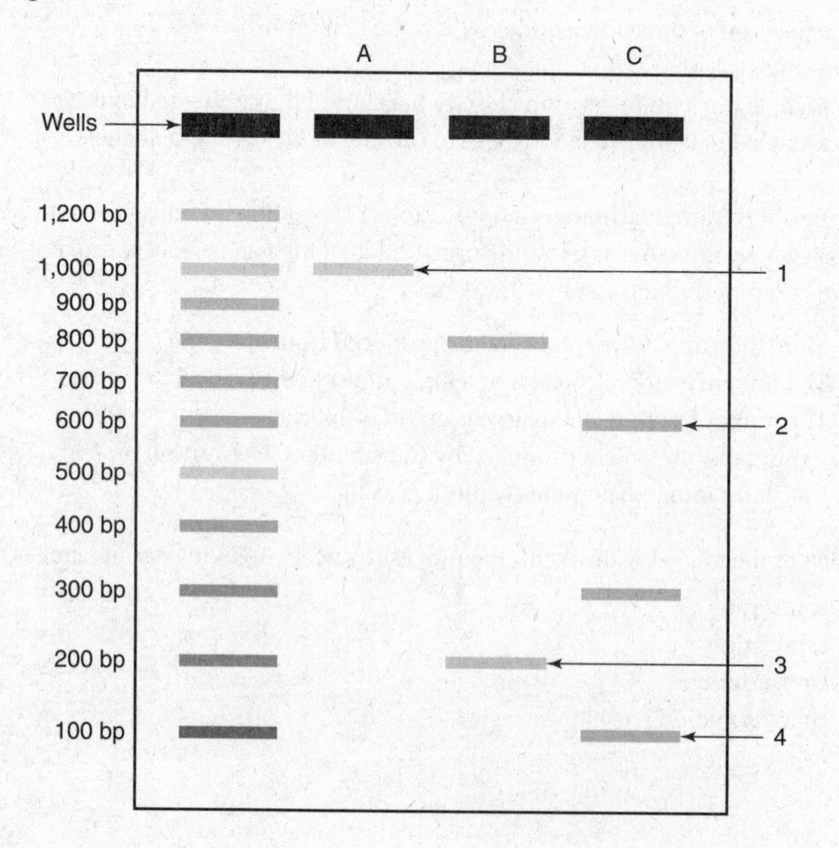

11. What is the function of a DNA ladder?
    (A) It points to the longest piece of DNA.
    (B) It points to the shortest piece of DNA.
    (C) It represents all the possible restriction fragments that can be found on that gel.
    (D) It can be used to identify the size of each restriction fragment in the gel by comparing the fragment to the ladder.

12. Which of the DNA fragments is the shortest on this gel?
    (A) 1
    (B) 2
    (C) 3
    (D) 4

13. There are three bands in Lane C. Which of the following is correct about the number of restriction sites in the DNA placed in the well in Lane C?
    (A) There were no restriction sites on this piece of DNA.
    (B) There were only two restriction sites.
    (C) There were only three restriction sites.
    (D) There were somewhere between two and three restriction sites.

---

14. Which of the following is correct about *Eco*RI?
    (A) It is one of the most-studied bacteriophages.
    (B) It is one of the many species of bacteria that live in the human intestine.
    (C) It is a type of DNA that is used extensively in bioengineering.
    (D) It is an enzyme that cuts DNA into fragments at specific recognition sites.

15. Which of the following statements is correct about the role of DNA polymerase in DNA replication?
    (A) DNA polymerase brings two separated strands of DNA back together after new strands form.
    (B) After the parental strands of DNA separate, DNA polymerase prevents them from reattaching.
    (C) DNA polymerase adds nucleotides to the 3 end of a growing strand of DNA.
    (D) DNA polymerase is an enzyme that untwists the double helix at the replication fork.

16. Which of the following is an example of wobble?
    (A) amino acids are carried to the ribosome to form a polypeptide chain
    (B) the excision of introns from mRNA
    (C) the binding of a primer to DNA
    (D) four codons can all code for the same amino acid

17. Which of the following is correct regarding sickle cell anemia?
    (A) It is caused by a chromosome mutation that resulted from nondisjunction.
    (B) It is common in people from the Middle East.
    (C) It is caused by a point mutation.
    (D) A person with sickle cell anemia is resistant to many other genetic disorders.

18. If AUU is the codon, what is the anticodon?

(A) AUU
(B) TAA
(C) UUA
(D) UAA

19. In 1952, Alfred Hershey and Martha Chase carried out a now famous experiment that proved that DNA, not protein, is the heritable material. Their model organism was the T2 bacteriophage, a phage that infects *E. coli* bacteria that are normally present in the human intestine.

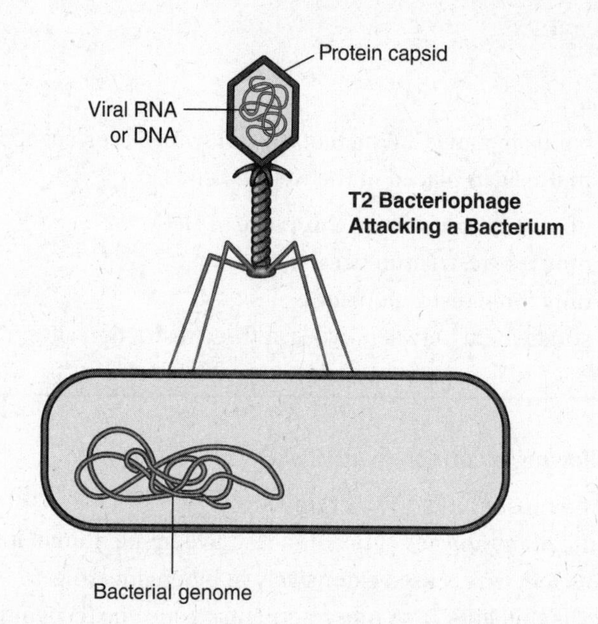

Hershey and Chase knew that when the T2 bacteriophage infected an *E. coli* cell, the phage viruses somehow reprogrammed the bacteria to produce thousands of T2 viruses. It was clear that the T2 bacteriophage transmitted some material into the bacteria. However, Hershey and Chase had to demonstrate how this transfer was accomplished. Here is essentially what they did. They set up two phage cultures: one growing in radioactive sulfur (^{35}S), and the other growing in radioactive phosphorus (^{32}P). The protocol is shown in the following chart.

	Step 1	Step 2	Step 3	Step 4
Group 1	Grow bacteriophage in radioactive sulfur (^{35}S).	Allow culture to infect *E. coli.*	Separate the viruses remaining on the bacterial cell surfaces from the cells themselves by using a blender.	Spin the bacteria into a pellet, and analyze it for the presence of radioactive substance.
Group 2	Grow bacteriophage in radioactive phosphorus (^{32}P).	Allow culture to infect *E. coli.*	Separate the viruses remaining on the bacterial cell surfaces from the cells themselves by using a blender.	Spin the bacteria into a pellet, and analyze it for the presence of radioactive substance.

Using the description of this experiment and what you know about DNA and proteins, which of the following statements correctly describes what radioactive substance will be identified inside the bacterial cells and provides the correct reasoning for that claim?

(A) ^{32}P will be found in the bacteria cells because protein from the T2 bacteriophage coat entered the bacteria.
(B) ^{32}P will be found in the bacteria cells because DNA from the T2 bacteriophage interior entered the bacteria.
(C) ^{35}S will be found in the bacteria cells because protein from the T2 bacteriophage coat entered the bacteria.
(D) ^{35}S will be found in the bacteria cells because protein from the T2 bacteriophage interior entered the bacteria.

20. If a segment of DNA is 5′-TGA AGA CCG-3′, the RNA that results from the transcription of this segment will be

(A) 5′-TGA AGA CCG-3′
(B) 3′-ACU UCU GGC-5′
(C) 3′-ACT TCT GGC-5′
(D) 3′-CGG UCU UCA-5′

21. Once transcribed, eukaryotic RNA normally undergoes substantial alteration that results primarily from

(A) removal of exons
(B) removal of introns
(C) addition of introns
(D) combining of RNA strands by a ligase

22. Which of the following contain one pyrimidine and one purine?

(A) adenine and guanine
(B) uracil and thymine
(C) cytosine and uracil
(D) adenine and cytosine

23. Which of the following acts as a primer that initiates the synthesis of a new strand of DNA?

(A) single-strand binding protein
(B) RNA
(C) DNA
(D) topoisomerases

24. If guanine makes up 28% of the nucleotides in a sample of DNA from an organism, then thymine would make up ____ % of the nucleotides.

(A) 28
(B) 56
(C) 22
(D) 44

Questions 25–27

The following three questions refer to the codon table for mRNA that was deciphered in the 1960s by Marshall Nirenberg:

FIRST BASE (5' end)	SECOND BASE: U		C		A		G		THIRD BASE (3' end)
U	UUU	Phe	UCU	Ser	UAU	Tyr	UGU	Cys	U
U	UUC	Phe	UCC	Ser	UAC	Tyr	UGC	Cys	C
U	UUA	Leu	UCA	Ser	UAA	Stop	UGA	Stop	A
U	UUG	Leu	UCG	Ser	UAG	Stop	UGG	Trp	G
C	CUU	Leu	CCU	Pro	CAU	His	CGU	Arg	U
C	CUC	Leu	CCC	Pro	CAC	His	CGC	Arg	C
C	CUA	Leu	CCA	Pro	CAA	Gin	CGA	Arg	A
C	CUG	Leu	CCG	Pro	CAG	Gin	CGG	Arg	G
A	AUU	Ile	ACU	Thr	AAU	Asn	AGU	Ser	U
A	AUC	Ile	ACC	Thr	AAC	Asn	AGC	Ser	C
A	AUA	Ile	ACA	Thr	AAA	Lys	AGA	Arg	A
A	AUG	Met or start	ACG	Thr	AAG	Lys	AGG	Arg	G
G	GUU	Val	GCU	Ala	GAU	Asp	GGU	Gly	U
G	GUC	Val	GCC	Ala	GAC	Asp	GGC	Gly	C
G	GUA	Val	GCA	Ala	GAA	Glu	GGA	Gly	A
G	GUG	Val	GCG	Ala	GAG	Glu	GGG	Gly	G

25. Which is the correct amino acid sequence that would be produced from this section of DNA?

GUACUACCCAGUCGC

(A) Leu – Ala – Thy – Ser – Arg
(B) Val – Leu – Pro – Ser – Arg
(C) Val – Trp – Pro – Phe – Gly
(D) Glu – Arg – Gly – Pro – Val

26. The cell that carries this codon sequence (. . . AGU UUU GUU GGU . . .) sustains two mutations at the bold nucleotides. The new codon sequence reads as follows: . . . AGC UUU GUU GGG . . . Which of the following statements correctly explains the results from such mutations?

(A) The amino acid sequence will be altered and may be passed to offspring if the mutation occurred in gamete-forming cells.
(B) The amino acid sequence will be altered and will result in a condition like the sickle cell trait, which is caused by a point mutation.
(C) The amino acid sequence will remain the same because multiple codons code for the same amino acid.
(D) The amino acid sequence will remain the same because 15% of all codons code for the same amino acid.

27. The following is a codon sequence in a cell: . . .UGU CCC UAC AGU . . . What would happen to the amino acid sequence if the codon sequence sustains one mutation and the bold nucleotide becomes A?

   (A) The amino acid sequence will remain unchanged because multiple codons code for the same amino acid.
   (B) The amino acid sequence will remain unchanged because almost every mutation is corrected by DNA proofreading.
   (C) The cell will be transformed into a cancer cell.
   (D) No amino acid sequence will be produced because UAA is a Stop codon, and translation will not continue.

---

28. A particular triplet code on DNA is AAA. What is the anticodon for it?

   (A) AAA
   (B) TTT
   (C) UUU
   (D) CCC

29. Which of the following is correct about DNA?

   (A) Almost 97% of the human genome are genes that code for polypeptides.
   (B) The regions that code for polypeptides are called introns.
   (C) The parts of DNA that do not code for polypeptides are regulatory sequences that control when genes turn on and off.
   (D) The human genome consists of about 100,000 genes.

30. Which one of the following statements is correct regarding the replication of DNA?

   (A) It occurs during metaphase of the cell cycle.
   (B) It occurs during anaphase of the cell cycle.
   (C) It occurs during prophase of the cell cycle.
   (D) It occurs during interphase of the cell cycle.

31. Which one of the following statements is correct regarding mad cow disease?

   (A) It is caused by a misfolded, infectious protein called a prion.
   (B) It is caused by a mutated virus.
   (C) It is caused by a bacteriophage.
   (D) It is caused by bacteria similar to *E. coli*, that normally lives in the human gut.

32. Which of the following acts as an inducer in the *lac* operon?

   (A) lactose
   (B) repressor
   (C) regulator
   (D) promoter

33. Which of the following serves as the binding site for RNA polymerase?

(A) lactose
(B) repressor
(C) regulator
(D) promoter

34. Which of the following codes for the repressor?

(A) lactose
(B) repressor
(C) regulator
(D) promoter

35. Which of the following binds to the operator?

(A) lactose
(B) repressor
(C) regulator
(D) promoter

36. Which of the following statements is correct regarding biotechnology techniques?

(A) PCR is used to cut DNA molecules.
(B) A DNA probe consists of a radioactive single strand of DNA.
(C) Restriction enzymes were first discovered in bacteriophage viruses.
(D) *Eco*RI is a name for a DNA probe.

37. With respect to sex chromosomes, human females have two X chromosomes, while human males have an X chromosome and a Y chromosome. What prevents females from having twice as many X chromosome gene products as males, who only possess a single copy of the X chromosome?

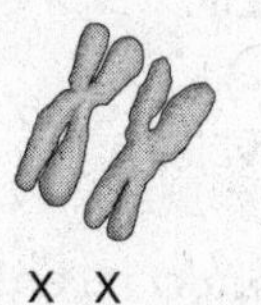

(A) Since they are larger, the male's Y chromosome contains two sets of X genes.
(B) The male's Y chromosome has evolved the necessary X genes.
(C) One of the female's X chromosomes in each body cell is inactivated.
(D) Even though the X chromosome is larger than the Y chromosome, one of the female's X chromosomes contains only recessive genes that do not exert much influence over the dominant genes of the other X chromosome.

38. Watson and Crick described DNA as a double helix that consists of two antiparallel strands. One strand runs 5′ to 3′. The other runs 3′ to 5′. Which of the following statements is correct about the 5′ end of the strand of DNA?

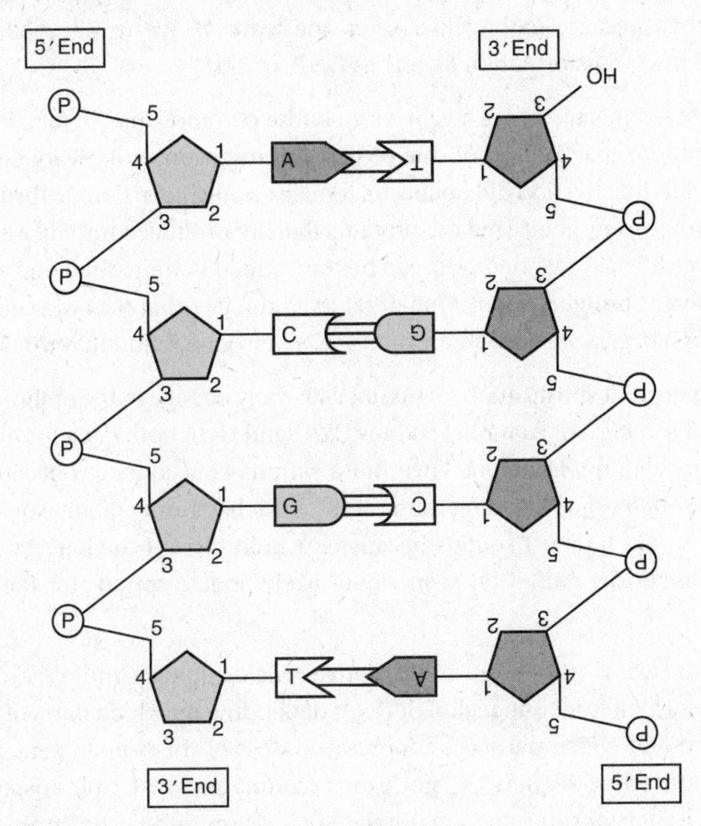

(A) It is the end to which the nucleotide is attached.
(B) It is the end to which the sugar is attached.
(C) It is the end to which the phosphate is attached.
(D) It is the end to which the hydroxyl is attached.

39. Which enzyme permanently seals together DNA fragments that have complementary sticky ends?

(A) DNA polymerase
(B) single-stranded binding protein
(C) reverse transcriptase
(D) DNA ligase

Answers Explained

1. **(D)** Choices A, B, and C describe events that suppress transcription and translation. In order for a gene to transcribe, RNA polymerase must bind to the promoter. (IST-1)

2. **(C)** Transcription occurs when RNA polymerase binds to the promoter, not the operator. Transcription is blocked when an active repressor binds to the operator. When allolactose binds to the repressor, it prevents the repressor from binding to the operator. This enables RNA polymerase to bind to the promoter, which turns *on* the operon. Choice D is incorrect because, although it is a true statement, that statement does not say anything about transcription. (IST-1)

3. **(D)** This question refers to the groundbreaking "transformation" experiment carried out by Frederick Griffith in 1928. He proved that some unknown heritable substance from the dead X-strain bacteria was transferred to the harmless Y-strain cells. The Y-strain then became pathogenic. Later, the work of Avery, MacLeod, and McCarty identified this "transformation factor" as DNA. (IST-1)

4. **(A)** Methylation suppresses a gene or an entire chromosome. In fact, it is responsible for the random inactivation of one of the X chromosomes in every cell in a human female. If this inactivation did not occur, females would have double the normal/necessary X chromosome genes and the proteins that are produced from it. Alternative splicing involves the fact that one gene can be transcribed in more than one way, producing many different proteins. Transduction refers to the fact that bacteria can uptake genes from other bacteria. Reverse transcriptase is an enzyme found in retroviruses. (IST-2)

5. **(C)** A single point substitution might not cause any change in the amino acid sequence because of *wobble.* For example, codons UUU and UUA both code for the same amino acid, phenylalanine. Deletions have more serious consequences because they cause a *reading frame shift.* A reading frame shift at the beginning of an exon or of a coding sequence would be most serious because it would disrupt the lengthy sequence that follows. One at the end of an exon would likely be less serious for the same reason. (IST-4)

6. **(A)** This answer is true because *alternative splicing* enables more than one polypeptide to be formed from one region of DNA, depending on which parts of the region are transcribed and which parts are suppressed. Most of the human genome consists of introns (noncoding sequences), not exons (coding regions). Spliceosomes are structures that remove introns and are unrelated to the question. Choice D is not a correct statement; mRNA gets broken down after it has been translated. This degradation is a mechanism that controls how much protein is produced. (IST-2)

7. **(C)** Transcription is the process by which DNA makes RNA. There are three types of RNA: mRNA, tRNA, and rRNA. Transcription occurs in three stages: initiation, elongation, and termination. (IST-1)

8. **(B)** DNA makes an exact copy of itself during replication. This process occurs in a semiconservative fashion, as proven by Meselson and Stahl. (IST-1)

9. **(D)** Griffith was the first to recognize the phenomenon of transformation while working with pneumococcus bacteria. (IST-1)

10. **(A)** Translation is the process by which the codons of mRNA sequence are changed into an amino acid sequence. Amino acids present in the cytoplasm are carried by tRNA molecules to the codons of the mRNA strand at the ribosome according to the base-pairing rules (A with U and C with G). (IST-1)

11. **(D)** Ladders are usually included in a gel because they are calibrated and identify the size of any fragment on the gel by comparing the fragment to the ladder. (IST-1 & SP 2)

12. **(D)** The shorter a fragment is, the farther and faster it will run from the well at the top of the gel in this image. (IST-1)

13. **(B)** Since there are three bands, that means that there were two restriction sites in the DNA before it was run through the gel. Review the following sketch of a stretch of DNA with two restriction sites.

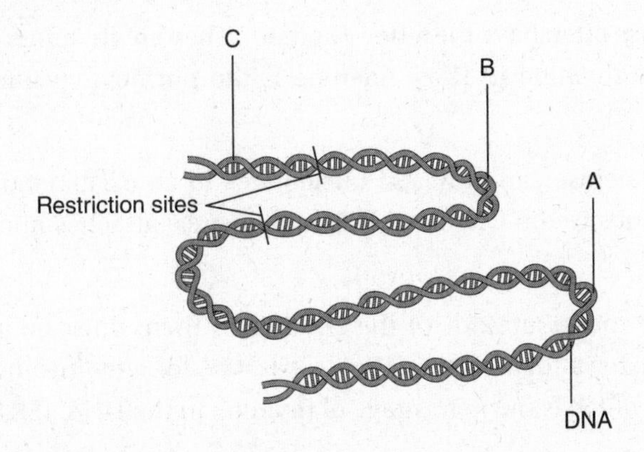

Each will produce one band on the gel. In this case, there is one large piece (A), one smaller piece (B), and one very small piece (C). (IST-1)

14. **(D)** *Eco*RI stands for *E. coli* restriction enzyme #1. It was the first restriction enzyme discovered. It acts as molecular scissors, cutting DNA at certain recognition sites. Choice A is not correct because that choice is referring to T2 bacteriophage. Choice B is not correct because that choice is referring to the bacterium *E. coli*, not the restriction enzyme. Choice C is not correct because *Eco*RI is not DNA. (IST-1)

15. **(C)** DNA polymerase adds new nucleotides to the 3′ end of a growing strand of DNA. It cannot add nucleotides to a naked strip of DNA. There must be an RNA primer there first. Choice A is referring to ligase. Choice B is referring to a single-stranded binding protein. Choice D is referring to a helicase enzyme. (IST-1 & SP 2)

16. **(D)** The pairing rules are not as strict for the third codon in mRNA as they are for the first two. One example is that UUU and UUA both code for phenylalanine. (IST-1)

17. **(C)** Sickle cell anemia is caused by a gene mutation in the gene that codes for hemoglobin. Sickle cell disease is common where malaria is endemic: in West Africa and Southeast Asia. People who are carriers for the sickle cell trait are resistant to malaria. Sickle cell disease does occur in the Middle East, but it is not common there. Sickle cell disease does occur in Caucasians, but only rarely. (IST-2 & IST-4)

18. **(D)** The codon is the nucleotide triplet associated with mRNA; the anticodon is the nucleotide sequence associated with tRNA. Codons and anticodons are complementary to each other. (IST-1)

19. **(B)** DNA consists of nucleotides made of phosphates, sugars, and nitrogen bases. It was the radioactive ^{32}P part of DNA from inside the bacteriophage that entered and infected the *E. coli*. That fact proved that DNA is the heritable material. The radioactive ^{35}S from the coat of the bacteriophage did not enter the *E. coli*. (SP 2)

20. **(B)** You are given a strand of DNA, which makes a strand of mRNA. Follow the base-pairing rules: T with A, C with G, C with G, and A with U. Remember, RNA contains uracil instead of thymine. If the DNA segment is 5′-TGA AGA CCG-3′, then the mRNA strand complementary to that is 3′-ACU UCU GGC-5′. (IST-1)

21. **(B)** Once transcription has occurred, the new RNA molecule undergoes RNA processing. During this process, introns (intervening sequences) are removed with the help of snRNPs, and a 5′ cap and poly(A) tail are added. (IST-1)

22. **(D)** Pyrimidines often have the letter *y* in them. They are thymine, cytosine, and uracil, which replaces thymine in RNA. Adenine is the purine; cytosine is the pyrimidine. (IST-1)

23. **(B)** DNA polymerase can only add nucleotides to an existing strand of nucleotides. RNA primer binds to the DNA, and DNA polymerase attaches nucleotides to the RNA primer. (IST-1)

24. **(C)** If guanine makes up 28% of the DNA, then there must be an equal amount of cytosine (28%), for a total of 56%. That leaves 44% for adenine and thymine. Dividing 44 by 2 = 22%, which is the percentage of thymine in the DNA. (SP 5)

25. **(B)** The sequence consists of codons. Read them off the chart provided to determine the amino acid for that codon. GUA = Val, CUA = Leu, etc. None of the other choices fit the correct pattern. (IST-1)

26. **(C)** First, establish if there is any change in the amino acids due to the mutation in the codon. AGU codes for the amino acid ser (serine), as does the mutant, AGC. Also, GGU codes for the amino acid gly (glycine), as does the mutated version, GGG. Therefore, there is no change in the amino acid sequence because multiple codons produce the same amino acid. (IST-4)

27. **(D)** In this case, UAC codes for Tyr, while UAA, the mutated version, codes for a Stop codon. The cell stops reading the mRNA strand, and no polypeptide will be produced. (IST-4)

28. **(A)** DNA (triplet code) makes RNA (codon), which makes protein (anticodon). If the triplet in DNA is AAA, then the codon on mRNA is UUU, and the anticodon on tRNA is AAA. (IST-1)

29. **(C)** Only 1.0% of DNA codes for proteins. Most of the remaining DNA, about 75% of it, are regulatory sequences that are involved with gene regulation. (IST-2)

30. **(D)** Replication of DNA occurs during the S (synthesis) phase of interphase. (IST-1)

31. **(A)** Prions are infectious proteins. They can turn normal proteins into misfolded ones. (IST-1)

32. **(A)** Lactose, or allolactose (an isomer of lactose), is the inducer that facilitates transcription by first binding to and deactivating the repressor. With the repressor deactivated, RNA polymerase can bind to the promoter, and transcription of the enzymes (to digest lactose) can occur. (IST-1)

33. **(D)** The promoter is the binding site for RNA polymerase. When RNA polymerase binds to the promoter, transcription occurs. (IST-1)

34. **(C)** The regulator gene codes for the repressor, a protein that inhibits transcription. In the *lac* operon, when the repressor binds to the operator, no transcription can occur. (IST-1)

35. **(B)** The repressor inhibits transcription by binding to the operator and preventing RNA polymerase from binding to the promoter. (IST-1)

36. **(B)** A DNA probe is a radioactive single strand of DNA used to tag and identify a specific sequence in a strand of DNA. PCR is a cell-free system that amplifies small pieces of DNA rapidly. Restriction enzymes are found in bacteria. *Eco*RI was the first restriction enzyme discovered. Every person has a unique set of RFLPs. (IST-1)

37. **(C)** In every cell of the human female, one of the X chromosomes is inactivated (probably by methylation). If this were not so, females would have double the amount of X chromosome gene products than males. The inactivated X chromosomes can be seen under a microscope as a condensed spot near the nuclear membrane called a Barr body. In any one cell, the X chromosome that is condensed is random. Overall, in the three trillion cells that make up the human body, half of them will have one active X chromosome and the other half will have an inactive X chromosome. (IST-2)

38. **(C)** Examine the figure provided with the question. You should notice that phosphate marks the 5 end and OH marks the 3′ end. (IST-1)

39. **(D)** DNA ligase seals together DNA fragments that have complementary sticky ends. (IST-1)

Biological Diversity 10

→ THE THREE-DOMAIN CLASSIFICATION SYSTEM
→ EVOLUTIONARY TRENDS IN ANIMALS
→ PHYLOGENETIC TREES

Big Idea: EVO
Enduring Understandings: EVO-1, EVO-2 & EVO-3
Science Practice: 4

For the complete list of Big Ideas, Enduring Understandings, and Science Practices, refer to the "AP Biology Course and Exam Description" from the College Board: *https://apcentral.collegeboard.org/pdf/ap-biology-course-and-exam-description .pdf?course=ap-biology*

INTRODUCTION

Taxonomy or **classification** is the naming and classification of species. It began in the eighteenth century when Carolus Linnaeus developed the system used today, the **system of binomial nomenclature**. This system has two main characteristics: a two-part name for every organism (for example, human is *Homo sapiens* and lion is *Panthera leo*), and a hierarchical classification of species into broader groups of organisms. These broader groups, or **taxa**, in order from the general to the specific are **kingdom**, **phylum**, **class**, **order**, **family**, **genus**, and **species**.

THE THREE-DOMAIN CLASSIFICATION SYSTEM

Over the years, the classification system that scientists use has changed several times. Currently, scientists use a system based on DNA analysis that accurately reflects evolutionary history, **phylogeny**, and the relationships among organisms. Our current system reflects the idea that all organisms descend from a common ancestor. This system is called the **three-domain system**. In it, all life is organized into three domains: **Bacteria**, **Archaea**, and **Eukarya**.

EVO-2
Our classification system reflects evolutionary relationships.

Figure 10.1 shows the organization of the current three-domain system of classification. (*The term **Monera** is no longer used in this system. Instead, prokaryotes are spread across two different domains: Archaea and Bacteria.*)

Table 10.1 shows a comparison of Bacteria, Archaea, and Eukaryotes. Notice, for some characteristics, the archaea resemble eukaryotes more than they resemble prokaryotes.

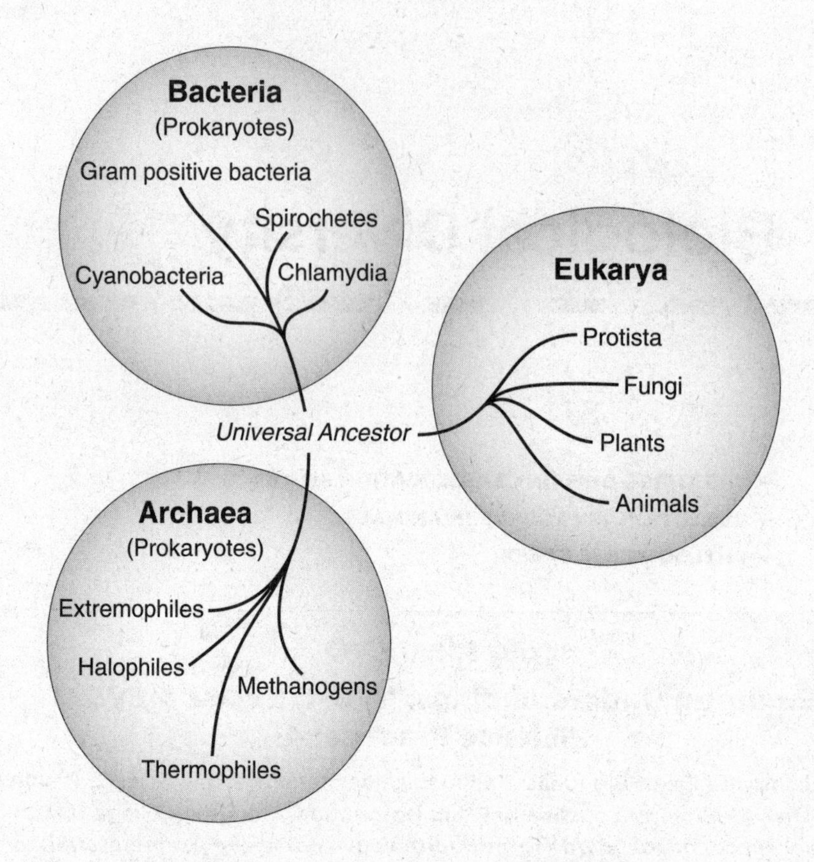

Figure 10.1 The Three-Domain Classification System

Table 10.1

Bacteria, Archaea, and Eukaryotes			
Feature	**Bacteria**	**Archaea**	**Eukaryotes**
Membrane-enclosed organelles	Absent	Absent	Present
Peptidoglycan in cell wall	Present	Absent	Absent
RNA polymerase	One type	Several kinds	Several kinds
Introns (noncoding regions of genes)	Absent	Present in some genes	Present
Antibiotic sensitivity to streptomycin, chloramphenicol	Inhibited	Not inhibited	Not inhibited

Domain Bacteria

- All are single-celled **prokaryotes** with no internal membranes (no nucleus, mitochondria, or chloroplasts).
- Some are anaerobes; some are aerobes.
- Bacteria play a vital role in the ecosystem as **decomposers** that recycle dead organic matter.
- Many are **pathogens**, causing disease.
- Bacteria play a vital role in **genetic engineering**. The bacteria from the human intestine, *Escherichia coli*, are transformed to manufacture human insulin.
- Some bacteria carry out **conjugation**, a primitive form of sexual reproduction where individuals exchange genetic material.
- They have a thick, rigid cell wall containing a substance known as **peptidoglycan**.
- Some carry out photosynthesis, but others do not.
- No introns (noncoding regions within the DNA).
- Corresponds roughly to the old grouping Eubacteria and includes blue-green algae, bacteria like *E. coli* that live in the human intestine, those that cause disease like *Clostridium botulinum* and *Streptococcus*, and those necessary in the nitrogen cycle, like nitrogen-fixing bacteria and nitrifying bacteria.
- Viruses are placed here because we do not know where else to place them.

Domain Archaea

- Unicellular
- Prokaryotic—no internal membranes such as a nucleus
- Includes **extremophiles**, organisms that live in extreme environments, like

  1. **METHANOGENS**—obtain energy in a unique way by producing methane from hydrogen
  2. **HALOPHILES**—thrive in environments with high salt concentrations like Utah's Great Salt Lake
  3. **THERMOPHILES**—thrive in very high temperatures, like in the hot springs in Yellowstone Park or in deep-sea hydrothermal vents

- Introns present in some genes
- No peptidoglycan

Domain Eukarya

- All organisms have a nucleus and internal organelles
- No peptidoglycan in cells
- Includes the four remaining kingdoms: Protista, Fungi, Plantae, and Animalia (see Table 10.2)

Today, taxonomy as a separate discipline is being replaced by the study of **systematics**, which includes taxonomy but considers biological diversity in an evolutionary context. Systematics focuses on tracing the ancestry of organisms. This is particularly important in light of current advances in DNA techniques that allow scientists to compare two species at the molecular level.

Table 10.2

The Four Kingdoms of Eukarya	
Kingdom	**Characteristics**
Protista	Includes the widest variety of organisms, but all are eukaryotes Includes organisms that do not fit into the Fungi or Plantae kingdoms, such as seaweeds and slime molds Consists of single and primitive multicelled organisms Includes heterotrophs and autotrophs Ameba and paramecium are heterotrophs Euglena are primarily autotrophic with red eyespots and chlorophyll to carry out photosynthesis Protozoans like ameba and paramecium are classified by how they move Mobility by varied methods: ameba—pseudopods; paramecium—cilia; euglena—flagella Some carry out conjugation, a primitive form of sexual reproduction Some cause serious diseases like amebic-dysentery and malaria
Fungi	All are heterotrophs and eukaryotes Secrete hydrolytic enzymes outside the body where extracellular digestion occurs; then the building blocks of the nutrients are absorbed into the body of the fungus by diffusion Are important in the ecosystem as decomposers Cell walls are composed of chitin, not cellulose Examples: yeast, mold, mushrooms, the fungus that causes athlete's foot
Plantae	All are autotrophic eukaryotes Some plants have vascular tissue (Tracheophytes); some do not have any vascular tissue (Bryophytes) Examples: mosses, ferns, cone-bearing and flowering plants
Animalia	All are heterotrophic, multicellular eukaryotes Are grouped in 35 phyla, but this book discusses 9 main phyla: Porifera, Cnidaria, Platyhelminthes, Nematoda, Annelida, Mollusca, Arthropoda, Echinodermata, and Chordata Most animals reproduce sexually with a dominant diploid stage In most species, a small, flagellated sperm fertilizes a larger, nonmotile egg This category is monophyletic, meaning all animal lineages can be traced back to one common ancestor Classified by anatomical features (homologous structures), DNA data, and embryonic development

EVOLUTIONARY TRENDS IN ANIMALS

> **EVO-2**
> Organisms are linked by lines of descent from common ancestry.

Life began as tiny, primitive, single-celled organisms in the oceans. The first **multicellular** eukaryotic **organisms** evolved about 1.5 billion years ago. The appearance of each phylum of animal represents the evolution of a new and successful body plan. These important trends include: specialization of tissues, germ layers, body symmetry, cephalization, and body cavity formation. Specifics of these trends are summarized in the sections that follow. Table 10.3 also summarizes this information.

Table 10.3

Trends in Animal Development from the Primitive to the Complex	
From the Primitive	**To the Complex**
No symmetry or radial symmetry with little or no sensory apparatus	Bilateral symmetry with a head end and complex sensory apparatus
No cephalization	Cephalization
Two cell layers: ectoderm and endoderm (diploblastic)	Three cell layers: ectoderm, mesoderm, and endoderm (triploblastic)
No true tissues	True tissues, organs, and organ systems
Life in water	Life on land and all the modification it requires
Sessile	Motile
Few organs, but no organ systems	Many organ systems and much specialization

Specialized Cells, Tissues, and Organs

We need to begin with some definitions.

- The **cell** is the basic unit of all forms of life. A neuron is a cell.
- A **tissue** is a group of similar cells that perform a particular function. The sciatic nerve is a tissue.
- An **organ** is a group of tissues that work together to perform related functions. The brain is an organ.

Sponges (Porifera) consist of a loose federation of cells, which are not considered tissue because the cells do not function as a unit. However, they possess cells that can sense and react to the environment.

Cnidarians, like the hydra and jellyfish, possess only the most primitive and simplest forms of tissue.

As larger and more complex animals evolved, specialized cells joined to form real tissues, organs, and organ systems. Flatworms have organs, but no organ systems.

More complex animals, like annelids (segmented worms) and arthropods, have organ systems.

Germ Layers

Germ layers are the main layers that form various tissues and organs of the body. They are formed early in embryonic development as a result of gastrulation. Complex animals are **triploblastic**. They consist of the ectoderm, endoderm, and mesoderm.

- The **ectoderm**, or outermost layer, becomes the skin and nervous system, including the nerve cord and brain.
- The **endoderm**, the innermost layer, becomes the viscera (guts) or the digestive system.
- The **mesoderm**, middle layer, becomes the blood and bones.

Primitive animals, like the **Porifera** and **Cnidarians**, have only two cell layers and are called **diploblastic**. Their bodies consist of ectoderm, endoderm, and **mesoglea** (middle glue), which connects the two layers together.

Bilateral Symmetry

Whereas primitive animals exhibit no symmetry or radial symmetry (see Figure 10.3), most sophisticated animals exhibit **bilateral symmetry** (see Figure 10.2). The echinoderms seem to be an exception because they exhibit bilateral symmetry only as larvae and revert to radial symmetry as adults. In bilateral symmetry, the body is organized along a **longitudinal axis** with right and left sides that mirror each other.

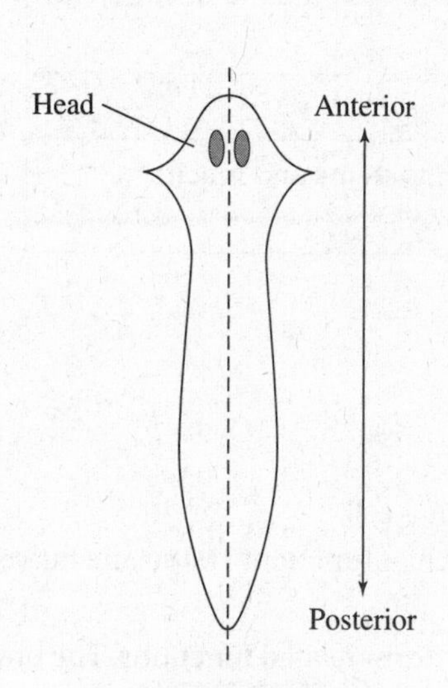

Figure 10.2
Bilateral Symmetry
in the Planaria Flatworm

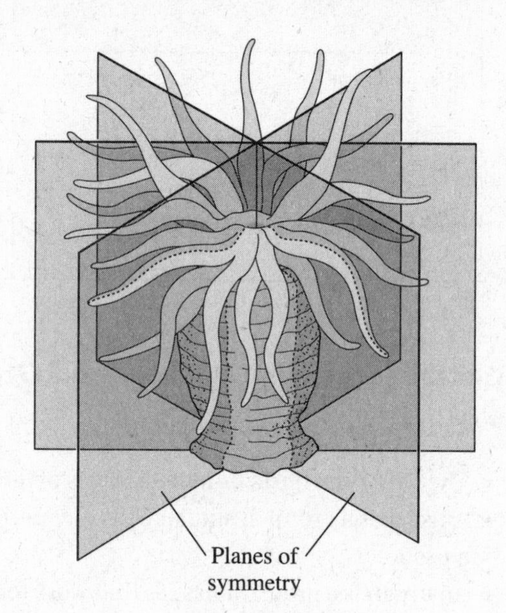

Figure 10.3
Radial Symmetry

Cephalization—Development of a Head End

Along with bilateral symmetry comes a front end (**anterior**) and a rear end (**posterior**). Sensory apparatus and a brain are clustered at the anterior, while digestive, excretory, and reproductive systems are located at the posterior. This enables animals to move faster to flee or to capture prey more effectively. Bilateral animals are all **triploblastic**, with an **ectoderm**, **mesoderm**, and **endoderm**.

PHYLOGENETIC TREES

All living things evolved from a common ancestor almost 4 billion years ago, which is why the principles of biology apply to all organisms. The evolutionary history of these relationships is known as **phylogeny**. A **phylogenetic tree** or **cladogram** is a diagrammatic reconstruction of that history. Phylogenetic trees used to be based on morphology and physical behaviors. Since more genomes have been sequenced, biologists are now constructing phylogenetic trees based on DNA and evolutionary relationships. Phylogenetic analysis of human mitochondrial DNA has allowed scientists to construct a history of human migration out of Africa. The growing database of DNA sequences enables researchers to study more species, but it also makes building phylogenetic trees more complex. A phylogenetic tree can usually be built in several different ways, and *every phylogenetic tree is a hypothesis.* Scientists narrow down the possibilities by using the principle of **maximum parsimony**, which states that one should follow the simplest explanation that coincides with the facts.

EVO-3
Phylogenetic trees and cladograms are constantly being revised based on biological data and emerging knowledge.

Table 10.4 shows shared and derived traits for building a phylogenetic tree. A " – " indicates that a trait is absent; a " + " indicates that a trait is present.

Table 10.4

Shared and Derived Traits for Building a Phylogenetic Tree

	Lancelet (Outgroup)	Lamprey	Bass	Frog	Turtle	Leopard
Vertebral column (backbone)	–	+	+	+	+	+
Hinged jaws	–	–	+	+	+	+
Four walking legs	–	–	+	+	+	+
Amnion	–	–	–	–	+	+
Hair	–	–	–	–	–	+

Figure 10.4 shows a **phylogenetic tree** for the animals listed in Table 10.4. It includes *ingroups*, the organisms of interest—lamprey, bass, frog, turtle, and leopard—and as a point of reference, the *outgroup*, lancelet. The outgroup is the group that diverged before the lineage evolved, in this case, vertebrates. When two lineages diverge, the split is depicted as a *node*. In this example, the nodes are development of a vertebral column, hinged jaws, four walking legs, amnion, and the development of hair. All animals share characteristics with their ancestors and also differ from them. In this tree, all the animals (except the lancelets)

EVO-3 & SP 4
Be able to create a simple cladogram that correctly represents evolutionary history from data provided. (See Table 10.4.)

> **EVO-3**
> Be able to evaluate evidence presented in a phylogenetic tree or simple cladogram.

share a vertebral column. That trait is known as a **shared ancestral trait** or **character**. In contrast, each animal in one **clade** or lineage has a trait that is not shared with their ancestors. That new trait is known as a **shared derived trait** or **character**. For example, hair in the leopard (and in all mammals) is a derived trait.

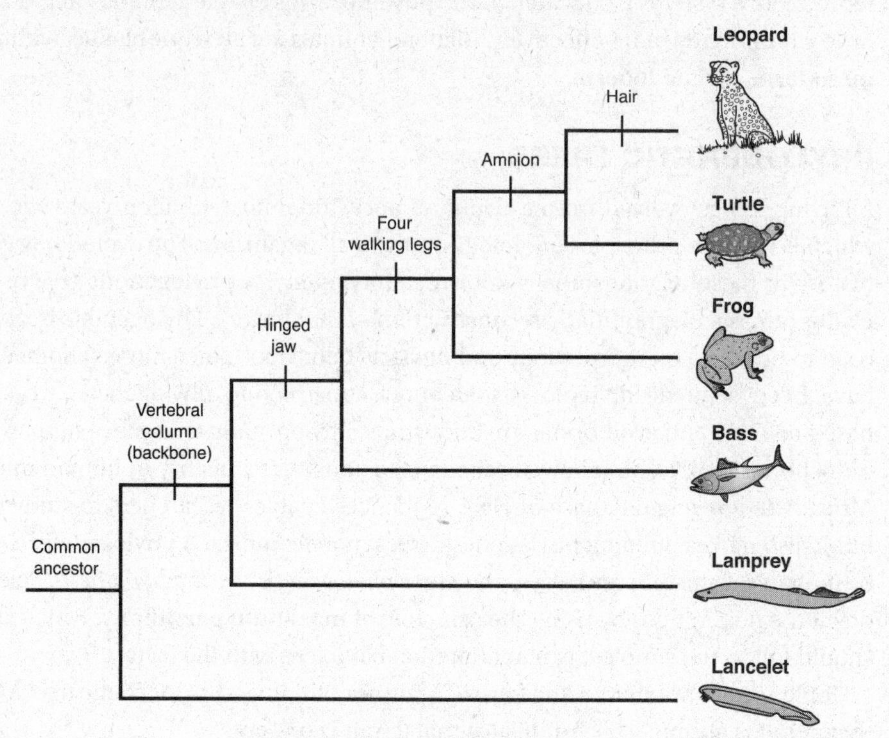

Figure 10.4 Phylogenetic Tree

CHAPTER SUMMARY

- This chapter was an introduction to Big Idea: EVO.
- Our current system of classification is based on DNA analysis and **phylogeny**, or evolutionary history. The fact that organisms share many **conserved** (unchanged) processes is evidence that all organisms are linked by lines of descent from a common ancestor. For example, DNA and RNA are universal. They carry the same genetic code across all domains.
- All organisms are placed into one of three **domains**: **Archaea**, **Bacteria**, and **Eukarya**. Organisms are grouped in a hierarchy of increasingly *exclusive* categories called **taxa**: **domain**, **kingdom**, **phylum**, **class**, **order**, **family**, **genus**, and **species**. The **taxon** (singular) species is the *least* diverse grouping. Our system of classification—**binomial nomenclature**—requires that the scientific name of every organism has two parts; the genus name and species name. Humans are called *Homo sapiens*; a red maple tree is called *Acer rubrum*.
- **Cladograms** or **phylogenetic trees** are diagrammatic reconstructions or models of evolutionary history. They can usually be built in different ways and are updated as new information is discovered. **Ingroups** is the term used for the organisms that are being compared, while **outgroups** are not closely related to the ingroups and are on a separate clade. Outgroups are included merely for comparison. All organisms share traits with their ancestors and also have evolved new and different traits. The traits or **characteristics** an organism shares with its ancestors are called **shared ancestral traits**, while new

traits are called **derived traits.** On a cladogram, all organisms that share a new *derived* trait are located on one branch or **clade.**

- Be able to analyze and evaluate a phylogenetic tree, and be able to construct your own. Study Table 10.4, which presents the data used to construct the phylogenetic tree in Figure 10.4.

PRACTICE QUESTIONS

1. The figure below depicts our system of classification of all organisms, presented in a hierarchical and nested pyramid.

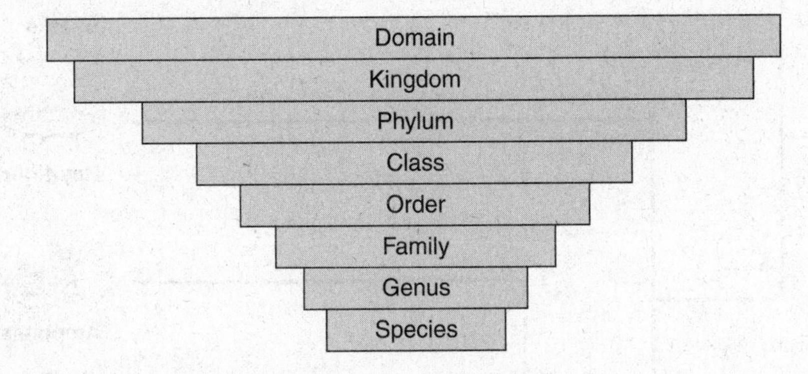

Which of the following is correct about our current system?

(A) Organisms are placed based on similarity of appearance.
(B) As you move up the pyramid, each level is more inclusive.
(C) This pyramid does not apply to Archaea.
(D) Organisms in the Bacteria domain differ from this system in that the scientific name for bacteria does not consist of two names.

2. Which of the following is correct about our system of classification?

(A) All prokaryotes are classified in the Bacteria domain.
(B) Protista, like ameba, are classified in the most varied Eukaryotic kingdom.
(C) Photosynthetic organisms are all classified in the Plantae kingdom.
(D) Fungi have cell walls made of cellulose, as do plants.

3. Which of the following contains prokaryote organisms capable of surviving extreme conditions of heat and salt concentration?

(A) archaea
(B) viruses
(C) protists
(D) fungi

4. Which of the following is best characterized as being eukaryotic and heterotrophic and having cell walls made of chitin?

(A) plants
(B) animals
(C) archaea
(D) fungi

5. A randomly selected group of organisms from a family would show more genetic variation than a randomly selected group from a
   (A) genus
   (B) kingdom
   (C) class
   (D) domain

6. Refer to the following phylogenetic tree.

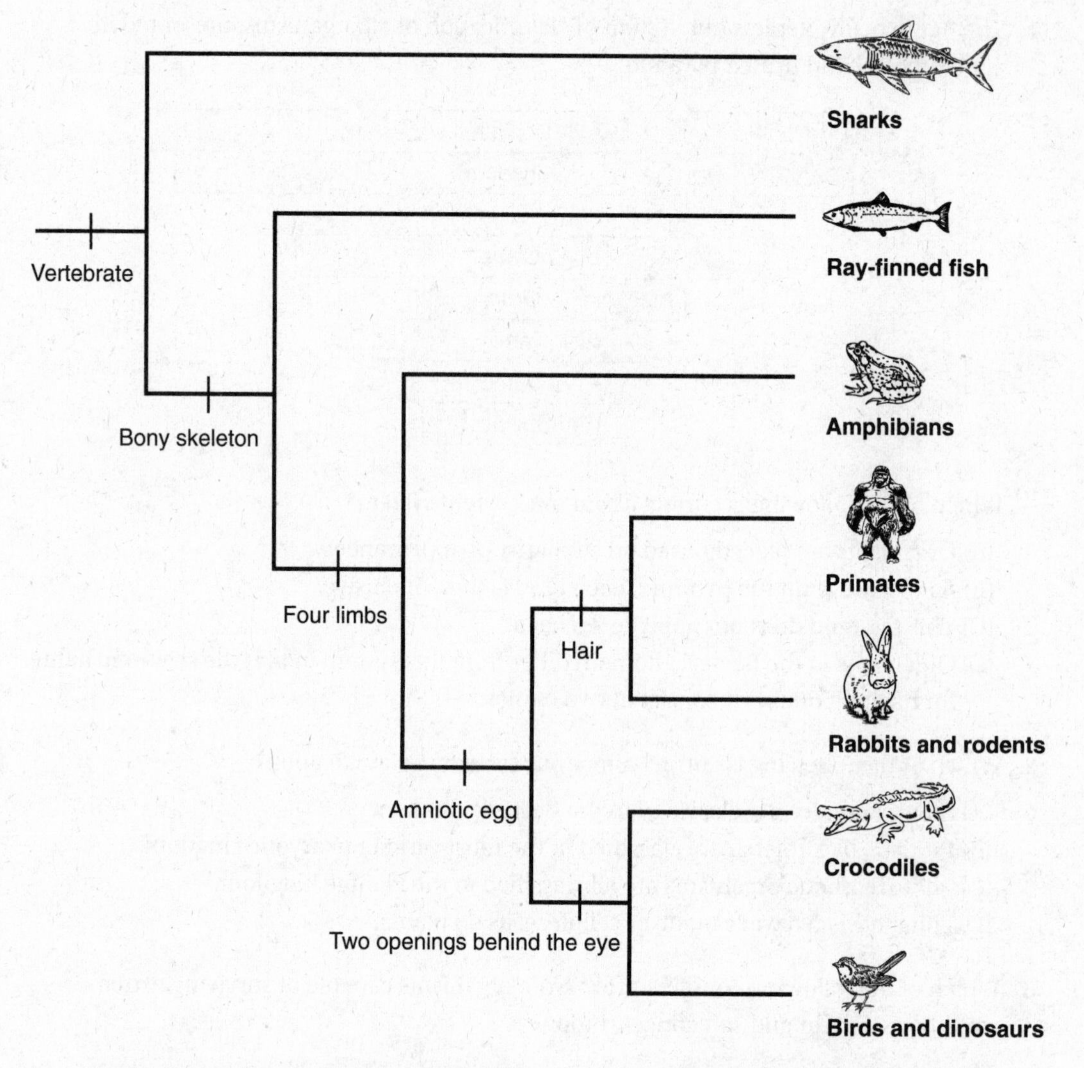

Which of the following statements is most consistent with this phylogenetic tree?
   (A) Ray-finned fish are a direct descendent of sharks because the shark skeleton is made of the more primitive cartilage, while the fish skeleton is more advanced, consisting of bone.
   (B) Crocodiles and rodents are closely related because they share a direct common ancestor.
   (C) Amphibians and primates are more closely related than amphibians and rabbits are because they are placed closer together on the cladogram.
   (D) The common ancestor of dinosaurs, crocodiles, and primates produced amniotic eggs.

7. Data regarding the presence (+) or absence (−) of five derived traits in four different species are shown in the table below.

	TRAIT				
SPECIES	1	2	3	4	5
A	+	+	−	−	−
B	+	+	+	−	−
C	−	−	−	−	−
D	+	−	−	+	+

Which of the following cladograms provides the simplest and most accurate representation of the data in this table?

(A)

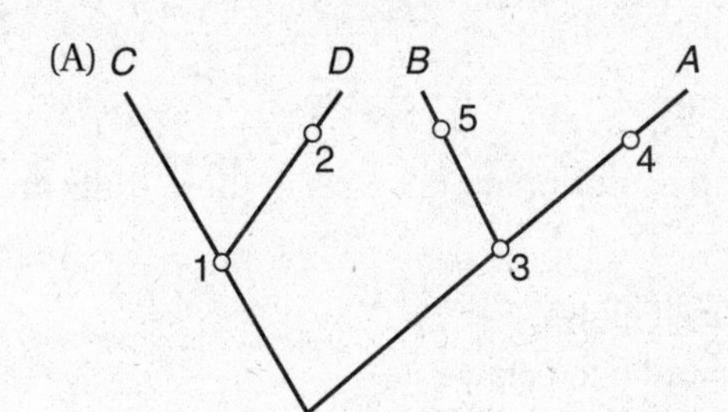

(B)

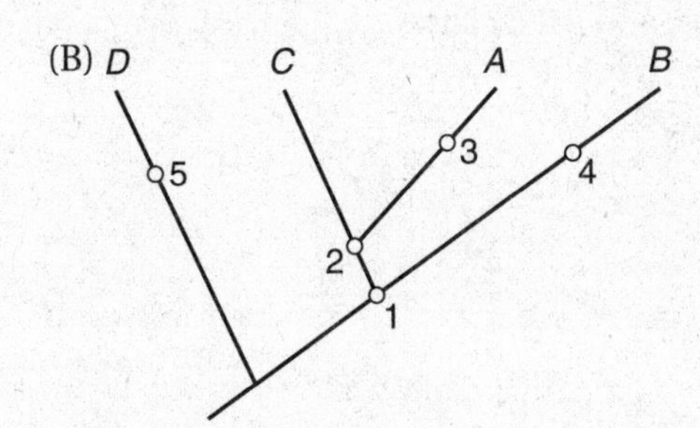

(C)

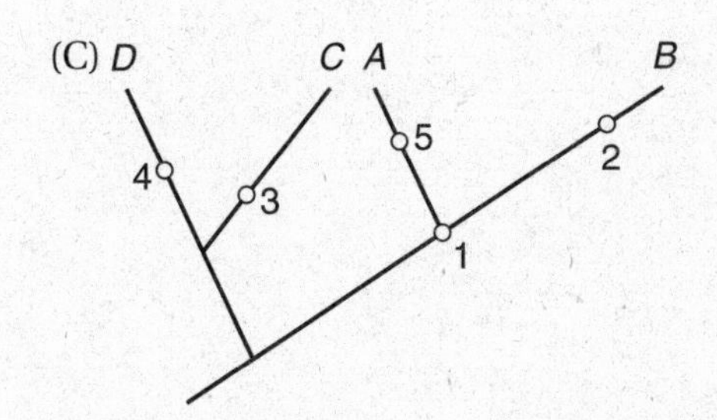

(D)

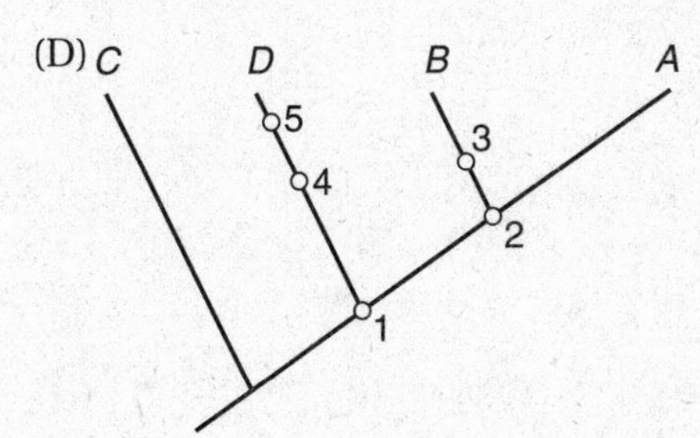

8. In the cladogram below, assume that the ancestor had a long tail, ear flaps, external testes, and fixed claws.

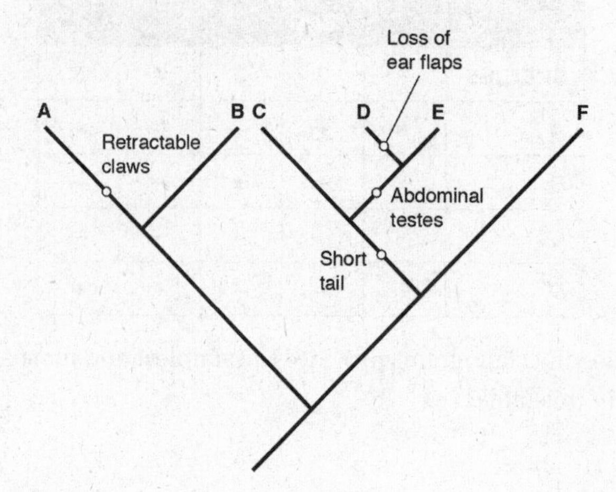

Based on the cladogram and assuming that all evolutionary changes in these traits are shown, what traits does organism D have?

(A) Long tail, ear flaps, external testes, and fixed claws
(B) Long tail, no ear flaps, abdominal testes, and fixed claws
(C) Short tail, no ear flaps, external testes, and fixed claws
(D) Short tail, no ear flaps, abdominal testes, and fixed claws

9. Which cladogram below most accurately represents the information provided in the table?

Organism	Derived Character		
	Backbone	**Legs**	**Hair**
Earthworm	Absent	Absent	Absent
Salmon	Present	Absent	Absent
Lizard	Present	Present	Absent
Cat	Present	Present	Present

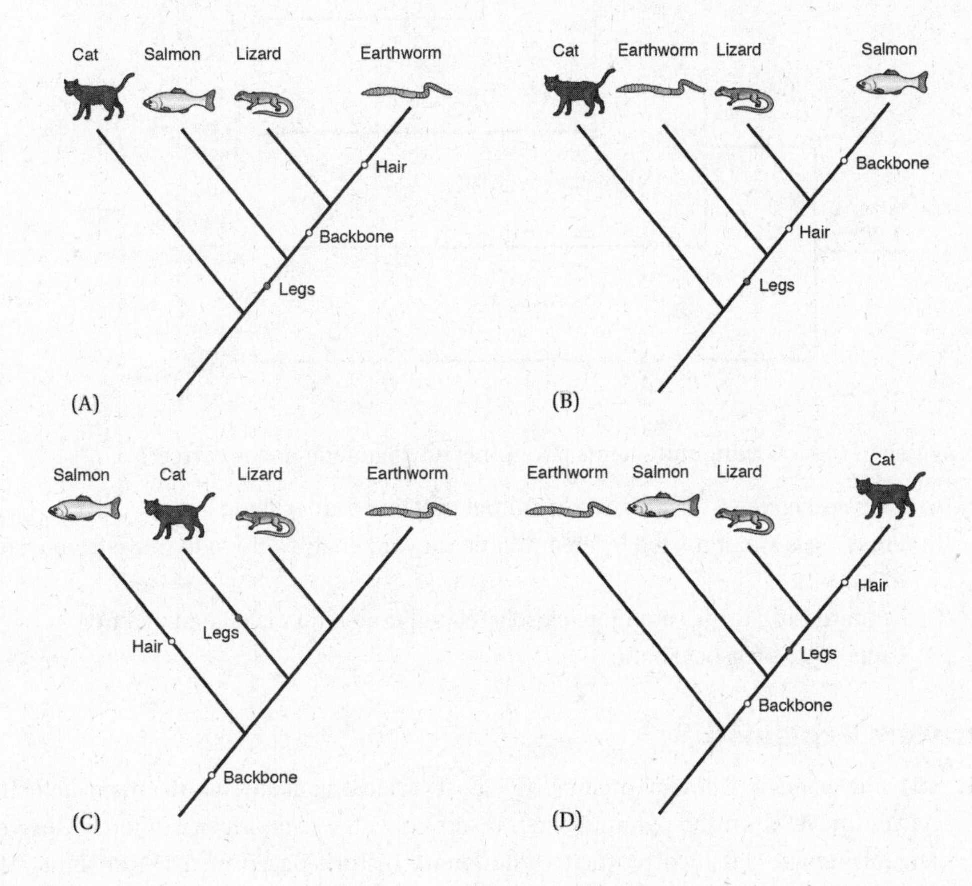

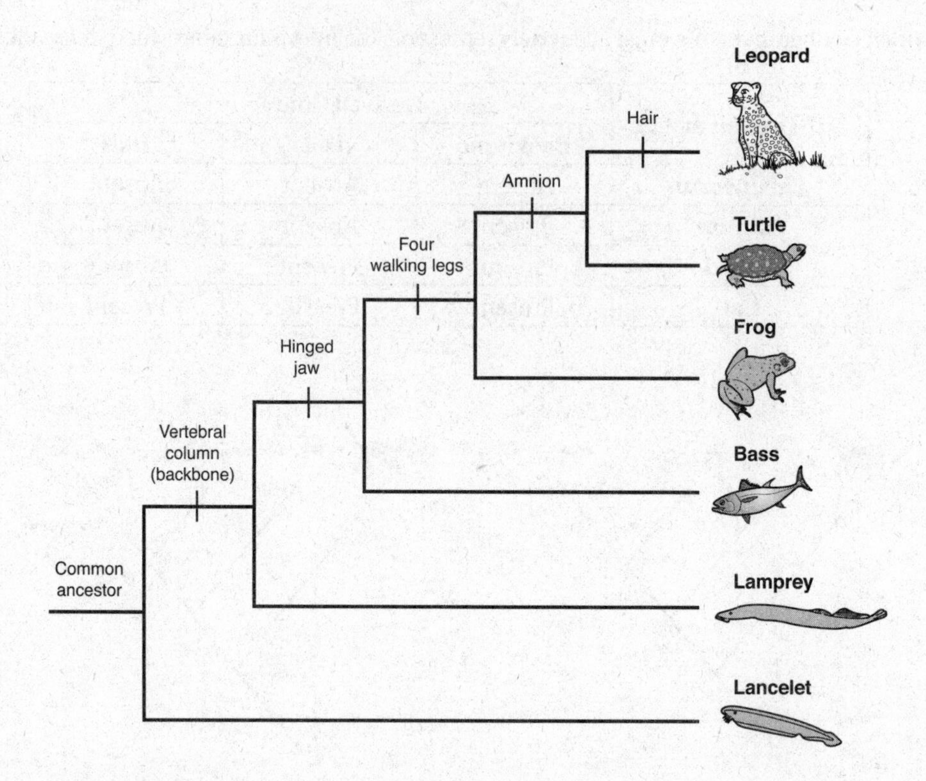

10. Which of the following statements about this phylogenetic tree is correct?

    (A) Bass descended from an ancient animal that had four walking legs.
    (B) Frogs' eggs are protected by an amniotic sac that enabled them to move from water to land.
    (C) Leopards and turtles are more closely related than either is related to a frog.
    (D) Lancelets have a backbone.

Answers Explained

1. **(B)** The taxon of "Domain" includes the most varied organisms. It is the most inclusive. As you move down the pyramid, each taxon holds fewer and fewer different species. Organisms are classified by their evolutionary history, based on DNA analysis. The pyramid holds every type of organism. Bacteria are named in the same manner as every other organism: with two names (i.e., *Escherichia coli* [*E. coli*]). (EVO-3)

2. **(B)** The kingdom Protista holds the most varied species. Prokaryotes—organisms without internal membranes (like a nucleus)—are grouped into two domains: Bacteria and Archaea. Photosynthetic organisms are not limited to plants. There are photosynthetic Protista and Bacteria. The cell walls of fungi are made of chitin, not cellulose. (EVO-3)

3. **(A)** Archaea is the group that includes the extremophiles. Most biologists believe that the prokaryotes and the archaea diverged from each other in very ancient times. One basic way in which the two differ is in their nucleic acids. (EVO-1 & EVO-2)

4. **(D)** Fungi are all heterotrophs. They secrete digestive enzymes and digest food outside the organism. This is called extracellular digestion. Once the food is digested and broken down into building blocks, it is absorbed into the body of the organism by diffusion. The cell walls of mushrooms consist of chitin. (EVO-2)

5. **(A)** The genus is a narrower grouping than the family, so it contains organisms with fewer differences. (EVO-3)

6. **(D)** Dinosaurs, crocodiles, and primates share a common ancestor. Ray-finned fish do not descend from sharks. They have a very distant common ancestor in vertebrates. Crocodiles and rodents are not close relatives and do not share a direct common ancestor. The fact that the amphibians and primates are near each other on the cladogram is not reflective of a relationship. Remember that on this (or any) cladogram, primates and rodents and rabbits can change position, rotating around the upright line. (EVO-3)

7. **(D)** Begin with the species with no traits: that is Species *C*. It must be placed on a line separated from the others. It is an "outgroup." Then, notice that all the other Species (*A*, *B*, and *D*) have Trait 1. Thus, Trait 1 must appear below all of them on the cladogram. Next, notice that only Species *A* and *B* have Trait 2. Thus, they must be off on a line and above Trait 2. Then, you'll see that only Species *B* has Trait 3, so it must be diverted off on a line above Trait 3. Finally, Species *D* has Trait 4 and Trait 5, but no other species has those traits. Thus, Species *D* must be above Trait 1 (because it has that trait), below Trait 2 and Trait 3 (because it does not have those traits), and above Trait 4 and Trait 5 (because it does have those traits). (EVO-3)

8. **(D)** Organism D has all the traits of the organisms on the same clade. In this case, that is short tail and abdominal testes. The node below D is loss of ear flaps—that is a new or "derived" trait—one that is not shared with the previous organisms. Also, organism A has the derived trait of retractable claws, which implies that all other organisms have fixed claws. (EVO-3)

9. **(D)** One can read a cladogram from left to right or right to left. The earthworm has none of the traits listed in the table, so it must be placed in a clade separate from the other animals; either on extreme left or extreme right. The three remaining animals all have the trait, "backbone," but only the salmon has no other trait listed in the table. Thus, the salmon will be on its own clade after the trait "backbone." The lizard and cat both have legs, so their clades must emanate from that trait. Only the cat has "hair" so it is on its own clade. (EVO-3 & SP 4)

10. **(C)** Both leopards and turtles have an amnion. As a result, they are the most closely related animals on the cladogram. Frogs do not have an amnion. They must lay their eggs in a moist or watery area so the eggs do not dry out. Bass descended from an ancestor that had a hinged jaw, as did frogs, turtles, and leopards. Bass did not descend from an animal that had four walking legs, but frogs, turtles, and leopards did. The lancelet descended from an animal that did not have a backbone. (EVO-3)

Evolution

11

Big Ideas: EVO & SYI

Enduring Understandings: EVO-1, EVO-2 & EVO-3; SYI-3

Science Practices: 1, 3 & 5

For the complete list of Big Ideas, Enduring Understandings, and Science Practices, refer to the "AP Biology Course and Exam Description" from the College Board: *https://apcentral.collegeboard.org/pdf/ap-biology-course-and-exam-description.pdf?course=ap-biology*.

INTRODUCTION

Evolution is the change in allelic frequencies in a population. Sometimes populations evolve rapidly, as in the case of the development and spread of antibiotic-resistant bacteria. However, evolution mostly takes place over hundreds, thousands, or millions of years. The theory of evolution is supported by scientific evidence from many disciplines.

EVO-3
Speciation and extinction have occurred throughout Earth's history.

HISTORY OF EARTH

Events in Earth's history can be dated by several methods. Studying undisturbed *sedimentary rock layers* and the fossils within them reveals the *relative age* of rocks and fossils. The *absolute age* of rocks can be accurately measured by *radiometric dating*, which is based on the decay of radioactive isotopes and half-life. Although radiometric dating is a powerful tool, sometimes there is inadequate rock or fossils available to measure, so scientists must use other techniques. *Paleomagnetic dating* uses the fact that Earth's magnetic poles shift and sometimes even reverse. These changes are recorded in rock layers. Other scientific methods of dating Earth use changes in sea levels, molecular clocks, and measurements of continental drift.

By employing all these methods, scientists have learned that Earth has radically changed since it formed 4.5 billion years ago. Continents have shifted. The climate has gone through periods of extreme cooling and warming. The oceans have risen and lowered repeatedly. Volcanic activity has drastically changed Earth and killed off untold numbers of species. Meteorites bombarding Earth from outer space were responsible for the dinosaurs going extinct and giving mammals a chance to expand around the globe. **Five major extinctions** have occurred. Additionally, 4 billion years ago, the atmosphere contained no free oxygen. However, cyanobacteria, oxygen-generating organisms that form rocklike structures called **stromatolites**, provided free oxygen for the oceans and atmosphere. At first, this high oxygen level killed off most anaerobic prokaryotes for which oxygen is toxic. Then the high oxygen levels made possible a rapid diversification of life on land and in the seas. A relatively short period known as the *Cambrian explosion* (535–525 million years ago) was characterized by the sudden appearance of many present-day animal phyla.

EVIDENCE FOR EVOLUTION

Many areas of scientific study provide evidence for evolution. Here are six.

1. Fossil Record

The fossil record documents the existence of species that have become **extinct** or have evolved into other species. **Radiometric dating** and **half-life** accurately measure the age of fossils. Prokaryotes were the first organisms to develop on Earth, and they are the oldest fossils. Paleontologists have discovered many transitional forms that link older fossils to modern species, such as the transition from ***Eohippus*** to the modern horse, ***Equus***. ***Archaeopteryx*** is a fossil that links reptiles and birds. The fossil record also indicates that all the organisms alive today are only a tiny fraction of all the organisms that ever lived. In other words, most life that existed on Earth went extinct.

EVO-1 & EVO-2
Scientific evidence supports the idea that evolution has occurred in all species.

2. Comparative Anatomy

The study of different structures contributes to scientists' understanding of the evolution of anatomical structures and of evolutionary relationships.

- The wing of a bat, the lateral fin of a whale, and the human arm all have the same internal bone structure, although the function of each varies. These structures, known as **homologous structures**, have a common origin and reflect a common ancestry. See Figure 11.1.

- **Analogous structures**, such as a bat's wing and a fly's wing, have the same function. However, the similarity is superficial and reflects an adaptation to similar environments, not descent from a recent common ancestor.
- **Vestigial structures**, such as the appendix, are evidence that structures have evolved. The appendix is a vestige of a structure needed when human ancestors ate a very different diet.

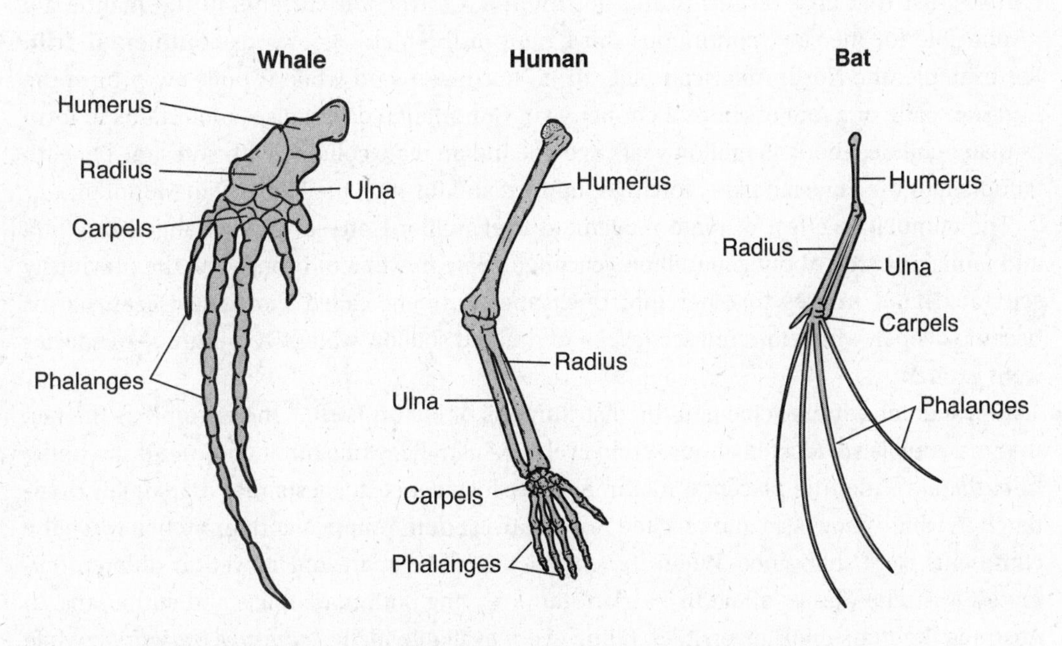

Figure 11.1 Comparison of Homologous Structures in Three Mammals

3. Comparative Biochemistry

Organisms that have a common ancestor will have common biochemical pathways. The more closely related the organisms are to each other, the more similar their biochemistry is. Humans and mice are both mammals. This close relationship is the reason that medical researchers can test new medicines on mice and extrapolate the results to humans. The protein **cytochrome *c*** is one of several electron carriers in the electron transport chain in cellular respiration and photosynthesis. It is present in all aerobic organisms. Analysis of its amino acid sequence is used to determine ancestral relationships and build cladograms. The amino acid sequence of cytochrome *c* in humans and chimps is almost identical.

4. Comparative Embryology

Closely related organisms go through similar stages in their embryonic development. For example, all vertebrate embryos go through a stage in which they have gill pouches on the sides of their throats. In fish, the gill pouches develop into gills. In mammals, they develop into eustachian tubes in the ears.

5. Molecular Biology

Since all aerobic organisms contain cells that carry out aerobic cell respiration, they all contain the polypeptide **cytochrome *c***. A comparison of the amino acid sequence of cytochrome *c* among different organisms shows which organisms are most closely related. The cytochrome *c* in human cells is almost identical to that of our closest relatives, the chimpanzee and gorilla, but differs from that of a pig.

6. Biogeography

According to the theory of **plate tectonics**, continents and oceans rest on giant plates of Earth's crust that float on top of the hot **mantle**. Convection currents in the mantle are responsible for the slow, continuous movement of the plates known as **continental drift**. For example, the North American plate drifts northwestward while it pulls away from the Eurasian plate at a rate of about 2 cm per year. Continental drift causes mountains to form as plates collide. About 45 million years ago, the Indian plate collided with and sank beneath (subducted) the Eurasian plate, forcing it upward and forming the Himalayan Mountains.

The cumulative effect of plate movement over millions of years has changed the flora and fauna of Earth. About 250 million years ago, plate movement brought all the previously separated land masses together into one supercontinent called **Pangaea**. Ocean basins became deeper, which lowered sea level and drained shallow waters. A multitude of species went extinct.

Continental drift has changed the distributions of life on Earth. One example is the fact that marsupials are located almost exclusively in Australia, while other continents are home to *eutherians*, the true placental mammals. Fossil evidence suggests that marsupials originated in what is now Asia and reached Australia via South America and Antarctica while the continents were still joined. When the continents broke apart and moved to different climates, Australia was set afloat like a giant raft, carrying both marsupials and eutherians. In Australia the marsupials diversified, filling every available niche (*adaptive radiation*), while the true placental mammals, which were not adapted to the Australian environment and climate, went extinct. On other continents the opposite occurred; only the true placental mammals survived and diversified, while marsupials went extinct.

HISTORICAL CONTEXT FOR EVOLUTIONARY THEORY

NOTE

Darwin's theory of evolution was developed over time and influenced by the work of other scientists.

Aristotle spoke for the ancient world with his theory of *Scala Natura*. According to this theory, all life forms can be arranged on a ladder of increasing complexity, each with its own allotted rung. The species are permanent and do not evolve. Humans are at the pinnacle of this ladder of increasing complexity.

Carolus Linnaeus or **Carl von Linné** (1707–1778) specialized in **taxonomy**, the branch of biology concerned with naming and classifying the diverse forms of life. He believed that scientists should study life and that a classification system would reveal a divine plan. He developed the naming system used today: **binomial nomenclature**. In this system, every organism has a unique name consisting of two parts: a **genus** name and a **species** name. For example, the scientific name of humans is *Homo sapiens*.

Georges Cuvier, who died in 1832 before Darwin published his thesis, studied fossils and realized that each stratum of Earth is characterized by different fossils. He advocated **catastrophism**, that a series of events in the past occurred suddenly and were caused by mechanisms different from those operating in the present. These events were responsible for the changes in life on Earth. Cuvier's detailed study of fossils was very important in the development of Darwin's theory.

James Hutton, one of the most influential geologists of his day, published his theory of **gradualism** in 1795. He stated that Earth had been molded, not by sudden, violent events, but by slow, gradual change. The effects of wind, weather, and the flow of water that he saw in his lifetime were the same forces that formed the various geologic features on Earth, such as mountain ranges and canyons. His theories were important because they were based on the idea that Earth had a very long history and that change is the normal course of events.

Charles Lyell was a leading geologist of Darwin's era. He stated that geological change results from slow, continuous actions. He believed that Earth was much older than the 6,000 years thought by early theologians. His text, *Principles of Geology*, was a great influence on Darwin.

Lamarck was a contemporary of Darwin who also developed a theory of evolution. He published his theory in 1809, the year Darwin was born. His theory relies on the ideas of **inheritance of acquired characteristics** and **use and disuse**. He stated that individual organisms change in response to their environment. According to Lamarck, the giraffe developed a long neck because it ate leaves from the tall acacia tree for nourishment and had to stretch to reach them. The animals stretched their necks and passed the acquired trait of an elongated neck onto their offspring. Although this theory may seem funny today, it was widely accepted in the early nineteenth century.

Alfred Russell Wallace, a naturalist and author, published an essay discussing the process of natural selection that was identical to Darwin's work, which had not yet been published. Many people credit Wallace, along with Darwin, for the theory of natural selection.

Charles Darwin was a naturalist and author who, when he was 22, left England aboard the HMS Beagle to visit the Galapagos Islands, South America, Africa, and Australia. By the early 1840s, Darwin had worked out his **theory of natural selection** or **descent with modification** as the mechanism for how populations evolve, but he did not publish them. Perhaps he was afraid of the furor his theories would cause. He finally published "**On the Origin of Species**" in 1859 when he was spurred on to publish by the appearance of a similar treatise by Wallace. Darwin's theory challenged the traditional view of a young Earth (about 6,000 years old) inhabited by unchanging species.

DARWIN'S THEORY OF NATURAL SELECTION

Natural selection is a major mechanism of evolution; it acts on phenotypic variation in populations. Here are the tenets of **Darwin's theory of natural selection**:

> **EVO-1**
> Evolution is measured by reproductive success.

- **Populations tend to** grow exponentially, **overpopulate**, and exceed their resources. Darwin developed this idea after reading Malthus's work, a treatise on population growth, disease, and famine published in 1798.
- **Overpopulation results in competition and a struggle for existence**.
- **In any population, there is variation and an unequal ability of individuals** to survive and reproduce. Darwin, however, could not explain the origin of variation in a population. (Mendel's theory of genetics, published in 1865, would have given Darwin an understanding of why variation occurs in a population. However, the ramifications of Mendel's theories were not understood until many years after he presented them.)
- **Only the best-fit individuals survive and get to pass on their traits to offspring**.
- **Evolution occurs as advantageous traits accumulate in a population**. No individual organism changes in response to pressure from the environment. Rather, the frequency of an allele within a population changes.

How the Giraffe Got Its Long Neck

According to **Darwin's theory**, ancestral giraffes were short-necked animals although neck length varied from individual to individual. As the population of animals competing for the limited food supply increased, the taller individuals had a better chance of surviving than those with shorter necks. Over time, the proportion of giraffes in the population with longer necks increased until only long-necked giraffes existed.

How Insects "Become" Resistant to Pesticides

Insects do not actually become resistant to pesticides. Instead, some insects are naturally resistant to a particular chemical insecticide. When the environment is sprayed with that insecticide, the resistant insects have the **selective advantage**. All the insects not resistant to the insecticide die, and the remaining resistant ones breed quickly with no competition. The entirely new population is resistant to the insecticide. The natural selection toward insecticide resistance is an example of **directional selection**.

> **EVO-3**
> Species extinction rates are rapid at times of ecological stress.

TYPES OF SELECTION

Natural selection can alter the frequency of inherited traits in a population in five different ways, depending on which phenotypes in a population are favored. Five types of selection are **stabilizing**, **disruptive** or **diversifying**, **directional**, **sexual**, and **artificial**; see Figure 11.2 for a few of these types of selection.

> **SP 3**
> Be able to describe a model (graph) that represents evolution within a population.

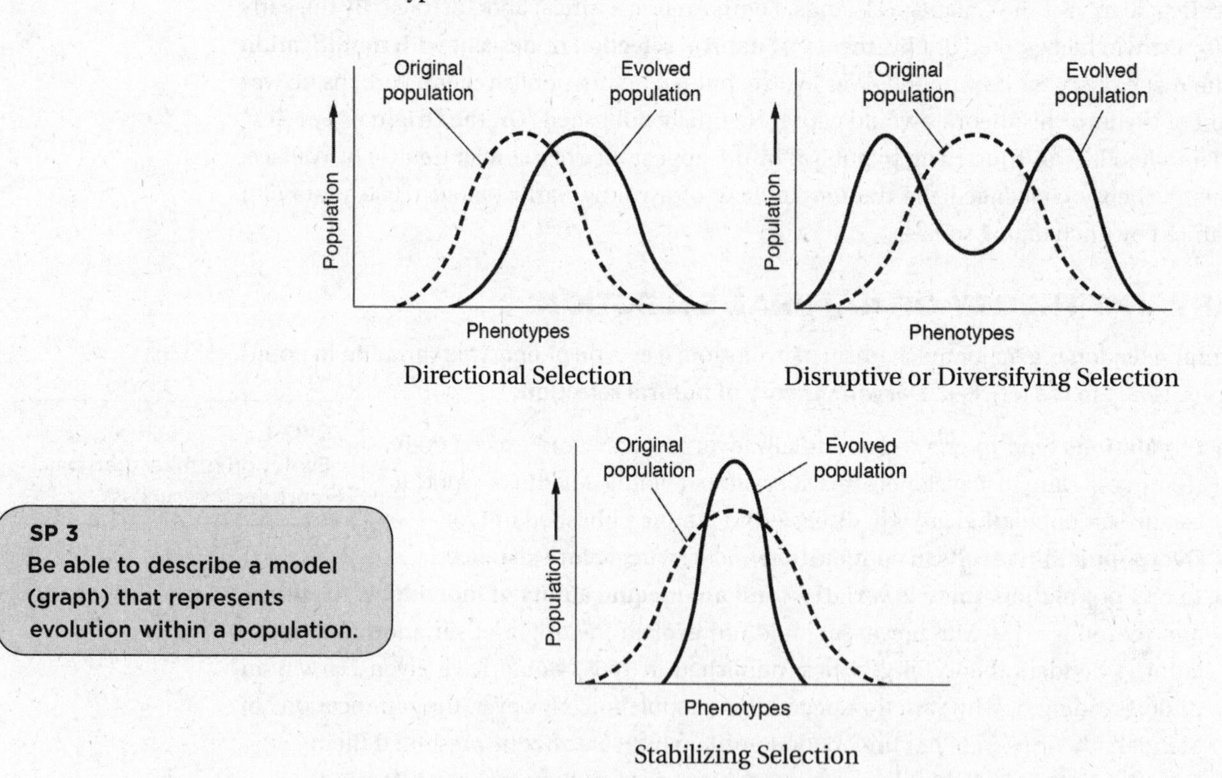

Figure 11.2 Three Types of Selection

1. Stabilizing Selection

Stabilizing selection, sometimes called purifying selection, eliminates the extremes and favors the more common intermediate forms. Many mutant forms are weeded out in this way.

- In humans, stabilizing selection keeps the majority of birth weights in the 6–8 pound (2.7–3.6 kg) range. For babies much smaller and much larger, infant mortality is greater.
- In Swiss starlings, genotypes that lead to a clutch size (the number of eggs a bird lays) of up to five will have more surviving young than birds of the same species that lay a larger or smaller number of eggs.

2. Disruptive or Diversifying Selection

Disruptive or **diversifying selection** increases the extreme types in a population at the expense of intermediate forms. What may result is called **balanced polymorphism**: one population divided into two or more distinct types. Over great lengths of time, disruptive selection may result in the formation of two new species.

Imagine that an environment with very light rocks and very dark soil is colonized by light, intermediate-colored, and dark mice. The frequency of very light and very dark mice, which are both camouflaged, would increase, while the intermediate-colored mice would die out because of predation. Pressure from the environment selects for two extreme characteristics.

3. Directional Selection

Changing environmental conditions give rise to **directional selection**, where one phenotype replaces another in the gene pool. Here are two examples of directional selection.

> **EVO-1**
> Environments change and act as a selective mechanism on populations.

- One example of **directional selection** is **industrial melanism** in **peppered moths**, *Biston betularia.* Until 1845 in England, most peppered moths were light; a few individuals were found to be dark. With increasing industrialization, smoke and soot polluted the environment, making all the plants and rocks dark. By 1900, all moths in the industrialized regions were dark; only a few light-colored individuals could be found. Before the industrial revolution, white moths were camouflaged in their environment and dark moths were easy prey for predators. After the environment was darkened by heavy pollution, dark moths were camouflaged and had the selective advantage. Within a relatively short time, dark moths replaced the light moths in the population.
- **Directional selection** can produce rapid shifts in allelic frequencies. For example, soon after the discovery of **antibiotics**, bacteria appeared that were resistant to these drugs. Scientists now know that the genes for antibiotic resistance are carried on **plasmids**, small DNA molecules, which can be transferred from one bacterial cell to another and which can spread the mutation for **antibiotic resistance** very rapidly within the bacterial population. The appearance of antibiotics themselves does not induce mutations for resistance; it merely selects against susceptible bacteria by killing them. Since only resistant individuals survive to reproduce, the next generation will all be resistant. Joshua Lederberg carried out the experiment that proved that some bacteria are resistant to antibiotics prior to any exposure.

4. Sexual Selection

Sexual selection is selection based on variation in secondary sexual characteristics related to competing for and attracting mates. In males, the evolution of horns, antlers, large stature, and great strength are the result of sexual selection. Male elephant seals fight for supremacy of a harem that may consist of as many as fifty females. In baboons, long canines are important for male-male competition. Differences in appearance between males and females are known as **sexual dimorphism**. In many species of birds, the females are colored in a way to blend in with their surroundings, thus protecting them and their young. The males, on the other hand, have bright, conspicuous plumage because they must compete for the attention of the females.

> **EVO-1**
> Some phenotypic variations significantly increase or decrease the fitness of an organism or a population.

5. Artificial Selection

Humans breed plants and animals by seeking individuals with desired traits as breeding stock. This is known as **artificial selection**. Racehorses are bred for speed, and laying hens are bred to produce more and larger eggs. Humans have bred cabbage, brussels sprouts, kale, kohlrabi, cauliflower, and broccoli all from the wild mustard plant by selecting for different traits.

PRESERVING VARIATION IN A POPULATION

Variation in a population is necessary in order for a population to evolve as the environment changes. Although Darwin could not explain the origin of variation, he knew that variation exists in every population. A good example of this is the existence of hundreds of breeds of dogs. All dogs belong to one species, *Canus familiaris.*

While it would seem that natural selection would tend to reduce genetic variation by removing unfavorable genotypes from a population, nature has many mechanisms to preserve it. Here are eight mechanisms that *preserve* diversity or variation in a gene pool or population: balanced polymorphism, geographic variation, sexual reproduction, outbreeding, diploidy, heterozygote advantage, frequency-dependent selection, and evolutionary neutral traits.

1. Balanced Polymorphism

Balanced polymorphism is the presence of two or more phenotypically distinct forms of a trait in a single population of a species. The shells of one genus of land snail exhibit a wide range of colors and banding patterns. Banded snails living on dark, mottled ground are less visible than unbanded ones and therefore are preyed upon less frequently. In areas where the background is fairly uniform, unbanded snails have the selective advantage. Each **morph** is better adapted in a different area, but both varieties continue to exist.

2. Geographic Variation

Two different varieties of rabbit continue to exist in two different regions in North America. Rabbits in the cold, snowy northern regions are camouflaged with white fur and have short ears to conserve body heat. Rabbits living in warmer, southern regions have mottled fur to blend in with surrounding woodsy areas and long ears to radiate off excess body heat. Such a graded variation in the phenotype of an organism is known as a **cline**. Because the variation in rabbit appearance is due to differences in northern and southern environments, this is an example of a **north-south cline**.

3. Sexual Reproduction

Sexual reproduction provides variation due to the shuffling and recombination of alleles during meiosis and fertilization.

- **Independent assortment of chromosomes** during metaphase I results in the recombination of unlinked genes.
- **Crossing-over** is the exchange of genetic material of homologous chromosomes and occurs during meiosis I. It produces individual chromosomes that combine genes inherited from two parents. In humans, two or three crossover events occur per homologous pair.
- The **random fertilization** of one ovum by one sperm out of millions results in enormous variety among the offspring.

EVO-1
The more genetic variation in a population, the greater the capacity for change or evolution and survival.

4. Outbreeding

Outbreeding is the mating of organisms within one species that are not closely related. It is the opposite of inbreeding, the mating of closely related individuals. Outbreeding maintains both variation within a species and a strong gene pool. Inbreeding weakens the gene pool because if organisms that are closely related interbreed, detrimental recessive traits tend to appear in homozygous recessive individuals. Many mechanisms have evolved to promote outbreeding.

In lions, the dominant male of a pride chases away the young maturing males before they become sexually mature. This ensures that these young males will not inbreed with their female siblings. These young males roam the land, often over great distances, looking for another pride to join. If one of these young male lions can successfully overthrow the king of another pride, he will inseminate all the females of that new pride and develop his own lineage.

5. Diploidy

Diploidy, the $2n$ condition, maintains and hides a huge pool of alleles that may be harmful in the present environment but that could be advantageous when conditions change in the future.

6. Heterozygote Advantage

Heterozygote advantage preserves multiple alleles in a population. It is a phenomenon in which the hybrid individual is selected for because it has greater reproductive success. The hybrids are sometimes better adapted than the homozygotes. Notice that heterozygote advantage is defined in terms of genotype, not phenotype.

In the case of **sickle cell anemia** in West Africa, people who are hybrid (*Ss*) for the sickle cell trait have the selective advantage over other individuals. Those who are hybrid have normal hemoglobin and do not suffer from sickle cell disease. However, they are resistant to malaria, which is endemic in West Africa. Those individuals who are homozygous for the sickle cell trait (*ss*) are at a great disadvantage because they have abnormal hemoglobin and suffer from and may die of sickle cell disease. People who are homozygous for normal hemoglobin (*SS*) do not have sickle cell disease but are susceptible to and may die of **malaria**. Thus, the mutation for the sickle cell trait is retained in the gene pool.

7. Frequency-Dependent Selection

Another mechanism that preserves variety in a population is known as **frequency-dependent selection** or the **minority advantage**. This acts to decrease the frequency of the more common phenotypes and increase the frequency of the less common ones. In predator-prey relationships, predators develop a **search image**, or standard representation of prey, that enables them to hunt a particular kind of prey effectively. If the prey individuals differ, the most common type will be preyed upon disproportionately while the less common individuals will be preyed upon to a lesser extent. Since these rare individuals have the selective advantage, they will become more common for a time, will lose their selective advantage, and will eventually be selected against.

8. Evolutionary Neutral Traits

Evolutionary neutral traits are traits that seem to have no selective advantage. One example is the different **blood types** in humans. Scientists do not understand where they evolved from or why they have remained (been conserved) in the human population. Perhaps they actually influence survival and reproductive success in ways that are difficult to perceive or measure.

CAUSES OF EVOLUTION OF A POPULATION

The agents of change for a population, that is, those things that cause **evolution**, are **genetic drift**, **gene flow**, **mutations**, **nonrandom mating**, and **natural selection**.

EVO-1
In addition to natural selection, chance and random events can influence evolution, especially for small populations.

Genetic Drift

Genetic drift is change in the gene pool due to chance. It is a fluctuation in the frequency of alleles from one generation to another and is unpredictable. It tends to limit diversity. There are two examples: the **bottleneck effect** and the **founder effect**.

- **Bottleneck effect**: Natural disasters, such as fires, earthquakes, and floods, reduce the size of a population *unselectively*, resulting in a loss of genetic variation. The resulting population is much smaller and not representative of the original one. Certain alleles may be underrepresented or overrepresented compared with that of the original population. This is known as the **bottleneck effect**. Here are two examples.

  The high rate of Tay-Sachs disease among Eastern European Jews is attributed to a population bottleneck experienced by Jews in the Middle Ages. During that period, many Jews were persecuted and killed, and the population was reduced to a small fraction of its original size. Of the individuals who remained alive, there happen to have been a disproportionate percentage of people who carried the Tay-Sachs gene. Since Jews in Europe remained isolated and did not intermarry with other Europeans to any great extent, the incidence of the trait remained unusually high in that population.

  From the 1820s to the 1880s along the California coast, the northern elephant seal was hunted almost to extinction. Since 1884, when the seal was placed under government protection, the population has increased to about 35,000; all are descendants from that original group and have little genetic variation.

- **Founder effect**: When a small population breaks away from a larger one to colonize a new area, it is most likely not genetically representative of the original larger population. Rare alleles may be overrepresented. This is known as the **founder effect** and occurred in the Old Order Amish of Lancaster, Pennsylvania. All of the colonists descended from a small group of settlers who came to the United States from Germany in the 1770s. Apparently one or more of the settlers carried the rare but dominant gene for **polydactyly**, having extra fingers and toes. Due to the extreme isolation and intermarriage of the close community, this population now has a high incidence of polydactyly.

Notice that whether a trait is dominant or recessive merely determines if it is expressed or remains hidden. It does not determine how common the trait is in a population. The increase or decrease in allelic frequency of a trait results from genetic drift or from the trait being advantageous or disadvantageous.

Gene Flow

Gene flow is the movement of alleles into or out of a population. It can occur as a result of the migration of fertile individuals or gametes between populations. For example, pollen from one valley can be carried by the wind across a mountain to another valley. Gene flow tends to increase diversity.

Mutations

Mutations are changes in genetic material and are the raw material for evolutionary change. They increase diversity. A single point mutation can introduce a new allele into a population. *Although mutations at one locus are rare, the cumulative effect of mutations at all loci in a population can be significant.*

Nonrandom Mating

Individuals choose their mates for a specific reason. The selection of a mate *serves to eliminate the less-fit individuals.* Snow geese exist in two phenotypically distinct forms: white and blue. Blue geese tend to mate with blue geese, and white geese tend to mate with white geese. If, for some reason, the blue geese became more attractive and both blue and white geese began mating with only blue geese, the population would evolve quickly, favoring blue geese.

Natural Selection

Natural selection is the major mechanism of evolution in any population. Those individuals who are better adapted in a particular environment exhibit *better reproductive success.* They have more offspring that survive and pass their genes on to more offspring.

EVO-1
Natural selection acts on phenotypic variations in populations.

HARDY-WEINBERG EQUILIBRIUM—CHARACTERISTICS OF STABLE POPULATIONS

Hardy and **Weinberg**, two scientists, described the characteristics of a **stable, nonevolving population**, that is, one in which allelic frequencies do not change. For example, if the frequency of an allele for a particular trait is 0.5 and the population is not evolving, in 1,000 years the frequency of that allele will still be 0.5.

According to Hardy-Weinberg, if the population is stable, the following must be true:

1. **THE POPULATION MUST BE VERY LARGE.** In a large population, a small change in the gene pool will be diluted by the sheer number of individuals and no change in the frequency of alleles will occur. (In a small population, the smallest change in the gene pool will have a major effect in allelic frequencies.)
2. **THE POPULATION MUST BE ISOLATED FROM OTHER POPULATIONS.** There must be no migration of organisms into or out of the gene pool because that could alter allelic frequencies.
3. **THERE MUST BE NO MUTATIONS IN THE POPULATION.** A mutation in the gene pool could cause a change in allelic frequency by introducing a new allele.
4. **MATING MUST BE RANDOM.** If individuals select mates, then those individuals that are better adapted will have a reproductive advantage and the population will evolve.
5. **THERE MUST BE NO NATURAL SELECTION.** Natural selection causes changes in relative frequencies of alleles in a gene pool.

The Hardy-Weinberg Equation

The Hardy-Weinberg equation enables us *to calculate frequencies of alleles in a population.* Although it can be applied to complex situations of inheritance, for the purpose of explanation here, we will discuss a simple case—a gene locus with only two alleles. Scientists use the letter ***p*** to stand for the frequency of the **dominant allele** and the letter ***q*** to stand for the frequency of the **recessive allele**.

The Hardy-Weinberg equation is

$$p^2 + 2pq + q^2 = 1 \quad \text{or} \quad p + q = 1$$

The monohybrid cross is the basis for this equation.

	A	*a*
A	*AA*	*Aa*
a	*Aa*	*aa*

$p^2 = AA =$ homozygous dominant

$2pq = 2(Aa) =$ hybrid

$q^2 = aa =$ homozygous recessive

As of 2020, you are allowed to use your calculator on the exam. Even if the Hardy-Weinberg problems are not so simple, the principle is the same. Here are three sample problems.

PROBLEM 1

If 9% of the population has blue eyes, what percent of the population is hybrid for brown eyes? What percent of the population is homozygous for brown eyes?

To solve this problem, follow these steps:

1. The trait for blue eyes is homozygous recessive, *bb*, and is represented by q^2.

   $q^2 = 9\%$ Converting to a decimal: $q^2 = 0.09$

2. To solve for q, find the square root of 0.09, which is 0.3.
3. Since $p + q = 1$, if $q = 0.3$, then $p = 0.7$.
4. The hybrid brown condition is represented by $2pq$.
5. To solve for the percent of the population that is hybrid, substitute values for $2(p)(q)$.

   The percentage of the population that is hybrid brown is $2(.7)(.3) = 42\%$.

6. Homozygous dominant is represented by p^2.

   The percentage of the population that is homozygous brown is $p^2 = (.7)^2 = 49\%$.

BE CAREFUL

The square root of 0.09 is 0.3, not 0.03.

PROBLEM 2

Determine the percent of the population that is homozygous dominant if the percent of the population that is homozygous recessive is 16%.

1. Homozygous recessive $= q^2 = 16$. Therefore, $q^2 = .16$ and $q = 0.4$.
2. If $q = 0.4$, then $p = 0.6$.
3. Therefore, the percentage of the population that is homozygous dominant $= p^2 = (.6)^2 = .36 = 36\%$.

REMEMBER

No single organism changes in response to a change in the environment. Rather, the frequency of an allele in the population may change.

PROBLEM 3

Determine the percent of the population that is hybrid if the allelic frequency of the recessive trait is 0.5.

1. In this example, you are given the value of q (not q^2). You only need to subtract from 1 to get the value of p.
2. If $p + q = 1$ and $q = 0.5$, then $p + 0.5 = 1$.
3. Since both $p = 0.5$ and $q = 0.5$, the percentage of the population that is hybrid is $2pq = 0.5 \times 0.5 \times 2 = 50\%$.

SPECIATION AND REPRODUCTIVE ISOLATION

The definition of a **species** is a population whose members have the potential to interbreed in nature and produce viable, fertile offspring. Lions and tigers can be induced to interbreed in captivity but would not do so naturally. Therefore, they are considered separate species. Horses and donkeys can interbreed in nature and produce a mule that is not fertile. Therefore, the horse and donkey belong to different species.

A **species** is defined in terms of **reproductive isolation**, meaning that one group of genes becomes isolated from another to begin a separate evolutionary history. Once separated, the two isolated populations may begin to diverge genetically under the pressure of different selective forces in different environments. If enough time elapses and differing selective forces are sufficiently great, the two populations may become so different that, even if they were brought back together, interbreeding would not naturally occur. At that point, **speciation** is said to have taken place. *Anything that fragments a population and isolates small groups of individuals may cause speciation.* The following describes different modes of speciation due to different modes of isolation, and Figure 11.3 shows diagrams of **allopatric** and **sympatric speciation**.

> **EVO-1**
> Speciation may occur when two populations become reproductively isolated from each other.

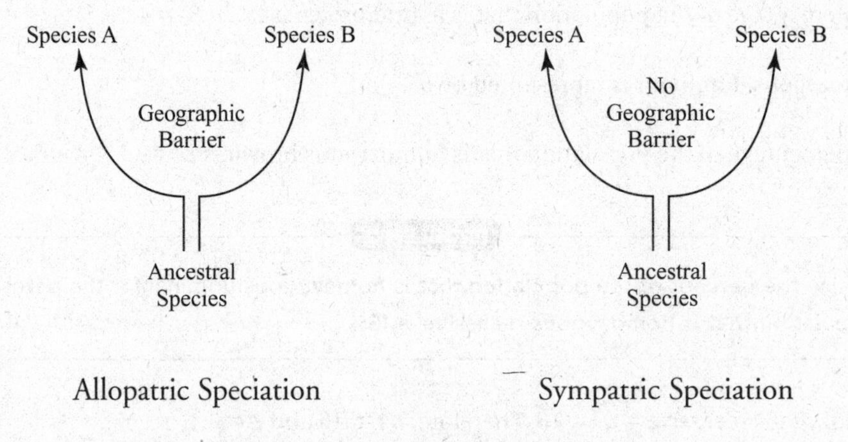

Figure 11.3 Two Types of Speciation

Allopatric Speciation

Allopatric speciation is caused by **geographic isolation**: separation by mountain ranges, canyons, rivers, lakes, glaciers, altitude, or longitude.

Sympatric Speciation

Under certain circumstances, speciation may occur without geographic isolation, in which case the cause of the speciation is **sympatric**. Examples of **sympatric speciation** are **polyploidy**, **habitat isolation**, **behavioral isolation**, **temporal isolation**, and **reproductive isolation**.

- **Polyploidy**: This is the condition where a cell has more than two complete sets of chromosomes ($4n$, $8n$, etc.). It is common in plants and can occur naturally or through breeding. It results from nondisjunction during meiosis when gametes with the $2n$ chromosome number are fertilized by another abnormal ($2n$) gamete, resulting in a daughter cell with

$4n$ chromosomes. Plants that are polyploid cannot breed with others of the same species that are not polyploid and are functionally isolated from them. See Figure 11.4.

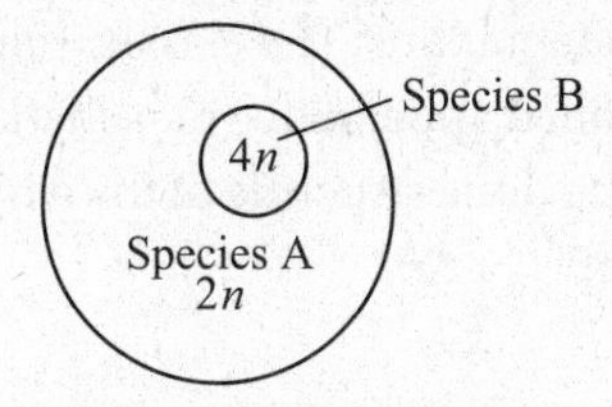

Figure 11.4 Polyploidy

- **Habitat isolation**: Two organisms live in the same area but encounter each other rarely. Two species of one genus of snake can be found in the same geographic area, but one inhabits the water while the other is mainly terrestrial.
- **Behavioral isolation**: Sticklebacks, small saltwater fish that have been studied extensively, have elaborate mating behavior. At breeding time, in response to increased sunlight, the males change in color and develop a red underbelly. The male builds a nest and courts the female with a dance that triggers a complex set of movements between the partners. If either partner fails in any step of the mating dance, no mating occurs and no young are produced.

  Male fireflies of various species signal to females of their kind by blinking the lights on their tails in a particular pattern. Females respond only to characteristics of their own species, flashing back to attract males. If, for any reason, the female does not respond with the correct blinking pattern, no mating occurs. The two animals become isolated from each other.
- **Temporal isolation**: Temporal refers to time. A flowering plant colonizes a region with areas that are warm and sunny and areas that are cool and shady. Flowers in the regions that are warmer become sexually mature sooner than flowers in the cooler areas. This separates flowers in the two different environments into two separate populations.
- **Reproductive isolation**: Closely related species may be unable to mate because of a variety of reasons. Differences in the structure of genitalia may prevent insemination. Differences in flower shape may prevent pollination. Things that prevent mating are called **prezygotic barriers**. For example, a small male dog and a large female dog cannot mate because of the enormous size differences between the two animals. Things that prevent the production of fertile offspring, once mating has occurred, are called **postzygotic barriers**. One example might be that a particular zygote is not viable. Both prezygotic and postzygotic barriers result in **reproductive isolation**.

PATTERNS OF EVOLUTION

The evolution of different species is classified into five patterns: divergent, convergent, parallel, coevolution, and adaptive radiation; see Figure 11.5.

Divergent Evolution

Divergent evolution occurs when a population becomes isolated (for any reason) from the rest of the species, becomes exposed to new selective pressures, and evolves into a new species. All the examples of allopatric and sympatric speciation that were discussed previously are examples of divergent evolution.

Convergent Evolution

When unrelated species occupy the same environment, they are subjected to similar selective pressures and show similar adaptations. The classic example of **convergent evolution** is the whale, which has the streamlined appearance of a shark because the two evolved in the same environment. The underlying bone structure of the whale, however, reveals an ancestry common to mammals, not to fish.

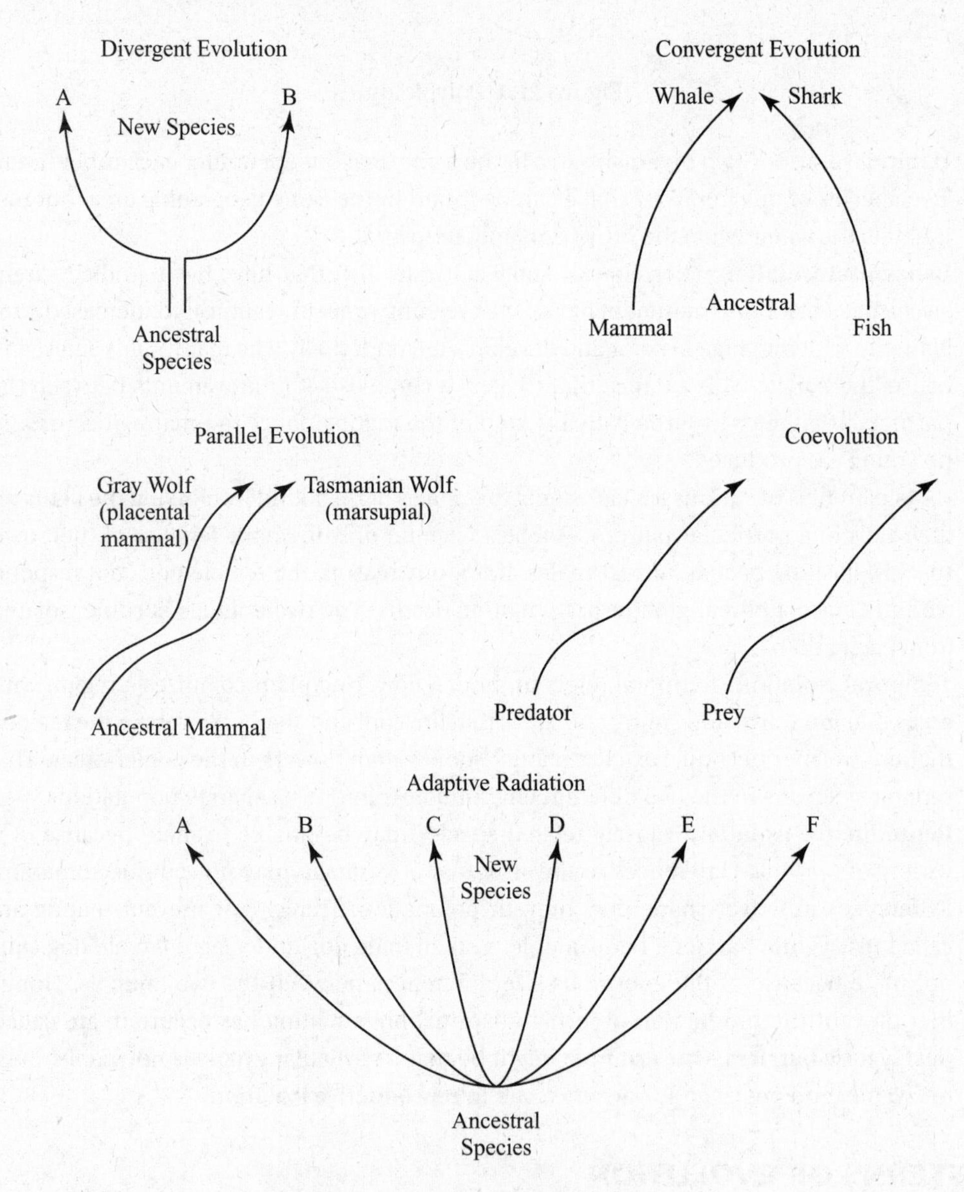

Figure 11.5 Patterns of Evolution

Parallel Evolution

Parallel evolution describes two related species that have made similar evolutionary adaptations after their divergence from a common ancestor. The classic example of this includes the marsupial mammals of Australia and the placental mammals of North America. (The only mammals in Australia are the ones that have been introduced from abroad, like rabbits.) There are striking similarities between some placental mammals like the gray wolf of North

America and the marsupial Tasmanian wolf of Australia because they share a common ancestor and evolved in similar environments.

Coevolution

Coevolution is the reciprocal evolutionary set of adaptations of two interacting species. All predator-prey relationships are examples and the relationship between the **monarch butterfly** and **milkweed plant** is another. The **milkweed plant** contains poisons that deter herbivores from eating them. The butterfly lays its eggs in the milkweed plant and when the larvae (caterpillars) hatch, they feed on the milkweed and absorb the poisonous chemicals from the plant. They store the poison in their tissues. This poison, which is present in the adult butterfly, makes the butterfly toxic to any animal who tries to eat it. (The butterfly exhibits bright conspicuous warning colors that deter predators.)

Adaptive Radiation

Adaptive radiation is the emergence of numerous species from a common ancestor introduced into an environment. Each newly emerging form specializes to fill an ecological niche. All 14 species of **Darwin's finches** that live on the **Galapagos Islands** today diverged from a single ancestral species perhaps 10,000 years ago. There are currently six ground finches, six tree finches, one warbler finch, and one bud eater.

MODERN THEORY OF EVOLUTION

Gradualism

Gradualism is the theory that organisms descend from a common ancestor gradually, over a long period of time, in a linear or branching fashion. Big changes occur by an accumulation of many small ones. According to this theory, fossils should exist as evidence of every stage in the evolution of every species with no missing links. However, the fossil record is at odds with this theory because scientists rarely find **transitional forms** or **missing links**.

EVO-1 & EVO-2
Populations of organisms continue to evolve.

Punctuated Equilibrium

The favored theory of evolution today is called **punctuated equilibrium** and was developed by **Stephen J. Gould** and **Niles Eldredge** after they observed that the gradualism theory was not supported by the fossil record. This theory proposes that new species appear suddenly after long periods of stasis. A new species changes most as it buds from a parent species and then changes little for the rest of its existence. The sudden appearance of the new species can be explained by the **allopatric model** of speciation. A new species arises in a different place and expands its range, outcompeting and replacing the ancestral species. See Figure 11.6, which illustrates gradualism and punctuated equilibrium.

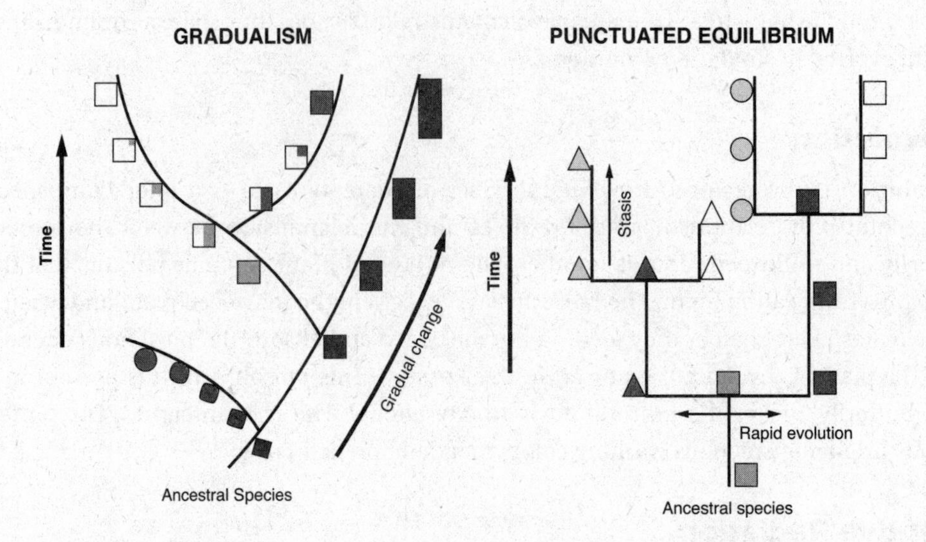

Figure 11.6 Gradualism and Punctuated Equilibrium

EVO-DEVO

How is it that humans and chimps are so different when our DNA is almost identical? (We share 98.9% of our DNA sequence.) We are bipedal (walk upright) and can send rockets to the moon, while chimps still walk on four legs in the rainforest. We don't know all the answers. However, the field of **evo-devo**, *evolutionary developmental biology*, provides some answers. In essence, major changes in body form and function can come about when some genes regulate other genes.

Certain Genes Alter Other Genes to Influence Development

The DNA sequence that codes for a particular structure might be identical in different species. However, these genes might be **upregulated** or **downregulated** by other genes, resulting in different outcomes in the different species. Two examples are homeotic genes and heterochrony.

HOMEOTIC GENES

Homeotic genes are *master regulatory genes* that control spatial organization of body parts. They determine such features as where wings or legs will develop on an insect or where the petals go on a flower. One particular homeotic gene, called a **Hox gene**, provides positional information in developing embryos. Among crustaceans, a change in the location of two Hox genes will transform antennae into legs. (See Figure 11.7.)

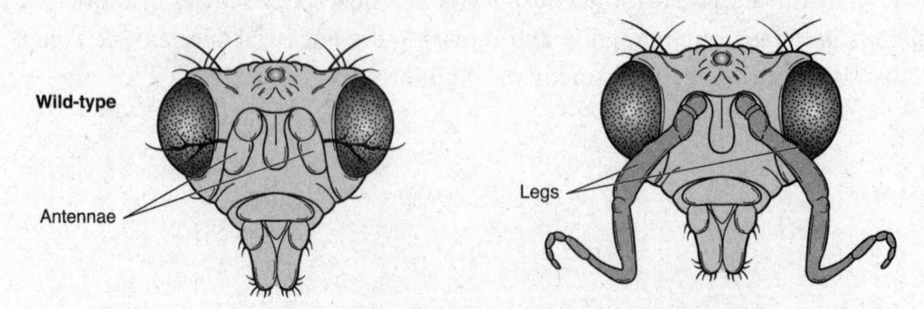

Figure 11.7 Homeotic Genes Control the Placement of Body Parts

HETEROCHRONY

Heterochrony is an evolutionary change in the *rate or timing of the development of body parts.* Figure 11.8 shows how the chimp and human skulls compare as they develop. As infants, they are very similar. As the chimp matures, the face becomes more prognathic (sticks out) and the jaw and teeth become larger and more powerful. In contrast, the development in the human skull seems to stop before those same changes happen. Perhaps the genetic code for the morphology of the skull in chimps and humans is identical. However, something signals the human skull genes to stop developing sooner.

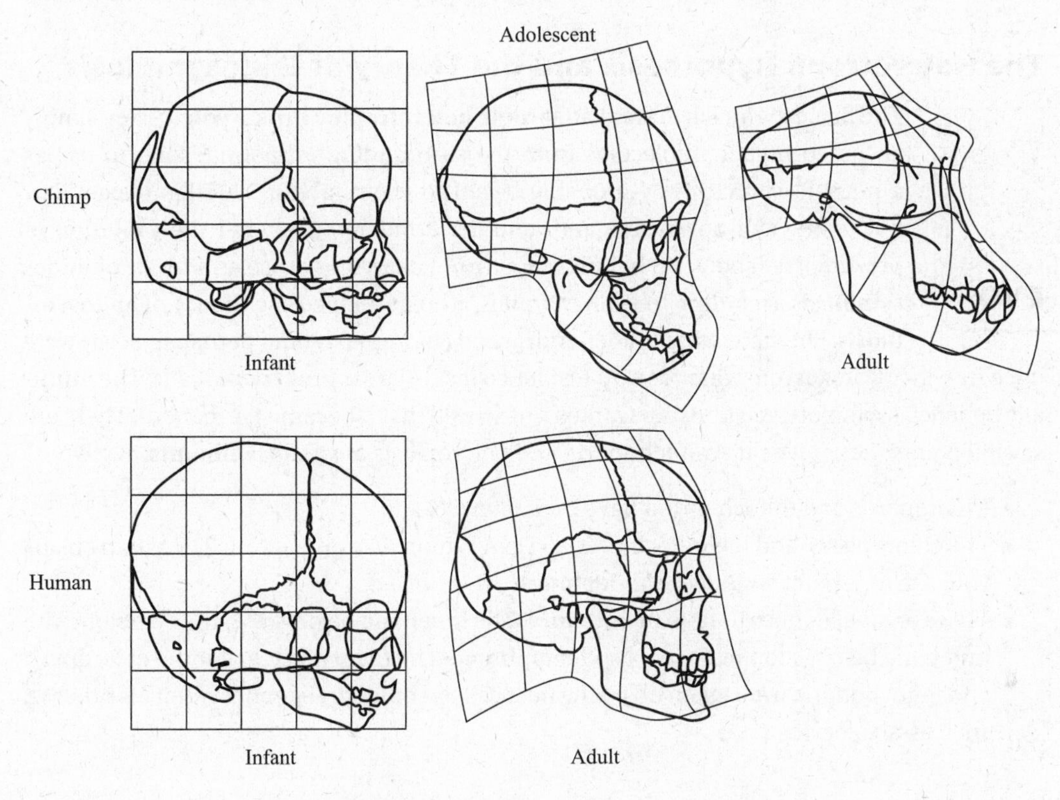

Figure 11.8 Relative Skull Growth Rate

Heterochrony is also responsible for the slow growth of leg and pelvic bones that led to the reduction and eventual loss of hind limbs in whales. It can also alter the timing of reproductive development compared to the rest of the body. One example is the adult aquatic salamander, the *axolotl,* which retains juvenile structures—gills—into adulthood.

REMEMBER

There was no free oxygen in the ancient atmosphere.

THE ORIGIN OF LIFE

The ancient atmosphere consisted of CH_4, NH_3, CO, CO_2, N_2 and H_2O, but lacked free O_2. There was probably intense lightning and ultraviolet (UV) radiation that penetrated the primitive atmosphere, providing energy for chemical reactions. Scientists have tried to mimic this early atmosphere to determine how the first organic molecules and earliest life developed. Here is a synopsis of those classic experiments.

SYI-3
Chemical experiments have shown that it is possible to form complex organic molecules from inorganic ones in the absence of life.

A. I. Oparin and **J. B. S. Haldane**, in the 1920s, hypothesized separately that under the conditions of early Earth, organic molecules could form.

Without corrosively reactive molecular oxygen present to react with and degrade them, organic molecules could form and remain.

Stanley Miller and **Harold Urey**, in the 1950s, tested the Oparin-Haldane hypothesis and proved that almost any energy source would have converted the molecules in the early atmosphere into a variety of organic molecules, including **amino acids**. They used electricity to mimic lightning and UV light that must have been present in great amounts in the early atmosphere.

Sidney Fox, in more recent years, carried out similar experiments. However, he began with organic molecules (not the original inorganic ones) and was able to produce membrane-bound, cell-like structures he called proteinoid microspheres that would last for several hours.

The Heterotroph Hypothesis and the Theory of Endosymbiosis

EVO-2
There are several hypotheses about the natural origin of life on Earth, each with supporting evidence.

The first cells on Earth were **anaerobic heterotrophic prokaryotes**. They simply absorbed organic molecules from the surrounding primordial soup to use as a nutrient source. They probably began to evolve about 3.5 billion years ago. Eukaryotes did not evolve until another 2 billion years after the evolution of prokaryotes (about 1.5 billion years ago). Eukaryotes arose as a result of **endosymbiosis** according to Lynn Margulis, who developed the **theory of endosymbiosis**. She stated that mitochondria and chloroplasts (and perhaps nuclei) were once free-living prokaryotes that took up residence inside larger prokaryotic cells. The mutually beneficial symbiotic relationship worked out so well that it became permanent. There are several points that prove that mitochondria and chloroplasts are **endosymbionts**.

- Chloroplasts and mitochondria have their own DNA.
- The chloroplasts' and the mitochondria's DNA is more like prokaryotic DNA than eukaryotic DNA. It is not wrapped with **histones**.
- These organelles have double membranes. The inner one belonged to the symbiont; the outer one belongs to the host plasma membrane. The theory states that the chloroplast and mitochondria were taken up by the host cell by some sort of endosymbiosis process, such as phagocytosis.

RNA WORLD

The **RNA World** concept hypothesizes that about 4 billion years ago, the first genetic substance on Earth was not the large complex DNA molecule but the small, single-stranded RNA molecule.

It has been proven that **ribozyme**, a type of RNA, can catalyze reactions like enzymes do, as well as transmit information to the next generation like DNA does. In 1989, the Nobel Prize was awarded to Professors Thomas Cech and Sidney Altman for their discovery of ribozyme, which has several surprising functions:

- Functions as an enzyme to catalyze reactions
- Splices RNA by itself, without the need for proteins
- Removes its own *introns* during RNA processing
- Joins amino acids together to form a polypeptide during translation in ribosomes

In the past 50 years since the development of the RNA World hypothesis and the discovery of ribozymes, the concept of RNA World has gone from speculation to a prevailing idea.

Note that the existence of ribozymes is an exception to the idea that all biological catalysts are proteins. RNA is not a protein.

EXAPTATION

We have learned that natural selection is a major mechanism in the evolution of any population—the best adapted traits will be selected. However, not every trait results because it is adaptive.

Traits that evolve by natural selection in one context but are then co-opted for another purpose are called **exaptations**. For example, feathers might have originally arisen as adaptations in the context of selection for insulation, and only later were they co-opted for flight. In this case, the general form of feathers is an *adaptation* for insulation and an *exaptation* for flight.

HALF-LIFE

Being able to date fossils accurately is important for reconstructing the history of life. One of the most common techniques to determine the *absolute* (not errorless) age of a fossil is **radiometric dating**. It is based on the half-life of **radioisotopes**. **Half-life** is the time required for half of the nucleus of a radioactive sample to decay into its products. After one half-life has passed, half the original material has decayed into atoms of a new element. The other half of the element remains intact. Half-lives can be as short as a fraction of a second or as long as billions of years. Carbon-14 has a half-life of 5,730 years. Uranium-238 has a half-life of 4.5 billion years. Half-life is *not* affected by temperature, pressure, or any other environmental conditions.

Fossils contain isotopes of elements that accumulated in the organisms when they were alive. For example, living organisms contain mostly the stable carbon isotope carbon-12 (C-12), as well as a small amount of radioactive carbon-14 (C-14). When the organism dies, it stops accumulating carbon and the amount of C-12 in its tissues remains the same. However, the C-14 that it contained at the time of death slowly decays into another element, nitrogen-14 (N-14). Therefore, by measuring the ratio of C-14 to C-12 in a fossil, you can determine the fossil's age.

This method of carbon dating works for fossils up to age 75,000 years. If a fossil is older than that, it contains too little carbon-14 to be detected by current equipment.

Dating Fossils Older than 75,000 Years

We may be able to date fossils older than 75,000 years *indirectly* if the ancient fossils are embedded in volcanic rock. Lava, as it cools, may absorb and trap radioisotopes from the surrounding environment. If fossils are sandwiched between two layers of volcanic rock *and if* we can date the two layers by a similar means as discussed above, then we know the age range of the fossils. If two volcanic layers surrounding fossils are determined to be 490 and 495 million years old (myo), then we know the age of the fossils to be between 490 and 495 myo.

Calculate Half-Life Problem

A sample contains 100.0 g of thorium-234. After 96.4 days, there is 6.25 g remaining of Th-234. Calculate the half-life of Th-234. Follow this procedure.

First, determine how many half-lives have passed to get from 100.0 g to 6.25 g: 100.0 g divided by 2 = 50.0 g; divided by 2 again = 25.0 g; divided again by 2 = 12.5 g; and divided again by 2 = 6.25 g. You divided 4 times to get from 100.0 g to 6.25 g. This means that 4 half-lives passed to leave 6.25 g remaining from the original 100.0 g.

Now, divide your original 96.4 days by 4 (number of half-lives) = 24.1 days. Therefore, the half-life of thorium-234 is 24.1 days.

CHAPTER SUMMARY

This chapter covered Big Ideas: EVO and SYI.

- Evidence for evolution comes from various scientific fields: **fossil record**, **comparative anatomy**, **comparative biochemistry**, **comparative embryology**, **molecular biology**, and **biogeography**.
- According to Darwin's theory of natural selection, populations tend to grow and exceed their limited resources. Competition for those limited resources results in differential survival. Individuals with more favorable phenotypes are more likely to survive to produce offspring, passing their genes to subsequent generations. **Fittest** is defined only in terms of reproductive success, not necessarily the biggest or fastest.
- A diverse gene pool is important for the survival of a species in a changing environment.
- In addition to natural selection, *chance* and *random, nonselective* events (**genetic drift**, including **bottleneck** and **founder effects**) can influence the evolutionary process, especially for small populations.
- Five conditions must be satisfied for a population to be in Hardy-Weinberg (H-W) equilibrium: 1. large population; 2. absence of migration; 3. no mutations; 4. random mating; and 5. absence of natural selection. These conditions are seldom met. Be able to calculate simple H-W problems.
- Some phenotypes significantly increase or decrease the **fitness** of an organism and of a population, such as sickle cell anemia, peppered moth, DDT resistance, and antibiotic resistance in bacteria.
- Humans alter the environment and affect the survival of other species, such as artificial selection, planting only one variety of a particular crop, overuse of antibiotics, and habitat destruction.
- Fossils can be dated by a variety of methods. See the section entitled "Half-Life."
- **Morphological homologies** provide evidence for evolution and common ancestry, such as similarity of front limb bones in whales, humans, and bats.
- **Biochemical similarities** across all kingdoms provide evidence for evolution and common ancestry. Examples include cytochrome *c* and the fact that major features of the genetic code are shared by all organisms.
- **Cladograms** can be constructed to demonstrate common ancestry and speciation using traits that are **shared** and **derived.** (See Chapter 10.)
- The rate of evolution can vary (**gradualism** versus **punctuated equilibrium**). Evolution is rapid in times of stress, such as peppered moths, bacterial antibiotic resistance, and insect DDT resistance.
- There are several hypotheses about the natural origin of life on Earth: **RNA World hypothesis**, **heterotroph hypothesis**, **theory of endosymbiosis**, the experiments of **Miller** and **Urey**, and so on. Primitive Earth provided inorganic precursors from which organic molecules could have been synthesized. There was no free oxygen present when Earth was formed 4.6 billion years ago.
- There have been five major extinctions.

PRACTICE QUESTIONS

1. Cystic fibrosis (CF) is the most common inherited disease in Caucasian populations. One out of 2,500 newborn infants has the disease, with a calculated carrier frequency of 1 in 25. The disease is caused by an autosomal recessive allele. Until recently, cystic fibrosis was nearly always lethal at a young age. The frequency of the CF gene has caused people to speculate for many years as to why it persists at such high levels in the gene pool. When you have a gene that codes for a fatal disorder, you expect the numbers in the population to decrease. However, this has not happened with CF. In terms of evolution, which of the following is the best explanation for why the CF gene has persisted at such high frequency?

   (A) The gene is caused by a mutation in adults who subsequently are diagnosed with the disease.
   (B) Oddly, people who carry the CF allele have more children than people who do not.
   (C) CF is a dominant gene.
   (D) Carriers of the CF gene have some sort of selective advantage over people who do not have the condition.

2. The graph below shows evolution in a population over time. Which of the following statements best describes what is shown in the graph?

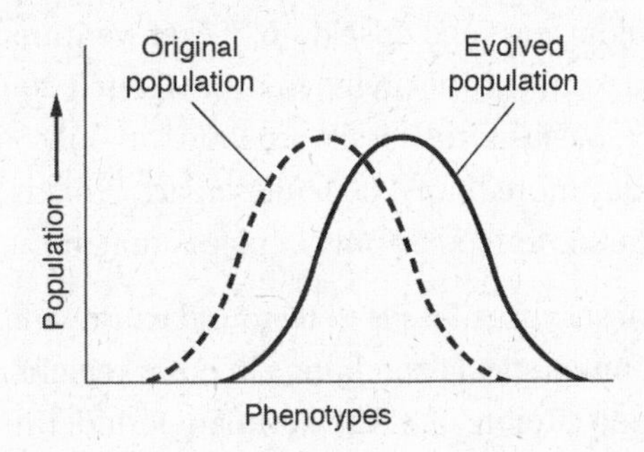

   (A) It is an example of artificial selection because the transformation from one population to the other is so sudden.
   (B) It describes a population where the extremes have been eliminated because they are selected against by rapid changes in the environment.
   (C) This describes a situation in which one phenotype replaces another, such as when a population of bacteria becomes resistant to antibiotics.
   (D) This is an example of diversifying selection in which one population is divided into two distinct populations, perhaps because of pressure from a predator.

3. Which of the following is part of Darwin's theory of evolution?

   (A) The largest, fastest, or smartest animal will survive.
   (B) One polymorphic population of snail exhibited a range of shell colors, from white to brown shells. Where the surrounding gravel was light, snail populations developed white shells because it gave them an advantage.
   (C) Through many years of training, a chimpanzee can learn to use sign language. As a result, her offspring will be able to learn sign language shortly after birth.
   (D) The male elk with the largest antlers will mate with all the females in the herd.

4. Which of the following would be the best *outgroup* for a cladogram about apes?

   (A) gorilla
   (B) Old World monkey
   (C) chimpanzee
   (D) human

5. Antibiotic-resistant bacteria are becoming more common worldwide. According to the CDC, over 80,000 invasive MRSA (methicillin-resistant *Staphylococcus aureus*) infections and 11,285 related deaths occur every year. Which of the following best provides selective pressure from the environment responsible for an increase in antibiotic resistance?

   (A) There must have been a one-time mutation in *S. aureus* bacteria that caused them to become resistant to methicillin.
   (B) Changes in the DNA of resistant *S. aureus* have caused them to become less resistant.
   (C) There is excessive use of antibiotics in humans and animals.
   (D) There was a change in the drug methicillin that weakened it. Scientists must find a way to make the drug stronger.

6. The extinct freshwater reptile Mesosaurus lived during the early Permian period from 299 million to 271 million years ago. It seldom, if ever, ventured onto land. Because it had modifications as a result of only living in freshwater, it could not have swum across the Atlantic Ocean. Yet, Mesosaurus fossils are found in only South Africa and Brazil, two places that are today more than 4,800 miles apart. What is the best explanation for why the Mesosaurus fossils are found only in places that are so far apart?

   (A) There are many Mesosaurus fossils to be found worldwide. There have simply not been enough paleontological expeditions in other regions to find them.
   (B) Because of changing climate after the Permian period, the animal evolved into something that looked quite different. These fossils have not yet been recognized as Mesosaurus.
   (C) The original Mesosaurus range spanned across a small part of the supercontinent Pangaea, which began to break apart 175 million years ago. The area that included the Mesosaurus range is now part of Brazil and South Africa.
   (D) The Mesosaurus expanded out from its original range in Brazil to South Africa by genetic drift.

7. Phage viruses acquire bits of bacterial DNA as they infect one bacterium after another. In some cases, the phage transfers these bits of host bacterial DNA as it infects another bacterium. Which of the following best explains the evolutionary significance of this process for bacteria?

   (A) It increases genetic variation in a population of bacteria and may speed up the population's rate of evolution.
   (B) It explains why the viruses evolve so rapidly.
   (C) It clarifies how bacteria reproduce asexually.
   (D) It accurately describes the process of transformation.

8. Earth has witnessed five major extinctions, when most of Earth's organisms disappeared. Which of the following choices would most likely be the cause of these mass extinctions?

(A) Spontaneous mutations in the most common species made that species suddenly maladapted for their environment.
(B) A diminishing of the sun's power caused the deaths of many plants and the subsequent collapse of the global food web.
(C) Rapid climate change left formerly well-adapted organisms suddenly unfit in the new environment.
(D) A series of massive meteorites caused all these extinctions.

9. In a population of Netherland Dwarf rabbits, the allele for agouti (A) is dominant, while the allele for solid black coloration (a) is recessive. If 65% of the population is agouti in color, what is the frequency of the dominant allele?

(A) 0.2
(B) 0.4
(C) 0.6
(D) 0.8

10. According to the Hardy-Weinberg equation, the dominant allele is represented by

(A) p
(B) q
(C) q^2
(D) p^2

11. The fact that the population of peppered moths in England changed from white to black in fifty years is an example of which type of selection?

(A) stabilizing selection
(B) disruptive selection
(C) directional selection
(D) sexual selection

12. The fact that human newborns usually weigh between 6 and 8 pounds (2.7 and 3.6 kg) is an example of which type of selection?

(A) stabilizing selection
(B) disruptive selection
(C) directional selection
(D) sexual selection

13. The fact that in one region of New Jersey there are two distinct types (two different colors) of one species of snake is an example of which type of selection?

(A) stabilizing selection
(B) disruptive selection
(C) directional selection
(D) sexual selection

14. The fact that large horns and giant antlers are characteristic of males is an example of which type of selection?
   (A) stabilizing selection
   (B) disruptive selection
   (C) directional selection
   (D) sexual selection

15. In 1953, Stanley Miller set up a closed glass system (as shown below), containing a mixture of gases, including H_2 (hydrogen gas), CH_4 (methane), NH_3 (ammonia), and H_2O (water vapor).

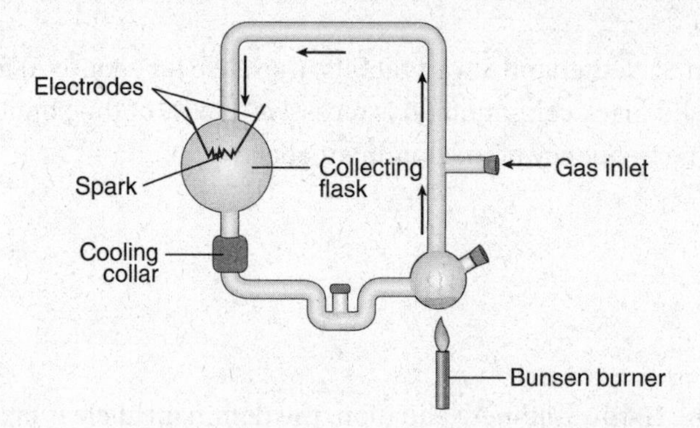

   He discharged sparks to mimic lightning. A condenser cooled the atmosphere, raining water and dissolved molecules into the collecting flask. As material cycled through the apparatus, Miller periodically collected samples for analysis.

   What molecules did Miller find in his collecting flask, and what hypothesis did this result demonstrate?
   (A) He collected simple sugars, which demonstrated that organisms that photosynthesized their own food could have been the first organisms on Earth.
   (B) He collected simple sugars, which demonstrated that hydrogen gas, methane, ammonia, and water vapor were all present on early Earth.
   (C) He collected simple organic molecules under conditions that mimic early Earth, which demonstrated how life could have evolved abiotically on Earth.
   (D) He collected simple organic molecules under conditions that mimic the early solar system, which demonstrated that life may have come to Earth from outer space (panspermia).

16. In the nineteenth century, a large lava flow spread across one of the Galapagos Islands, dividing it into two parts: dry highlands and lowlands that often become inundated by the Pacific Ocean. As a result, a new species of finch evolved on those islands. Which of the following choices provides the correct cause of the evolution of the new species of finch and also includes a correct explanation of that cause?
   (A) Reproductive isolation is the cause of this evolution because the males and females of the two populations can no longer identify each other's mating calls.
   (B) Adaptive radiation is the cause of this evolution because the two populations of finches feed in different niches.
   (C) Geographic isolation is the cause of this evolution because something has physically separated the two populations.
   (D) Temporal isolation is the cause of this evolution because the birds in one region begin nesting at a different time than the birds in the other region.

17. A population of moths has inhabited a region of central Massachusetts for hundreds of years. The moths exist in a wide variety of colors that match the color of tree trunks on which they settle to hide from predators. A large limestone mining quarry recently opened in that same region. The limestone dust from the operation reduces sunlight and pollutes the air with white dust.

The graphs below illustrate four possible changes to the moth population as a result of the change in the environment (due to the presence of the mining operation).

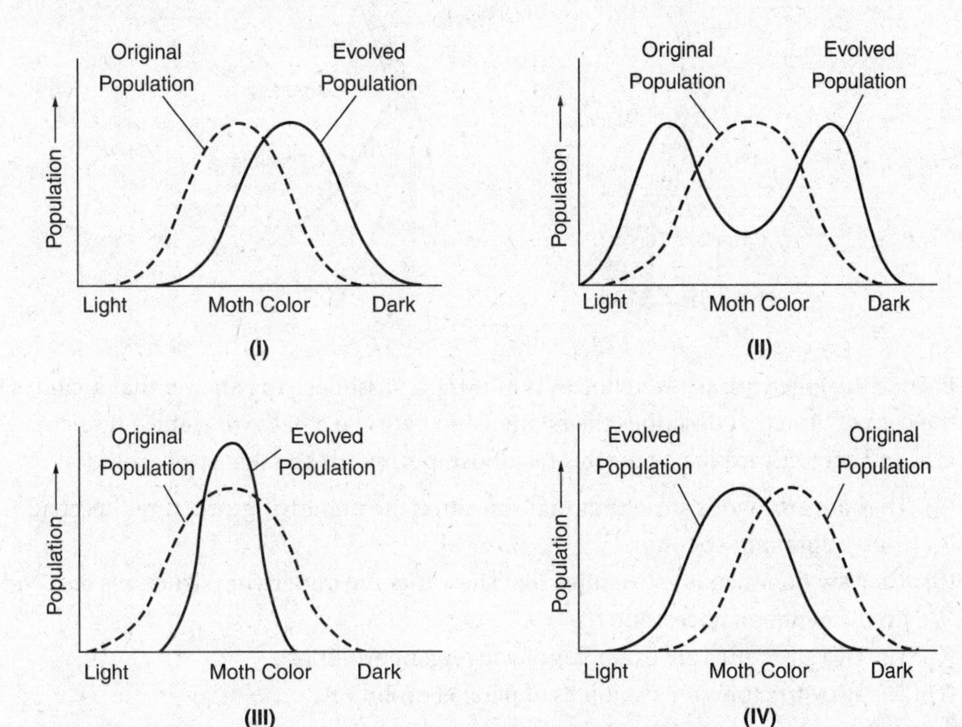

Which of the following predictions includes the most likely change that will occur in the coloration of the moth population in the future and provides the correct reasoning for that answer?

(A) The coloration range will shift toward the darker colors as in graph I because the predators will be able to find the darker moths more easily.
(B) The coloration range will shift toward the extremes as in graph II because more mid-range colored moths will be camouflaged.
(C) The coloration range will become narrower as in graph III because both the lighter and darker moths will be eaten more often.
(D) The coloration range will shift toward the lighter colors as in graph IV because the predators will select moths that are darker.

18. A bat's wing and a human's arm appear very different and have very different functions. However, bats and humans are closely related.

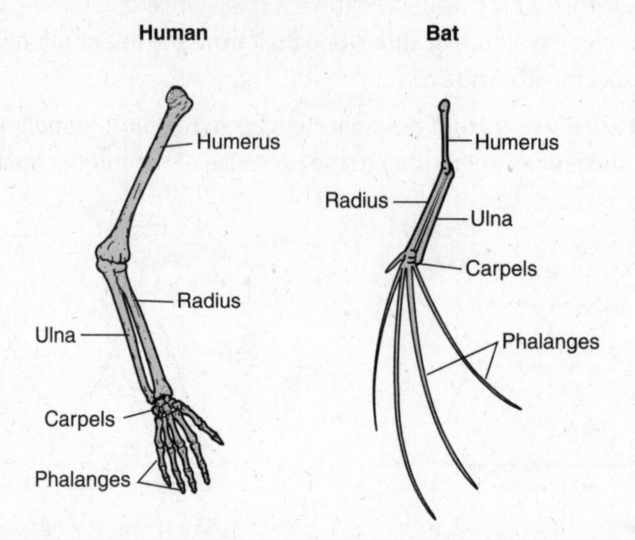

Each of the following answer choices is in two parts. Choose the answer that includes the correct term that describes the relationship between a bat's wing and a human's arm and correctly explains how this relationship supports the theory of evolution.

(A) They are analogous structures that show that the underlying structures descend from a common ancestor.
(B) They are homologous structures that show that the underlying structures descend from a common ancestor.
(C) The two structures are examples of convergent evolution.
(D) The two structures are examples of parallel evolution.

19. The expansion of species of finches in the Galapagos Islands, from 1 species to 14 different species, is an example of which of the following?

(A) founder effect
(B) parallel evolution
(C) adaptive radiation
(D) convergent evolution

20. The establishment of a genetically unique population through genetic drift is an example of which of the following?

(A) founder effect
(B) parallel evolution
(C) adaptive radiation
(D) convergent evolution

21. The independent development of similarities between unrelated groups resulting from adaptation to similar environments is an example of which of the following?
(A) founder effect
(B) parallel evolution
(C) adaptive radiation
(D) convergent evolution

22. The fact that the Tasmanian wolf of Australia is a marsupial but looks very similar to the gray wolf (a placental mammal of North America) is an example of which of the following?
(A) founder effect
(B) parallel evolution
(C) adaptive radiation
(D) convergent evolution

23. The sickle cell trait is recessive. In order to have the disease, a person must have two recessive alleles. Researchers were collecting data on the incidences of sickle cell anemia in children under the age of 14 in Touba, Senegal, in West Africa. They gathered data from 9,851 children, and they discovered that 3,942 had the disease, while 5,909 did not. Assuming that the population is in equilibrium, what percentage of those without the disease are carriers?
(A) 32%
(B) 40%
(C) 47%
(D) 63%

24. Two populations of finches on Isla Santa Cruz, in the Galapagos Islands in Ecuador, are isolated from each other. Although their population density is the same, one population is significantly larger than the other. A new bacterial infection has rapidly spread across the entire island, attacking, among others, both populations of finches.

Which of the following statements best predicts how the new infection will affect the two populations and provides the correct reasoning for that prediction?
(A) The smaller population will be less affected because there will be fewer individuals in that population that could become ill.
(B) The larger population will be more affected because, as a result of more individuals in that population, there will be more mutations that will make more individuals vulnerable.
(C) The larger population will be less affected because there will be greater genetic variation in a large population and because it will be more likely that some individuals will be immune to the infection.
(D) The two populations will be equally affected because they both evolved from one ancestral population and will have similar responses to the infection.

25. Lake Washington is a glacial lake that parallels Washington State's Pacific coast. In the 1950s, it was contaminated with 20 million gallons of phosphorus-laden sewage from surrounding farms. By the 1960s, it had become a 300,000-acre cesspool with water clarity of less than 2 feet. However, in 1968, a historic cleanup helped clear the waters of Lake Washington and transformed the lake into a pristine boaters' paradise. After the cleanup was complete, the lake's water clarity had increased to 15 feet.

    The three-spined stickleback is a small fish (3–4 cm in length) that lives in both marine water and freshwater. In 1955, the saltwater sticklebacks in Lake Washington tended to have up to 30 plates on their bodies to protect them from large marine predators. However, the freshwater species of sticklebacks in Lake Washington tended to have little protective armor in 1955, when scientists began to study the fish in the lake.

    Which of the following statements predicts what you would expect to happen to the freshwater sticklebacks in Lake Washington after the cleanup in 1968 and provides the correct reasoning for that prediction?

    (A) The number of plates on any individual freshwater stickleback fish would increase if the threat from predation were great enough.
    (B) The average number of plates in the freshwater stickleback fish population would not change, even though the water became clearer, because the number of plates on a fish is genetically determined.
    (C) The average number of plates in the freshwater stickleback fish population would increase as the water became clearer because having more protection from predation would be an advantage.
    (D) The average number of plates in the freshwater stickleback fish population would increase as the water became clearer as a result of increased UV radiation and an increase in mutations in the fish population.

Answers Explained

1. **(D)** The fact that the CF gene has not disappeared is a puzzle. However, experimentation with mice who carry the gene has demonstrated that a CF carrier is resistant to cholera. Published work also shows that the CF gene may confer resistance to typhoid and tuberculosis. Choices A and B do not make sense given the information in the question. The fact that any gene is dominant has nothing to do with how common it is in a population. Generally, a gene remains in a population if it is somehow adaptive or, at least, not harmful. (EVO-1)

2. **(C)** This graph shows one population at one time changing (moving to the right) into another population. Choice A is predicated on sudden and extreme change. Nothing in the graph tells you about how quickly the change happened. Choice B is not correct because it describes stabilizing selection, whereas the graph shows directional selection. Choice D describes a graph where one population divides into two. (EVO-1 & SP 1)

3. **(D)** Evolutionary success is measured by reproductive success. The organism that produces the most offspring is considered the most successful. Choice A is not correct because the *best-adapted* organism survives; that is not always the largest, fastest, or smartest organism. Choice B is not correct because the wording implies that the snail needed to be a certain color and so it became that color. Organisms do not evolve because they "need" to. If they are not adapted, they migrate or die. Choice C is not correct because acquired traits are not inherited. (EVO-1)

4. **(B)** Monkeys have tails and are only very distantly related to apes. The other choices are all apes and are closely related. (EVO-3)

5. **(C)** Overexposure to antibiotics provided the selective pressure for antibiotic-resistant bacteria to take over. When exposed to antibiotics, all nonresistant bacteria die leaving only resistant ones to proliferate. Drug companies did not change the formula of the antibiotic. They did not make it weaker. (EVO-1)

6. **(C)** There is lots of fossil evidence about the breakup of Pangaea. This is merely one example. There has been an enormous amount of paleontological evidence supporting the existence of Pangaea. (SYI-3)

7. **(A)** Bacteria are able to evolve rapidly because of tremendous variation within their gene pool. (EVO-1)

8. **(C)** Rapid changes in the environment can cause rapid changes in species—or even mass extinctions. (EVO-3)

9. **(B)** If 65% of the population is agouti, then 35% of the population is black, which means that $q^2 = 0.35$. If $q^2 = 0.35$, then $q = 0.591$ or 0.6. If $q = 0.6$, then the frequency of the p allele, which is the dominant allele, is 0.4 because $p + q = 1$. (EVO-1 & SP 5)

10. **(A)** According to Hardy-Weinberg equilibrium, p is the dominant allele and q is the recessive allele. (EVO-1)

11. **(C)** The black peppered moths replaced the white peppered moths. Since one characteristic replaced another, this is directional selection. (EVO-1)

12. **(A)** Stabilizing selection tends to eliminate the extremes in a population. (EVO-1)

13. **(B)** Disruptive selection tends to select for the extremes. Originally, there was probably a range of coloration of snakes in the area in question. Over time, pressure from the environment selected against different colorations until only two remained. (EVO-1)

14. **(D)** Sexual selection has to do with the selection for traits that attract a mate. (EVO-1)

15. **(C)** The purpose of this now-famous experiment was to demonstrate that life on Earth could have begun from simple inorganic molecules. (SYI-3)

16. **(C)** Choice A is not correct because the stem of the question says nothing about bird calls. Choice B is not correct because the stem of the question says nothing about feeding. Choice D is not correct because, again, the stem of the question says nothing about nesting. (EVO-1)

17. **(D)** Notice the labels for "Light" and "Dark" on the x-axis of each of the four graphs. The *population* of moths will change in color due to pressure from the environment. *No individual* will change. Within the moth population, there are both dark and light moths. In the new environment, the dark moths will be seen more readily by predators and get eaten. The remaining ones (to be counted by researchers) will be lighter. (EVO-1 & SP 5)

18. **(B)** The two structures are homologous because they have the same underlying structure, which reflects shared ancestry. Analogous structures do not have similar underlying structures and do not reflect common ancestry. These two structures are not examples of convergent evolution nor are they examples of parallel evolution. (EVO-3 & SP 1)

19. **(C)** Adaptive radiation is the emergence of numerous species from one common ancestor introduced into a new environment. Today, 14 different species of finches are on the Galapagos Islands where originally there was only 1 species. Each species fills a different niche. (EVO-1)

20. **(A)** Genetic drift is evolution through chance. The founder effect is one example of genetic drift. Another is the bottleneck effect. (EVO-1)

21. **(D)** The classic example of convergent evolution can be seen in the whale and the shark. The two animals are unrelated; the whale is a mammal and the shark is a fish. However, they look alike because they experience the same environmental pressures. They both have a streamlined appearance with fins because that design is best for living in the ocean, not because they are related or have a recent common ancestor. (EVO-3)

22. **(B)** Eutherians (placental mammals) and marsupials are closely related although they diverged several million years ago. Although they live thousands of miles apart, these two animals live in similar environments and are under the same selective pressures from their respective environments. As a result, they have evolved along similar parallel lines. (EVO-3)

23. **(C)** 3,942 children had the disease out of a total of 9,851 children. Therefore, 40% of the children are afflicted, and 40% = $0.40 = q^2$. If $q^2 = 0.40$, then $q = 0.63$. If $q = 0.63$, then $p = 0.37$. To determine what percentage of those without the disease are carriers, that is $2(p)(q) = 2(.37)(.63) = 0.47$ or 47%. (EVO-1 & SP 5)

24. **(C)** The essence of this question is the fact that the more genetic variation there is in any population, the more that population will be able to withstand changes or pressure from the environment. (EVO-1)

25. **(C)** Choice C is the most reasonable statement and is in accordance with the theory of natural selection. Choice A is not correct because individual organisms do not change because they need to. What changes is the rate of occurrence of a trait in a population. If a trait is advantageous, it will likely spread within a population over time. Choice B is not correct because it denies that populations change over time, which is the essence of the theory of evolution. Choice D is not correct because a mutation could increase or decrease the number of plates on a fish. (EVO-1)

Ecology

12

→ PROPERTIES OF POPULATIONS
→ POPULATION GROWTH
→ COMMUNITY ECOLOGY
→ ENERGY FLOW AND PRIMARY PRODUCTION
→ ENERGY FLOW AND THE FOOD CHAIN
→ ECOLOGICAL SUCCESSION
→ BIOMES
→ CHEMICAL CYCLES
→ HUMANS AND THE BIOSPHERE

Big Ideas: EVO, ENE, IST & SYI

Enduring Understandings: EVO-1; ENE-1, ENE-3 & ENE-4; IST-5; SYI-1, SYI-2 & SYI-3

Science Practice: 2

For the complete list of Big Ideas, Enduring Understandings, and Science Practices, refer to the "AP Biology Course and Exam Description" from the College Board: *https://apcentral.collegeboard.org/pdf/ap-biology-course-and-exam-description.pdf?course=ap-biology*.

INTRODUCTION

Ecology is the study of the interactions of organisms with their physical environment and with each other. Here is some introductory vocabulary for the topic.

1. A **population** is a group of individuals of one species living in one area that have the ability to interbreed and interact with each other.
2. A **community** consists of all the organisms living in one area.
3. An **ecosystem** includes all the organisms in a given area, as well as the abiotic (nonliving) factors with which they interact.
4. **Abiotic factors** are nonliving and include temperature, water, sunlight, wind, rocks, and soil.
5. **Biosphere** is the global ecosystem.

PROPERTIES OF POPULATIONS

Here are 5 properties of populations you should know.

1. Size

Size is the total number of individuals in a population and is represented by *N*.

2. Density

Density is the number of individuals per unit area or volume. Counting the number of organisms inhabiting a certain area is often very difficult, if not impossible. For example, imagine trying to count the number of ants in 1 acre (0.5 ha) of land. Instead, scientists use **sampling techniques** to estimate the number of organisms living in one area. One sampling technique commonly used to estimate the size of a population is called **mark and recapture**. In this technique, organisms are captured, tagged, and then released. Some time later, the same process is repeated and the following formula is used for the collected data,

$$N = \frac{\text{(number marked in first catch)} \bullet \text{(total number in second catch)}}{\text{number of recaptures in second catch}}$$

Suppose 50 zebra mussels are captured, marked, and released. One week later, 100 zebra mussels are captured and 10 are found to have markings already. When using the formula, the total population would be about 500 zebra mussels.

3. Dispersion

Dispersion is the pattern of spacing of individuals within the area the population inhabits; see Figure 12.1. The most common pattern of dispersion is **clumped**. Fish travel this way in schools because there is safety in numbers. Some populations are spread in a **uniform** pattern. For example, certain plants may secrete toxins that keep away other plants that would compete for limited resources. **Random** spacing occurs in the absence of any special attractions or repulsions. Trees can be spaced randomly in a forest.

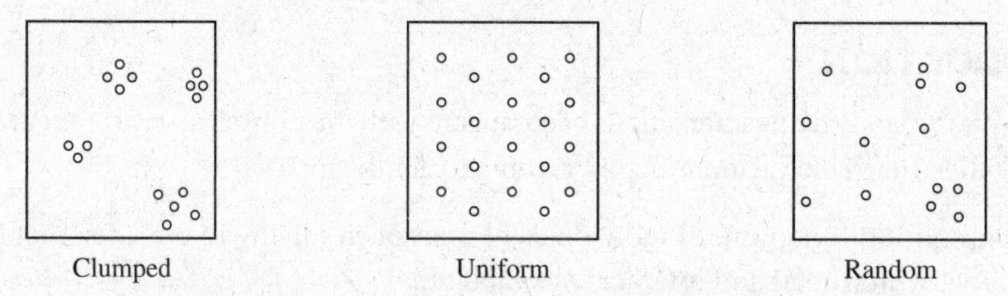

Figure 12.1 Dispersion Pattern of Individuals in One Area

4. Survivorship Curves

Survivorship or **mortality curves** show the size and composition of a population. There are three types of survivorship curves.

- **Type 1** curves show organisms with low death rates in young and middle age and high mortality in old age. There is a great deal of parenting, which accounts for the high survival rates of the young. This is characteristic of humans.

- **Type 2** curves describe a species with a death rate that is constant over the life span. This describes the hydra, reptiles, and rodents.
- **Type 3** curves show a very high death rate among the young but then show that death rates decline for those few individuals that have survived to a certain age. This is characteristic of fish and invertebrates that release thousands of eggs, have external fertilization, and have no parenting; see Figure 12.2.

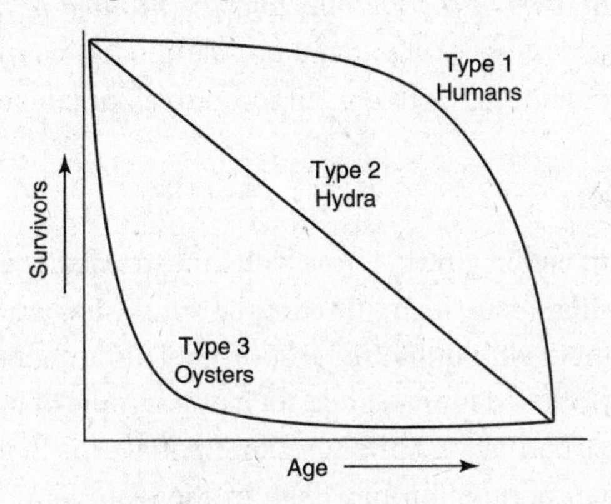

Figure 12.2 Survivorship Curves

SYI-1
The stability of populations, communities, and ecosystems is affected by interactions with both biotic and abiotic factors.

5. Age Structure Diagrams

Another important parameter of populations is age structure. An age structure diagram shows the relative numbers of individuals at each age. Figure 12.3 shows two age structure diagrams. Country I shows the age structure of the human population of India; the pyramidal shape is characteristic of developing nations with half the population under the age of 20. Even after taking into account the disease, famine, natural disasters, and emigration that will occur, the population in 20 years will be enormous. Country II shows an age structure for a developed nation like the United States with a stable population, **zero population growth**, where the number of people in each age group is about the same and the birth rates and the death rates are about equal.

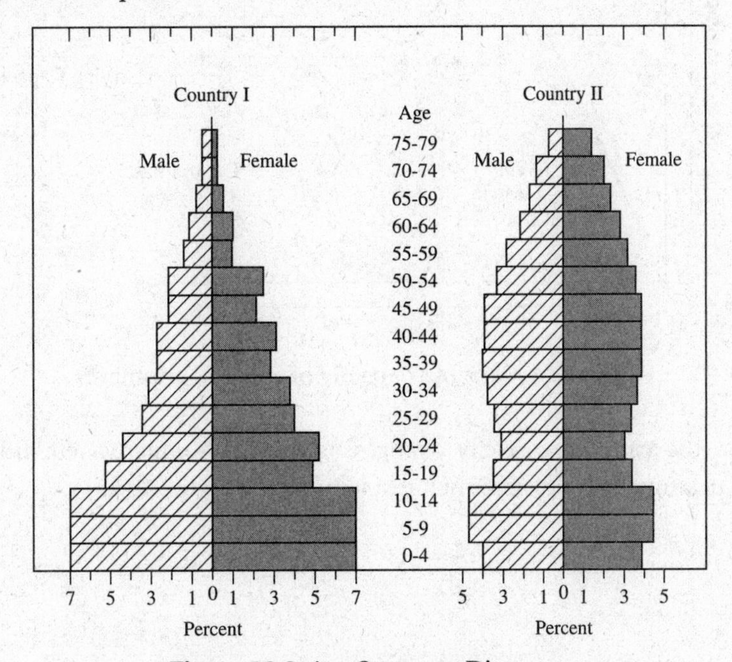

Figure 12.3 Age Structure Diagrams

POPULATION GROWTH

Every population has a characteristic **biotic potential**, the maximum rate at which a population could increase under ideal conditions. Different populations have different biotic potentials, which are influenced by several factors. These factors include *the age at which reproduction begins, the life span during which the organisms are capable of reproducing, the number of reproductive periods in the lifetime, and the number of offspring the organism is capable of reproducing.* Regardless of whether a population has a large or small biotic potential, certain characteristics about growth are common to all organisms.

Exponential Growth

The simplest model for population growth is one with unrestrained or **exponential growth**. This population has no predation, parasitism, or competition. It has no immigration or emigration and is in an environment with unlimited resources. This is characteristic of a population that has been recently introduced into an area, such as a sample of bacteria newly inoculated onto a petri dish. Although exponential growth is usually short-lived, the human population has been in the exponential growth phase for over 300 years.

> **SYI-1**
> A population can produce a density of individuals that exceeds the system's resources.

Carrying Capacity

Ultimately, there is a limit to the number of individuals that can occupy one area at a particular time. That limit is called the **carrying capacity (K)**. Each particular environment has its own carrying capacity around which the population size oscillates; see Figure 12.4.

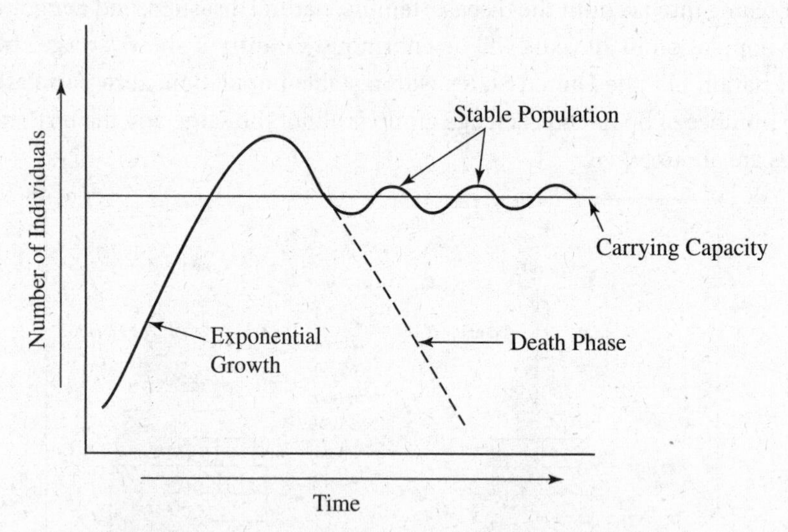

Figure 12.4 Carrying Capacity of One Environment

In addition, the carrying capacity changes as the environmental conditions change. Perhaps a fire destroyed several acres of forest habitat. See Figure 12.5.

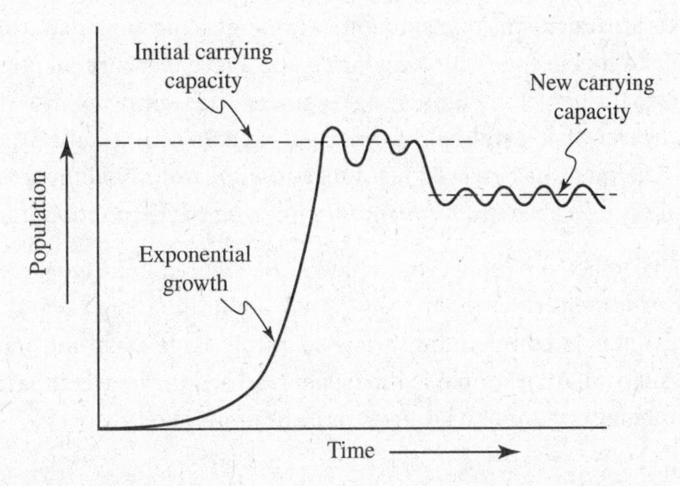

Figure 12.5 Carrying Capacity Changes as Environmental Conditions Change

Limiting Factors

Limiting factors are those factors that limit population growth. They are divided into two categories: **density-dependent** and **density-independent** factors.

- **Density-dependent factors** are those factors that increase directly as the population density increases. They include competition for food, the buildup of wastes, predation, and disease.
- **Density-independent factors** are those factors whose occurrence is unrelated to the population density. These include earthquakes, storms, and naturally occurring fires and floods.

Growth Patterns and Life Histories

Some species are opportunistic; they reproduce rapidly when the environment is uncrowded and resources are vast. They are referred to as **r-strategists**. Other organisms, the **K-strategists**, live in a dense population near the carrying capacity (K), where the competition is fierce. Table 12.1 compares the two life strategies.

Table 12.1

Comparison of Two Life Strategies	
r-strategists	**K-strategists**
Many young	Few young
Little or no parenting	Intensive parenting
Rapid maturation	Slow maturation
Small young	Large young
Reproduce once	Reproduce many times
Example: insect	Examples: mammals and birds

- **r-strategists** are organisms that sometimes undergo a big bang pattern of reproduction. An example of this in the plant kingdom is the agave (century plant), which grows for many years in the desert, accumulating resources and waiting for the "right moment." In an unusually wet year, it suddenly sends up a large flowering stalk, produces seeds, and then dies. The plant has evolved a strategy to increase probability of success through sheer numbers *when the environment is in flux and the population has decreased.*
- **K-strategists**, which are repeat reproducers, are favored in a *stable environment where competition for resources is intense*, and where adults are more likely to survive to breed again. With intense competition, a few relatively large offspring may have a better chance at survival. An example is the loggerhead sea turtle, which lays about 300 eggs per year and may continue to lay eggs for 30 or more years.

A Case Study—The Hare and the Lynx

An excellent study in population growth involves the populations of **snowshoe hare** and **lynx** at the Hudson Bay Company, which kept records of the pelts sold by trappers from 1850–1930. The data reveal fluctuations in the populations of both animals. The hare feeds on the grass, and the lynx feeds on the hare. So the cycles in the lynx population are probably caused by cyclic fluctuations in the hare population. The hare population experiences cycles of **exponential growth** and **crashes**. Additionally, cycles in the hare population are probably due to a **limited food supply** for the hare due to a combination of malnutrition from cyclical overcrowding and overgrazing and of predation by the lynx; see Figure 12.6.

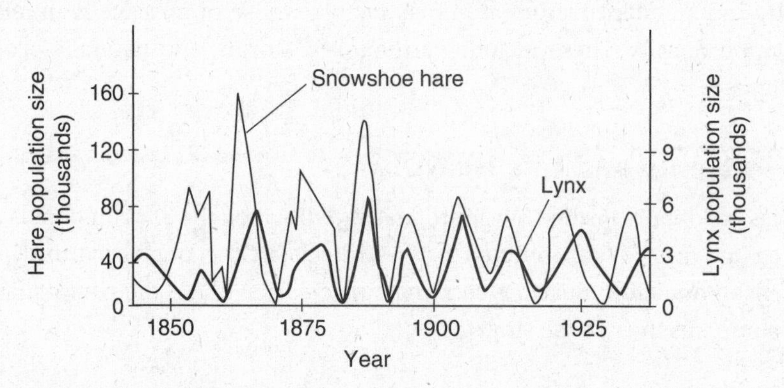

Figure 12.6 Fluctuations in Hare and Lynx Populations from 1850 to 1930

COMMUNITY ECOLOGY

Communities are made of populations that interact with the environment and with each other. They are characterized by how diverse and dense they are. Species diversity has two components. One is **species richness**, the number of different species in the community. The other is **relative abundance** of the different species. In general, diverse communities are *more productive* because they are more stable and survive for longer periods of time. They are also better able to withstand and recover from environmental stresses such as a drought or an incursion by an invasive species.

SYI-3
The structure of a community is measured and described in terms of species composition and species diversity.

Interactions within a community are complex but can be divided into five categories: **competition**, **predation**, **herbivory**, **symbiosis**, and **facilitation**.

1. Competition

The Russian scientist **G. F. Gause** developed the **competitive exclusion principle** after studying the effects of interspecific competition in a laboratory setting. He worked with two very similar species: *Paramecium caudatum* and *Paramecium aurelia.* When he cultured them separately, each population grew rapidly and then leveled off at the carrying capacity. However, when he put the two cultures together, *P. aurelia* had the advantage and drove the other species to extinction. His principle states that *two species cannot coexist in a community if they share a* ***niche****, that is, if they use the same resources.*

In nature, there are two related outcomes, besides extinction, if two species *inhabit the same niche and therefore compete for resources.* One of the species will evolve through natural selection to exploit different resources. This is called **resource partitioning**. Another possible outcome is what occurred on the **Galapagos Islands**. Finches evolved different beak sizes through natural selection and were able to eat different kinds of seeds and avoid competition. This divergence in body structure is called **character displacement.**

2. Predation

Predation can refer to one animal eating another animal, or it can also refer to animals eating plants. For their protection, animals and plants have evolved defenses against predation.

SYI-1
Interactions among populations affect the distribution and abundance of populations.

Animals have evolved **active defenses** such as **hiding**, **fleeing**, or **defending** themselves. These, however, can be very costly in terms of energy. Animals have also evolved **passive defenses** such as **cryptic coloration** or **camouflage** that make the prey difficult to spot. Here are three examples:

- **Aposematic coloration** is the very bright, often red or orange, coloration of poisonous animals as a warning that possible predators should avoid them.
- **Batesian mimicry** is copycat coloration where one harmless animal mimics the coloration of one that is poisonous. One example is the **viceroy butterfly**, which is harmless but looks very similar to the **monarch butterfly**, which stores poisons in its body from the milkweed plant.
- In **Müllerian mimicry**, two or more poisonous species, such as the cuckoo bee and the yellow jacket, resemble each other and gain an advantage from their combined numbers. Predators learn more quickly to avoid any prey with that appearance.

3. Herbivory

Herbivory generally refers to an interaction (+/–) in which an organism eats part or parts of a plant or alga. Cattle graze on grass. Invertebrates like beetles and grasshoppers also eat vegetation. In the oceans, herbivores include snails, sea urchins, and manatees. Like predators, herbivores have special adaptations for grazing the right plant body or flower, such as specialized teeth or a modified digestive system. **Plants** have evolved **spines** and **thorns** and chemical **poisons**, such as **strychnine**, **mescaline**, **morphine**, and **nicotine**, to fend off attacks by animals.

4. Symbiosis

When two or more species live in direct and intimate contact with each other, their relationship is called **symbiotic**. These relationships can be helpful, neutral, or harmful.

A. MUTUALISM is a symbiotic relationship where both organisms benefit (+/+). An example is the bacteria that live in the human intestine and produce vitamins. Another example of mutualism is **mycorrhizae**, symbiotic relationships between roots and fungi that enhance nutrient uptake and were critical to the development of vascular plants.

B. COMMENSALISM is a symbiotic relationship where one organism benefits and one is unaware of the other organism (+/o). Barnacles that attach themselves to the underside of a whale benefit by gaining access to a variety of food sources as the whale swims into different areas. In addition, the whale is unaware of the barnacles.

C. PARASITISM is a symbiotic relationship where one organism, the parasite, benefits while the host is harmed (+/–). A tapeworm in the human intestine is an example of a parasite.

5. Facilitation

Organisms can have positive effects (+/+ or 0/+) on the survival and reproduction of other species without living in direct and intimate contact with them. A good example of facilitation is the black rush *Juncus gerardi*, which makes the soil more hospitable for other species in New England salt marshes. It helps prevent salt buildup in the soil by shading the soil surface and reducing evaporation.

ENERGY FLOW AND PRIMARY PRODUCTION

Every day, Earth is bombarded with enough sunlight to supply the needs of the entire human population for the next 25 years. Most solar radiation, though, is absorbed, scattered, or reflected by the atmosphere. Only a small fraction actually reaches green plants, and less than 1% is actually converted to chemical bond energy by photosynthesis. However, that energy is the basis for almost all of Earth's food chains and fuels all life on Earth. (An example of a food chain that does not rely on solar energy is one located around deep-ocean thermal vents.) Ecologists use two terms when they discuss energy flow on Earth: gross primary productivity and net primary productivity. **Gross primary productivity (GPP)** is the amount of light energy that is converted to chemical energy by photosynthesis per unit time. **Net primary productivity (NPP)** is equal to the GPP minus the energy used by producers for their own cellular respiration.

ENE-1
All living systems require the constant input of free energy.

Different ecosystems vary in their NPP as well as what they contribute to the total or **global NPP** of Earth. Tropical rainforests are among the most productive terrestrial ecosystems and contribute a large portion of Earth's overall net primary production. (Unfortunately, that number is shrinking as we cut down rainforests to make way for farming.) Coral reefs, on the other hand, have a very high NPP but contribute relatively little to the global NPP because they occupy such a tiny part of the planet. The open oceans are just the opposite of coral reefs. Their NPP is very low per unit area. Because they occupy three-fourths of the globe, however, their contribution to the global NPP is higher than that of any other biome.

ENERGY FLOW AND THE FOOD CHAIN

The **food chain** is the pathway along which food is transferred from one **trophic** or feeding **level** to another. Energy, in the form of food, moves from the **producers** to the **herbivores** to the **carnivores**. Only about **10 percent** of the energy stored in any **trophic level** is converted to organic matter at the next trophic level. This means that if you begin with 10,000 kJ of plant matter, the food chain can support 1,000 kJ of herbivores (primary consumers), 100 kJ of secondary consumers, and only 10 kJ of tertiary consumers. As a result of the loss of energy from one trophic level to the next, food chains are rather short. They never have more than four or five trophic levels. As you might expect, long food chains are less stable than short ones. This is because population fluctuations at lower trophic levels are magnified at higher levels, causing local extinction of top predators. A good model to demonstrate the interaction of organisms in the food chain and the loss of energy is the **food pyramid**; see Figure 12.7.

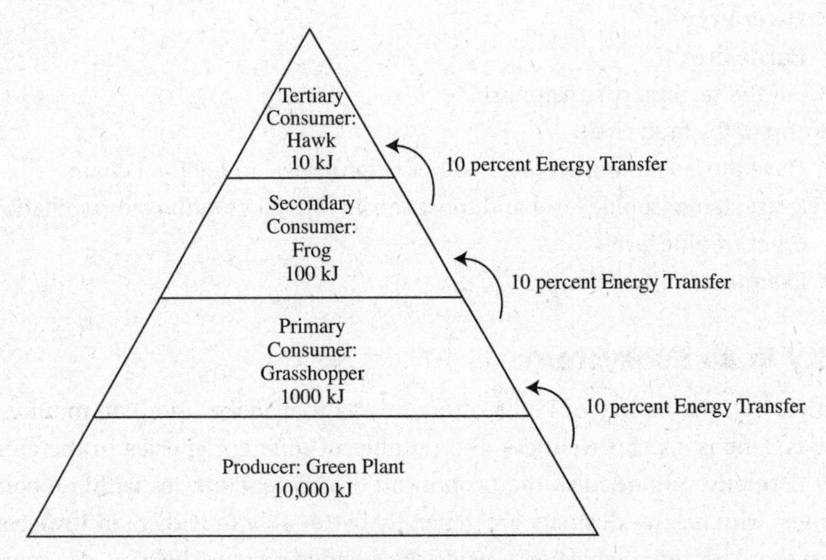

Figure 12.7 Food Pyramid

Food chains are not isolated; they are interwoven with other food chains into a **food web**. An animal can occupy one trophic level in one food chain and a different trophic level in another food chain. Humans, for example, can be primary consumers when eating vegetables but are tertiary consumers when eating a steak. Here are two sample food chains, each with four trophic levels:

Producers → Primary Consumers → Secondary Consumers → Tertiary Consumers

Terrestrial Food Chain
Green Plant → Grasshopper → Frog → Hawk

Marine Food Chain
Phytoplankton → Zooplankton → Small Fish → Shark

- **Producers**
  - ✔ **Autotrophs**
  - ✔ **Green plants**
  - ✔ Convert light energy to chemical bond energy
  - ✔ Have the greatest biomass of any trophic level
  - ✔ Examples: diatoms and phytoplankton
- **Primary consumers**
  - ✔ **Heterotrophs**
  - ✔ **Herbivores**
  - ✔ Eat the producers
  - ✔ Examples: grasshoppers, zooplankton

- **Secondary consumers**
  - ✔ **Heterotrophs**
  - ✔ **Carnivores**
  - ✔ Eat the primary consumers
  - ✔ Examples: frogs, small fish
- **Tertiary consumers**
  - ✔ **Heterotrophs**
  - ✔ **Carnivores**
  - ✔ Eat the secondary consumers
  - ✔ Top of the food chain
  - ✔ Have the **least biomass** of any other trophic level in the food chain
  - ✔ Least stable trophic level and most sensitive to fluctuations in populations of the other trophic levels
  - ✔ Examples: hawks, sharks

Diversity in an Ecosystem

Species diversity, the variety of kinds of organisms that make up a community, has two components. One is **species richness**, the number of different species in the community. The other is **relative abundance**, the proportion of different species within a community. Communities with greater diversity are generally better able to withstand **invasive species** (organisms that become established outside their native range). They are also more able to recover from environmental stresses, such as drought or disease. Here is one unfortunate present-day example of the importance of diversity in a community.

> **ENE-4 & SYI-1**
> Communities are composed of populations of organisms that interact in complex ways.

Although there are several thousand varieties of banana in the world, plantations only cultivate the one variety we eat, the Cavendish. Unfortunately, a fungus, *Fusarium oxysporum*, is attacking the Cavendish worldwide. It has already destroyed crops across Taiwan, Indonesia, and Malaysia, and threatens plantations in South and Central America. This lack of diversity in banana culture has left the banana vulnerable to disease, and may leave us with no bananas to eat.

Dominant species in a community are the species that are the most abundant or that collectively have the highest biomass. They exert control over the abundance and distribution of other species. Sugar maples in North American forests are an example. They affect the abiotic factors, such as shade and soil nutrients (from rotting leaves), which in turn provide special habitats for many other species.

Keystone species are not abundant in a community. However, they exert major control over other species in the community. Sea otters in the North Pacific are a perfect example. They are high in the food chain and feed on sea urchins, which feed mainly on kelp. Where the sea otters are abundant, there are few sea urchins and kelp forests are abundant. In contrast, where orcas feed on sea otters, sea urchins are abundant and kelp is rare.

BOTTOM-UP AND TOP-DOWN CONTROL OF A COMMUNITY

There are two simplified models for the structure of a community based on direction of influence. The **bottom-up model** focuses on influence from lower to higher trophic levels. For example, an increase of minerals available in the environment will increase the biomass of the producers, and will increase the biomass up to and including the highest trophic level.

However, if you add or remove predators to and from the *bottom-up community,* the effect will not extend down to the bottom levels.

In contrast, the **top-down model** (developed in 1963) states the opposite. Removing the top carnivores from a community increases the abundance of lower primary carnivores, which decreases the numbers of herbivores, resulting in an increase in the mass of producers. This model is also called the **trophic cascade model**. An excellent example is what happened in Yellowstone Park. The U.S. National Park Service exterminated the wolf population in the park in 1926, but reintroduced them in the 1990s after public sentiment shifted. During the years when there were no wolves, the aspen tree population in the park decreased from 6% to 1% because of intense browsing by Rocky Mountain elk. When wolves were reintroduced into the park, the aspen trees looked healthier and their density increased substantially. Presumably this resulted from the top-down pressure on the elk population by the wolf population.

Biological Magnification

Organisms at higher trophic levels have greater concentrations of accumulated toxins stored in their bodies than those at lower trophic levels. This phenomenon is called **biological magnification**. The bald eagle almost became extinct because Americans sprayed heavily with the pesticide DDT in the 1950s, which entered the food chain and accumulated in the bald eagle at the top of the food chain. Because DDT interferes with the deposition of calcium in eggshells, the thin-shelled eggs were broken easily and few eaglets hatched. DDT is now outlawed, and the bald eagle was saved from extinction by human intervention.

Decomposers

Decomposers—**bacteria** and **fungi**—are usually not depicted in any diagram of a food chain. However, without decomposers to recycle nutrients back to the soil to nourish plants, there would be no food chain and no life.

ECOLOGICAL SUCCESSION

Most communities are dynamic, not stable. The size of a population increases and decreases around the carrying capacity. Migration of a new species into a habitat can alter the entire food chain. Major disturbances, whether natural or human-made, like volcanic eruptions, strip mining, clear-cutting a forest, and forest fires, can suddenly and drastically destroy a community or an entire ecosystem. What follows this destruction is the process of sequential rebuilding of the ecosystem called **ecological succession**. See Figure 12.8.

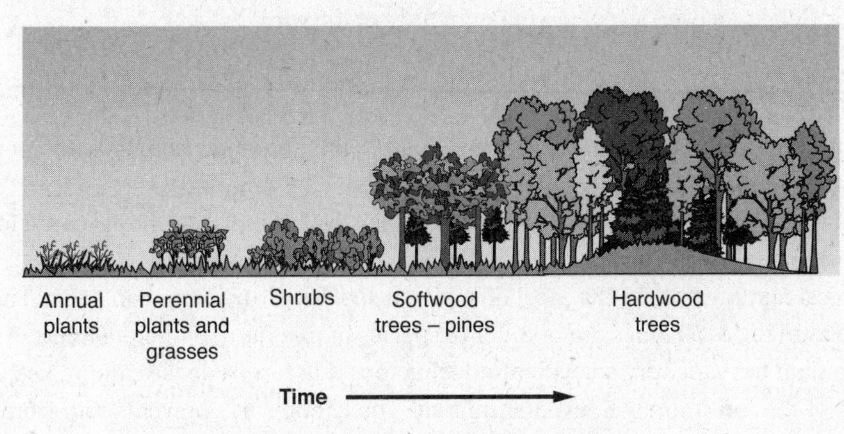

Figure 12.8 Ecological Succession

If the rebuilding begins in a lifeless area where even soil has been removed, the process is called **primary ecological succession**. *The essential and dominant characteristic of primary succession is soil building.* After an ecosystem is destroyed, the first organisms to inhabit a barren area are **pioneer organisms** like **lichens** (a symbiont consisting of algae and fungi) and **mosses**, which are introduced into the area as spores by the wind. Soil develops gradually as rocks weather and organic matter accumulates from the decomposed remains of the pioneer organisms. Once soil is present, pioneer organisms are overrun by other larger organisms: grasses, bushes, and then trees. The final stable community that remains is called the **climax community**. It remains until the ecosystem is once again destroyed by a **blowout**, a disaster that destroys the ecosystem once again.

One example of **primary succession** that was studied in detail occurred at the southern edge of Lake Michigan. As the lakeshore gradually receded northward after the last ice age (10,000 years ago), it left a series of new beaches and sand dunes exposed. Today, someone who begins at the water's edge and walks south for several miles will pass through a series of communities that were formed in the last 10,000 years. These communities represent the various stages beginning with bare, sandy beach and ending with a climax community of old, well-established forests. In some cases, the climax community is a beech-sugar maple forest; in other areas the forest is a mix of hickory and oak.

The process known as **secondary succession** occurs when an existing community has been cleared by some disturbance that leaves the soil intact. This is what happened in 1988 in Yellowstone National Park when fires destroyed all the old growth that was dominated by lodgepole pine but left the soil intact. Within one year, the burned areas in Yellowstone were covered with new vegetation.

While it is unlikely that the AP exam will specifically test you on biomes, they are an important biological phenomena that you should be familiar with.

BIOMES

Biomes are very large regions of the earth whose distribution depends on the amount of **precipitation** and **temperature** in an area. Each biome is characterized by **dominant vegetation** and **animal** life. There are many biomes, including freshwater, marine, terrestrial, and aquatic. If you begin at the equator and travel north, you will pass through several terrestrial biomes: tropical rainforest, desert, temperate grassland, temperate deciduous forest, taiga, and finally, tundra in the north. Changes in altitude produce effects similar to changes in latitude. On the slopes of the Appalachian Mountains in the east and the Rockies and coastal ranges in the west, there is a similar trend in biomes. As elevation increases and temperatures and humidity decrease, one passes through temperate deciduous forests to taigas to tundras. Here is an overview of the major biomes of the world.

Tropical Rainforest

- Tropical rainforests are found near the equator with abundant rainfall, stable temperatures, and high humidity.
- Although these forests cover only 4 percent of Earth's land surface, they account for more than 20 percent of Earth's net carbon fixation (food production).
- Tropical rainforests have the most diversity of species of any biome on Earth. They may have as many as 50 times the number of species of trees as a temperate forest.
- Dominant trees are very tall with interlacing tops that form a dense canopy, keeping the floor of the forest dimly lit even at midday. The canopy also prevents rain from falling directly onto the forest floor, but leaves drip rain constantly.

- Many trees are covered with **epiphytes**, photosynthetic plants that grow on other trees rather than supporting themselves. They are not parasites but may kill the trees inadvertently by blocking the light.
- Tropical rainforests house the most diverse animal species of any biome, including birds, reptiles, mammals, and amphibians.
- Some are **biodiversity hotspots**, meaning that many species are endangered.

Desert

- Deserts have less than 10 inches (25 cm) of rainfall per year; not even grasses can survive.
- Deserts experience the most extreme temperature fluctuations of any biome. Daytime *surface* temperatures can be as high as 158°F (70°C). With no moderating influence of vegetation, heat is lost rapidly at night. Shortly after sundown, temperatures drop drastically.
- Characteristic plants are the drought-resistant cactus with shallow roots to capture as much rain as possible during hard and short rains, which are characteristic of the desert.
- Other plants include sagebrush, creosote bush, and mesquite.
- There are many small annual plants that germinate only after a hard rain, send up shoots and flowers, produce seeds, and die, all within a few weeks.
- Most animals are active at night or during a brief early morning period or late afternoon, when the heat is not so intense. During the day, they remain cool by burrowing underground or hiding in the shade.
- Cacti can expand to hold extra water and have modified leaves called spines that protect against animals attacking the cactus for its water.

SYI-3
The distribution of local and global ecosystems changes over time.

- As an example of how severe conditions in a desert can be, in the Sahara Desert, there are regions hundreds of miles across that are completely barren of any vegetation.
- Characteristic animals include rodents, kangaroo rats, snakes, lizards, arachnids, insects, and a few birds.

Temperate Grasslands

- Temperate grasslands cover huge areas in both the temperate and tropical regions of the world.
- Temperate grasslands are characterized by low total annual rainfall or uneven seasonal occurrence of rainfall, making conditions inhospitable for forests.
- Principal grazing mammals include bison and pronghorn antelope in the United States and wildebeest and gazelle in Africa. Also, burrowing mammals, such as prairie dogs and other rodents, are common.

Temperate Deciduous Forest

- Temperate deciduous forests are found in the northeast of North America, south of the taiga, and are characterized by trees that drop their leaves in winter.
- Temperate deciduous forests include many more plant species than the taiga does.
- Temperate deciduous forests show **vertical stratification** of plants and animals; that is, there are species that live on the ground, on the low branches, and on the treetops.
- The soil is rich due to decomposition of leaf litter.
- Principal mammals include squirrels, deer, foxes, and bears, which are dormant or hibernate through the cold winter.

Conifer Forest—Taiga

- Taigas are located in northern Canada and much of the world's northern regions.
- They are dominated by conifer (evergreen) forests, like spruce and fir.
- The landscape is dotted with lakes, ponds, and bogs.
- They have very cold winters.
- This is the largest terrestrial biome.
- Taigas are characterized by heavy snowfall; the trees are shaped with branches directed downward to prevent heavy accumulations of snow from breaking their branches.
- Principal large mammals include moose, black bears, lynx, elks, wolverines, martens, and porcupines.
- Flying insects and birds are prevalent in summer.
- Taigas have greater variety in species of animals than tundras do.

Tundra

- Tundras are located in the far northern parts of North America, Europe, and Asia.
- They are characterized by **permafrost**, permanently frozen subsoil found in the farthest point north including Alaska.
- The tundra is commonly referred to as the **frozen desert** because it gets very little rainfall and what rainfall that does occur cannot penetrate the frozen ground.
- The tundra has the appearance of gently rolling plains with many lakes, ponds, and bogs in depressions.
- Insects, particularly flies, are abundant. As a result, vast numbers of birds nest in the tundra in the summer and migrate south in the winter.
- Principal mammals include reindeer, caribou, Arctic wolves, Arctic foxes, Arctic hares, lemmings, and polar bears.
- Though the number of individual organisms in the tundra is large, the number of species is small.
- The high Arctic tundra—a subset of permafrost—remains a strong carbon sink.

ENE-3, IST-5
The activities of organisms are affected by interactions with biotic and abiotic factors.

Aquatic Biomes

Aquatic biomes cover about 75% of Earth. Unlike terrestrial biomes, they are not characterized by a single dominant group of organisms. The primary distinction among aquatic biomes is *salinity*. There are freshwater, estuary, and marine biomes. **Freshwater biomes** have a salinity of less than 0.1% and include rivers, streams, ponds, and wetlands. Some of our freshwater reserves are stored in groundwater. Freshwater makes up less than 4% of Earth's aquatic biomes. **Estuaries** are located at the mouths of rivers where saltwater and freshwater mix. Salt marshes and mangrove forests are estuaries that support enormous populations of animal life. However, the largest biome on Earth is the marine biome with a salinity of 3% on average. Here are the characteristics of marine biomes:

- The marine biomes are the largest biomes, covering three-fourths of Earth's surface.
- They are the most stable biome with temperatures that vary little because water has a high heat capacity and there is such an enormous volume of water.
- They provide most of Earth's food and oxygen.
- The marine biomes are divided into different regions classified by the amount of sunlight they receive, the distance from the shore, and the water depth and whether it is open water or ocean bottom.

CHEMICAL CYCLES

Although Earth receives a constant supply of energy from the sun, chemicals must be recycled. Here are three chemical cycles: **the water cycle**, **the carbon cycle**, and **the nitrogen cycle**.

The Water Cycle

Remember: Living systems depend on the properties of water that result from its polarity and from hydrogen bonding.

Water evaporates from Earth, forms clouds, and rains over the oceans and land. Some rain percolates through the soil and makes its way back to the seas. Some evaporates directly from the land, but most evaporates from plants by **transpiration**.

The oceans contain 97% of the water in the biosphere. About 2% is locked in glaciers and polar ice caps, and the remaining 1% is in lakes, rivers, and groundwater. A negligible amount is in the atmosphere.

The Carbon Cycle

Remember: Carbon moves from the environment to organisms, where it is used to build carbohydrates, proteins, lipids, or nucleic acids. Carbon is used in storage compounds and cell formation in all organisms.

The basis of the carbon cycle are the reciprocal processes of **photosynthesis** and **respiration**.

> **ENE-3**
> **Organisms must exchange matter with the environment to grow, reproduce, and maintain organization.**

- Cell respiration by animals and bacterial decomposers adds CO_2 to the air and removes O_2.
- The burning of fossil fuels adds CO_2 to the air.
- Photosynthesis removes CO_2 from the air and adds O_2.

The major reservoirs of carbon are fossil fuels and plant and animal biomass. Carbon is also found in the soil, in dissolved carbon compounds in the oceans, in sediments in aquatic ecosystems, and in the atmosphere as CO_2 (carbon dioxide) and CO (carbon monoxide).

The Nitrogen Cycle

Remember: Nitrogen moves from the environment to organisms, where it is used in building proteins and nucleic acids.

Very little nitrogen enters ecosystems directly from the air. Most of it enters ecosystems by way of bacterial processes.

> **ENE-1**
> **Energy flows, but matter is recycled.**

- **Nitrogen-fixing bacteria** live in the nodules in the roots of legumes and convert **free nitrogen** into the **ammonium ion** (NH_4^+).
- **Nitrifying bacteria** convert the ammonium ion into **nitrites** and then into **nitrates**.
- **Denitrifying bacteria** convert **nitrates** (NO_3) into **free** atmospheric **nitrogen**.
- **Bacteria of decay** decompose **organic matter** into **ammonia**.

The main reservoir of nitrogen is the atmosphere, which contains about 79% nitrogen gas (N_2). Nitrogen is also found bound in the soil and in lake, river, and ocean sediments. It is also fixed into animal and plant biomass. See Figure 12.9.

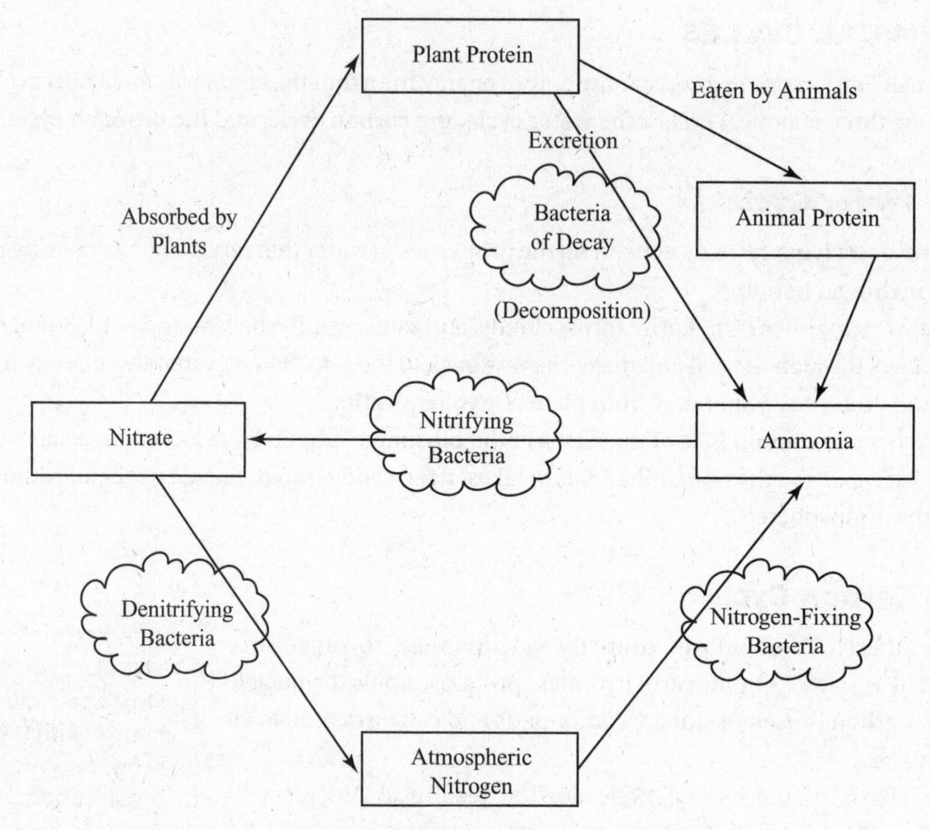

Figure 12.9 The Nitrogen Cycle

HUMANS AND THE BIOSPHERE

Humans threaten to make Earth uninhabitable as the population increases exponentially and as people waste natural resources, destroy animal habitats, and pollute the air and water. Here are several examples of this.

Eutrophication of the Lakes

Humans have disrupted freshwater ecosystems, causing a process called **eutrophication**. Runoff from sewage and manure from pastures increase nutrients in lakes and cause excessive growth of algae and other plants. Shallow areas become choked with weeds, and swimming and boating become impossible. As these large populations of photosynthetic organisms die, two things happen. First, organic material accumulates on the lake bottom and reduces the depth of the lake. Second, **detritivores** use up oxygen as they decompose the dead organic matter. Lower oxygen levels make it impossible for some fish to live. As fish die, decomposers expand their activity and oxygen levels continue to decrease. The process continues, more organisms die, the oxygen levels decrease, more decomposing matter accumulates on the lake bottom, and ultimately, the lake disappears. See Figure 12.10.

> **EVO-1 & SYI-2**
> Humans impact the environment and hasten change at both local and global levels.

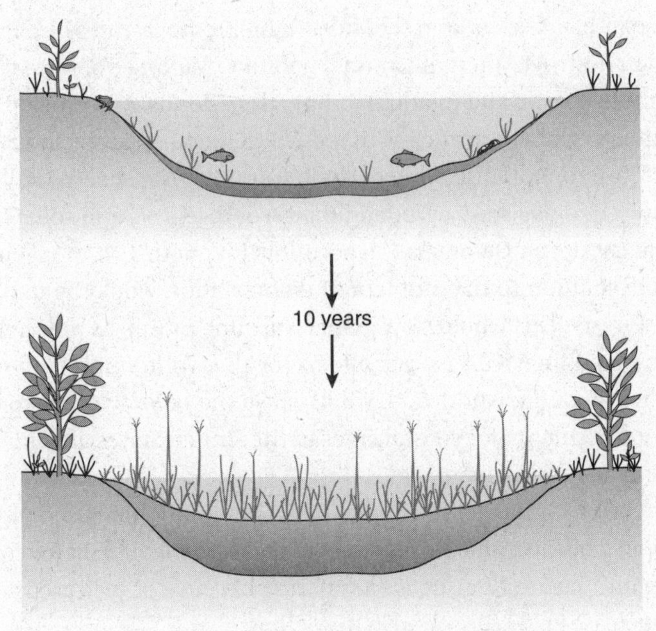

Figure 12.10 Process of Eutrophication

Acid Rain

Acid rain is caused by pollutants in the air from the **combustion of fossil fuels**. Nitrogen and sulfur pollutants in the air turn into **nitric, nitrous, sulfurous**, and **sulfuric acids**, which cause the pH of the rain to be less than 5.6. This kills the organisms in lakes and damages ancient stone architecture.

Toxins

Toxins from industry have gotten into the **food chain**. Most cattle and chicken feed contain **antibiotics** and **hormones** to accelerate animal growth but may have serious ill effects on humans who eat the chicken and beef. Any **carcinogens** or **teratogens** (causing birth defects) that get into the food chain accumulate and remain in the human body's fatty tissues because we occupy the top of the food chain. This process is called biological magnification.

Global Warming

To understand global warming, we first must talk about the **greenhouse effect**. CO_2 and water vapor in the atmosphere absorb and retain much of the light and heat that comes to Earth from the sun. If the greenhouse effect did not occur, the average temperature on the surface of Earth would be much colder and life as we know it would not exist. However, atmospheric CO_2 levels have increased by more than 40% during the last 150 years due to the burning of fossil fuels and deforestation. Scientists link increased CO_2 levels to **global warming**. According to NASA, the past five years are, collectively, the warmest years in the modern record (since 1880), and 2018 had the fourth highest mean temperature yet.

The region where global warming has already had a great impact is in the far north, which includes the Arctic tundra and northern coniferous forests. As temperatures rise, snow and ice melt, uncovering darker and more absorbent surfaces. As a result, more radiation is absorbed and Earth is warmed even more. In the summer of 2007, Arctic sea ice covered the smallest area on record. Melting ice and the resulting decrease in habitat endanger Arctic animals such as polar bears, seals, and seabirds. Higher temperatures also increase the likelihood of fires, which destroy even more animal habitats. Melting polar ice causes the seas to rise, resulting in more coastal and inland flooding, flash floods, and erosion.

The only solution to global warming is to reduce CO_2 emissions by industrialized nations and to reduce deforestation, particularly in the tropics. Forests absorb CO_2 from the atmosphere as they grow. They also store carbon in their wood, leaves, and soil. Forests release the CO_2 when they are cut down. Deforestation accounts for about 12% of greenhouse gas emissions. One possible solution to the problem of deforestation would be to pay countries *not* to cut down their forests. This would slow global warming as well as preserve biodiversity.

Global warming could have disastrous effects for the world's population. An increase of 1.0°C on average temperature worldwide would cause the polar ice caps to melt, raising the level of the seas. As a result, major coastal cities in the United States, including New York, Los Angeles, and Miami, would be under water.

Coral reefs like the Great Barrier Reef in Australia are under increased physiological stress due to an increase in global warming. This stress makes it more difficult for coral to build their skeletons. Oysters and sea urchins are also suffering because of increased acidification of the oceans due to an increase in CO_2 dissolved in the oceans.

Acidification of the Oceans

Carbon dioxide from the atmosphere normally dissolves in the oceans by combining with H_2O to form carbonic acid. With increased atmospheric carbon dioxide from burning fossil fuels, the oceans are rapidly becoming more acidic. The acidification of the oceans also directly results in a decrease in the concentration of carbonate ions (CO_3^{2-}) in the oceans. This compound is required by many marine organisms, including reef-building corals and animals that build shells. Since coral reefs provide shoreline protection and support a great diversity of commercial fish species, their destruction would be a great loss. In addition to affecting the shell and reef-building organisms directly, many food chains that include these shell-building animals are also negatively affected.

Depleting the Ozone Layer

The accumulations in the air of **chlorofluorocarbons**, chemicals used for refrigerants and aerosol cans, have caused the formation of a hole in the protective **ozone layer**. This allows more ultraviolet (UV) light to reach Earth, which is responsible for an increase in the incidences of **skin cancer** (melanoma) worldwide.

Introducing New Species

The introduction of any new species into an environment can threaten native wildlife and ***decrease biodiversity***. Nonnative species with no natural predators or parasites can grow, reproduce, and spread without limit. These **invasive species** can outcompete, displace, or even kill native species.

Invasive species do not have to come from another country. For example, lake trout are native to the Great Lakes. However, they are considered to be invasive in Yellowstone Lake, Wyoming, because they compete with native cutthroat trout for resources and habitat.

Invasive species are primarily spread by human activities, often unintentionally. Ships carry aquatic organisms in their ballast water. Insects are carried in wood, in shipping pallets, and in crates around the world. Some imported ornamental plants, like the strangler fig in Florida, escape into the wild and become invasive. People who are unaware of the consequences release "pets" they no longer want into the wild. This happened with Burmese pythons, which have become a frightening problem in the Everglades. They grow as long as 19 feet and have hunted native marsh rabbits almost to extinction.

Invasive species cost billions of dollars each year in removal and/or rehabilitation. Here are two more examples:

- The **"killer" honeybee**: The African honeybee is a very aggressive subspecies of honeybee that was brought to Brazil in 1956 to breed a variety of bee that would produce more honey in the tropics than the Italian honeybee. The African honeybees escaped by accident and have been spreading throughout the Americas. By the year 2000, ten people were killed by these bees in the United States.
- The **zebra mussel**: In 1988, the zebra mussel, a fingernail-sized mollusk native to Asia, was discovered in a lake near Detroit. No one knows how the mussel got transplanted there, but scientists infer it was accidentally carried by a ship from a freshwater port in Europe to the Great Lakes. Without any local natural predator to limit its growth, the mussel population exploded. They were first discovered when they were found to have clogged the water intake pipes of those cities whose water is supplied by Lake Erie. To date, the zebra mussel has caused millions of dollars of damage. In addition, the influx of the zebra mussel threatens several native species with extinction by outcompeting indigenous species.

EVO-1 & SYI-2
An introduced species can exploit a new niche free of predators or competitors, thus exploiting new resources. It can devastate native species.

Pesticides vs. Biological Control

Scientists have developed a variety of pesticides, chemicals that kill organisms that we consider to be undesirable. These pesticides include insecticides, herbicides, fungicides, and mice and rat killers. On the one hand, these pesticides save lives by increasing food production and by killing animals that carry and cause diseases like bubonic plague (diseased rats) and malaria (Anopheles mosquitoes). On the other hand, exposure to pesticides can cause cancer in humans. Moreover, spraying with pesticides ensures the development of resistant strains of pests through natural selection. The pests come back stronger than before. This problem requires that we spray more and more, which means more people will be exposed to these toxic chemicals.

An alternative to widescale spraying with pesticides is called biological control. The following are some biological methods of getting rid of pests without using dangerous chemicals:

1. Use crop rotation—change the crop planted in a field.
2. Introduce natural enemies of the pests—you must be careful, however, that you do not disrupt a delicate ecological balance by introducing an invasive species.
3. Use natural plant toxins instead of synthetic ones.
4. Use insect birth control—male insect pests can be sterilized by exposing them to radiation and then releasing them into the environment to mate unsuccessfully with females.

CHAPTER SUMMARY

This chapter provided information for Big Ideas: EVO, ENE, IST, and SYI.

- Although there are no complicated chemical equations and molecular pathways in this chapter, you must connect some of the concepts in this chapter with those from earlier chapters, namely, photosynthesis, cellular respiration, and evolution.
- **Ecology** is the study of the *interactions* of organisms and populations with each other and with their physical environment. The properties of **populations** are **size**, **density**, **dispersion**, **survivorship curves**, and **age structure diagrams**.
- A **population** is a group of individuals of one species living in one area that have the ability to interact and interbreed with each other. There are two categories of factors that limit the growth of any population. One is **density-dependent factors** such as competition for food, the buildup of wastes, predation, and disease. The other is **density-independent factors**, which include earthquakes, storms, naturally occurring fires and floods. If a population experiences few or none of these limiting factors, it will grow *exponentially* and exceed the **carrying capacity** of its **ecosystem**. As a result, the population will decrease or die.
- A **community** consists of all the organisms that live in one area. They are characterized by their **diversity** and **density**. The most successful communities are those that have the greatest diversity and are not so dense that they exceed the carrying capacity of the ecosystem. A diverse community has a better chance for survival in a changing environment. This idea connects with the concepts of **natural selection** and **evolution**. Interactions within a community are complex. However, they can be divided into five categories: **competition**, **predation**, **herbivory**, **symbiosis**, and **facilitation**.
- All living things require a *constant input of free energy*. Organisms must exchange matter with the environment to grow, reproduce, and maintain homeostasis. When you discuss the flow of energy in a **food chain**, remember that **producers** are **autotrophs**; they capture light energy and convert it into food by **photosynthesis**. When energy travels through a food chain, only 10% of the energy is actually converted into organic matter at the next **trophic level**. The rest is lost to **heat** and **cellular respiration** in animals along the way. Toxins that enter at the bottom of a food chain accumulate and concentrate in the higher trophic levels. This phenomenon is called **biological magnification**.
- The distribution of local and global ecosystems changes over time. This refers to **ecological succession** and **biomes**.
- Lastly, humans threaten Earth in many ways. Our population is increasing exponentially. We waste natural resources, destroy animal habitats, and pollute the air and water. Be able to discuss other specific examples of how humans negatively impact our world, such as **acid rain** and **acidification of the oceans**, **eutrophication of lakes**, **global warming**, and the **introduction of invasive species**. Also, reflect on remedies for these problems.

PRACTICE QUESTIONS

1. In 2011, a giant earthquake caused a tsunami that flooded the Fukushima Daiichi Nuclear Power Plant in Japan. The flooding caused the cores of the nuclear reactors to melt down, releasing tons of radioactive materials into the environment. Although most sea life in the nearby Pacific Ocean survived the initial exposure to massive amounts of radiation, scientists are uncertain about what will happen in the future.

   Which of the following statements describes the most likely consequence of such a release of radioactive material into the environment and includes appropriate evidence to support that description?

   (A) All sea life will be wiped out in five years because radiation kills living things.
   (B) Plant-eating animals will die within five years, which will cause the food chain to collapse because upper trophic levels feed on lower ones.
   (C) Radioactive materials will enter the aquatic food web and hurt the top predators because toxins become more concentrated in successive levels up the food chain.
   (D) Radioactive materials will enter the aquatic food web and most affect the lowest trophic levels because the lowest trophic levels have the greatest biomass.

2. Which of the following best explains why there are usually no more than five trophic levels in a food chain?

   (A) There are not enough organisms to fill more than five levels.
   (B) There is too much competition among the organisms at the lower levels to support more animals at higher levels.
   (C) The statement is not true; there can be unlimited trophic levels.
   (D) Energy is lost at each trophic level.

3. Below is a climograph that shows the mean temperature and precipitation for the temperate forests in the United States for the past 50 years.

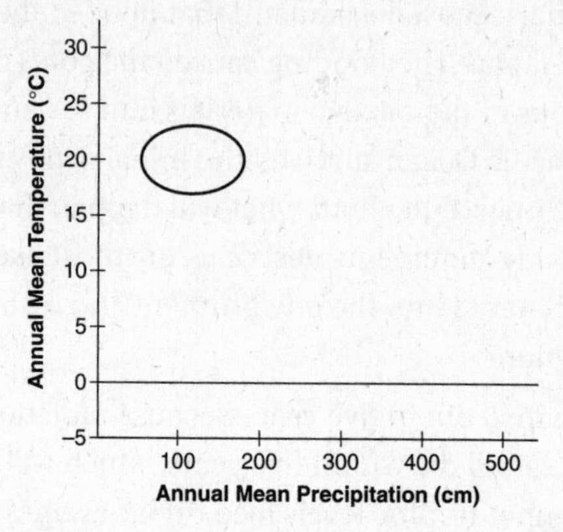

Predict how the climograph would appear if carbon dioxide levels continue to rise and the United States continues to become warmer and dryer over the next 50 years.

(A)

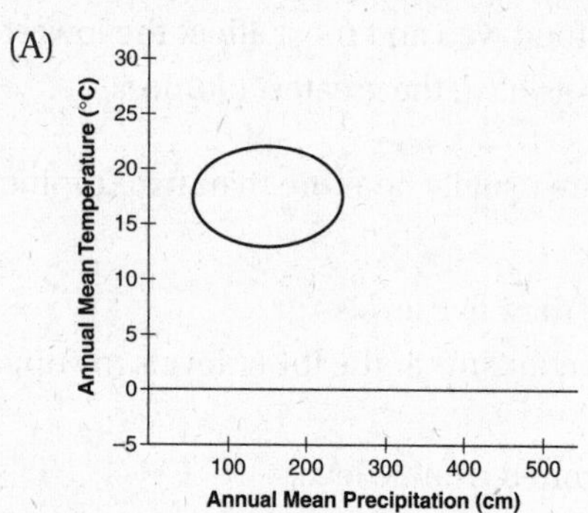

(B)

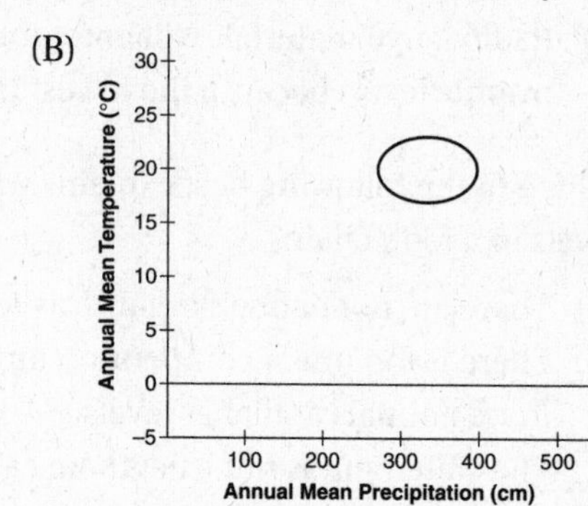

(C)

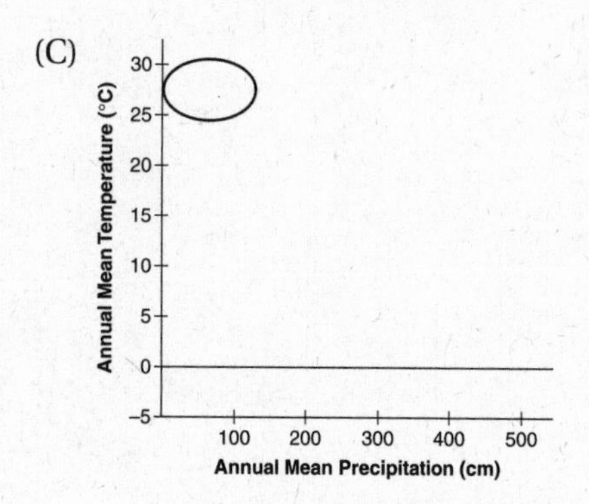

(D)

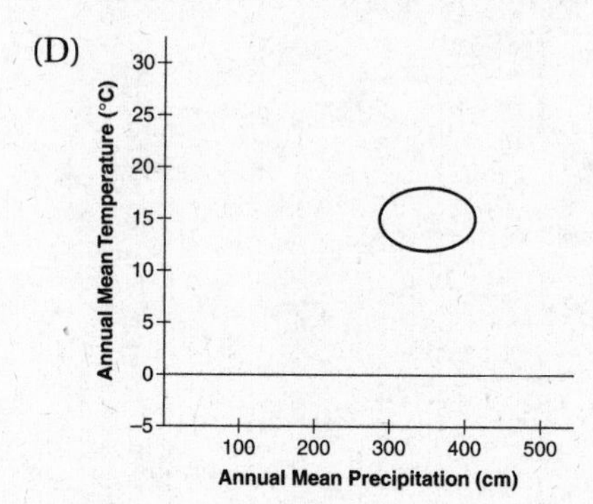

4. The steppes of Siberia are part of a biome that consists of a dry, cold grassland. If you were looking elsewhere in the world for a similar biome, which of the following variables would you consider most important?
   (A) rainfall and temperature
   (B) amount of sunlight and human population size
   (C) variation in seasons (i.e, summer vs. winter)
   (D) altitude and water supply

5. A biodiversity hotspot is a relatively small region on Earth that contains many species that are not found anywhere else in the world. For example, nearly 30% of all bird species inhabit hotspots that make up only 2% of Earth's total landmass. In addition, hotspots contain species that are endangered by ongoing human activities, like deforestation, that cause habitat destruction.
   Which of the following is correct about biodiversity hotspots?
   (A) Most biodiversity hotspots are located in tropical and subtropical regions.
   (B) Many biodiversity hotspots are located across Europe.
   (C) As a result of tight regulations in countries that contain identified biodiversity hotspots, endangered species in those hotspots are safe and assured of protection in the future.
   (D) Each biodiversity hotspot is characterized by the presence of specific, identified, common animals and plants.

6. In Yellowstone National Park, the reintroduction of wolves has decreased the deer and elk populations. Studying this dynamic is an example of ____________ -level ecology.
   (A) ecosystem
   (B) community
   (C) population
   (D) individual

7. If climate change is warming the North and South Poles faster than it is warming the temperate latitudes, which of the following biomes would likely face the most significant change?
   (A) taiga
   (B) temperate deciduous forest
   (C) tropical rainforest
   (D) tundra

Questions 8–10

Questions 8–10 refer to the idealized survivorship graph shown below.

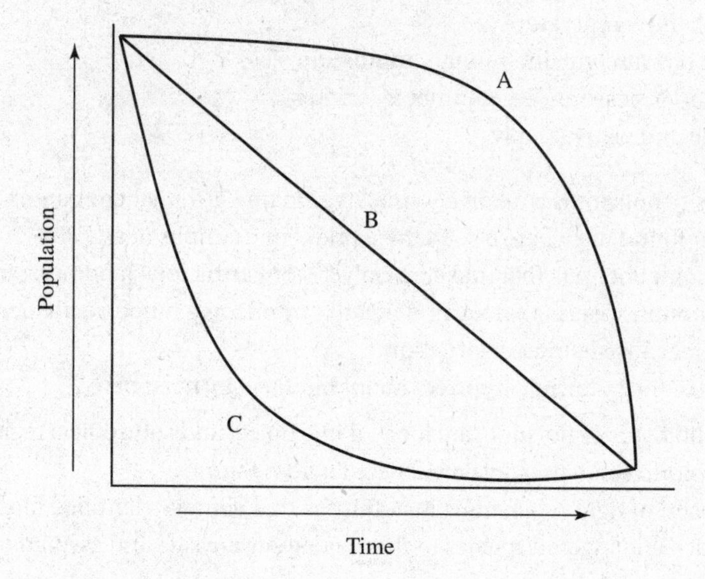

8. Some animal species are classified as K-strategists. They have few young at one time, but they reproduce more than once. They are slow to mature, and they engage in intensive parenting. Which of the following statements correctly identifies the line in the graph that represents a K-strategist and provides the correct reasoning for that answer?
   (A) Line A represents a K-strategist because, as a result of intensive parenting, this species has a high survival rate of offspring.
   (B) Line B represents a K-strategist because it is characteristic of either a K-strategist or an r-strategist under different conditions.
   (C) Line C represents a K-strategist because there is a large die off of the young, which is made up for by having hundreds or thousands of offspring at one time.
   (D) None of these lines fits the characterization of a K-strategist.

9. Animals that are classified as r-strategists have many small young at one time and engage in little or no parenting. They mature rapidly and reproduce once. An insect is an example of an r-strategist. Which of the following statements correctly identifies the line in the graph that represents an r-strategist and provides the correct reasoning for that answer?
   (A) Line A represents an r-strategist because, as a result of intensive parenting, this species has a high survival rate of offspring.
   (B) Line B represents an r-strategist because it is characteristic of either a K-strategist or an r-strategist under different conditions.
   (C) Line C represents an r-strategist because there is a large die off of the young, which is made up for by having hundreds or thousands of offspring at one time.
   (D) None of these lines fits the characterization of an r-strategist.

10. Which line on the graph characterizes an animal that releases thousands of eggs at once into the sea to be fertilized externally?

    (A) Line A characterizes this type of animal because, as a result of intensive parenting, this species has a high survival rate of offspring.
    (B) Line B characterizes this type of animal because it is characteristic of either a K-strategist or an r-strategist under different conditions.
    (C) Line C characterizes this type of animal because there is a large die off of the young, which is made up for by having hundreds or thousands of offspring at one time.
    (D) None of these lines fits the characterization of this type of animal.

---

11. The African savanna is a tropical grassland with warm to hot temperatures year-round, even during the summer rainy season. The savanna is characterized by grasses, dispersed bushes, and acacia trees. It contains a diverse community of organisms. Large cats (like lions and cheetahs) feed on herbivores (like impalas and warthogs). Herbivores consume leaves and grasses, which are plentiful in the rainy season. In the dry season, rivers and streams become limited or disappear completely, and fierce competition among the wildlife results. In a weakened state, an animal can easily succumb to disease. During the dry season, forest fires become more prevalent. They can destroy what little food is available. Both density-dependent and density-independent factors play a role in regulating life on the African savanna. Which of the following statements is correct about these factors?

    (A) Naturally occurring forest fires are density-dependent factors because they harm the environment and kill many animals.
    (B) Disease is a density-independent factor because its cause has nothing to do with the density of the population.
    (C) Starvation is a density-dependent factor because, if there are too many animals, there will not be enough food for all of them.
    (D) Lack of rainfall is a density-independent factor because, if there were fewer animals, there would be adequate water and not so many fires.

12. Dutch elm disease (DED) is caused by a fungus that kills elm trees. The fungus is carried by the elm bark beetle that was originally native to Asia. That beetle was introduced into Ohio in the 1930s. These beetles and this fungus infection have since spread to Australia and New Zealand, killing native elms at a rapid pace. At one time, elm trees provided shade on streets throughout America. Now, elm trees can only occasionally be found across the United States, Europe, and the United Kingdom.

    Which of the following best identifies the most significant danger caused by an invasive species like the elm bark beetle that carries the fungus that causes Dutch elm disease?

    (A) It can wipe out a single species.
    (B) It reduces biodiversity in an ecosystem.
    (C) The parasite can spread to humans.
    (D) It infects trees that have not evolved resistance.

13. Consider two different ecosystems. One is a stream-fed environment, with a constant water supply and stable year-round temperatures. This environment can support many large, well-established populations. The other ecosystem is a desert, which floods and dries out at unpredictable times. Due to the unfavorable conditions, the sizes of resident populations fluctuates.

    Which of the following choices accurately describes which of the two ecosystems would be more likely to support r-strategists and provides the correct reasoning for that answer?

    (A) The stable stream-fed environment would better support r-strategists because r-strategists can always outcompete any other population by having hundreds of offspring at once.
    (B) The stable stream-fed environment would better support r-strategists because r-strategists are less vulnerable to floods or drought than K-strategists are.
    (C) The changing desert environment would better support r-strategists because the r-strategists can lay hundreds of eggs after other animals have been killed off by floods or drought.
    (D) The changing desert environment would better support r-strategists because the r-strategists can produce a small number of large offspring that can complete against other populations.

14. What would most likely be the cause of bushes of one species growing in one area in a uniform spacing pattern?

    (A) random distribution of seeds
    (B) interactions among individuals in the population
    (C) chance
    (D) the varied nutrient supplies in that area

15. Animals from two different species utilize the same source of nutrition in one area. It is most accurate to say that the animals

    (A) will learn to get along
    (B) will compete for food
    (C) will die because there will not be enough food for both of them
    (D) will learn to eat different foods

16. The Japanese seastar is native to the coasts of Japan, Korea, and China. It is a large seastar that is a voracious eater and can be as wide as 20 inches across. Accidentally transported in the ballasts of ships, more than twenty years ago, the Japanese seastar has spread throughout Southeast Asia and the United States, where it has no natural predators. When it invades a new area, it feeds on native populations.

    Which of the following statements best explains what will happen when the Japanese seastar invades an ecosystem?

    (A) The seastar will enhance the quality of that ecosystem by only devouring nuisance native plants and animals.
    (B) The seastar will be a new food source for that ecosystem, and the biodiversity of that ecosystem will remain the same.
    (C) The biodiversity of that ecosystem will be increased because another species will be added to it.
    (D) The biodiversity of that ecosystem will be reduced because the seastar will feed on or outcompete native organisms.

Questions 17–20

Questions 17–20 refer to the following depiction of a food web for a terrestrial ecosystem.

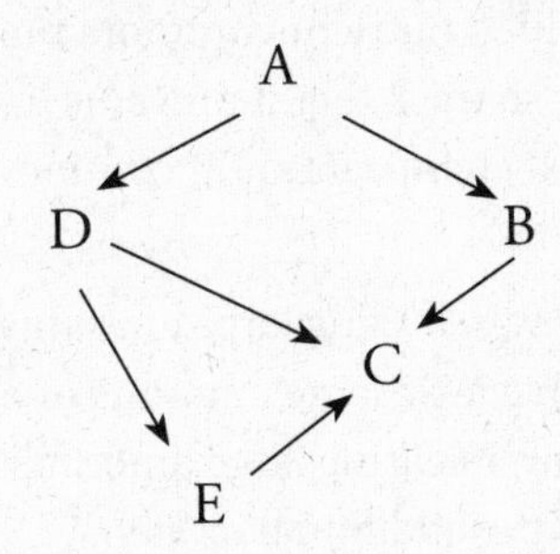

17. Which species is the producer?
    (A) A
    (B) B
    (C) C
    (D) D

18. A toxic pollutant would be found in highest concentrations in which species?
    (A) A
    (B) B
    (C) C
    (D) D

19. Which species would have the greatest biomass?
    (A) A
    (B) B
    (C) C
    (D) D

20. Which species would have the smallest biomass?
    (A) A
    (B) B
    (C) C
    (D) D

21. In the 1970s, scientists began to learn and report about a process called *eutrophication*, a phenomenon found in freshwater lakes. The process begins as leaky septic systems and fertilizer runoff from surrounding farms and lawns adds chemical pollutants and/or nutrients to lakes. These added nutrients promote the growth of primary producers, like algae. The algae can grow so thick that it prevents light from penetrating the water.

    Predict what will happen if chemical runoff continues to add nutrients to these lakes.

    (A) Algae will continue to grow, releasing large amounts of oxygen into the water. This will cause the fish to die because large amounts of oxygen are lethal to fish.
    (B) Algae will continue to grow, releasing large amounts of oxygen into the water. The fish will thrive because of the additional oxygen.
    (C) Layers of algae will die and will be decomposed by bacteria, using up oxygen in the water. This oxygen depletion will cause the death of many fish and other aquatic life.
    (D) Layers of algae will die, leaving the lake healthy and the water clear.

Questions 22–23

Questions 22–23 refer to the graph below that shows changes in population over time.

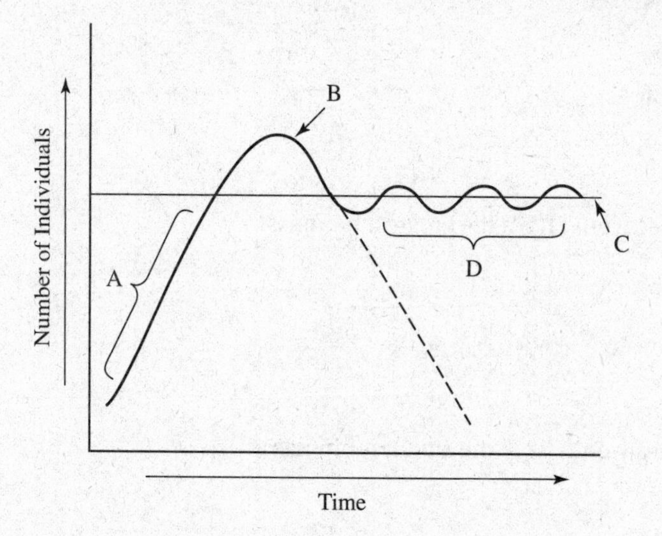

22. Which letter shows a mature, well-established population in favorable conditions?

    (A) A
    (B) B
    (C) C
    (D) D

23. Which letter shows the carrying capacity of the environment?

    (A) A
    (B) B
    (C) C
    (D) D

Questions 24–25

Questions 24–25 refer to the following age structure pyramid for a developing country.

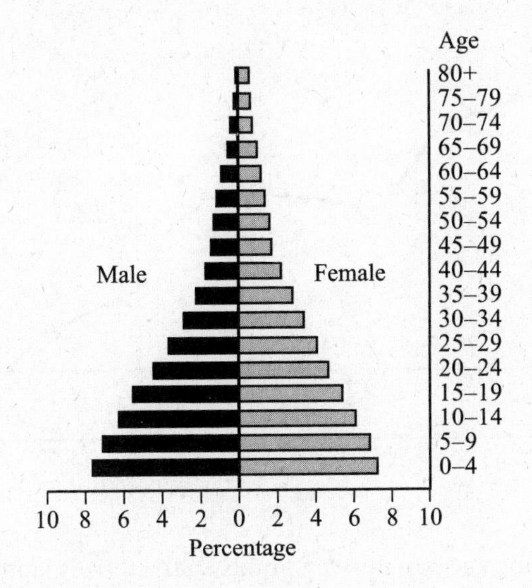

24. Which of the following statements regarding population growth would be most likely to occur over the next 20 years?

    (A) The population will decrease slowly.
    (B) The population will decrease rapidly.
    (C) The population will increase slowly.
    (D) The population will increase rapidly.

25. Which of the following situations would be most likely to occur in this country in the near future?

    (A) With increased numbers of people entering higher education, advances in technology will supply better health care.
    (B) With an increase in the number of people employed, the overall wealth of the country will increase.
    (C) There will be an increase in the demand for resources.
    (D) There will be a decline in the number of frail and elderly people.

26. Below is a graph showing the change in population (in millions) for Japan from 1950 to 2016.

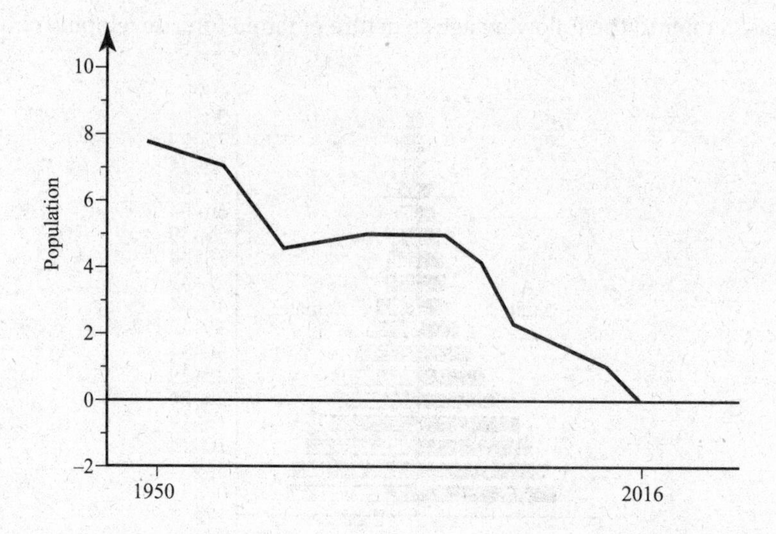

Which of the following age structure pyramids characterizes Japan's growth?

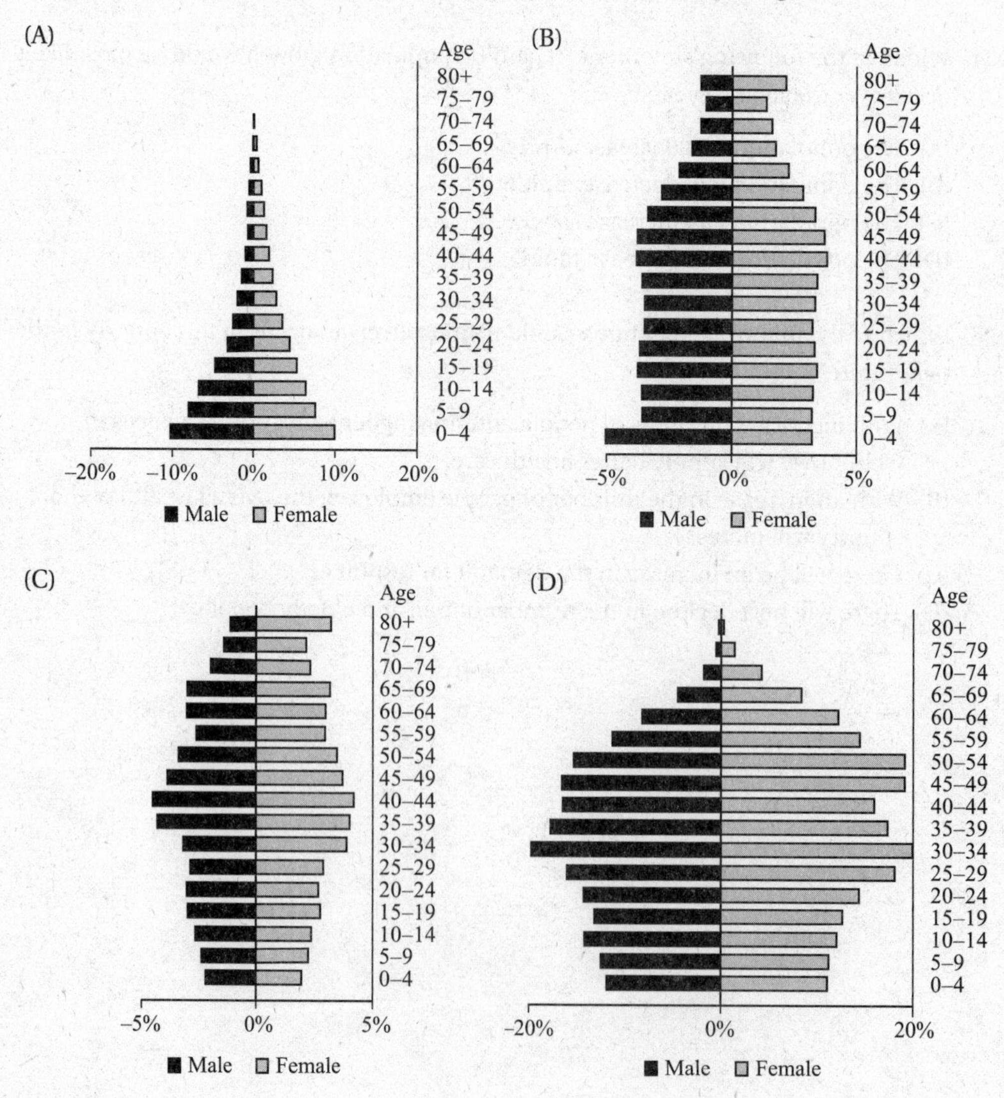

Questions 27–29

For questions 27–29, read this paragraph about an invasive species and then answer the questions that follow.

> Eurasian watermilfoil was accidentally introduced into North America from Europe. It spread westward into inland lakes primarily by boats and water birds. It reached midwestern states between the 1950s and 1980s. In nutrient-rich lakes, it can form thick, underwater strands of tangled stems and vast mats of vegetation at the water's surface. In shallow areas, the plant can interfere with water recreation such as boating, fishing, and swimming. The plant's floating canopy can also crowd out important native water plants. A key factor in the plant's success is its ability to reproduce through stem fragmentation and runners. A single segment of stem and leaves can take root and form a new colony. Fragments clinging to boats and trailers can spread the plant from lake to lake. The mechanical clearing of aquatic plants for beaches, docks, and landings creates thousands of new stem fragments and new plants.

27. Which of the following nutrients running off from the land would have the greatest effect on the growth of Eurasian watermilfoil?
    (A) carbon and hydrogen
    (B) nitrogen and phosphorus
    (C) hydrogen and phosphorus
    (D) carbon and phosphorus

28. Which of the following graphs would most accurately show what will happen to the number of different native species in the affected lakes in the next 5 years if the watermilfoil remains?

(A)

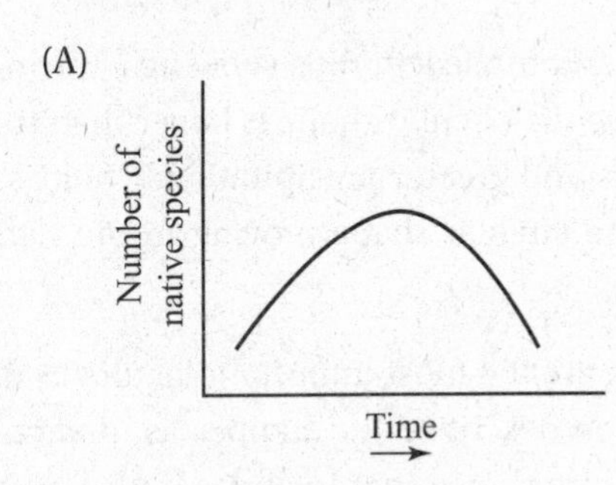

(B)

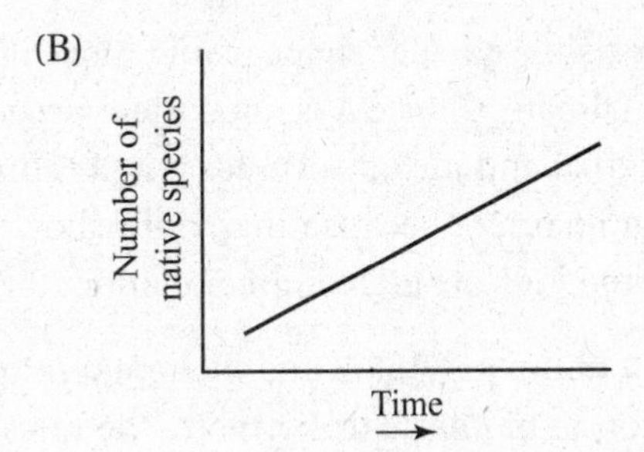

(C)

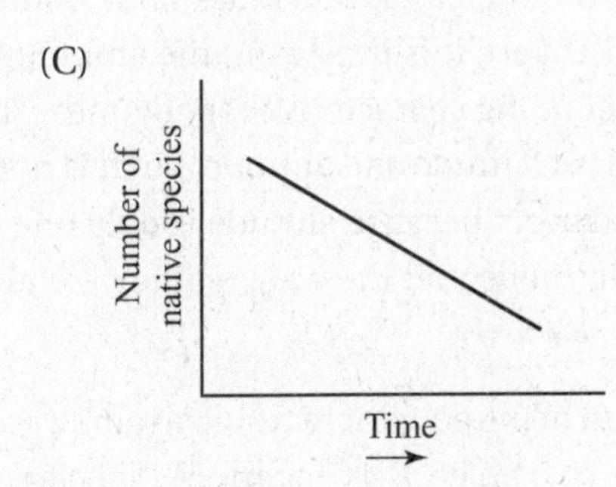

(D)

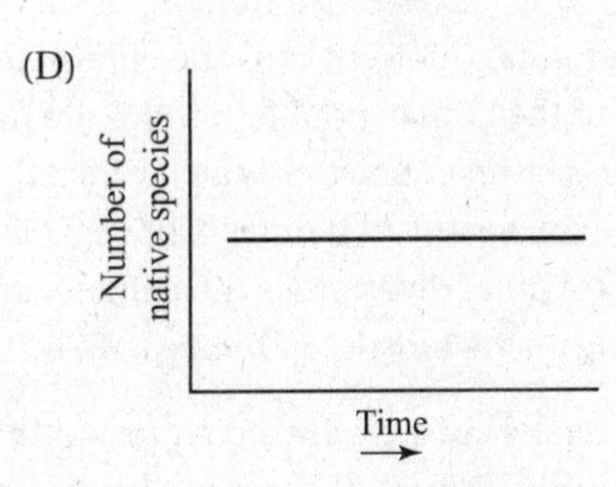

29. A midwestern town decided to treat their watermilfoil-infested shallow lake with a chemical that would target watermilfoil and leave any native plants unaffected. The treatment was carried out late in the growing season when people could no longer swim or boat in the lake because the watermilfoil had grown so dense. The treatment worked well. Within one week, all the watermilfoil had turned brown and died, and had fallen to the bottom of the lake. Which of the following would be the most likely immediate consequence of the lake treatment?

(A) Oxygen levels would decrease because of increased bacterial decomposition, causing fish to die.
(B) Native species would also die because of residual toxins in the lake.
(C) Oxygen levels would increase because with watermilfoil gone, photosynthesis would increase.
(D) CO_2 levels would decrease because of increased photosynthesis carried out by a sudden increase in the native plant population.

Answers Explained

1. **(C)** Choices A and B are not correct because, while it is true that radiation kills cells, there is no information about any five-year interval given in the question. Choice D is not correct because the danger that results from contamination of a food web most affects the highest trophic levels, not the lowest trophic levels. It is true that the lowest trophic level (the producers) has the greatest biomass, and it is also true that radioactive contamination to that level is horrible. However, the initial radiation contamination of the lowest trophic level (the producers) is a tiny fraction of the amount of radiation that works its way through the food chain from that level to the highest level. (ENE-3)

2. **(D)** Only about 10 percent of the energy from one trophic level is transferred to the next level. The other choices do not make any sense. (ENE-1)

3. **(C)** The existing circular shape would move up and to the left, thus showing a warmer and drier climate. Choice A is not correct because the circular shape is larger than that of the original and includes cooler temperatures and greater precipitation. Choices B and D are incorrect because they both show the circular shape moving to the right, which would indicate greater precipitation. (SP 2)

4. **(A)** The amount of rainfall and the temperature are the most important factors in the distribution of biomes. Furthermore, the question described the steppes as "dry, cold grassland" so it makes sense to look for an answer choice that includes both rainfall and temperature. Choice B must be eliminated because it is irrelevant; the amount of sunlight and the human population size are not defining characteristics of biomes. The variation in seasons, choice C, is an indirect result of latitude and altitude, but it is not a defining characteristic of biomes. Choice D is incorrect because altitude is only one of several factors that determines the climate, and altitude and the water supply are also not defining characteristics of biomes. (SYI-3)

5. **(A)** Biodiversity hotspots are mostly located south of the equator in tropical rainforests. There are no biodiversity hotspots in Europe, so choice B is incorrect. Species in biodiversity hotspots are constantly facing the threats posed by climate change and harmful human activities (like deforestation), so choice C is incorrect. Choice D is not correct because the species in hotspots are rare, not common. (SYI-3)

6. **(B)** This question is really a definition question. The definition of "community" is "all the organisms or animals in one place or ecosystem." Since this question focuses on wolves, deer, and elk (a community) in Yellowstone National Park, the answer must be choice B. Choice A is incorrect because ecosystems involve all living and abiotic factors in an area. Choice C is incorrect because the question focuses on a community, not just a population (which is defined as one kind of organism in an area). Choice D is incorrect because the question does not solely focus on individuals; rather, it discusses a community. (ENE-4)

7. **(D)** Of the four choices listed, the tundra biome is located closest to the North and South Poles and includes a layer of permafrost. The taiga biome is located just south of the tundra. The biome that is south of the taiga is the temperate deciduous forest, and the tropical rainforest biome is closest to the equator. (SYI-3)

8. **(A)** There is a high survival rate of the young, and death occurs in old age. The curve is flat in the beginning, and then drops at old age. (SYI-1)

9. **(C)** Line C represents an r-strategist because it is an initial steep downward curve. This line represents that the highest mortality occurs early in life due to a lack of protective parenting, which is characteristic of external fertilization. Choice A describes a K-strategist because there is an initial flat line (representing a low death rate among the young as a result of intensive parenting) followed by a quick drop off (which represents death during old age). Choice B is incorrect because r-strategists and K-strategists each have their own characteristic survivorship curves that do not vary. Choice D is incorrect because there is one line that represents the survivorship curve of an r-strategist. (ENE-1)

10. **(C)** Fertilization is external with no parenting, so survival of the young is poor. This is why the curve dips steeply initially. (SYI-1)

11. **(C)** Density-dependent factors have to do with population density and all things related to that, such as disease and starvation. Density-independent factors are factors that affect a population but do not have to do with population density, such as flooding, precipitation, and wind. (SYI-1)

12. **(B)** The most important problem caused by an invasive species is that it reduces biodiversity. Every species plays a role in an ecosystem, and the loss of a single species can destroy an entire ecosystem. Choices A and D are incorrect because, although each is a correct statement, neither describes the *most significant* danger caused by an invasive species. Choice C is incorrect because the stem of the question does not give any information about the fungus spreading to humans. (EVO-1 & SYI-2)

13. **(C)** Due to the unfavorable conditions in the changing and harsh desert, many organisms die, leaving a small population. r-strategists reproduce large numbers of offspring at once when there is room (i.e., when the population is low). K-strategists favor stable environments, which are apt to be at carrying capacity (since organisms are not being killed off by unfavorable conditions). (IST-5)

14. **(B)** Plants may secrete toxins that keep other plants from growing nearby. This minimizes competition for limited resources. (ENE-4)

15. **(B)** This is a restatement of Gause's competitive exclusion principle. (ENE-4)

16. **(D)** The greatest danger posed by an invasive species is that it will feed on or outcompete native organisms, thus reducing the biodiversity of that ecosystem. Choice A is incorrect because there is no way to know that the invader will only devour the nuisance native plants and animals. Choice B is incorrect because the seastar will be the voracious invader; it will not be a food source. Choice C is incorrect because the invading species may kill off some native species and reduce the number of native populations. (IST-5)

17. **(A)** Species A is the producer because both B and D feed on it. (ENE-1)

18. **(C)** Species C would accumulate the most toxins because it is at the top of the food chain. The longest chain runs from A to D to E to C. (ENE-1)

19. **(A)** The producer always has the greatest biomass. (ENE-1)

20. **(C)** The top consumer always has the least biomass. (ENE-1)

21. **(C)** Eutrophication is caused by runoff from surrounding shorelines, particularly after a storm or from snowmelt. Oils, soaps, sewage, and chemicals (like nitrogen and phosphorus) that run into the lake foster the growth of algae. The algae can grow so thick that light is blocked, and algae at the lower depths cannot survive. The lower algae die and are decomposed by bacteria, which use up available oxygen. The situation can get bad enough to kill fish and other living things in the water. (ENE-1 & SYI-2)

22. **(D)** The population at D fluctuates around the carrying capacity (C) for that environment. (SYI-1)

23. **(C)** The carrying capacity is the maximum population size that can be supported by the available resources. It is symbolized as K. (ENE-1)

24. **(D)** This diagram is an age structure pyramid. It shows a distribution of the population by age and percentage of the entire population. There are very few elderly people, and the population of young people approaching reproductive age is very large and getting even larger. One can assume that barring some disaster or law limiting the number of children a couple can have, a larger and larger number of people in the future will have children. (ENE-1)

25. **(C)** The main point here is that with an increased population, there will be increase in the demand for resources, regardless of any other variable. (ENE-4)

26. **(D)** The graph shows a population whose size is rapidly decreasing. Choice A shows a population that will get much larger in the future because there is a large number of children who have yet to reach procreation age. Choice B is a population with an even age distribution. One can expect that the population size will not change in the near future. Choice C shows a population with a fairly even distribution, except for people in their 30s and 40s, which is larger. It seems that this group has not yet had children, although it may in the future. Choice D shows a large population of people in their 50s who passed childbearing age and did not have children. In addition, there is a large cohort of people in their mid-30s who have not yet had children. So the population will continue to get smaller, as shown in the graph. (SP 2)

27. **(B)** Nitrogen and phosphorus are the main ingredients in fertilizer because they fuel the growth of plants. All plants need nitrogen to make proteins and need phosphorus to make proteins and DNA. Carbon and hydrogen are everywhere and are not contaminants. (ENE-3)

28. **(C)** The result of an infestation of invasive or alien species is that they outcompete all other species and reduce diversity. Without diversity, a population or an ecosystem has less chance of survival if the environment changes. Without diversity, there can be no evolution. (EVO-1 & SYI-2)

29. **(A)** From the description, you learn that watermilfoil represents a huge biomass in the lake. Once it is killed, there will be massive decomposition by bacteria in the lake. Decomposition uses up oxygen which, in turn, threatens the fish population. Also, consider that this change is sudden, brought about by human intervention. There is no time for fish to acclimatize. (EVO-1 & SYI-2)

13 Animal Behavior

→ FIXED ACTION PATTERN
→ MIGRATION
→ ANIMAL SIGNALS AND COMMUNICATION
→ LEARNING MODIFIES BEHAVIOR
→ SOCIAL BEHAVIOR
→ NATURAL SELECTION AND REPRODUCTIVE SUCCESS
→ EVOLUTION OF BEHAVIOR

Big Ideas: EVO & SYI
Enduring Understandings: EVO-1; SYI-3
Science Practice: 5

For the complete list of Big Ideas, Enduring Understandings, and Science Practices, refer to the "AP Biology Course and Exam Description" from the College Board: *https://apcentral.collegeboard.org/pdf/ap-biology-course-and-exam-description.pdf?course=ap-biology*.

INTRODUCTION

Animal behavior is carried out in response to internal or external *stimuli*. It can be solitary or social, fixed or variable. Whatever the behavior, it enables organisms to hide, search for food, or find a mate. It evolved because of substantial evolutionary pressures. When we speak about behavior, we refer to either the proximate or the ultimate causes of behavior. The **proximate causes** are the immediate, genetic, physiological, neurological, and developmental mechanisms that determine how an individual behaves. The **ultimate causes** result from the evolutionary pressures that have fashioned an animal's behavior.

The study of behavior and its relationship to its evolutionary origins is called **ethology**. Foremost in the field of ethology are three scientists who shared the Nobel Prize in Physiology and Medicine in 1973: Karl von Frisch, Niko Tinbergen, and Konrad Lorenz. **Karl von Frisch** is known for his extensive studies of communication in honeybees and his famous description of their **waggle dance**. **Niko Tinbergen** is known for his elucidation of the **fixed action pattern**. **Konrad Lorenz** is famous for his work with **imprinting**.

EVO-1
Innate behaviors are inherited.

As you study each topic, think of how a particular behavior benefits one animal and makes it more fit than other animals. Also think about how the behavior enhances the animal's survival and how the behavior might have evolved.

FIXED ACTION PATTERN

A **fixed action pattern (FAP)** is an innate, seemingly *highly stereotypic behavior* that once begun is continued to completion, no matter how useless it is. FAPs are initiated by external stimuli called **sign stimuli**. When these stimuli are exchanged between members of the same species, they are known as **releasers**. An example of an FAP studied by **Tinbergen** involves the **stickleback fish**, which attacks other males that invade its territory. The **releaser** for the attack is the red belly of the intruder. The stickleback will not attack an invading male stickleback that lacks a red underbelly, but it will readily attack a non-fish-like wooden model as long it has a splash of red visible.

MIGRATION

EVO-1
Behavior in animals is triggered by environmental stimuli and is vital to reproduction, natural selection, and survival.

Animals **migrate** in response to environmental stimuli, like changes in day length, precipitation, and temperature. The environment also provides cues for navigation.

- Some migrating animals track their position relative to the sun. Although the sun's position changes throughout the day, animals can monitor changes in the position of the sun against an internal **circadian clock** to keep track of where they are.
- Nocturnal animals keep track of their position using the North Star, which has a fixed position in the sky.
- Pigeons track their positions relative to Earth's magnetic field.
- Gray whales migrate seasonally between the Bering Sea and the coastal lagoons of Mexico. They do this by knowing and remembering elements in their environment. Orienting by landmarks is called **piloting**.

ANIMAL SIGNALS AND COMMUNICATION

REMEMBER
An important theme in AP Biology is communication and signaling. Pheromones are a good example of how cells communicate with one another.

Animals mark their territory with chemical signals called **pheromones**. These chemicals can also act as alarm signals. For example, if a catfish is injured, a substance is released from its skin that disperses in the water and induces a fright *response* in other fish. The frightened fish become hypervigilant and form into tightly packed schools for protection at the lake or river bottom.

Visual signals are effective in open places in daylight. They provide information about many factors including the sex, strength, and social status of an individual. An example of symbolic communication is the waggle dance in bees that gives details about the location of a food source.

LEARNING MODIFIES BEHAVIOR

Learning is a sophisticated process in which the responses of the organism are modified as a result of experience. The capacity to learn can be tied to length of life span and complexity of the brain. If the animal has a very short life span, like a fruit fly does, it has no time to learn, even if it had the ability to do so. It must therefore rely on fixed action patterns. In contrast, if the animal lives for a long time and has a complex brain, then a large part of its behavior depends on prior experience and learning.

Habituation

Habituation is one of the simplest forms of learning. An animal comes to ignore a persistent stimulus so it can go about its business. If you tap a dish that contains a hydra, it will quickly shrink and become immobile. If you keep tapping, after a while the hydra will begin to ignore the tapping, elongate, and continue moving about. It has become **habituated** to the stimulus.

Associative Learning

Associative learning is one type of learning in which one stimulus becomes linked to another through experience. Examples of associative learning are classical conditioning and operant conditioning.

- **Classical conditioning**, a type of **associative learning**, is widely accepted because of the ingenious work of **Ivan Pavlov** in the 1920s. Normally, dogs salivate when exposed to food. Pavlov trained dogs to associate the sound of a bell with food. The result of this conditioning was that dogs would salivate upon merely hearing the sound of the bell, even though no food was present.
- **Operant conditioning**, also called **trial and error** learning, is another type of **associative learning**. An animal learns to associate one of its own behaviors with a reward or punishment and then repeats or avoids that behavior. The best-known studies involving operant conditioning were done by **B. F. Skinner** in the 1930s. In one study, a rat was placed into a cage that contained a lever that released a pellet of food. At first, the rat would depress the lever only by accident and would receive food as a reward. The rat soon learned to associate the lever with the food and would depress the lever at will. Similarly, an animal can learn to carry out a behavior to avoid punishment. Such systems of rewards and punishment are the basis of most animal training.

Imprinting

Imprinting is learning that occurs during a **sensitive** or **critical period** in the early life of an individual and is **irreversible** for the length of that period. When you see ducklings following closely behind their mother, you are seeing the result of successful imprinting. Mother-offspring bonding in animals that depend on parental care is critical to the safety and development of the offspring. If the pair does not bond, the parent will not care for the offspring and the offspring will die. At the end of the juvenile period, when the offspring can survive without the parent, the response disappears.

Classic imprinting experiments were carried out by **Konrad Lorenz** with geese. Geese hatchlings will follow the first thing they see that moves. Although the object is usually the mother goose, it can be a box tied to a string or, in the case of the classic experiment, it was Konrad Lorenz himself. Lorenz was the first thing the hatchlings saw, and they became **imprinted** on the scientist. Wherever he went, they followed.

Examples of Learning and Problem Solving

If a chimpanzee is placed into a room with several boxes spaced out on the floor and a banana hanging high out of reach, the chimp will stack the boxes on top of each other until he reaches the banana. This problem-solving behavior is highly developed in some mammals, especially in primates. It has also been observed in some birds, especially ravens. In one study, a raven was able to fly to a branch and step on a string in order to bring some hanging food within reach. Interestingly enough, some ravens were not able to solve the problem, which shows that *problem-solving ability varies with individual ability and experience.*

EVO-1
Natural selection favors innate or learned behaviors that increase survival and reproductive success.

Learning Sometimes Happens in Stages

Some birds learn songs in stages from other members of their species. The white-crowned sparrow hears the species song within the first 50 days of life. Although it does not sing during this *sensitive period*, the young bird memorizes the song and chirps in response to hearing it. This sensitive period is followed by a second learning phase when the juvenile bird sings tentative notes of the song and compares it with what he hears around him. Once its own song matches what he has heard, the song is "crystallized" as the final white-crowned sparrow song.

Learning from Others

Many animals learn to solve problems by observing the behavior of other individuals. Young wild chimps learn how to crack open oil palm nuts with stones by copying experienced chimps. Wild vervet monkeys in Kenya learn to make alarm calls. At first, they make indiscriminate calls in response to danger and later fine tune the call as they learn from older monkeys in the group.

SOCIAL BEHAVIOR

Social behavior is any kind of interaction among two or more animals, usually of the same species. It is a relatively new field of study, developed in the 1960s. Types of social behaviors are **cooperation**, **agonistic behavior**, **dominance hierarchies**, **territoriality**, and **altruism**.

EVO-1
Cooperative behaviors tend to increase the fitness of the individual and the group.

Cooperation

Cooperation enables the individuals to carry out a behavior, such as hunting, that they can do as a group more successfully than they can do separately. Lions or wild dogs will hunt in a pack, enabling them to bring down a larger animal than an individual could ever bring down alone.

Agonistic Behavior

Agonistic behavior is aggressive behavior. It involves a variety of threats or actual combat to settle disputes among individuals. These disputes are commonly held over access to food, mating, or shelter. This behavior involves both real aggressive behavior as well as ritualistic or symbolic behavior. One combatant does not have to kill the other. The use of symbolic behavior often prevents serious harm. A dog shows aggression by baring its teeth and erecting its ears and hair. It stands upright to appear taller and looks directly at its opponent. If the aggressor succeeds in scaring the opponent, the loser engages in submissive behavior that says, "You win, I give up." Examples of submissive behavior are looking down or away from the winner. Dogs or wolves put their tail between their legs and run off. Once two individuals have settled a dispute by agonistic behavior, future encounters between them usually do not involve combat or posturing.

Dominance Hierarchies

Dominance hierarchies are pecking order behaviors that dictate the social position an animal has in a culture. This is commonly seen in hens where the alpha animal (top-ranked) controls the behaviors of all the others. The next in line, the beta animal, controls all others except the alpha animal. Each animal threatens all animals beneath it in the hierarchy. The top-ranked animal is assured of first choice of any resource, including food after a kill, the best territory, or the most fit mate.

Territoriality

A **territory** is an area an organism defends and from which other members of the community are excluded. Territories are established and defended by *agonistic behaviors* and are used for capturing food, mating, and rearing young. The size of the territory varies with its function and the amount of resources available.

Altruism

Altruism is described as a behavior that reduces an individual's reproductive fitness (the animal might die) but increases the fitness of the colony or family. Scientists do not agree on how the trait of altruism remains in a population if the individual carrying the gene or genes that is responsible for the behavior dies.

> **EVO-1**
> **Organisms have a variety of signaling behaviors that can result in reproductive success.**

One explanation can be seen in honeybees, where worker females share 75% of the same genes. When a worker honeybee stings an intruder while defending a hive, the worker sacrifices itself for its relatives, individuals who carry most of the same genes as that particular worker. The individual dies, but relatives survive to pass on their genes, including the gene for altruistic behavior. This concept is known as **kin selection** or **inclusive fitness**.

NATURAL SELECTION AND REPRODUCTIVE SUCCESS

Foraging Behavior: Cost-Benefit Analysis

Benefits are measured in terms of **fitness enhancement**—the more fit an animal is, the more likely it will get to pass more genes to the next generation. However, there must be a balance between the benefits of a behavior and the costs. Think about **foraging behavior** (all the behaviors involved in food gathering and eating). The benefits of eating are obvious. A robust, well-fed animal will have an advantage in competition for a mate. However, foraging has costs that might not be so obvious. Besides all the energy expended in foraging, it might be dangerous. For example, an animal might have to chase away another predator for the remains of a kill. Time spent foraging is also time lost from defending one's territory or from protecting one's young. *Ultimately, natural selection will favor behavior that minimizes the costs of foraging and maximizes its benefits.*

Mating Behavior and Mate Choice

Beyond the act of copulation and fertilization, mating behavior involves mating rituals, whether the animals are monogamous or polygamous, and the extent of parental care. Mating behavior even dictates morphological characteristics. Among **polygynous species**, such as elk, one male inseminates many females and the males are larger and more highly ornamented. This difference in appearance between males and females is called **sexual dimorphism**. In animal species where the female mates with more than one male, known as **polyandrous species**, the female is the showier of the two.

EVOLUTION OF BEHAVIOR

Some behaviors are simply controlled by a gene or a set of genes. Think of spiderwebs. Each web is characteristic of a particular species. Since the adult spider dies before the eggs hatch, a spider cannot learn to build a particular web from its parent. So the control of the behavior—spinning a web—must be genetic.

To understand the *ultimate cause* of a behavior, one must look at the behavior in terms of what **selective advantage** it confers on the individual or group. An interesting behavior that exemplifies this is altruism.

CHAPTER SUMMARY

The information in this chapter supported Big Ideas: EVO and SYI.

Behavior enables an individual to do all the things it needs to do to be successful: stay alive and pass on its genes to the next generation. Animal behaviors evolved under *strong evolutionary pressures.*

When you look at a particular behavior, analyze it in two ways:

1. What systems are responsible for that behavior (**proximate causes**)? Is the behavior genetic or is it learned?
2. How does the behavior make the animal more fit? How or why did that behavior evolve? What are the **ultimate causes** for that behavior?

PRACTICE QUESTIONS

1. Which of the following scientists studied communication in bees?
   (A) B. F. Skinner
   (B) Karl von Frisch
   (C) Niko Tinbergen
   (D) Ivan Pavlov

2. Which of the following scientists developed the idea of classical conditioning?
   (A) B. F. Skinner
   (B) Karl von Frisch
   (C) Niko Tinbergen
   (D) Ivan Pavlov

3. Which of the following scientists developed the idea of operant conditioning?
   (A) B. F. Skinner
   (B) Karl von Frisch
   (C) Niko Tinbergen
   (D) Ivan Pavlov

4. Which of the following scientists developed the idea of fixed action pattern?
   (A) B. F. Skinner
   (B) Karl von Frisch
   (C) Niko Tinbergen
   (D) Ivan Pavlov

5. Pavlov's dogs are an example of which of the following?
   (A) fixed action pattern
   (B) habituation
   (C) classical conditioning
   (D) imprinting

6. The fact that one stimulus becomes linked to another is an example of which of the following?
   (A) fixed action pattern
   (B) associative learning
   (C) classical conditioning
   (D) imprinting

7. Geese hatchlings following their mother is an example of which of the following?
   (A) fixed action pattern
   (B) associative learning
   (C) classical conditioning
   (D) imprinting

8. Innate, highly stereotypic behavior that must continue until it is completed is an example of which of the following?

   (A) fixed action pattern
   (B) associative learning
   (C) classical conditioning
   (D) imprinting

9. The scrub jay is a bird that is indigenous to North America. "Helper" scrub jays that have not mated and have no territory of their own often assist mated pairs in food gathering and raising their offspring. Which of the following statements best explains how this behavior fits into the theory of natural selection?

   (A) This behavior can be explained by genetic drift and is not supported by the theory of natural selection.
   (B) The "helper" birds can increase their own reproductive success by helping relatives who share some of their same genes.
   (C) The "helper" birds have a common genetic mutation that directs this behavior, even though it decreases their own reproductive fitness.
   (D) Natural selection will select against these "helper" birds, and there will be fewer and fewer of them.

10. Groundhogs are solitary animals that prefer to live in transitional areas where forests meet well-vegetated open fields; these are the areas where they can spot potential predators. They are diggers: they excavate tunnels underground in which they live and raise their young. The tunnel may have several entrances and can be up to 50 feet long. When groundhogs sense danger, they retreat into their tunnel.

    Based on this information, which of the following statements is correct in regard to a groundhog's tunneling behavior?

    (A) It is innate and is controlled by genes.
    (B) It is a learned behavior (trial and error).
    (C) It is a cooperative behavior—several adults will dig one tunnel.
    (D) When frightened, the groundhog will become paralyzed and will be unable to dig a tunnel.

11. When a worker honeybee stings an intruder to protect the hive, the worker honeybee dies. This phenomenon is known as altruism: an individual sacrifices itself for the group. What scientists do not fully understand is how this genetic trait remains in the gene pool if the individual that carries it dies. Which of the following statements offers the best explanation for how the trait persists in the gene pool?
    (A) The genetic trait of altruism arises frequently as a mutation because the colony must be protected.
    (B) The worker honeybee that dies shares most of her genes with her sister worker honeybees. When she dies, others that share most of the same genes will get to pass the altruistic trait on to their offspring.
    (C) Only some worker honeybees carry the altruism trait. They have the selected advantage over other worker honeybees and get to pass the trait on to their progeny after stinging the intruder.
    (D) This is a learned behavior that the queen honeybee teaches to the worker honeybees.

12. Among hens, the alpha (top-ranked) male controls the behavior of all the other animals in the group. The next in line, the beta animal, controls all the other animals except for the alpha male. Each animal can threaten all animals beneath it in the "pecking order." The alpha individual gets first choice of any resource: the most food and the best territory, as well as the opportunity to mate with all the females.
    Which of the following statements explains how this hierarchy benefits the individual and/or entire population?
    (A) The fittest individual will pass on the fittest genes to the next generation, thus maintaining the best gene pool for the entire population.
    (B) The fittest individual will gain the most resources, which will benefit itself and its offspring.
    (C) The weakest individuals will die, but their genes will persist in the population gene pool only by chance.
    (D) Successful reproduction is limited to only the fittest individuals.

13. Agonistic behavior is aggressive behavior. It involves a variety of threats to settle disputes among individuals. It also involves ritualistic or symbolic threatening behavior that is meant to avoid a "battle to the death." Which of the following is an example of agonistic behavior?
    (A) Two lions preserve their vast territory by protecting it together.
    (B) A young male lion attacks and kills the head of the pride and mates with all the females in the pride.
    (C) Male song birds sing to protect their territory.
    (D) Silverback (alpha male) gorillas communicate their power by hooting, chest pounding, leg kicking, and sideways running when approached by another male.

14. Habituation is one of the simplest forms of learning. For example, if a student taps the side of a Petri dish that contains a hydra (a cnidarian), at first the animal will shrink into a ball, becoming immobile. After several minutes of repeated tapping, however, the hydra will extend its tentacles, begin to move, and continue feeding. Which of the following accurately describes the advantage of this behavior, known as habituation?
    (A) The animal conserves energy and can remain immobile for days without feeding, until the threat ceases.
    (B) The hydra that can remain immobile for the longest time are considered the fittest, have the fittest genes, and will produce the fittest offspring.
    (C) The hydra that can remain immobile for the longest time will be able to live the longest.
    (D) The animal can continue feeding almost uninterrupted.

15. Animals that help other animals are expected to be
    (A) stronger than other animals
    (B) related to the animals they help
    (C) male
    (D) female

16. When it is time for them to spawn (lay eggs), salmon return to the very stream where their lives began. Once there, over the course of 1–2 days, female salmon lay hundreds of eggs that are fertilized by a male. Within a few days or weeks, all of the adults die. The fertilized eggs hatch, and all of the baby salmon migrate downstream to the ocean, where they grow to adulthood and live for years. Some salmon migrate thousands of miles into the ocean. When it is time to spawn, the salmon return to their place of origin.

    In one experiment, to determine where exactly salmon return to spawn, salmon eggs were randomly separated into two groups. Group A's eggs were left where those eggs had been laid in the stream. Group B's eggs were moved to a hatchery in a nearby stream. As adults, Group A returned to spawn in the stream where they had hatched. Group B returned to the hatchery to spawn.

    Using the information provided and your knowledge of animal behavior, which of the following choices correctly identifies the behavior demonstrated by the salmon and explains how that behavior helped the salmon determine where to return to spawn?
    (A) The behavior displayed here is cooperation. Using information stored in their brains during a sensitive period of life, the salmon collectively navigated to the place where they were fry (young fish).
    (B) The behavior displayed here is imprinting. Using information stored in their brains during a sensitive period of life, the salmon correctly navigated to the place where they had hatched to spawn.
    (C) The behavior displayed here is dominance hierarchy. A dominant female led the other salmon to work collectively to locate the site where they were fry (young fish).
    (D) The behavior displayed here is agonistic behavior. A dominant female led the others to the site where they had hatched.

17. Peter Berthold at the Max Planck Institute for Ornithology in Germany has been studying migratory patterns in birds for more than 30 years. Here is a description of one of his studies with birds called blackcaps (*Sylvia atricapilla*). He used three sets of birds for the study.

Experimental Design:
Group I: Blackcaps captured while wintering in Britain and then bred in Germany in an outdoor cage
Group II: The offspring of the birds captured in Britain; kept in outdoor cages
Group III: Young birds collected from nests near the lab and then raised in outdoor cages

In autumn during the migratory period, Berthold placed all three sets of birds into large glass-covered, funnel-shaped cages lined with carbon-coated paper. When the funneled cages were placed outdoors at night, the birds made marks with their beaks on the paper that indicated in which direction they were trying to fly (migrate).

Results of the Study:
The wintering adult birds captured in Britain and their lab-raised offspring both attempted to migrate to the west (toward Britain). The young birds collected from nests in southern Germany attempted to migrate to the southwest.

Which statement best describes what this experiment demonstrated about migratory patterns in blackcaps?

(A) Migratory patterns in blackcaps are controlled by Earth's magnetic field.
(B) Migratory patterns in blackcaps are controlled by environmental cues and hormones.
(C) Migratory patterns in blackcaps are not fixed but can be altered by relocating the nests.
(D) Migratory patterns in blackcaps have a genetic basis.

18. A study on kin selection and altruistic behavior was carried out using Belding's ground squirrels (*Spermophilus beldingi*). Pups are nursed for about a month and then weaned, at which point they are free to move away from their birthplace.

An adult female Belding's ground squirrel acts as a sentinel, standing up on her hind feet to sound loud alarm calls, trills, and single-note whistles when an aerial or a terrestrial predator is sensed nearby. Upon hearing the warning call, squirrels quickly try to escape into their underground nests. By sending out an alarm, the young may be protected, but the whistling adult female clearly puts herself in danger. Data were collected comparing the distance in meters that females and males move from their birthplace after weaning. Choose the graph that best demonstrates kin selection by females.

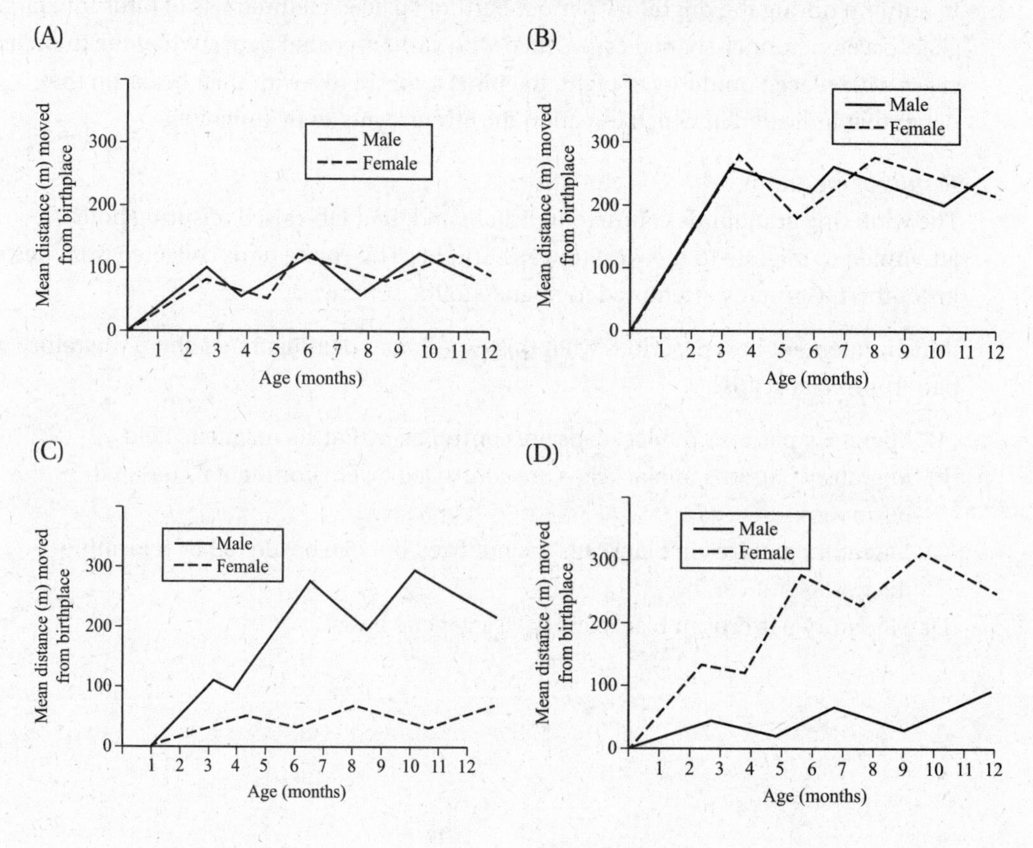

Answers Explained

1. **(B)** Karl von Frisch studied and named the waggle dance in bees. (EVO-1)

2. **(D)** Classical conditioning is a form of associative learning. Ivan Pavlov is famous because he "conditioned" dogs to salivate at the sound of a bell. (EVO-1)

3. **(A)** Operant conditioning is also called trial and error learning. An animal learns to associate one of its own behaviors with a reward or punishment and then tends to repeat or avoid that behavior. The best-known lab studies in operant conditioning were done by B. F. Skinner in the 1930s. (EVO-1)

4. **(C)** A fixed action pattern enables an animal to engage in complex behavior automatically without having to "think" about it. An FAP is a sequence of behaviors that is unchangeable and usually carried out to completion once initiated. Niko Tinbergen is most associated with FAPs. (EVO-1)

5. **(C)** An example of classical conditioning is how Pavlov trained his dogs to salivate at the sound of a bell. (EVO-1)

6. **(B)** When one stimulus becomes associated with one response, it is associative learning. (EVO-1)

7. **(D)** Konrad Lorenz imprinted his geese onto himself. They thought he was their mother and followed him everywhere. (EVO-1)

8. **(A)** This is the definition of a fixed action pattern. (EVO-1)

9. **(B)** This question revolves around **altruism** and the concept of "inclusive fitness." Choice B explains these ideas perfectly. There are many examples in nature of individual animals sacrificing themselves for the well-being of the entire group or population. Scientists assume that they do it because it increases the overall fitness of the entire group. Also, the altruistic individuals sacrifice themselves for individuals who have some similar genes and who will pass those genes onto the next generation. Choice A is incorrect because genetic drift is due to an accidental change in a population and is not relevant to this question. Choices C and D do not accurately describe this behavior and how it fits into the theory of natural selection. (SYI-3)

10. **(A)** Digging tunnels is a complex behavior. Complex behaviors in animals other than primates are generally innate. Groundhogs are solitary animals (as stated in the question), which rules out learning or teaching opportunities. If the animal learned to dig a tunnel by trial and error, it would not survive the learning process. Thus, since groundhogs are solitary animals, that means that choices B and C are incorrect. Choice D is inconsistent with what is stated in the question. (SYI-3)

11. **(B)** Choice B describes the theory for the occurrence of this phenomenon that is most accepted by scientists. Choice A is not correct because mutations do not arise as a result of being needed. Choice C does not make sense because the honeybee dies after stinging an intruder. Choice D is not correct because one honeybee cannot "teach" another how to die in order to save the group. (EVO-1)

12. **(A)** The alpha male, the fittest individual, gets to pass on his genes because he mates with every, or mostly every, female. Since his genes may be the only ones that get inherited, he benefits and so does the entire group or population. (EVO-1)

13. **(D)** Choice A describes cooperation, not agonistic behavior. Choice B describes a "battle to the death." Agonistic behavior is meant to avoid any killing. Choice C describes territoriality. (EVO-1)

14. **(D)** Hydra spend most of their time feeding. If the hydra ceased to feed for any extended length of time, it would die. (EVO-1)

15. **(B)** Animals that help other animals are engaging in altruistic behavior. Altruistic behavior is seemingly selfless behavior that may save kin that are carrying genes similar to those in the individual that sacrificed itself. (EVO-1)

16. **(B)** This experiment demonstrates the idea of imprinting. The salmon in both groups swam back to where they had hatched years before to spawn. They did not collectively swim to where they were fry, nor did a dominant female lead them to where they were fry, which rules out choices A and C. Choice D is incorrect because agonistic behavior is real or symbolic aggressive behavior, but there is no aggressive behavior described in the question. (EVO-1)

17. **(D)** The offspring of the birds raised in Britain attempted to fly toward Britain even though they had never migrated from or to Britain. This proves that the blackcaps' ability to navigate must have been inborn. (SP 5)

18. **(C)** According to the graph for choice C, females stay near their birthplace while the males move away. This means that the females are staying close to their relatives because when they send out an alarm and risk their lives, they are doing it for close relatives. This phenomenon is known as kin selection. (SP 5)

PART 3: ILLUSTRATIVE EXAMPLES REVIEW

Illustrative examples are topics that are suggested by the College Board, but are not required for the AP Biology exam. However, they are important for you to know because studying these topics will help you understand material that is presented throughout the course, and these topics can also serve as examples that you can use to explain your answers to the free-response questions on the exam. Using the illustrative examples from these chapters will give you a way to demonstrate a greater understanding, which will help your score on the exam.

The five chapters in this part of the book are filled with illustrative examples. Think about how to integrate these topics into the broader, required themes of the AP Biology course and exam. Here are two examples of what these chapters have to offer:

- They will explain how the timing and coordination of growth enables organisms to respond to environmental cues. Some information for those ideas can be found in the chapter on **plants** (Chapter 14), the chapter on **human physiology** (Chapter 15), and the chapter on **animal reproduction and development** (Chapter 18).
- They will expand your knowledge of cell-to-cell communication, with specific examples about chemical signaling from the **endocrine and nervous systems** (Chapter 16) and **the human immune system** (Chapter 17).

You may remember some of these topics from a previous Biology course, and/or many of them may be new to you. Either way, here is an opportunity for you to broaden your understanding of Biology and master topics that will help you better understand the concepts covered in your AP Biology course and on your AP exam. Dig in!

Plants 14

→ STRATEGIES THAT ENABLED PLANTS TO MOVE TO LAND
→ THE LEAF
→ TRANSPORT IN PLANTS
→ WATER POTENTIAL AND PASSIVE TRANSPORT
→ PLANT RESPONSES TO STIMULI

INTRODUCTION

The material in this chapter is not required for the AP Biology exam as per the new College Board curriculum. That is why this chapter is placed in Part 3 of this book and why there are no classification references to the curriculum frameworks (such as ENE-1). However, four of the thirteen investigative labs—#1, #4, #5, and #11—involve plants. In addition, your teacher may present some of this material to be used for what the College Board calls "illustrative examples" to help you understand the material. The most important topics in this chapter are how plant cells communicate, how plant cells respond to their environments, and how increased surface area increases absorption rates.

Remember, this chapter consists of illustrative examples. Although these topics are not required for the AP exam, studying them will broaden your understanding of Biology and provide you with examples to use as you write your answers to the free-response questions.

Plants are defined as multicelled, eukaryotic, photosynthetic **autotrophs**. Their cell walls are made of cellulose, and their surplus carbohydrate is stored as starch.

Plants evolved from aquatic green algae about 500 million years ago. Along with fungi and animals, they gradually colonized the land as they evolved adaptations to a dry environment. Today, most plants live on land. They have diversified into almost 300,000 different species inhabiting all but the harshest environments. Plants stabilize the soil they live in and provide a home for billions of insects and larger animals. They release oxygen into the atmosphere and absorb carbon dioxide. Most of the world depends on rice, beans, soy, corn, and wheat for survival.

Focus on mycorrhizae, leaf and stomate structure, signal transduction pathways, phototropism, and photoperiodism.

STRATEGIES THAT ENABLED PLANTS TO MOVE TO LAND

Plants began life in the seas and moved to land as competition for resources increased. The biggest problems a plant on land faces are supporting the plant body and absorbing and conserving water. Here are some modifications that evolved that enable plants to live on land:

- **Cell walls** made of cellulose lend support to the plant whose cells, unsupported by a watery environment, must maintain their own shape.
- **Roots** and **root hairs** absorb water and nutrients from the soil. Root hairs are finger-like extensions of root epidermal cells near the tip of each root. By increasing the root's surface area, they greatly enhance the absorption of water and minerals from the soil into the plant.
- **Transport tissue** (xylem and phloem) can move fluids great distances.
- **Stomates** open to exchange photosynthetic gases and close to minimize excessive water loss.
- The waxy coating on the leaves, **cutin**, helps prevents excess water loss from the leaves.
- In some plants, gametes and zygotes form within a protective jacket of cells called gametangia that prevents drying out.
- Sporopollenin, a tough polymer, is resistant to almost all kinds of environmental damage and protects plants in a harsh terrestrial environment. It is found in the walls of spores and pollen.
- **Seeds** and **pollen** are a means of dispersing offspring. They also have a protective coat that prevents desiccation.
- Xylem and phloem vessels enable plants to grow tall.
- Lignin embedded in xylem and other plant cells provides support.

Mycorrhizae are a type of fungus that live in plant roots and form a symbiotic relationship with the plants. Nearly all plants on Earth rely on mycorrhizae for enhanced uptake of nutrients and water from the soil. In return, the plants provide the fungi with carbohydrates they formed during photosynthesis. Fossil evidence demonstrates that mycorrhizae colonized the land before plants 405 million years ago, perhaps enabling plants to colonize land in the first place.

THE LEAF

The leaf is organized to maximize sugar production while minimizing water loss. The epidermis is covered by a waxy cuticle made of cutin to minimize water loss. **Guard cells** are modified epidermal cells that contain chloroplasts, are photosynthetic, and control the opening of the stomates. The inner part of the leaf consists of palisade and spongy mesophyll cells whose function is photosynthesis. The cells in the palisade layer are packed tightly, while the spongy cells are loosely packed to allow for diffusion of gases into and out of these cells. Vascular bundles or veins are located in the mesophyll and carry water and nutrients from the soil to the leaves and also carry sugar, the product of photosynthesis, from the leaves to the rest of the plant. Specialized mesophyll cells called bundle sheath cells surround the veins and separate them from the rest of the mesophyll. Figure 14.1 is an illustration of a leaf.

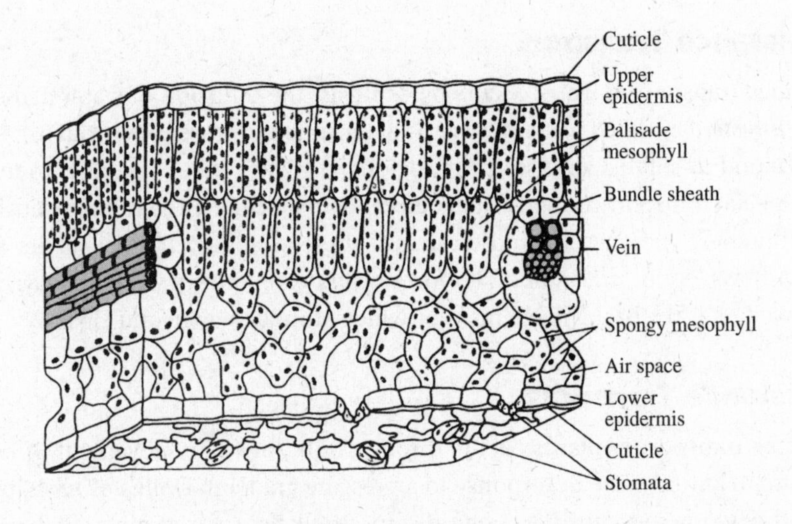

Figure 14.1 Structure of a Leaf

Stomates

Plants lose water by transpiration. About 90 percent of that water escapes through stomates, which account for only about 1 percent of the surface of the leaf. Guard cells are modified epithelium that contain chloroplasts that control the opening and closing of the stomates by changing their shape. The cell walls of guard cells are not uniformly thick. Cellulose microfibrils are oriented in such a direction (radially) that when the guard cells absorb water by osmosis and become **turgid**, they curve like hot dogs, causing the stomate to open. When guard cells lose water and become **flaccid**, the stomate closes. See Figure 14.2. In addition, some plants have stomates nestled in **stomatal crypts** that further minimize exposure of the stomate to air. This reduces water loss even more. (Review Chapter 7, "**Photosynthesis**.")

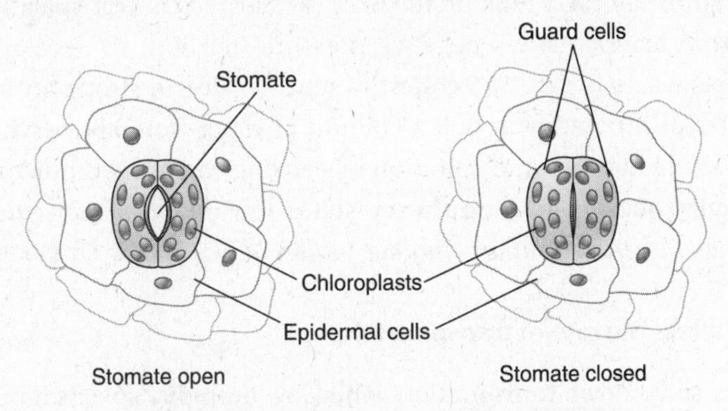

Figure 14.2 Stomates and Guard Cells

TRANSPORT IN PLANTS

Plants employ a variety of mechanisms to transport substances short and long distances. The selective permeability of plasma membranes controls short-distance movement of substances into and out of cells. Both active and passive transport mechanisms occur in plants, and plant cell membranes are equipped with the same types of pumps, channels, and transport proteins that other cells have.

Short-Distance Transport

Short-distance transport of water occurs by **osmosis**, the diffusion of water across a membrane. The *direction* in which water diffuses is determined by **water potential** (ψ). Free water, water not bound to solutes, diffuses from regions of higher water potential to lower water potential. Special transport proteins called **aquaporins** facilitate the rapid osmosis of water across membranes. The opening and closing of these selective channels affect the *rate* at which water flows, not its *direction.* As it moves, water can perform work, such as cell expansion. Cell swelling is the basic mechanism by which **stomates** open and close.

Long-Distance Transport

Long-distance transport in plants occurs through **bulk flow**, the movement of liquid from high pressure to low pressure in response to a pressure gradient. Unlike *osmosis*, bulk flow is *independent of solute concentration.* Long-distance bulk flow carries material through xylem and phloem vessels.

Phloem transports sugar throughout the plant by a process called **translocation**. The sugar moves from the **source**, where it is produced in the leaves, to the **sink**, where it is used or stored. The sink includes the roots and fruits. Proton pumps create an *electrochemical gradient* that powers the loading of sugar into sieve tubes of the phloem. **Translocation** requires an input of energy.

Xylem fluid (water and minerals from the soil) rises in a plant against gravity but requires no energy. The fluid in the xylem can be *pushed up* by root pressure or *pulled up* by **transpirational pull**. **Root pressure** results from water flowing into the stele from the soil as a result of the high mineral content in the root cells. Root pressure can push xylem sap upward only a few yards (meters). Droplets of water that appear in the morning on the leaf tips of some herbaceous dicots, like strawberries, are due to root pressure. This is known as **guttation**.

Transpirational pull can carry fluid up the world's tallest trees. **Transpiration**, the **evaporation of water from leaves**, causes negative pressure (tension) to develop in the xylem tissue from the roots to the leaves. The **cohesion** of water due to strong attraction between water molecules makes it possible to pull a column of water from above within the xylem. The absorption of sunlight drives transpiration by causing water to evaporate from the leaf. **Transpirational pull–cohesion tension theory** states that *for each molecule of water that evaporates from a leaf by transpiration, another molecule of water is drawn in at the root to replace it.*

Several factors affect the rate of transpiration:

- High humidity slows down transpiration, while low humidity speeds it up.
- Wind can reduce humidity near the stomates and thereby increase transpiration.
- Increased light intensity will increase photosynthesis and thereby increase the amount of water vapor to be transpired and increase the rate of transpiration.
- Closing stomates stops transpiration.

WATER POTENTIAL AND PASSIVE TRANSPORT

Water potential (**Ψ** or psi) is measured in megapascals (MPa) by plant biologists. By definition, Ψ of pure water in an open container under standard conditions of temperature (273 K or 0°C) and pressure (1 atmosphere) is 0 (zero) MPa. Water potential (Ψ) results from solute potential (Ψ_s) plus pressure potential (Ψ_p): $\Psi = \Psi_s + \Psi_p$.

Solute Potential (Ψ_s)

- Is the *osmotic potential.*
- Is directly proportional to molarity.
- The Ψ_s of pure water is 0 (zero).
- The Ψ_s is always a negative value; as the solute concentration increases, the Ψ_s decreases (for example, a 0.1 M solution of sugar has a Ψ_s of -0.23 MPa).

Pressure Potential (Ψ_p)

- Is the physical pressure on a solution.
- Can have a positive or negative value relative to atmospheric pressure.
- The water in living cells is usually under positive pressure due to the uptake of water by osmosis (i.e., turgor pressure in plants); the cell contents are *pressed* up against the cell wall.
- The water in xylem vessels in plants (which transport water up a plant) is under negative pressure; the water and nutrients are *pulled* upward in a vascular plant because the pressure decreases inside the xylem as water transpires from the leaves.

Remember: *Water moves from higher water potential to lower water potential.*

PLANT RESPONSES TO STIMULI

Hormones

Plant hormones help coordinate growth, development, and responses to environmental stimuli. They are produced in very small quantities, but they have a profound effect on the plant because the hormone signal is amplified. **Signal transduction pathways** amplify the hormonal signal and connect it to specific cell responses. A plant's response to a hormone usually depends not so much on absolute quantities of hormones but on relative amounts. Hormones can have multiple effects on a plant, and they can work synergistically with other hormones or in opposition to them. Auxins, cytokinins, gibberellins, abscisic acid, and ethylene gas are examples of plant hormones.

AUXINS

- An unequal distribution of auxins is responsible for phototropisms.
- They enhance apical dominance, which is the preferential growth of a plant upward (toward the sun) rather than laterally.
- Auxins are the main ingredient in *rooting powder,* which helps develop roots quickly.

CYTOKININS

- Cytokinins stimulate cytokinesis and cell division.
- They delay *senescence* (aging) by inhibiting protein breakdown.
- Spraying cytokinins on flowers keeps them fresh.

GIBBERELLINS

- Gibberellins promote stem and leak elongation.
- They induce "bolting," the rapid growth of a floral stalk on which a flower and fruit will develop.

ABSCISIC ACID (ABA)

- ABA inhibits growth and promotes seed dormancy.
- It enables plants to withstand drought.
- It closes stomates during times of water stress.

ETHYLENE GAS

- Ethylene gas promotes ripening, which in turn triggers increased production of ethylene gas.
- Commercial producers pick fruit before it is ripe and spray it with ethylene gas when the fruit gets to market to ripen it.

Tropisms

A tropism is the growth of a plant toward or away from a stimulus. Examples are **thigmotropisms** (touch), **geotropisms** or **gravitropisms** (gravity), and **phototropisms** (light). A growth of a plant toward a stimulus is known as a **positive tropism**, while a growth away from a stimulus is a **negative tropism**.

Phototropisms result from an *unequal distribution* of one category of plant hormones called auxins, which accumulate on the side of the plant away from the light. Since auxins cause growth, the cells on the shady side of the plant enlarge and the stem bends toward the light.

Geotropisms result from an interaction of auxins and **statoliths**, specialized plastids that contain dense starch grains.

Photoperiodism

IMPORTANT TIP

The timing and coordination of physiological events are regulated by multiple mechanisms.

Plants detect the presence of light as well as its direction, intensity, and wavelength. The environmental stimulus a plant uses to detect the time of year is the **photoperiod**, the relative lengths of day and night. Plants have a biological clock set to a 24-hour day, known as a **circadian rhythm**. The physiological response to the photoperiod, such as flowering, is known as **photoperiodism**. Some plants will flower only when the light period is longer than a certain number of hours. These plants are called **long-day plants**. Some plants are **short-day plants**, and some are **day-neutral** and will flower regardless of the length of day. Plants actually respond to the length of darkness, not the length of light. So despite its name, a *long-day plant* is actually a *short-night plant.*

Early summer has short periods of darkness that induce flowering in **long-day plants** (such as an iris), but not in **short-day plants** (such as a goldenrod). **Late fall** has long periods of darkness that induce flowering in **short-day plants**, but not in **long-day plants**.

The photoreceptor responsible for keeping track of the length of day and night is the pigment **phytochrome**. There are two forms of phytochrome: **Pr (red light–absorbing)** and **Pfr (infrared light–absorbing)**. Phytochrome is synthesized in the Pr form. When the plant is exposed to light, Pr converts to Pfr. In the dark, Pfr reverts back to Pr. The conversion from one to the other enables the plant to keep track of time. The plant is able to sense the concentrations of the two phytochromes and respond accordingly.

$$\text{Pr} \underset{\substack{\text{IR light} \\ \text{(Slow conversion at night)}}}{\overset{\substack{\text{(daylight)} \\ \text{Red light}}}{\rightleftarrows}} \text{Pfr (triggers germination)}$$

CHAPTER SUMMARY

- Think of this chapter as a source of illustrative examples that can help you on your free-response questions on the AP exam. One thing the College Board is emphatic about is that it wants students to demonstrate that they can make connections between topics and across various domains.
- Knowing the structure of **leaves** and of **stomates** will help you understand the important process of **photosynthesis**. Understanding the role of **mycorrhizae** in plant growth will give you an example of the interdependence of organisms within an **ecosystem**. How plants load and carry sucrose inside **phloem vessels** is an important example of **active transport**. Understanding how water and nutrients travel through xylem will help you to understand **Lab #11**. The material about hormones in this chapter is a good example of one of the College Board's favorite topics—**cell communication**. Plants use **signal transduction pathways** for the same reason as do organisms in other kingdoms. These pathways amplify a specific signal from inside or outside a cell. Another important concept put forth by the College Board is that the timing and coordination of physiological events are regulated by multiple mechanisms. **Photoperiodism** is a great example of that.

PRACTICE QUESTIONS

1. In most of the Rocky Mountains, red squirrels feed on lodgepole pine seeds. They harvest pine cones from the trees and store them through the winter. They are most successful at harvesting seeds from wide, heavy pine cones. Some forests that are inhabited by red squirrels are also inhabited by crossbill birds, which compete with the squirrels for lodgepole pine seeds. In those areas where the two animals have been competing for lodgepole pine seeds for thousands of years, the crossbills have evolved a deeper and wider bill than the common bill found in the rest of its species. This enables them to compete for seeds more effectively against the red squirrels. In four isolated forest ranges east and west of the Rocky Mountains, red squirrels have been absent for 10,000 to 12,000 years, and red crossbills are the main lodgepole pine seed predator.

   Which of the following would best describe the crossbill in these isolated mountain forest ranges?

   (A) The crossbills would have a deeper and wider bill.
   (B) The crossbills would have evolved a bill that would help them compete against red squirrels, although what that bill would look like is unclear.
   (C) The crossbills would have the common bill.
   (D) The crossbills would have evolved into a larger bird in order to compete more favorably with the red squirrels.

2. If grown individually under short-day conditions, a short-day plant will flower, but a long-day plant will not. However, both will flower if they are grafted together and are exposed to short-day conditions, as depicted in the figure below:

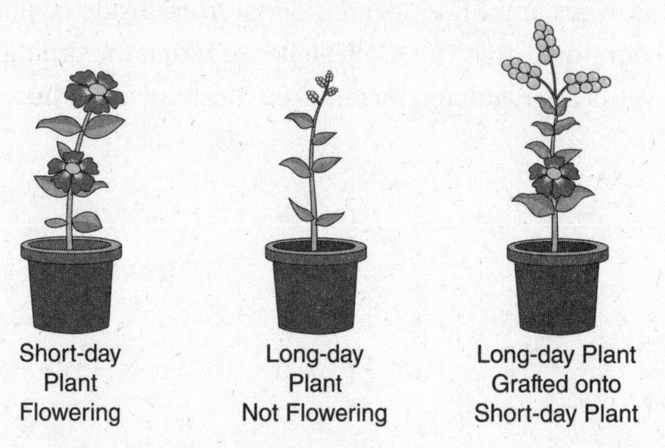

   What can be deduced from this phenomenon?

   (A) The genes of the long-day plant mutated.
   (B) The genes of the short-day plant mutated.
   (C) There was an error in the procedure because short-day and long-day blooming are both genetically controlled.
   (D) There must have been some chemical substance that was transmitted from the lower flower to the upper one up the graft.

3. Which of the following is the most accurate explanation for how plants uptake water from the roots and carry it to the leaves?
   (A) Water is pulled up to the leaves by a combination of cohesion forces, adhesion forces, and transpiration of water from leaves.
   (B) Water is pushed up the xylem using energy it absorbs from the sun.
   (C) Water is pushed up the phloem using energy it absorbs from the sun.
   (D) Hormones are directly responsible for moving water from the soil to the leaves.

4. Which of the following is most likely to happen when stomates are open?
   (A) CO_2 will leave the leaf.
   (B) Guard cells will lose water and become flaccid.
   (C) Transpiration will shut down.
   (D) Water will be released.

5. In plants, special transport proteins called aquaporins facilitate the rapid diffusion of water across membranes. Scientists engineered a mutation in a type of mustard plant so that the plant developed normally except for the fact that the cell membranes had no aquaporins. Given the function of aquaporins described above, which of the following statements describes the most likely result of such a mutation?
   (A) The fruit from this plant would not ripen.
   (B) Flowering in the plant would be altered because the plant's circadian rhythm would be altered.
   (C) The plant would not be able to produce pollen.
   (D) The stomates in the leaves would not open and close properly.

6. Which of the following is most important for the normal functioning of stomates?
   (A) nerve impulses being sent to the stomates from the leaf stem
   (B) the osmosis of water across membranes in a leaf
   (C) the pigment phytochrome
   (D) the hormone ethylene gas

7. Which of the following is most responsible for the movement of sugars down a tree from its photosynthetic leaves to its roots?
   (A) the hormone abscisic acid (ABA)
   (B) ATP
   (C) the evaporation of water from leaves
   (D) active transport in xylem tissue

8. The grass *Dichanthelium lanuginosum* lives in hot soils and houses fungi of the genus *Curvularia* in their leaves in a symbiotic relationship. Researchers performed experiments in the field to test the impact of *Curvularia* on the heat tolerance of this grass. They grew plants with and without *Curvularia* in soils of different temperatures. After 3 weeks, they measured plant mass and the number of new shoots produced. (Endophytes are fungi that live inside a plant body part without harming the plant. Parasites harm the plant they live with.)

Here is a bar graph of the results.

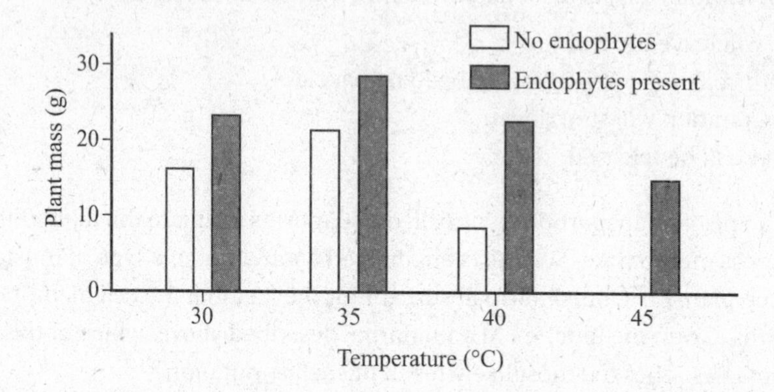

Which of the following best describes the role of *Curvularia* in this study and the reason for that role?

(A) *Curvularia* is an endophyte because plants with *Curvularia* grew better than those without it, except at high temperatures.

(B) *Curvularia* is an endophyte because plants with *Curvularia* grew better than those without it, especially at high temperatures.

(C) *Curvularia* is a parasite because plants with *Curvularia* did not grow as well as those without it, except at high temperatures.

(D) *Curvularia* is a parasite because plants with *Curvularia* grew less than those without it, regardless of the temperature.

9. The illustration below presents xylem and phloem vessels in a stem of a plant, with the direction of the flow of water and nutrients in the vessels. Also shown are the values for water potential in various cells (labeled A, B, C, and D).

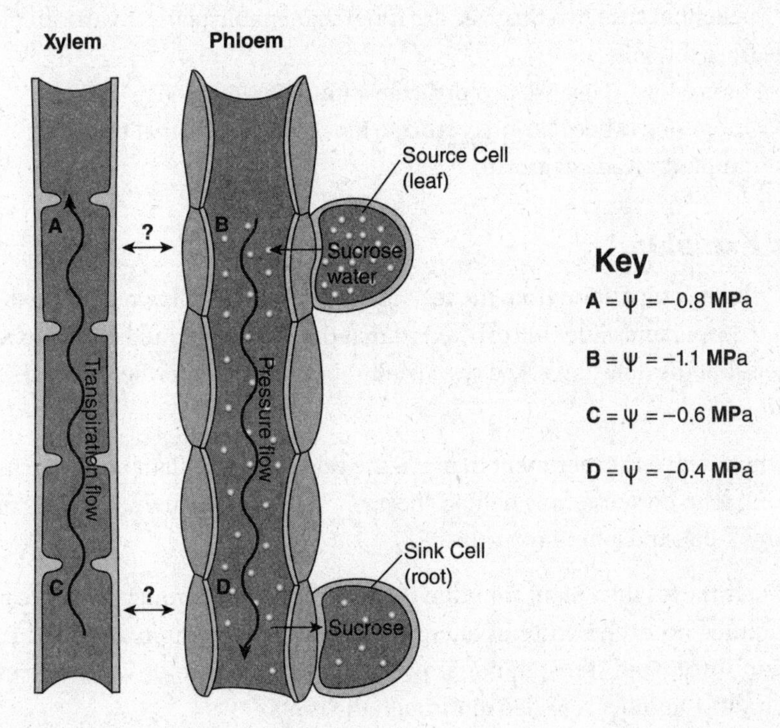

What is the correct direction of flow from cell to cell based on the water potential values in each area?

(A) A → B; C → D
(B) B → A; D → C
(C) A → B; D → C
(D) B → A; C → D

10. Compared with a cell that has few aquaporins in its membranes, a cell that contains many aquaporins will have a

(A) higher water potential
(B) lower water potential
(C) faster rate of active transport
(D) faster rate of osmosis

11. Compared to seedlings grown without mycorrhizae, seedlings grown with mycorrhizae will

    (A) grow taller because mycorrhizae are hormones that stimulate growth
    (B) grow taller because mycorrhizae are fungi that enable plants to absorb more nutrients from the soil
    (C) grow more slowly because mycorrhizae stimulate senescence
    (D) grow more slowly because mycorrhizae block a critical signal transduction pathway that ultimately leads to growth

Answers Explained

1. **(C)** Without competition from the red squirrels, the crossbills would have the common bill. The deeper and wider bill evolved so that the crossbills could compete with the red squirrels, but the question asked specifically about the region where no red squirrels are present.

2. **(D)** This is a classic experiment to prove the existence of a plant hormone that induces flowering. The hormone was named "florigen," and surprisingly, it causes flowering in both short-day and long-day plants.

3. **(A)** Water moves up a plant from the roots to the leaves through vessels called xylem, and requires no expenditure of energy. It occurs by a combination of adhesion and cohesion forces and transpirational pull. As one molecule of water leaves a leaf by transpiration, another is pulled in through the roots.

4. **(D)** When stomates open, CO_2 moves inward, not outward. As water vapor leaves the leaf through open stomates, other molecules of water are pulled in at the roots and into the xylem by transpirational pull. When light strikes a leaf, guard cells become rigid or turgid, which changes their shape, causing stomates to open. Thus, when stomates are open, the guard cells are rigid or turgid, not flaccid.

5. **(D)** Stomates open as the guard cells that control stomate function fill with water and become turgid. Stomates close as the guard cells that control stomate function lose water and become flaccid. Aquaporins allow large amounts of water to rapidly diffuse across membranes into cells. Choice A is incorrect because fruit ripening is caused by the hormone ethylene gas; it would be unaffected by the presence or absence of aquaporins. Choice B is incorrect because circadian rhythm and flowering in plants are controlled by phytochromes and hormones, respectively. Choice C is incorrect because aquaporins have nothing to do with pollen and reproduction.

6. **(B)** Changes in water potential cause water to flow into and out of guard cells that surround stomates on the surface of leaves. Stomates open as the guard cells that control stomate function fill with water and become turgid. Stomates close as the guard cells that control stomate function lose water and become flaccid. Choice A is incorrect because plants do not have nerves or neurons. Choice C is incorrect because phytochromes have to do with circadian rhythm and photoperiodism. Choice D is incorrect because the hormone ethylene gas is responsible for fruit ripening.

7. **(B)** The process by which sugars flow down a tree within phloem tissue requires energy. Cells of the phloem tissue utilize ATP for this process. Choice A is incorrect because the hormone ABA is related to the suppression of growth in a plant, as well as the promotion of seed dormancy. Choice C is incorrect because the evaporation of water from leaves (transpiration) powers the movement of water and nutrients up a plant in xylem vessels. Choice D is incorrect because sugars are transported in phloem, not xylem. Also, xylem flow does not require the expenditure of energy, whereas phloem transport does.

8. **(B)** Notice that the plants grown without *Curvularia* did not grow at all at 45 degrees. Clearly, *Curvularia* helps the plants grow and is not a parasite.

9. **(C)** Water flows from high water potential to low water potential. $\Psi = -0.8$ MPa is a higher value than $\Psi = -1.1$ MPa, so water flows from A to B. $\Psi = -0.4$ MPa is a higher value than $\Psi = -0.6$ MPa, so water flows from D to C. Notice the circle of flow from A to B and then from D to C.

10. **(D)** Aquaporins are channels that can rapidly open and close to allow large amounts of water across a plant membrane. The water flows down the gradient and requires no energy to move.

11. **(B)** Mycorrhizae enable plants to uptake greater amounts of water and nutrients.

15 Human Physiology

- → DIGESTION IN DIFFERENT ANIMALS
- → DIGESTION IN HUMANS
- → GAS EXCHANGE IN DIFFERENT ANIMALS
- → GAS EXCHANGE IN HUMANS
- → CIRCULATION IN DIFFERENT ANIMALS
- → HUMAN CIRCULATION
- → OSMOREGULATION
- → EXCRETION
- → MUSCLE

Remember, this chapter consists of illustrative examples. Although these topics are not required for the AP exam, studying them will broaden your understanding of Biology and provide you with examples to use as you write your answers to the free-response questions.

INTRODUCTION

The material in this chapter is not required for the AP Biology exam as per the new College Board curriculum. That is why this chapter is placed in Part 3 of this book and why there are no classification references to the curriculum frameworks (such as ENE-1). The College Board assumes that you already studied these topics in the first Biology course you took in high school. However, your teacher may cover some of these topics to give you a broader understanding from which to write your responses to the free-response questions on the AP Biology exam.

DIGESTION IN DIFFERENT ANIMALS

Hydra

In the **hydra** (cnidarians), digestion occurs in the **gastrovascular cavity**, which has only one opening. Cells of the **gastrodermis** (lining of the gastrovascular cavity) secrete digestive enzymes into the cavity for **extracellular digestion**. Some specialized nutritive cells have flagella that move the food around the gastrovascular cavity, and some have pseudopods that engulf food particles.

Earthworm

The digestive tract of the **earthworm** is a long, straight tube. As the earthworm burrows in the ground, creating tunnels that aerate the soil, the mouth ingests decaying organic matter along with soil. From the mouth, food moves to the esophagus and then to the **crop** where it is stored. Posterior to the crop, the **gizzard**, which consists of thick, muscular walls, grinds

STUDY TIP

Organs work together to do one job. One example of this is digestion.

up the food with the help of sand and soil that were ingested along with the organic matter. The rest of the digestive tract consists of the intestines where chemical digestion and absorption occur. Absorption is enhanced by the presence of a large fold in the upper surface of the intestine, called the **typhlosole**, which greatly increases the surface area of the intestine.

Grasshopper

Like the earthworm, the **grasshopper** has a digestive tract that consists of a long tube consisting of a **crop** and **gizzard**. However, there are several differences. The grasshopper has specialized mouth parts for tasting, biting, and crushing food and has a gizzard that contains **plates** made of **chitin** that help in grinding the food. In addition, in the grasshopper, the digestive tract is also responsible for removing nitrogenous waste (**uric acid**) from the animal.

DIGESTION IN HUMANS

The human digestive system has two important functions: **digestion** (breaking down large food molecules into smaller usable molecules) and **absorption** (the diffusion of these smaller molecules in the body's cells). **Fats** get broken down into glycerol and fatty acids, **starch** into monosaccharides, **nucleic acids** into nucleotides, and **proteins** into amino acids. **Vitamins** and **minerals** are small enough to be absorbed without being digested. The digestive tract is about 30 feet (9 m) long and made of **smooth** (**involuntary**) **muscle** that pushes the food along the digestive tract by a process called **peristalsis**.

Mouth

In the mouth, the **tongue** and differently shaped **teeth** work together to break down food mechanically. Form relates to function, and the type of teeth a mammal has reflects its dietary habits. Humans are **omnivores** and have three different types of teeth: **incisors** for cutting, **canines** for tearing, and **molars** for grinding. **Salivary amylase** released by **salivary glands** begins the chemical breakdown of **starch**.

Esophagus

After swallowing, food is directed into the esophagus, and not the windpipe, by the **epiglottis**, a flap of cartilage in the back of the **pharynx** (throat). No digestion occurs in the esophagus.

Stomach

The **stomach** churns food mechanically and secretes **gastric juice** that begins the digestion of **proteins**. The lining of the stomach contains **gastric pits** that are themselves lined with cells. **Chief cells** secret pepsinogen, the inactive form of **pepsin** that becomes activated by acid. **Parietal cells** secrete the hydrochloric acid that keeps the pH of gastric juices at 2–3 and activates pepsinogen. HCl also kills ingested microorganisms and breaks down protein.

The stomach of all mammals also contains **rennin** to aid in the digestion of the protein in milk. The **lower esophageal sphincter** at the top of the stomach keeps food in the stomach from backing up into the esophagus and burning it. The **pyloric sphincter** at the bottom of the stomach keeps the food in the stomach long enough to be digested.

Excessive acid can cause an **ulcer** to form in the esophagus, the stomach, or the **duodenum** (the first 12 inches [30 cm] of the small intestine). However, scientists now know that a common cause of ulcers is a particular bacterium, ***Helicobacter pylori***, which can be effectively treated with antibiotics.

Small Intestine

STUDY TIP

High surface area-to-volume ratios enhance an organism's ability to obtain necessary resources and eliminate waste.

Digestion is completed in the **duodenum**. Intestinal enzymes and pancreatic amylases hydrolyze starch and glycogen into maltose. **Bile**, which is produced in the **liver** and stored in the **gallbladder**, is released into the small intestine as needed and acts as an **emulsifier** to break down fats, creating greater surface area for digestive enzymes. **Peptidases**, such as trypsin and chymotrypsin, continue to break down proteins. Nucleic acids are hydrolyzed by **nucleases**, and **lipases** break down fats. Once digestion is complete, the lower part of the small intestine is the site of **absorption**. Millions of fingerlike projections called **villi** absorb all the nutrients that were previously released from digested food. Each villus contains capillaries (that absorb amino acids, vitamins, and monosaccharides), and a **lacteal**, a small vessel of the **lymphatic system**, which absorbs fatty acids and glycerol. Each epithelial cell of the villus has many microscopic cytoplasmic appendages called **microvilli** that greatly increase the rate of nutrient absorption by the villi. Figure 15.1 shows a villus with a lacteal, capillaries, and microvilli.

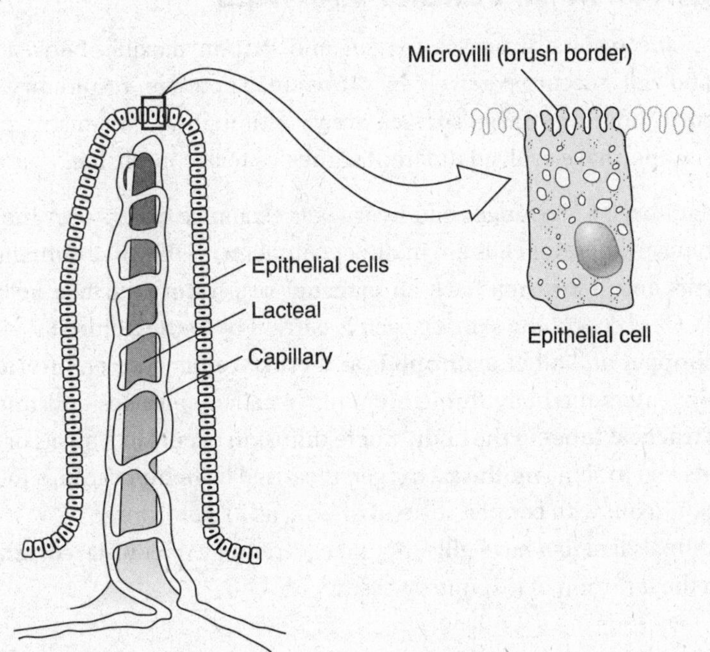

Figure 15.1 A Villus that Contains Blood Vessels and a Lacteal

Large Intestine

The **large intestine** or **colon** serves three main functions: **egestion**, the removal of undigested waste; **vitamin production**, from bacteria symbionts living in the colon; and the **removal of excess water**. Together, the small intestine and colon reabsorb 90 percent of the water that entered the alimentary canal. If too much water is removed from the intestine, **constipation** results; if inadequate water is removed, **diarrhea** results. The last 7–8 inches (18–20 cm) of the gastrointestinal tract is called the **rectum**. It stores **feces** until their release. The opening at the end of the digestive tract is called the **anus**.

Hormones that Regulate the Digestive System

Hormones are released as needed as a person sees or smells food or as food moves along the gut. Table 15.1 summarizes the hormones involved in regulating digestion.

Table 15.1

Hormones that Regulate Digestion		
Hormone	**Site of Production**	**Effect**
Gastrin	Stomach wall	Stimulates sustained secretion of gastric juice
Secretin	Duodenum wall	Stimulates pancreas to release bicarbonate to neutralize acid in duodenum
Cholecystokinin (CCK)	Duodenum wall	Stimulates pancreas to release pancreatic enzymes and gallbladder to release bile into small intestine

GAS EXCHANGE IN DIFFERENT ANIMALS

An exchange of the respiratory gases oxygen and carbon dioxide, between the external environment and cells, occurs passively by **diffusion**. Therefore, respiratory surfaces must be **thin**, be **moist**, and have **large surface areas**. Although all organisms must exchange respiratory gases, they have evolved different strategies to accomplish it.

- In simple animals, like **sponges** and **hydra**, gas exchange occurs over the entire surface of the organism wherever cells are in direct contact with the environment.
- **Earthworms** and **flatworms** have an **external respiratory surface** because diffusion of O_2 and CO_2 occurs at the skin. Oxygen is carried by **hemoglobin** dissolved in blood.
- The **grasshopper** and other **arthropods** and **crustaceans** have an **internal respiratory surface**. Air enters the body through openings called **spiracles** and travels through a system of **tracheal tubes** in the body, where diffusion occurs in sinuses or hemocoels. In **arthropods** and in some **mollusks**, oxygen is carried by **hemocyanin**, a molecule similar to hemoglobin but with **copper**, instead of iron, as its core atom.
- Aquatic animals like **fish** have **gills** that take advantage of **countercurrent exchange** to maximize the diffusion of respiratory gases.

GAS EXCHANGE IN HUMANS

In **humans**, air enters the nasal cavity and is **moistened**, **warmed**, and **filtered**. From there, air passes through the **larynx** and down the **trachea** and **bronchi** into the tiniest **bronchioles**, which end in microscopic air sacs called **alveoli** where diffusion of respiratory gases occurs; see Figure 15.2. Humans have an **internal respiratory surface**. As the rib cage expands and the **diaphragm** contracts and lowers, the chest cavity expands, making the internal pressure lower than atmospheric pressure. Thus, air is drawn into the lungs by **negative pressure**.

The **medulla** in the brain, which contains the breathing control center, sets the rhythm of breathing and monitors **CO_2 levels** in the blood by sensing changes in the pH of the blood. CO_2, the by-product of cell respiration, dissolves in blood to form **carbonic acid**. Therefore, the higher the CO_2 concentration in the blood, the lower the pH. Blood pH lower than 7.4 causes the medulla to increase the rate of breathing to rid the body of more CO_2.

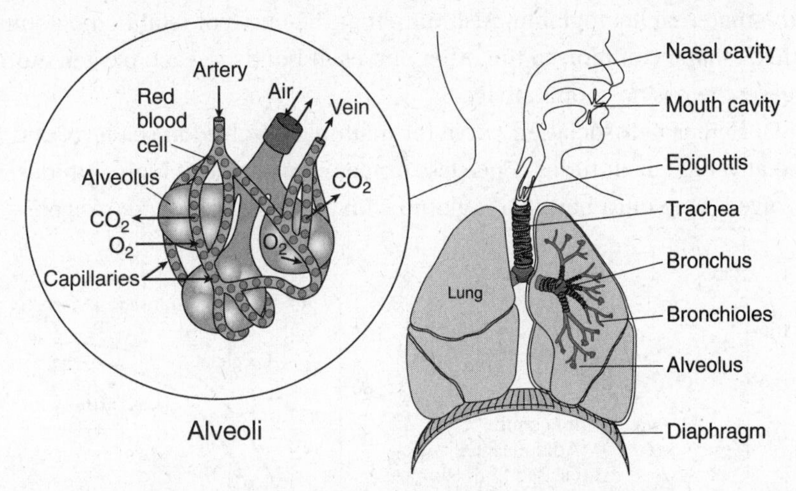

Figure 15.2 Adaptations for Human Respiration

The body is also sensitive to low O_2 levels but to a much lesser degree. O_2 sensors—chemoreceptors—are located in the nodes of neural tissue in the aorta and carotid arteries leaving the heart. If O_2 levels drop drastically, these chemoreceptors become activated and send nerve impulses to the medulla.

Hemoglobin

Oxygen is carried in the human blood by the respiratory pigment **hemoglobin**, which can combine loosely with four oxygen molecules, forming the molecule **oxyhemoglobin**. To function in the transport of oxygen, hemoglobin must be able to bind with oxygen in the lungs and unload it at the body cells. The more tightly the hemoglobin binds to oxygen in the lungs, the more difficult it is to unload the cells. Hemoglobin is an **allosteric** molecule and exhibits **cooperativity**. This means that once it binds to one oxygen molecule, hemoglobin undergoes a **shape change** and binds more easily to the remaining three oxygen molecules. In addition, hemoglobin's conformation is sensitive to pH. A drop in pH lowers the affinity of hemoglobin for oxygen (known as the Bohr shift). Because CO_2 dissolves in water to form carbonic acid, actively respiring tissue, which releases large quantities of CO_2, will lower the pH of its surroundings and induce hemoglobin to release its oxygen at the cells where needed.

STUDY TIP

On these graphs, the curve on the left shows greater affinity for oxygen.

Figure 15.3 consists of four graphs showing saturation-dissociation curves for hemoglobin (Hb). *The farther to the right the curve is, the less affinity the hemoglobin has for oxygen.*

- Graph A: Here is a dissociation curve of adult hemoglobin at normal and low blood pH showing the Bohr shift. At a lower pH, the hemoglobin has less affinity for oxygen.
- Graph B: Here is a dissociation curve for hemoglobin of two different mammals: a mouse and an elephant. The mouse has a much higher metabolism, and its body cells have a correspondingly higher oxygen requirement. To accommodate the animal's oxygen needs, the mouse hemoglobin has a dissociation curve located to the right of the elephant hemoglobin dissociation curve. In other words, mouse hemoglobin drops off O_2 at cells more easily than does elephant hemoglobin.
- Graph C: Here is a dissociation curve for fetal and maternal hemoglobin. Fetal hemoglobin has a higher affinity for oxygen than adult hemoglobin does so it can take oxygen from the maternal hemoglobin. Also note that the curve of fetal hemoglobin does not have the S-shape common to the other curves. It bonds to each oxygen atom with the same ease; there is no cooperativity.
- Graph D: Here is a dissociation curve for mammals evolved at sea level and mammals evolved at very high altitudes. Since less oxygen is available at high altitudes, mammals that evolved there must have hemoglobin with a greater affinity for oxygen.

REMEMBER

Being able to interpret graphs is important.

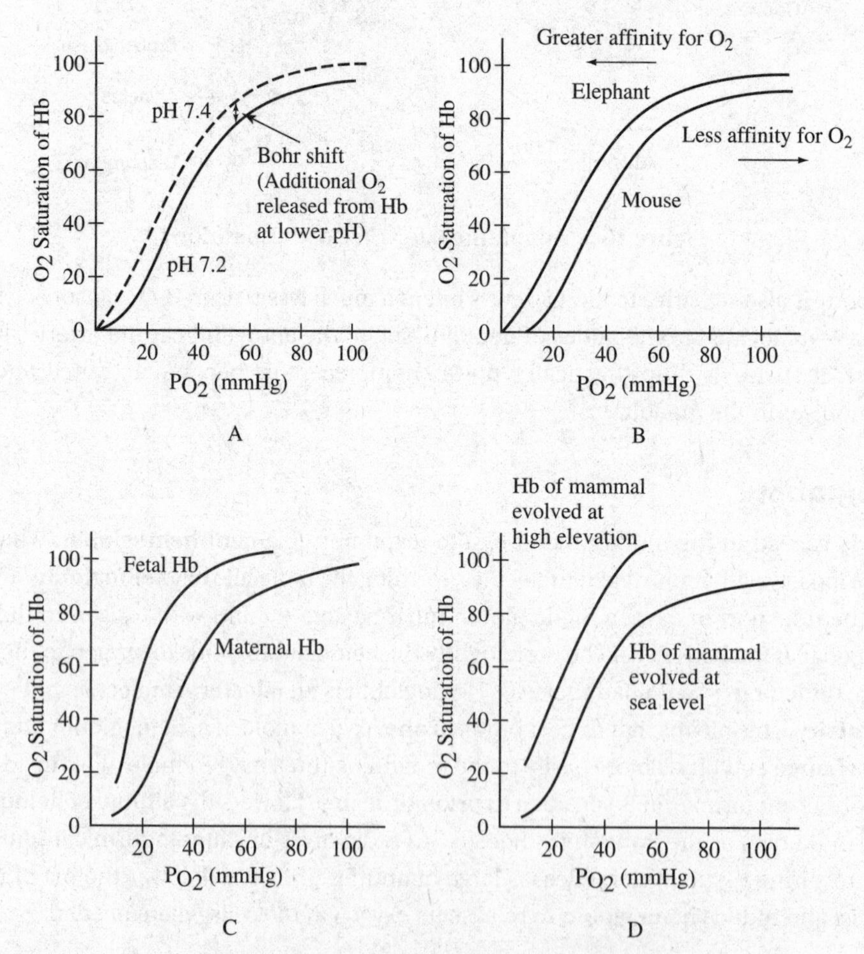

Figure 15.3 Affinity for Oxygen for Different Types of Hemoglobin or in Varying pH

Transport of Carbon Dioxide

Very little carbon dioxide is transported by hemoglobin. Most carbon dioxide is carried in the **plasma** as part of the reversible blood buffering **carbonic acid–bicarbonate ion system**, which maintains the blood at a constant pH of 7.4. The bicarbonate ion is produced in a two-stage reaction. First, carbon dioxide combines with water to form carbonic acid. This reaction is catalyzed by carbonic acid anhydrase found in red blood cells. Then carbonic acid dissociates into a bicarbonate ion and a proton. The protons can be given up into the plasma, which lowers the blood pH, or taken up by the **bicarbonate ion**, which raises the blood pH. Here is the equation:

$$\underset{\text{carbon dioxide + water}}{CO_2 + H_2O} \longleftrightarrow \underset{\text{carbonic acid}}{H_2CO_3} \longleftrightarrow \underset{\text{bicarbonate ion + proton}}{HCO_3^- + H^+}$$

CIRCULATION IN DIFFERENT ANIMALS

Primitive animals like the sponge and the hydra have no circulatory systems. All their cells are in direct contact with the environment, and such a system is unnecessary. The earthworm has a closed circulatory system where blood is pumped by the heart through arteries, veins, and capillaries. Oxygen is carried by hemoglobin that is dissolved in the blood. The grasshopper, as a representative animal of the arthropods, has an **open circulatory system**. After blood is pumped by the heart into an artery, it leaves the vessel and seeps through spaces called sinuses or hemocoels as it feeds body cells. The blood then moves back into a vein and circulates back to the heart. This system lacks capillaries. Arthropod blood is colorless and does not carry oxygen.

STUDY TIP

CO2 is carried in the plasma. O2 is carried by red blood cells.

HUMAN CIRCULATION

Human circulation consists of a **closed circulatory system** with arteries, veins, and capillaries. Table 15.2 shows the components of blood.

Table 15.2

Components of Blood		
Component	**Scientific Name**	**Properties**
Plasma	– –	Liquid portion of the blood Contains clotting factors, hormones, antibodies, dissolved gases, nutrients, and wastes Maintains proper osmotic potential of blood: 300 mOsm/L
Red Blood Cells	Erythrocytes	Carry hemoglobin and oxygen Do not have a nucleus and live only about 120 days Formed in the bone marrow and recycled in the liver
White Blood Cells	Leukocytes	Fight infections and are formed in the bone marrow Die fighting infections and are one component of pus One type of leukocyte—the B lymphocyte—produces antibodies
Platelets	Thrombocytes	These are not cells; rather, they are cell fragments that are formed in the bone marrow (from megakaryocytes) and they clot blood

Red blood cells, white blood cells, and platelets all develop in bone marrow from multipotent **stem cells**. These stem cells keep dividing and constantly replenish the population of blood cells throughout a person's life.

The Mechanism of Blood Clotting

Blood clotting is a complex mechanism that begins with the release of **clotting factors** from platelets and damaged tissue. It involves a complex set of reactions, including the activation of inactive **plasma proteins**. **Anticlotting factors** normally circulate in the plasma to prevent the formation of a clot or **thrombus**, which can cause serious damage even in the absence of an injury.

Here is the pathway of normal clot formation:

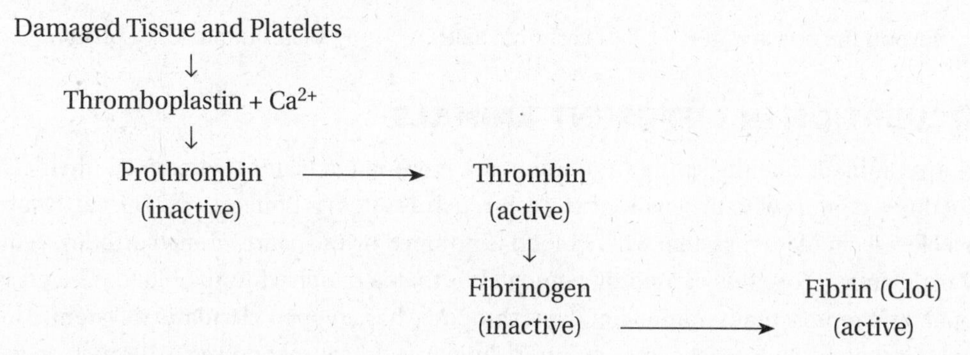

Structure and Function of Blood Vessels

Table 15.3 describes the various blood vessels.

Table 15.3

Blood Vessels		
Vessel	**Function**	**Structure**
Artery and Arteriole	Carry blood away from the heart under enormous pressures	Walls made of thick, elastic, smooth muscle
Vein and Venule	Carry blood back to the heart under very little pressure	Thin walls have valves to help prevent back flow; veins are located within skeletal muscle, which propels blood upward and back to the heart as the body moves
Capillary	Allows for diffusion of nutrients and wastes between cells and blood	Walls are one cell thick and so small that blood cells travel in single file

The Heart

The heart is located beneath the sternum and is about the size of a clenched fist. It beats about **70 beats per minute** and pumps about 5 quarts (5 L) of blood per minute, or the total volume of blood in the body each minute. Two **atria** receive blood from the body cells, and two **ventricles** pump blood out of the heart. Individual **cardiac muscle cells** have the ability to contract even when removed from the heart. The heart itself has its own innate **pacemaker**, the **sinoatrial (SA) node**, which sets the timing of the contractions of the heart. Located in the wall of the right atrium, it generates and sends electrical signals to the atrioventricular (AV) node. The action potential of pacemaker cells is generated by voltage-gated Ca^{2+} channels. From the pacemaker, impulses are sent to the bundle of His and Purkinje fibers, which trigger the ventricles to contract. Electrical impulses travel through the cardiac and body tissues to the skin, where they can be detected by an **electrocardiogram (EKG)**. The heart's pacemaker is influenced by a variety of factors: two sets of nerves that cause it to speed up or slow down, hormones such as adrenaline, and body temperature. **Blood pressure** is lowest in the veins and highest in the arteries when the ventricles contract. Blood pressure for all normal, resting adults is **120/80**. The **systolic number (120)** is a measurement of the pressure when the ventricles contract, while the **diastolic number (80)** is a measure of the pressure when the heart relaxes; see Figure 15.4.

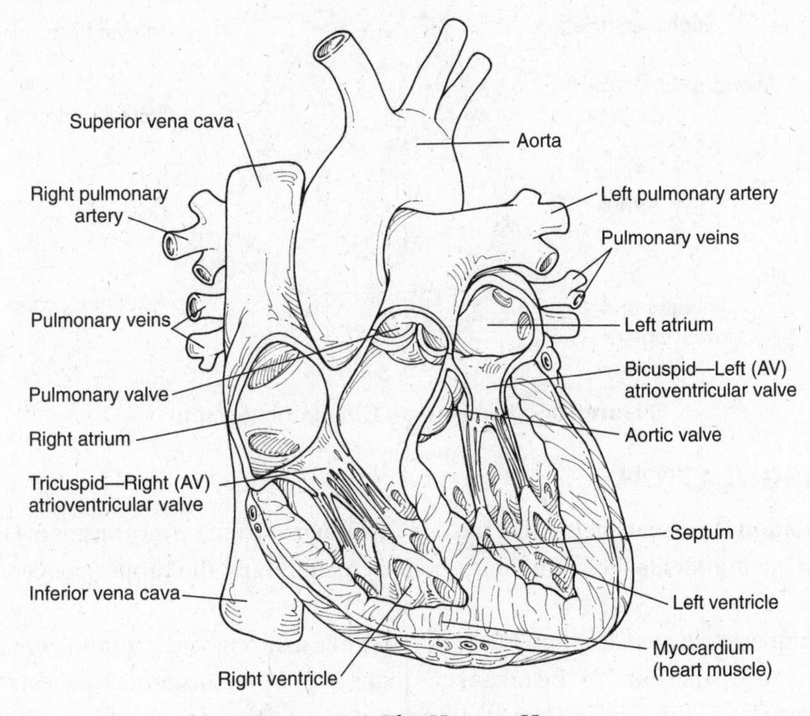

Figure 15.4 The Human Heart

Pathway of Blood

Blood enters the heart through the vena cava. From there it continues to the

Right atrium
Right atrioventricular (AV) valve—tricuspid valve
Right ventricle
Pulmonary semilunar valve
Pulmonary artery
Lungs
Pulmonary vein
Left atrium
Bicuspid (left AV) valve
Left ventricle
Aortic semilunar valve
Aorta
To all the cells in the body

STUDY TIP

Learn the pathway of the blood in the body.

Blood circulates through the **coronary circulation** (heart), **renal circulation** (kidneys), and **hepatic circulation** (liver). The **pulmonary circulation** includes the pulmonary artery, lungs, and pulmonary vein. Figure 15.5 is a drawing of the human circulatory system.

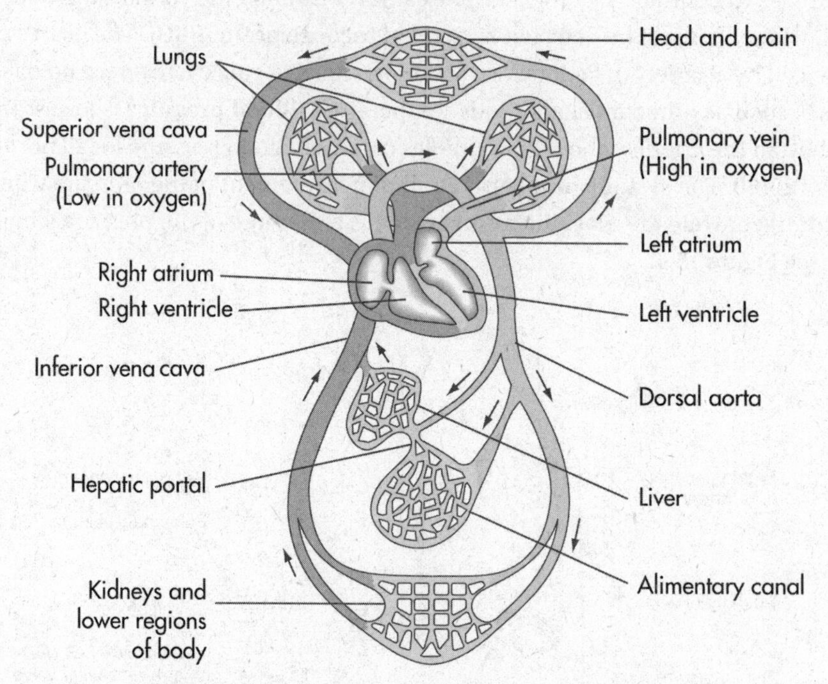

Figure 15.5 The Human Circulatory System

OSMOREGULATION

Osmoregulation is the management of the body's water and solute concentration. Organisms in different environments face different problems maintaining the proper concentration of body fluids.

For **marine vertebrates**, like bony fish, the ocean is a strongly dehydrating environment because it is very hypertonic to the organisms living in it. Fish constantly lose water through their gills and skin to the surrounding environment. To counteract the problem, they produce very little urine and drink large amounts of seawater. The extra salt that is taken in with the seawater is actively transported out through the gills.

The osmoregulation problems faced by **freshwater organisms** are the opposite of those living in salt water. The environment is hypotonic to the organisms, and they are constantly gaining water and losing salt. Freshwater fish excrete copious amounts of dilute urine.

Terrestrial organisms face entirely different problems. They have had to evolve mechanisms and structures that enable them to rid themselves of metabolic wastes while retaining as much water as possible.

EXCRETION

Different excretory mechanisms have evolved in various organisms for the purpose of osmoregulation and for the removal of metabolic wastes. Here are some organisms matched with their excretory structures:

Organism	Structures
Protista	Contractile vacuole
Platyhelminthes (planaria)	Flame cells
Earthworm	Nephridia (metanephridia)
Insects	Malpighian tubules
Humans	Nephrons

Excretion is the removal of metabolic wastes, which include **carbon dioxide** and **water** from cell respiration and **nitrogenous wastes** from protein metabolism. The organs of excretion in humans are the **skin**, **lungs**, **kidneys**, and the **liver** (where urea is produced). There are three nitrogenous wastes: **ammonia**, **urea**, and **uric acid**. Which waste an organism excretes is the result of the environment it evolved in and lives in. Here are some characteristics of each nitrogenous waste.

Nitrogenous Wastes

1. **AMMONIA**
   - Very soluble in water and highly toxic
   - Excreted generally by organisms that live in water, including the hydra and fish
2. **UREA**
   - Not as toxic as ammonia
   - Excreted by earthworms and humans
   - In mammals, it is formed in the liver from ammonia
3. **URIC ACID**
   - Pastelike substance that is not soluble in water and therefore not very toxic
   - Excreted by insects, many reptiles, and birds, with a minimum of water loss

The Human Kidney

The kidney functions as both an **osmoregulator** (regulates blood volume and concentration) and an organ of **excretion**. Humans have two kidneys supplied by blood from the **renal artery** and **renal vein**. The kidneys filter about 1,000–2,000 liters of blood per day and produce, on average, about 1.5 liters of urine (700–2,000 mL). As terrestrial animals, humans need to conserve as much water as possible. However, people must balance the need to conserve water against the need to rid the body of poisons. The kidney must adjust both the volume and the concentration of urine depending on the animal's intake of water and salt and the production of urea. If fluid intake is high and salt intake is low, the kidney will produce large volumes of **dilute (hypoosmotic) urine**. In periods of high salt intake where water is unavailable, the kidney can produce **concentrated (hyperosmotic) urine**.

The Nephron

The functional unit of the kidney is the **nephron**; see Figure 15.6. The nephron consists of a cluster of capillaries called the **glomerulus**, which sits inside a cuplike structure called **Bowman's capsule** and connects to a long narrow tube (the **renal tubule**). Each human kidney contains about 1 million nephrons. The nephron carries out its job in four steps: **filtration**, **secretion**, **reabsorption**, and **excretion**.

REMEMBER

Filtration is passive and nonselective.

Secretion is active and highly selective.

Reabsorption is passive, active, and selective.

Filtration occurs as blood pressure (about twice that of capillaries outside the kidney) forces fluid from the blood in the **glomerulus** into Bowman's capsule. Specialized cells of

Bowman's capsule are modified into **podocytes**, which, along with **slit pores**, increase the rate of filtration. Filtration occurs by *diffusion* and is *passive* and *nonselective*. The filtrate contains everything small enough to diffuse out of the glomerulus and into Bowman's capsule, including glucose, salts, vitamins, waste such as urea, and other small molecules. From Bowman's capsule, the filtrate travels into the **proximal tubule**.

Secretion occurs in the **proximal** and **distal tubules**. It is the *active, selective* uptake of certain drugs and toxic molecules that did not get filtered into Bowman's capsule. The proximal tubule also secretes ammonia to neutralize the acidic filtrate.

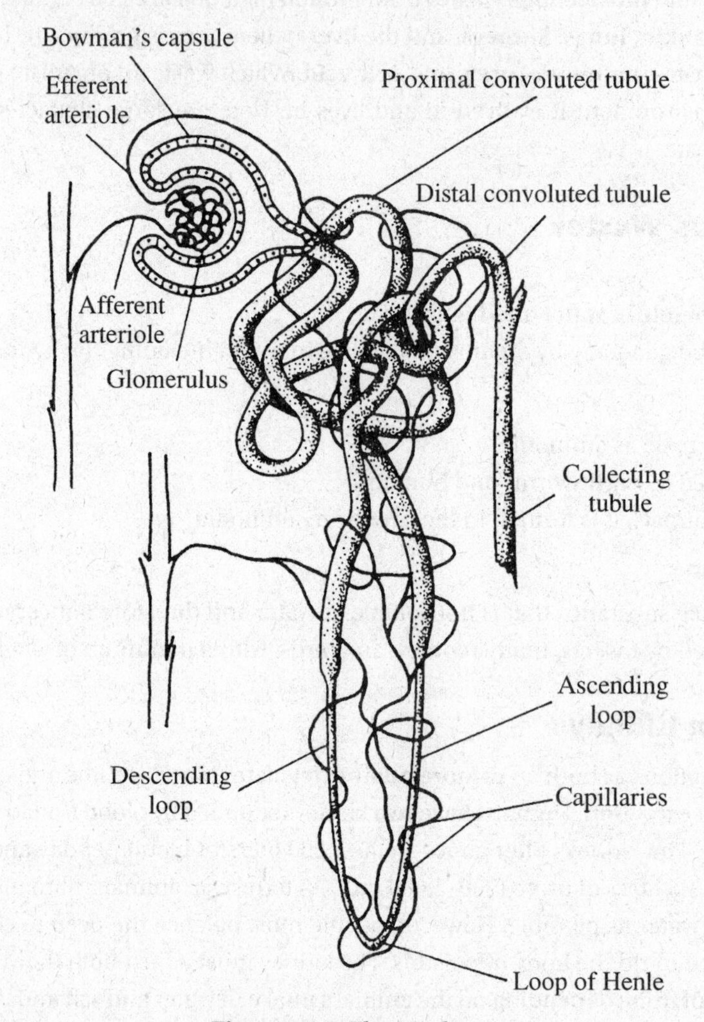

Figure 15.6 The Nephron

Reabsorption is the process by which most of the water and solutes (glucose, amino acids, and vitamins) that initially entered the tubule during filtration are transported back into the peritubular capillaries and, thus, back to the body. This process begins in the proximal convoluted tubule and continues in the loop of Henle and collecting tubule. The main function of the loop of Henle is to move salts from the filtrate and accumulate them in the medulla surrounding the loop of Henle and the collecting tubule. In this way, the loop of Henle acts as a **countercurrent exchange mechanism** (see Chapter 3, **"The Cell"**), maintaining a steep salt gradient surrounding the loop. This gradient ensures that water molecules will continue to flow out of the collecting tubule of the nephron, thus creating hypertonic urine and conserving water. *The longer the loop of Henle, the greater is the reabsorption of water.*

Excretion is the removal of metabolic wastes, for example, nitrogenous wastes. Everything that passes into the collecting tubule is **excreted** from the body. From the collecting tubule or duct, urine passes through the **ureter** to the **urinary bladder**. Urine is temporarily stored in the urinary bladder until it passes out of the body via the **urethra**. See Figure 15.7.

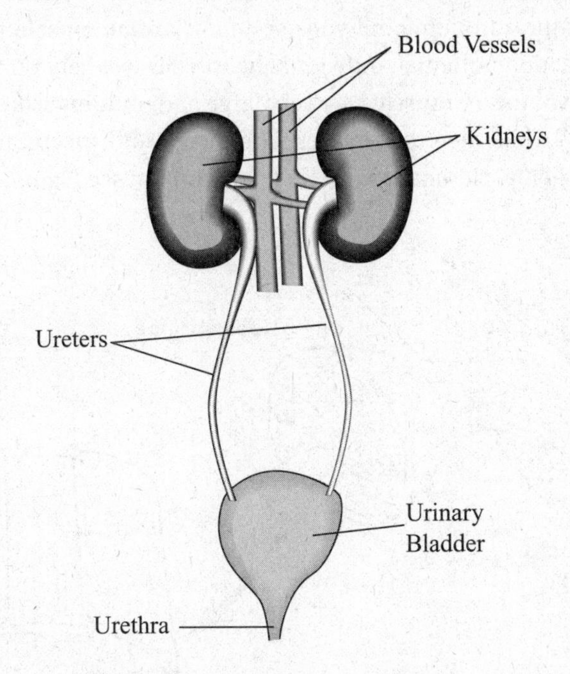

Figure 15.7 Urinary Tract

Hormone Control of the Kidneys

If blood pressure falls, the ability of the nephron to filter blood at Bowman's capsule is impaired, placing the body in danger. Therefore, the kidney must respond quickly by increasing blood pressure. It can do this because the kidney is under the control of three hormones: aldosterone, antidiuretic hormone (ADH), and renin.

IMPORTANT TIP

Hormone function is an important example of how cells use chemical signals to communicate.

Aldosterone is a hormone released by the adrenal glands in *response to a decrease in blood pressure or volume.* It acts on the distal tubules of the nephron to reabsorb more sodium ions and water, thus increasing blood volume and pressure. **ADH**, also known as **vasopressin**, is produced by the hypothalamus and is both stored in and released from the posterior pituitary. It is released in response to dehydration due to excessive sweating or inadequate water intake, which causes the blood to become too concentrated (osmolarity to increase). ADH increases the permeability of the collecting tubules to water by opening *renal aquaporins* (transmembrane proteins that function as water channels). This opening allows more water to be reabsorbed and urine volume to be reduced. **Renin**, a hormone released from the kidneys, converts an inactive protein into active **angiotensin**, which stimulates the adrenal cortex to release aldosterone.

Physicians prescribe several effective drugs to treat excessively high blood pressure. The most common drug prescribed for high blood pressure is a *diuretic,* commonly called a water pill, which reduces blood volume by increasing the amount of water released in urine. The person simply urinates more. Another commonly prescribed drug is an *angiotensin-converting enzyme (ACE) inhibitor,* which prevents the formation of angiotensin II. ACE inhibitors inhibit the formation of aldosterone and thereby decrease blood volume. Both of these drugs work in opposition to what the renin-angiotensin-aldosterone system normally does (which is raise blood pressure).

MUSCLE

There are three types of muscle: smooth, cardiac, and skeletal. **Smooth** or **involuntary** muscle makes up the walls of the **blood vessels** and the **digestive tract**. Because of the arrangement of its actin and myosin filaments, it does not have a striated appearance. It is under the control of the autonomic nervous system. **Cardiac muscle** makes up the **heart**. It generates its own action potential; individual heart cells will beat on their own in a saline solution. **Skeletal** or **voluntary muscles** are very large and multinucleate. They work in pairs: one muscle contracts while the other relaxes. The biceps and triceps are one example. The following discussion of muscles focuses on skeletal muscles; see Figure 15.8.

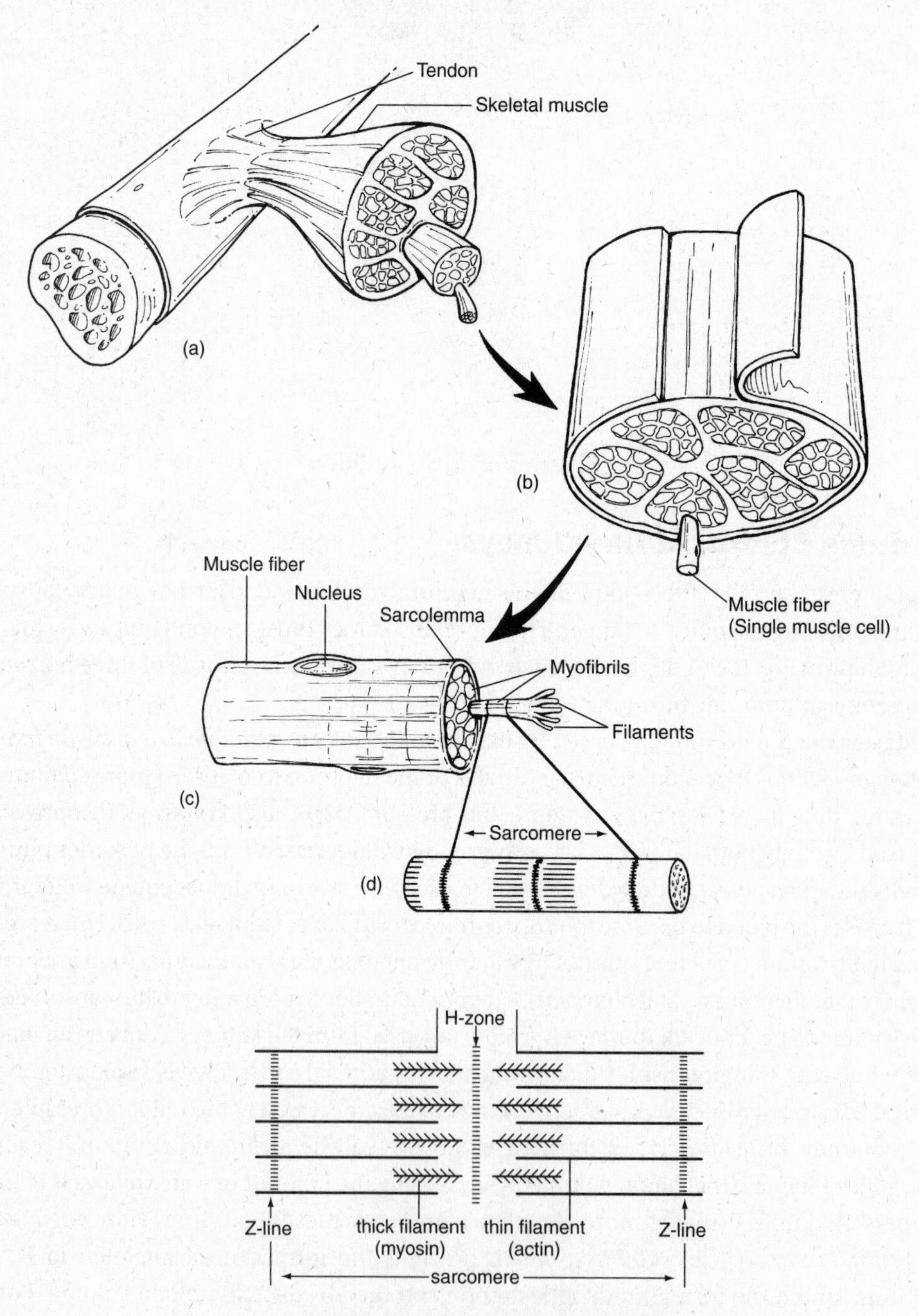

Figure 15.8 The Structure of Skeletal Muscle

Every muscle consists of bundles of thousands of **muscle fibers** that are individual cylindrical **muscle cells**. Each cell is exceptionally **large** and **multinucleate**. Skeletal muscle consists of modified structures that enable the cell to contract.

- The **sarcolemma** is modified **plasma membrane** that surrounds each muscle fiber and can propagate an **action potential**.
- The **sarcoplasmic reticulum (SR)** is modified **endoplasmic reticulum** that contains sacs of **Ca^{++}** necessary for normal muscle contraction.
- The **T system** is a system of tubules that runs perpendicular to the SR and that connects the SR to the extracellular fluid.
- The **sarcomere** is the functional unit of the muscle fiber (cell). Its boundaries are the **Z lines**, which give skeletal muscle its characteristic striated appearance.

STUDY TIP

Thin filaments = actin

Thick filaments = myosin

The Sliding Filament Theory

Within the cytoplasm of each muscle cell are thousands of fibers called **myofibrils** that run parallel to the length of the cell. Myofibrils consist of **thick** and **thin filaments**. Each **thin filament** consists of two strands of **actin proteins** wound around one another. Each **thick filament** is composed of two long chains of **myosin** molecules each with a globular head at one end. The contraction of the sarcomere depends on two other molecules, **troponin** and **tropomyosin**, in addition to **Ca^{++} ions**, to form and break cross-bridges. Muscle contracts as thick and thin filaments slide over each other.

The Neuromuscular Junction

The axon of a motor neuron synapses on a skeletal muscle at the **neuromuscular junction**. The neurotransmitter **acetylcholine**, released by vesicles from the axon, binds to **receptors** on the **sarcolemma**, depolarizes the muscle cell membrane, and sets up an **action potential**. The impulse moves along the **sarcolemma**, into the **T system**, and stimulates the **sarcoplasmic reticulum** to release Ca^{++}. The **Ca^{++} ions** then alter the troponin-tropomyosin relationship and the muscle contracts.

Summation and Tetanus

A single action potential in a muscle will cause the muscle to contract locally and minutely for a few milliseconds and then to relax. This brief contraction is called a **twitch**. If a second action potential arrives before the first response is over, there will be a **summation effect** and the contraction will be greater. If a muscle receives a series of overlapping action potentials, even further summation will occur. If the rate of stimulation is fast enough, the twitches will blur into one smooth, sustained contraction called **tetanus** (not related to the disease of the same name). Tetanus is what occurs when a large muscle such as the bicep muscle contracts. If the muscle continues to be stimulated without respite, it will eventually **fatigue** and relax; see Figure 15.9.

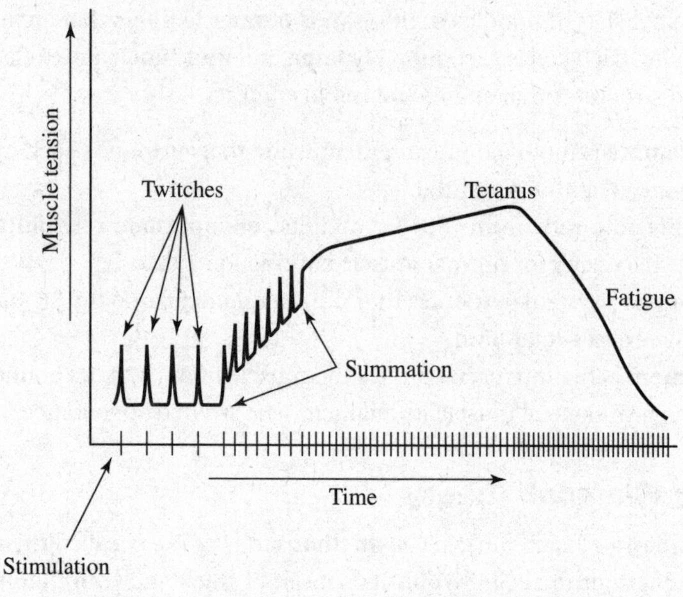

Figure 15.9 Progress of a Muscle Contraction

CHAPTER SUMMARY

Although you likely will not see any topics from this chapter covered in the multiple-choice questions on the AP Biology exam, you may draw on the topics discussed in this chapter for your answers to the free-response questions.

Homeostasis is a fundamental concept in Biology, and you should be able to give specific examples of it. Here are a few from this chapter:

- Hormones in the stomach and duodenal walls regulate digestion.
- Blood pH is maintained at 7.4 by the bicarbonate ion system.
- Endotherms use metabolic processes to maintain body temperature.

Another fundamental concept in Biology is **form relates to function**. There are several examples of that concept in this chapter:

- The extreme length of the small intestine relates to its function in digestion and absorption.
- Lungs have millions of microscopic, thin-walled, moist air sacs called alveoli that maximize diffusion of respiratory gases.
- The walls of arteries are made of thick, smooth muscle that can withstand the tremendous pressure of the blood they carry away from the heart. Veins, in contrast, have thin walls because they carry blood back to the heart under low pressure. The walls that make up capillaries are one cell thick so that respiratory gases can diffuse through them readily to and from the surrounding tissue.
- The left ventricle of the heart is the chamber that pumps blood away from the heart and around the body under high pressure. It has the thickest wall of any chamber.
- Skeletal muscles are ideally suited to contract. They consist of actin and myosin filaments that are organized so the filaments can slide past each other rapidly.
- The villi and microvilli expand the absorption function of the small intestine by greatly increasing the surface area for the diffusion of nutrients and waste.

PRACTICE QUESTIONS

DIGESTION

1. Gastric enzymes work best at a pH of
   (A) 2
   (B) 6
   (C) 8
   (D) 11

2. Which of the following is correct about the function of the small intestine?
   (A) Protein digestion begins here.
   (B) Both mechanical and chemical digestion occur here.
   (C) Sugars are digested by lipases here.
   (D) Fats are digested by lipases here.

3. Which of the following is correct about the large intestine?
   (A) The large intestine is the site of vitamin production.
   (B) The large intestine is the site of lipid digestion.
   (C) The large intestine is the site of the absorption of nutrients into the bloodstream.
   (D) The pH of the large intestine is acidic.

4. Which of the following is correct about the digestive system?
   (A) The digestion of proteins releases fatty acids.
   (B) The duodenum is the first 12 inches of the large intestine.
   (C) The secretion of digestive juice into the stomach is triggered by the release of a hormone.
   (D) The large intestine is longer than the small intestine.

5. Which of the following is correct about the digestive system?
   (A) The removal of too much water from the digestive system results in the condition known as diarrhea.
   (B) Pepsin is the enzyme that digests starch.
   (C) The type and position of human teeth reveals that we evolved to be carnivores.
   (D) The walls of the human digestive tract consist of smooth muscle tissue.

6. Which of the following correctly states the function of microvilli and includes the organ that contains microvilli?
   (A) Microvilli are found in the small intestine, and they complete the digestion of proteins.
   (B) Microvilli are found in the stomach, and they enhance the digestion of proteins.
   (C) Microvilli are found in the large intestine, and they enhance the absorption of nutrients.
   (D) Microvilli are found in the small intestine, and they enhance the absorption of nutrients.

7. Which is TRUE of the stomach?

   (A) The pyloric sphincter is at the top of the stomach.
   (B) The stomach lining releases lipases to begin fat digestion.
   (C) Hydrochloric acid activates the enzyme pepsinogen.
   (D) The pH of the stomach varies from acidic to basic depending on what must be digested.

8. The hormone gastrin is released by the ________ and has its effect on the ________.

   (A) duodenum; stomach
   (B) duodenum; pancreas
   (C) stomach; gastric lining
   (D) stomach; small intestine

9. The lacteal is found in the ________, and is involved with ________.

   (A) stomach; the release of hormones
   (B) duodenum; the hydrolysis of lipids
   (C) small intestine; the absorption of fatty acids
   (D) colon; the reabsorption of water

10. Absorption of nutrients occurs in the

    (A) duodenum of the colon
    (B) duodenum of the small intestine
    (C) latter part of the small intestine
    (D) latter part of the large intestine

GAS EXCHANGE

11. Which is CORRECT about gas exchange in humans?

    (A) As humans inhale, the pressure in the chest cavity decreases and air is drawn into the lungs.
    (B) Air is forced down the windpipe when a person inhales.
    (C) The breathing rate is controlled by the hypothalamus in the brain.
    (D) Hemoglobin carries carbon dioxide and oxygen in fairly equal amounts.

12. Tracheal tubes are found in

    (A) earthworms
    (B) hydra
    (C) fish
    (D) insects

13. Breathing in humans is usually regulated by

    (A) the number of red blood cells
    (B) the amount of hemoglobin in the blood
    (C) inherent genetic control
    (D) CO_2 levels and pH sensors

14. All of the following descriptions about the normal direction of the flow of blood are correct EXCEPT

(A) lungs → pulmonary artery
(B) right ventricle → tricuspid valve
(C) aorta → aortic semilunar valve
(D) vena cava → right atrium

15. The pacemaker of the heart is

(A) the sinoatrial node
(B) the atrioventricular node
(C) the diastolic node
(D) the semilunar node

16. The Bohr effect on the oxyhemoglobin dissociation curve is produced directly by changes in

(A) temperature
(B) pH
(C) CO_2 levels
(D) oxygen concentration

17. Which is TRUE of the human circulatory system?

(A) The right ventricle of the heart has the thickest wall.
(B) Veins have thick walls consisting of smooth muscle cells to assist in returning blood to the heart.
(C) Blood flow is slowest in capillaries to maximize the diffusion of nutrients and waste.
(D) The left and right ventricles contract alternately, which is responsible for the pulse sound.

18. In humans, the largest amount of the carbon dioxide produced by body cells is carried to the lungs as

(A) CO_2 attached to hemoglobin in the red blood cells
(B) the bicarbonate ion dissolved in the plasma
(C) the bicarbonate ion attached to hemoglobin
(D) CO_2 gas in solution in the plasma

19. During ventricular systole, the __________ valve(s) __________.

(A) semilunar; close
(B) semilunar; open
(C) AV; open
(D) AV and semilunar; close

20. All of the following are true about blood EXCEPT

(A) red blood cells live for about 120 days
(B) white blood cells are formed in the bone marrow
(C) platelets are not cells but are actually cell fragments
(D) platelets derive from specialized cells known as neutrophils

EXCRETION

21. Which of the following processes of the kidney is the LEAST selective?
    (A) secretion
    (B) filtration
    (C) reabsorption
    (D) the target of ADH

22. In humans, urea is produced in the
    (A) kidneys
    (B) urinary bladder
    (C) urethra
    (D) liver

23. The main nitrogenous waste excreted by birds is
    (A) ammonia
    (B) urea
    (C) uric acid
    (D) nitrite

24. Which of the following is INCORRECTLY paired?
    (A) nephridia-earthworm
    (B) flame cell-bird
    (C) Malpighian tubules-insect
    (D) nephron-human

25. Following a company picnic on a hot summer day, people who ate traditional fare of hot dogs and hamburgers, drank ice cold beer, and played softball for an hour became very thirsty. Which would probably contribute LEAST to their thirst?
    (A) eating salty food
    (B) drinking an ice cold beer
    (C) drinking lots of ice water
    (D) sweating

26. Which nitrogenous waste requires the least water for its excretion?
    (A) ammonia
    (B) urea
    (C) nitrites
    (D) uric acid

27. Where does filtration occur?
    (A) proximal and distal tubules
    (B) ascending loop of Henle
    (C) Bowman's capsule
    (D) collecting tubule

28. Where does secretion occur?
    (A) proximal and distal tubules
    (B) ascending loop of Henle
    (C) Bowman's capsule
    (D) collecting tubule

29. Which of the following is the area that is impermeable to the diffusion of water?
    (A) proximal and distal tubules
    (B) ascending loop of Henle
    (C) Bowman's capsule
    (D) collecting tubule

30. Which of the following is the target structure of ADH?
    (A) proximal and distal tubules
    (B) ascending loop of Henle
    (C) Bowman's capsule
    (D) collecting tubule

MUSCLE

31. Tendons connect
    (A) muscle to muscle
    (B) muscle to bone
    (C) ligaments to bones
    (D) ligaments to ligaments

32. The walls of arteries consist of
    (A) striated muscle and are under voluntary control
    (B) striated muscle and are not under voluntary control
    (C) smooth muscle and are controlled by the somatic nervous system
    (D) smooth muscle and are controlled by the autonomic nervous system

33. What is the basic unit of function of a skeletal muscle fiber?
    (A) myosin filaments
    (B) actin filaments
    (C) the sarcomere
    (D) Z line

34. What neurotransmitter at the synapse of a neuromuscular junction causes a muscle to contract?
    (A) GABA
    (B) norepinephrine
    (C) dopamine
    (D) acetylcholine

35. Which of the following is the name of a single muscle cell?

(A) sarcomere
(B) myofibril
(C) muscle fiber
(D) sarcolemma

36. All of the following are true about the contracting of skeletal muscle cells EXCEPT

(A) thick filaments are composed of actin
(B) Ca^{++} ions are necessary for normal muscle contraction
(C) the sarcolemma can propagate an action potential
(D) the T system connects the sarcolemma to the sarcoplasmic reticulum

Answers Explained

DIGESTION

1. **(A)** Gastric enzymes are strongly acidic and activate the gastric enzyme pepsinogen. Other enzymes, such as intestinal enzymes, are activated in an alkaline pH.

2. **(D)** Fats are digested by lipases in the small intestine. Choice A is incorrect because protein digestion begins in the stomach. Choice B is incorrect because mechanical and chemical digestion begin in the mouth. Sugars are digested by amylase, so choice C is incorrect as well.

3. **(A)** The large intestine carries out three functions: vitamin production, removal of undigested waste, and removal of excess water from the colon. Choices B and C are incorrect because digestion and nutrient absorption do not occur in the large intestine. Choice D is incorrect because the stomach is the only part of the digestive system with an acidic pH.

4. **(C)** The secretion of digestive juice into the stomach is triggered by the release of the hormone gastrin. Choice A is incorrect because the digestion of proteins releases amino acids, not fatty acids. Choice B is incorrect because the duodenum is the first 12 inches of the small intestine. Choice D is incorrect because the small intestine is much longer than the large intestine, but the large intestine has a wider diameter.

5. **(D)** Choice A is incorrect because the removal of too much water from the digestive system results in constipation. Choice B is incorrect because pepsin digests proteins. Choice C is incorrect because the presence of our different types of teeth reveals that we evolved to be omnivores.

6. **(D)** Microvilli are located in the small intestine. They increase the surface area to enhance the absorption of nutrients into the bloodstream. Choice A is incorrect because microvilli absorb nutrients *after* digestion is complete. Choices B and C are incorrect because they do not provide the correct organ that contains microvilli.

7. **(C)** The lower esophageal sphincter is at the top of the stomach; the pyloric sphincter is at the bottom of the stomach. The stomach releases the inactive form of pepsin (pepsinogen), which becomes activated by HCl acid, and digestion can begin. The pH of the stomach is very acidic. The stomach is stimulated by the hormone gastrin.

8. **(C)** This is a statement of fact. See Table 15.1, "Hormones that Regulate Digestion."

9. **(C)** The lacteal is inside the villi that line the inside of the small intestine. The duodenum is the first 12 inches (30 cm) of the small intestine and is the site of digestion. "Colon" is another name for large intestine.

10. **(C)** Digestion occurs in the duodenum—the first part of the small intestine, and absorption occurs in the latter part.

GAS EXCHANGE

11. **(A)** Humans breathe by negative pressure. When we inhale, the chest cavity expands as the diaphragm lowers. This increase in volume causes a decrease in internal pressure, and air is drawn into the lungs because the internal pressure is less than the external pressure.

12. **(D)** Spiracles, openings in the exoskeleton of the insects, connect to tracheal tubes that lead to the hemocoels where the diffusion of respiratory gases occurs. The respiratory surface is internal.

13. **(D)** The breathing rate is controlled by the medulla in the brain, which is primarily sensitive to CO_2 levels in the blood. The other choices are incorrect.

14. **(A)** From the lungs, blood flows into the left atrium via the pulmonary vein.

15. **(A)** The pacemaker of the heart is the sinoatrial node. From the sinoatrial node, the impulse passes to the atrioventricular node, then to the bundle of His.

16. **(B)** When the pH becomes more acidic, hemoglobin has less affinity for oxygen, so the hemoglobin will drop off some oxygen at the cells that have become slightly acidic due to an accumulation of CO_2 from cell respiration.

17. **(C)** Diffusion occurs in the thin-walled capillaries where the blood circulates slowly. The statements in choices A, B, and D are all false.

18. **(B)** Although a small amount of CO_2 is carried by the hemoglobin that is attached to red blood cells, most is dissolved in blood as the bicarbonate ion (HCO_3^-) and is not attached to anything. Carbon dioxide dissolves in water to produce the bicarbonate ion: $CO_2 + H_2O \rightarrow HCO_3^- + H^+$.

19. **(B)** Systole is the contraction of the ventricles of the heart. When the ventricles contract, blood is pushed out of the arteries through the semilunar valves while the bicuspid and tricuspid valves remain closed.

20. **(D)** Platelets are fragments of cells known as megakaryocytes. All the other choices are true.

EXCRETION

21. **(B)** During filtration, all substances small enough to diffuse out of the glomerulus will do so. It is the least selective process that occurs in the nephron.

22. **(D)** Urea is a nitrogenous waste that is produced in the liver; urine is produced in the kidneys.

23. **(C)** Uric acid is excreted as a crystal to conserve water by many terrestrial animals. However, humans and earthworms release nitrogenous waste in the form of urea.

24. **(B)** The flame cell is a primitive excretory structure in Platyhelminthes like planaria. All freshwater Protista have contractile vacuoles that pump out water that continually leaks in. The other pairings are all correct.

25. **(C)** All of the other options would greatly contribute to thirst. Alcohol, particularly, is dehydrating because it blocks the production of antidiuretic hormone, thus causing the release of extra water in the urine.

26. **(D)** Uric acid is not soluble in water; therefore, it does not require water in order to be excreted from the body.

27. **(C)** Filtration occurs in Bowman's capsule.

28. **(A)** Secretion occurs in the proximal and distal tubules.

29. **(B)** The ascending loop of Henle is impermeable to water.

30. **(D)** The target structure of ADH is the collecting tubule within the kidney.

MUSCLE

31. **(B)** This is a fact.

32. **(D)** The autonomic nervous system controls those things that are not consciously controlled. It controls smooth and cardiac muscles.

33. **(C)** The sarcomere is the basic unit of contraction of skeletal muscle.

34. **(D)** Acetylcholine is the neurotransmitter at a neuromuscular junction.

35. **(C)** A single muscle cell is called a muscle fiber. It is a very large, multinucleated cell.

36. **(A)** Thick filaments are composed of myosin, not actin. The other statements are correct about muscles.

Endocrine and Nervous Systems

16

→ **CHEMICAL SIGNALS**
→ **FEEDBACK MECHANISMS**
→ **HOW HORMONES TRIGGER A RESPONSE IN TARGET CELLS**
→ **TEMPERATURE REGULATION**
→ **NERVOUS SYSTEM**
→ **ORGANIZATION OF THE HUMAN BRAIN**

INTRODUCTION

The material in this chapter is not required for the AP Biology exam as per the new College Board curriculum. That is why this chapter is placed in Part 3 of this book and why there are no classification references to the curriculum frameworks (such as ENE-1). However, a major topic of Biology is cell signaling (how cells communicate). Thus, this chapter deals with just that: local and long-distance signaling between and among cells of the endocrine and nervous systems. Negative and positive feedback mechanisms are described as well.

Remember, this chapter consists of illustrative examples. Although these topics are not required for the AP exam, studying them will broaden your understanding of Biology and provide you with examples to use as you write your answers to the free-response questions.

CHEMICAL SIGNALS

Animals have two major regulatory systems that release chemicals: the **endocrine system** and the **nervous system**. The endocrine system secretes **hormones**, while the nervous system secretes **neurotransmitters**. Even though the two systems are separate, there is overlap between them, and together they work to regulate the body. **Epinephrine** (adrenaline), for example, functions as the **fight-or-flight** hormone secreted by the adrenal gland as well as a neurotransmitter that sends a message from one neuron to another.

Hormones

Hormones are produced in **ductless (endocrine) glands** and move through the blood to a specific **target** cell, tissue, or organ that can be far from the original endocrine gland. They can produce an immediate short-lived response, the way **adrenaline** (epinephrine) speeds up the heart rate and increases blood sugar. They can dramatically alter the development of an entire organism, the way **ecdysone** controls **metamorphosis** in insects. **Tropic hormones** have a far-reaching effect because they stimulate other glands to release hormones. For example, the anterior pituitary releases TSH (thyroid-stimulating hormone), which stimulates the thyroid to release thyroxine. Other types of chemical messengers reach their target

STUDY TIP

Signals released by one cell type can travel long distances to target cells of another type.

by special means. **Pheromones** in the urine of a dog carry a message between different individuals of the same species. In vertebrates, **nitric oxide** (NO), a gas, is produced by one cell and diffuses to and affects only neighboring cells before it is broken down. Table 16.1 is an overview of the hormones of the endocrine system.

Table 16.1

Overview of the Hormones of the Endocrine System		
Gland	**Hormone**	**Effect**
Anterior pituitary	Growth hormone (GH)	Stimulates growth of bones
	Luteinizing hormone (LH)	Stimulates ovaries and testes
	Thyroid-stimulating hormone (TSH)	Stimulates thyroid gland
	Adrenocorticotropic (ACTH) hormone	Stimulates adrenal cortex to secrete glucocorticoids
	Follicle-stimulating hormone (FSH)	Stimulates gonads to produce sperm and ova
	Prolactin	Stimulates mammary glands to produce milk
Posterior pituitary	Oxytocin	Stimulates contractions of the uterus and milk production by mammary glands
	Antidiuretic hormone (ADH)	Promotes retention of water by the kidneys
Thyroid	Thyroxine (T3 and T4)	Controls metabolic rate
	Calcitonin	Lowers blood calcium levels
Parathyroid	Parathormone	Raises blood calcium levels
Adrenal cortex	Glucocorticoids	Raises blood sugar levels
Adrenal medulla	Epinephrine (adrenaline) Norepinephrine (noradrenaline)	Raises blood sugar levels by increasing the rate of glycogen breakdown by the liver
Pancreas—islets of Langerhans	Insulin—secreted by β cells	Lowers blood glucose levels
	Glucagon—secreted by α cells	Raises blood glucose levels by causing the breakdown of glycogen into glucose
Thymus	Thymosin	Stimulates T lymphocytes
Pineal	Melatonin	Involved in biorhythms
Ovaries	Estrogen	Promotes uterine lining growth, promotes development and maintenance of primary and secondary characteristics of females
	Progesterone	Promotes uterine lining growth
Testes	Androgens	Support sperm production and promote secondary sex characteristics

The Hypothalamus

The **hypothalamus** plays a special role in the body; it is the *bridge between the endocrine and nervous systems.* The hypothalamus acts as part of the nervous system when, in times of stress, it sends electrical signals to the adrenal gland to release adrenaline. It acts like a nerve when it secretes **gonadotropin-releasing hormone (GnRH)** from neurosecretory cells that stimulate the anterior pituitary to secrete **FSH** and **LH.** It acts as an endocrine gland when it produces **oxytocin** and **antidiuretic hormone** that it stores in the posterior pituitary. The hypothalamus also contains the body's thermostat and centers for regulating hunger and thirst.

The hypothalamus receives information from nerves throughout the body, including the brain. In response, it initiates *endocrine* signaling appropriate to the environmental conditions. For example in many vertebrates, nerve signals from the brain pass sensory information about seasonal changes to the hypothalamus. The hypothalamus, in turn, regulates the release of reproductive hormones that are necessary during the breeding season. Signals from the hypothalamus travel to the posterior pituitary and anterior pituitary.

Posterior Pituitary

Two hormones are synthesized in the hypothalamus and reach the posterior pituitary by hypothalamic axons. These hormones are released from the posterior pituitary upon receiving nerve impulses from the hypothalamus. One hormone is **oxytocin,** which regulates milk secretion by mammary glands and uterine contractions during labor. In addition, oxytocin targets the brain, influencing behaviors related to trust, maternal care, pair bonding, and sexual activity. The other hormone, **antidiuretic hormone** (ADH), also known as **vasopressin,** targets the nephron and increases water retention by the kidneys, thus decreasing urine volume. This regulates blood osmolarity.

Anterior Pituitary

All hormones released by the anterior pituitary are controlled by at least one hormone from the hypothalamus, either a stimulating or an inhibiting hormone. For example, the hypothalamus releases **prolactin-releasing hormone** into tiny capillaries so it can immediately reach the anterior pituitary. This hormone stimulates the release of **prolactin** from the anterior pituitary, which stimulates milk production.

REMEMBER

Hormones can regulate behavior.

ADH also plays a role in the behavior of prairie voles, whose brains have large numbers of vasopressin receptors. The male prairie voles hover over their young pups, care for them, and act aggressively toward intruders. If, however, the males are injected with a drug that blocks the vasopressin receptor, the males fail to form pair-bonds after mating.

FEEDBACK MECHANISMS

STUDY TIP

Organisms use feedback mechanisms to regulate growth and reproduction and to maintain dynamic equilibrium.

A **feedback mechanism** is a self-regulating mechanism that increases or decreases the level of a particular substance. **Positive feedback** *enhances an already existing response.* During labor, for example, the pressure of the baby's head against sensors near the opening of the uterus stimulates more uterine contractions, which causes increased pressure against the uterine opening, which causes yet more contractions. This positive feedback loop brings labor *to an end* and the birth of a baby. This is very different from negative feedback. **Negative feedback** is a common mechanism in the endocrine system (and elsewhere) that *maintains homeostasis.* A good example is how the body maintains proper levels of thyroxine. When the level of thyroxine in the blood is too low, the hypothalamus releases **thyrotropin-releasing hormone (TRH)**, which stimulates the anterior pituitary to release a hormone, thyroid-stimulating hormone (TSH), which stimulates the thyroid to release more thyroxine. Thyroxine exerts negative feedback on the hypothalamus and the anterior pituitary. When the level of thyroxine is adequate, the hypothalamus stops stimulating the pituitary. (See "Positive and Negative Feedback of Menstrual Cycle" on page 392.)

REMEMBER

Signal transduction pathways link signal reception with cellular response.

HOW HORMONES TRIGGER A RESPONSE IN TARGET CELLS

There are two types of hormones, and they stimulate target cells in different ways. They are illustrated in Figures 16.1 and 16.2.

STUDY TIP

Know the two types of hormones and how each affects its target cells.

1. **Lipid** or **steroid hormones** diffuse directly through the plasma membrane and bind to a receptor inside the nucleus that triggers the cell's response.

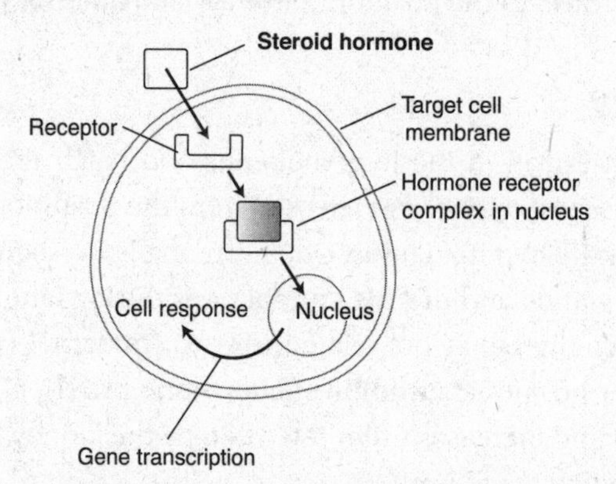

Figure 16.1 How Steroid Hormones Trigger a Response in a Cell

2. **Protein** or **polypeptide hormones** (nonsteroidal) cannot dissolve in the plasma membrane, so they bind to a receptor on the surface of the cell. Once the hormone (the first messenger) binds to a receptor on the surface of the cell, it triggers a secondary messenger, such as cAMP inside the cell, which converts the extracellular chemical signal to a specific response.

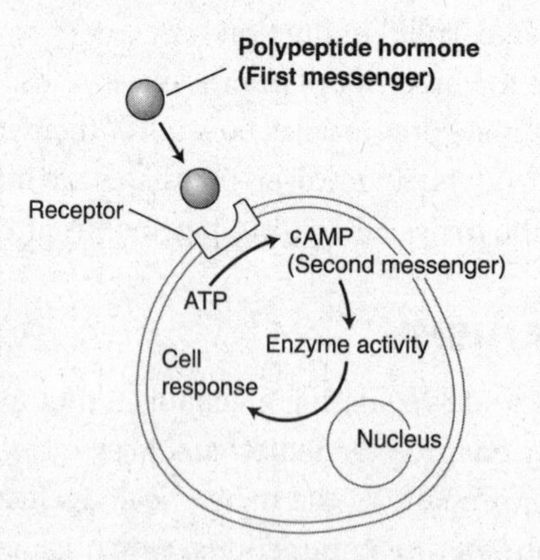

Figure 16.2 How Polypeptide Hormones Trigger a Response in a Cell

You may wish to refer back to Chapter 3, **"The Cell,"** for the basics about how hormones influence cells.

Details of How Two Hormones Function

TESTOSTERONE

Testosterone is a good example of how a steroid hormone functions. Testosterone is secreted by cells in the testes and travels through the blood, entering cells all over the body. It readily passes through the cell membrane and binds to a receptor in the cytoplasm, activating it. *Only cells with testosterone receptors can respond to testosterone.* With the hormone attached, the activated receptor enters the nucleus and acts as a **transcription factor**, turning on specific genes in the nucleus that control male characteristics. However, different genes are turned on in different cells. In muscle cells, the genes that increase muscle mass are turned on. In hair cells, genes for the protein keratin (what hair is made of) are turned on.

To stress the importance of receptors, let's look at an example. There is a condition in humans called *androgen insensitivity syndrome* (**AIS**). People with this condition are born with the XY genotype but lack testosterone receptors in all their cells. (They do, however, possess receptors for another hormone that inhibits normal development of internal female reproductive structures.) So although testosterone is produced in normal amounts, individuals with AIS do not develop male reproductive ducts, male genitals, or male secondary sexual characteristics. The result is a genetically XY individual who appears female but who has internal testes.

EPINEPHRINE

Epinephrine "tells" the liver to release glucose. The hormone epinephrine acts as the first messenger when it binds to and activates *G protein membrane receptors* on the surface of liver cells. The activated G protein then activates *adenylyl cyclase* in the membrane to produce cAMP (the second messenger). This begins a *kinase cascade* that does two things. First, it activates glycogen phosphorylase to convert glycogen into glucose. Second, it inhibits the conversion of glucose into glycogen. The cascade amplifies the original signal. For every molecule of epinephrine bound to a receptor on the cell surface, 20 molecules of cAMP are produced, and 10,000 molecules of glucose are released by the liver. The scientist who received the Nobel Prize in 1971 for discovering the role of cAMP in this pathway is Earl Sutherland.

BPA AND YOU

Bisphenol A (**BPA**) is an industrial compound used in the manufacturing of plastics, such as plastic food containers. The danger of this compound is that it mimics the feminizing hormone estrogen and can set up an estrogen *signal transduction pathway* even though no estrogen is present. So a person exposed to BPA experiences the same effect as if he or she received a dose of estrogen. Imagine the consequence if a baby or young child were exposed to this substance over the long term! Beginning in 2012 and continuing to today, manufacturers no longer use BPA in baby bottles and children's drinking cups.

TEMPERATURE REGULATION

Most life exists within a fairly narrow range, from 0°C (the temperature at which water freezes) to about 50°C. Temperatures on land fluctuate enormously. Therefore, temperature regulation, like water conservation, became a problem for animals when they moved to the land. To stay alive, animals must generate their own body heat, seek out a more suitable climate, and/or change behavior. Here are some examples of behavioral changes that alter body temperature:

- A snake can warm itself in the sun or cool off by hiding under a rock.
- Animals on a cold winter prairie huddle together to decrease heat loss.
- Bees swarming in a hive raise the temperature inside the hive.
- Dogs pant and sweat through their tongues to cool themselves off.
- Elephants lack sweat glands, but they cool themselves off by squirting water onto their skins and flapping their ears like fans.
- Humans shiver, jump around, and develop goosebumps to keep warm.
- Birds migrate to a warmer climate and a better source of food to feed themselves and their young.

Ectotherms

Ectotherms are animals that gain most of their body heat from their environment. They have such a low metabolic rate that the amount of heat they can generate is too small to have any effect on body temperature. The body temperature of an ectotherm equilibrates to the environment. They must maintain adequate body temperature through behavioral means, like those previously described. Ectotherms include fish, amphibians, and reptiles. The term ectotherm is closest in meaning to the common term cold-blooded, which has no real scientific meaning.

Endotherms

Endotherms are animals that use metabolic processes (oxidizing sugar) to produce body heat. The body temperature of endotherms remains constant. In a cold environment, an endotherm can keep its body warmer than the environment. The temperature of the human body fluctuates during the course of a day, but the average temperature is 37°C (98.6°F). All mammals and birds are endotherms. Some scientists believe that certain dinosaurs may have been endotherms. The term endotherm is closest in meaning to the common term warm-blooded, which has no real scientific meaning.

In terms of energy consumption, endothermy is very costly. Humans can use 60% of what we eat to maintain our body heat. The metabolic rate of a mammal is much higher than that of a reptile of similar size. As a consequence, mammals must take in many more calories than a reptile of similar size. When considering the metabolic requirements of mammals, in general, the smaller the mammal the faster the metabolism is. Flying birds have an even higher metabolic requirement than do mammals.

Despite the fact that the requirement for being an endotherm is so great, it may have given birds and mammals a critical advantage during the time when ancient Earth was dominated by reptiles. Perhaps being able to maintain a high body temperature made it possible for mammals and birds to invade and colonize colder environments that reptiles found uninhabitable.

Poikilotherms and Homeotherms

You may see the terms **poikilotherm** (having a body temperature that varies with the environment) and **homeotherm** (having a constant body temperature despite fluctuations in environmental temperature). However, these terms are not always meaningful. When discussing thermoregulation, the real issue is not whether an animal has variable or constant body temperature but, rather, the *source of heat* used to maintain its body temperature. For example, some scientists classify a fish as a poikilotherm. However, many fish inhabit water that has such a constant temperature that the fish also have a constant body temperature, just like a homeotherm. Although mammals are classified as homeotherms, their temperature sometimes varies greatly. During **torpor**, **estivation**, or **hibernation**, mammals save energy by drastically decreasing their metabolic rate and body temperature.

Problems of Living on Land

Through natural selection, animals have evolved various anatomical and physiological adaptations for life in different environments. The size of the ears in a jackrabbit can be correlated to the climate in which it lives. Jackrabbits living in cold, northern regions have small ears to minimize heat loss. In contrast, rabbits in warm, southern regions have long ears to dissipate heat from the many capillaries that make their ears appear pink. This anatomical difference across a geographic range is called a **north-south cline**.

Countercurrent heat exchange is a mechanism that has evolved in a variety of organisms. It helps to warm or cool extremities. See Figure 16.3. This can be seen in arctic animals such as the polar bear that must reach into icy waters to catch fish to eat. While its arm is in the frigid water, warm core blood is flowing out to the paw in the arteries, warming the chilled blood returning to the heart in veins, which lie directly next to the arteries.

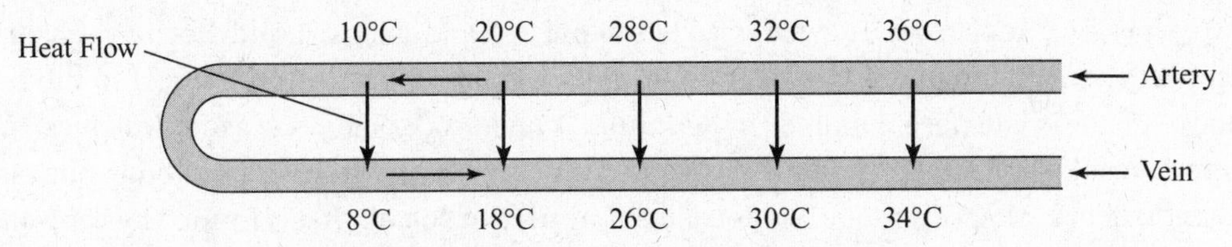

Figure 16.3 Countercurrent Heat Exchange

NERVOUS SYSTEM

The vertebrate nervous system consists of central and peripheral components.

1. **THE CENTRAL NERVOUS SYSTEM (CNS)** consists of the brain and spinal cord.
2. **THE PERIPHERAL NERVOUS SYSTEM (PNS)** consists of all nerves outside the CNS.

The peripheral nervous system is then further divided and subdivided into various systems. Table 16.2 gives an overview of the peripheral nervous system.

REMEMBER

Cells communicate with each other through direct contact with other cells or from a distance via chemical signaling.

Table 16.2

Components of the Peripheral Nervous System		
System		**Function**
SENSORY		Conveys information from sensory receptors or nerve endings
MOTOR		
■ Somatic System		Controls voluntary muscles
■ Autonomic System		Controls involuntary muscles
	Sympathetic	• **Fight-or-flight** response • Increases the heart rate and breathing rate • Liver converts glycogen to glucose • Bronchi of the lungs dilate and increase gas exchange • Adrenaline raises blood glucose levels
	Parasympathetic	• Opposes the sympathetic system • Calms the body • Decreases the heart rate and breathing rate • Enhances digestion

The Neuron

The neuron consists of a cell body, which contains the **nucleus** and other organelles, and two types of cytoplasmic extensions called **dendrites** and **axons**. **Dendrites** are **sensory**; they receive incoming messages from other cells and carry the electrical signal to the cell body. A neuron can have hundreds of dendrites. A neuron has only one axon, which can be several feet long in large mammals. **Axons** transmit an impulse from the cell body outward to another cell. Many axons are wrapped in a fatty **myelin sheath** that is formed by **Schwann cells.** Figure 16.4 shows a sketch of a neuron.

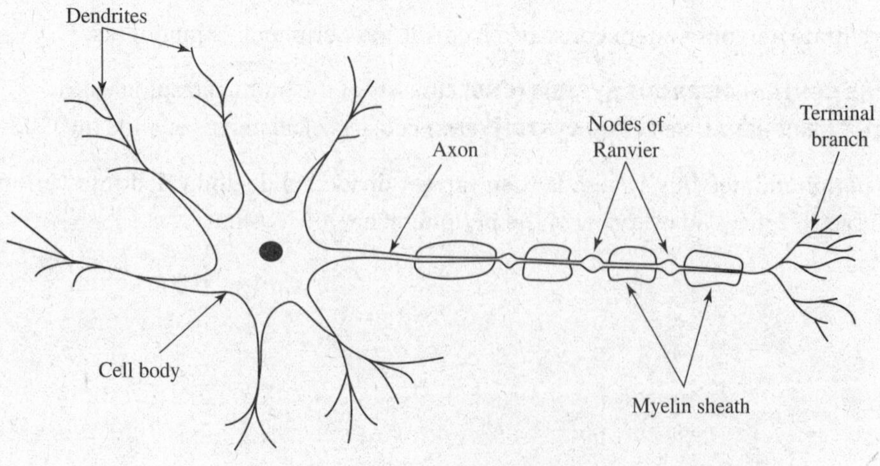

Figure 16.4 The Neuron

There are three types of neurons:

- **Sensory neurons** receive an initial stimulus from a sense organ, such as the eyes and ears, or from another neuron.
- A **motor neuron** stimulates **effectors** (**muscles** or **glands**). A motor neuron, for example, can stimulate a digestive gland to release a digestive enzyme or to stimulate a muscle to contract.
- The **interneuron** or **association neuron** resides within the spinal cord and brain, receives sensory stimuli, and transfers the information directly to a motor neuron or to the brain for processing.

The Reflex Arc

The simplest nerve response is a **reflex arc**. It is *inborn, automatic, and protective.* An example is the **knee-jerk reflex**, which consists of only two types of neurons: sensory and motor. A stimulus, a tap from a hammer, is felt in the sensory neuron of the kneecap, which sends an impulse to the motor neuron, which directs the thigh muscle to contract. A more complex reflex arc consists of three neurons: sensory, motor, and interneurons or association neurons; see Figure 16.5. A sensory neuron transmits an impulse to the interneuron in the **spinal cord**, which sends one impulse to the brain for processing and also one to the motor neuron to effect change immediately (at the muscle). This is the type of response that quickly jerks your hand away from a hot iron before your brain has figured out what occurred.

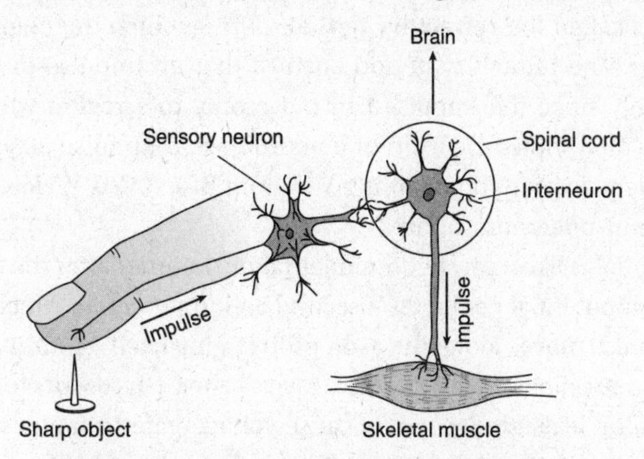

Figure 16.5 The Three-Neuron Reflex Arc

STUDY TIP

The neuron allows for the detection, generation, transmission, and integration of signal information.

Resting Potential

All living cells exhibit a **membrane potential**, a difference in electrical charge between the cytoplasm (negatively charged) and extracellular fluid (positively charged). Physiologists measure this difference in membrane potential using microelectrodes connected to a voltmeter. This potential should be between –50 mV and –100 mV. The negative sign indicates that the inside of the cell is negative relative to the outside of the cell. A neuron in its unstimulated or **polarized state** (**resting potential**) has a membrane potential of about **–70 mV**. The **sodium-potassium pump** maintains the membrane potential (the polarization) by actively pumping ions that leak across the membrane. Potassium ions leak passively down the gradient across the cell membrane through "leak channels" in resting neurons. In order for the nerve to fire, a stimulus must be strong enough to overcome the **resting threshold** or **resting potential**. *The larger the resting potential, the stronger the stimulus must be to cause the nerve to fire.*

Gated Channels

Neurons have **gated ion channels** that open or close in response to a stimulus and play an essential role in the transmission of electrical impulses. These channels allow only one kind of ion, such as sodium or potassium, to flow through them. If a stimulus triggers a **sodium ion gated channel** to open, sodium ions flow into the cytoplasm, resulting in a decrease in polarization to about –60 mV (from –70 mV). The membrane becomes somewhat **depolarized**, so it is easier for the nerve to fire. In contrast, if a **potassium ion gated channel** is stimulated, the membrane potential increases and the membrane becomes **hyperpolarized**, to about –75 mV, so that it is harder for the neuron to fire.

Action Potential

An **action potential**, or **impulse**, can be generated only in the **axon** of a neuron. When an axon is stimulated sufficiently to overcome the threshold, the permeability of a region of the membrane suddenly changes and the impulse can pass. **Sodium channels** open and sodium ions flood into the cell, down the concentration gradient. In response, **potassium channels open** and potassium ions flood out of the cell. This rapid movement of ions or **wave of depolarization** reverses the polarity of the membrane and is called an **action potential**. The action potential is localized and lasts a very short time. The **sodium-potassium pump** restores the membrane to its original polarized condition by pumping sodium and potassium ions back to their original position. This period of repolarization, which lasts a few milliseconds, is called the **refractory period**, during which the neuron *cannot respond to another stimulus.* The refractory period ensures that an impulse moves along an axon in one direction only since the impulse can move only to a region where the membrane is polarized. Figure 16.6 shows the axon of a neuron as an impulse action potential passes from left to right, depolarizing the membrane in front of it. (Also, review Figure 3.16, which illustrates the sodium-potassium pump.)

The action potential is like a row of dominoes falling in order after the first one is knocked over. The first action potential generates a second action potential, which generates a third, and so on. The impulse moves along the axon propagating itself *without losing any strength.* If the axon is large or myelinated, the impulse travels faster. This occurs for two reasons. First, the myelin reduces ion leakage. Second, because voltage-gated channels are located on the nodes, the impulse leaps from node to node (Ranvier) in saltatory (jumping) fashion.

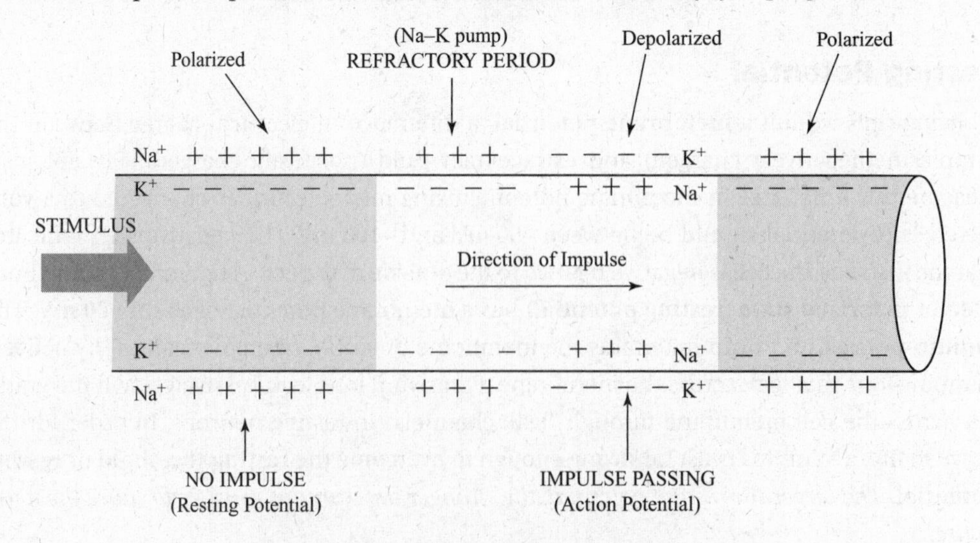

Figure 16.6 An Impulse Passing Along an Axon

The action potential is an *all-or-none* event; either the stimulus is strong enough to cause an action potential or it is not. The body distinguishes between a strong stimulus and a weak one by the **frequency** of action potentials. A strong stimulus sets up more action potentials than a weak one does.

The graph in Figure 16.7 traces the events of a membrane experiencing sufficient stimulation to undergo an action potential.

REMEMBER

Transmission of information along neurons and synapses results in either a stimulatory or an inhibitory response.

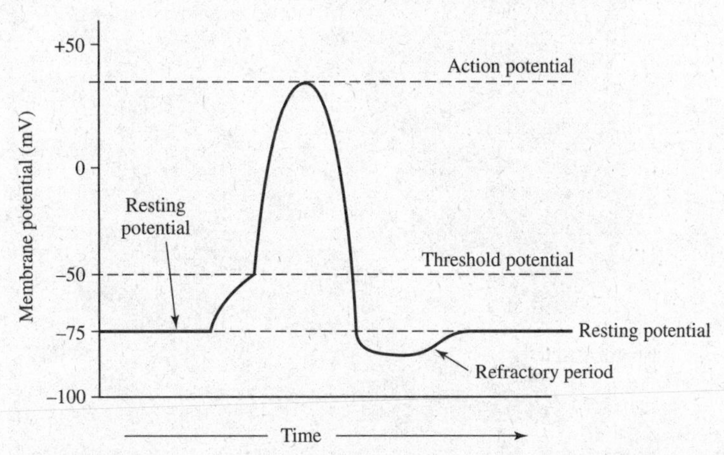

Figure 16.7 Axon Membrane Undergoing an Action Potential

The Synapse

Although an impulse travels along an axon electrically, it crosses a synapse **chemically**. The cytoplasm at the **terminal branch** of the **presynaptic neuron** contains many **vesicles**, each containing thousands of molecules of **neurotransmitter**. Depolarization of the presynaptic membrane causes **Ca^{++} ions** to rush into the terminal branch through **calcium gated channels**. This sudden rise in Ca^{++} levels stimulates the vesicles to fuse with the presynaptic membrane and release the neurotransmitter by **exocytosis** into the synaptic cleft. Once in the synapse, the neurotransmitter bonds with receptors on the **postsynaptic side**, altering the membrane potential of the postsynaptic cell and resulting in a response. Depending on the type of receptors and the ion channels they control, the postsynaptic cell will be either *inhibited* or *excited.* In vertebrates, synapses between motor neurons and muscle cells are always excitatory. However, synapses between neurons can also be inhibitory by causing hyperpolarization of the postsynaptic cell.

Shortly after the neurotransmitter is released into the synapse, it is destroyed by an enzyme called **esterase** and recycled by the presynaptic neuron. The neurotransmitter at all neuromuscular junctions is **acetylcholine**. Other neurotransmitters are **serotonin**, **epinephrine**, **norepinephrine**, and **dopamine**. In addition, the neurotransmitter acetylcholine stimulates some cells to release the gas **nitric oxide** (**NO**), which, in turn, stimulates other cells. A single postsynaptic neuron may receive 1,000 impulses from presynaptic neurons, releasing a variety of different neurotransmitters. All of these impulses are integrated and summed up into either hyperpolarization (inhibited) or hypopolarization (stimulated).

STUDY TIP

Cells communicate over short distances by using local regulators that target cells in the vicinity of the emitting cell.

Figure 16.8 shows the terminal branch of the neuronal axon and the synapse.

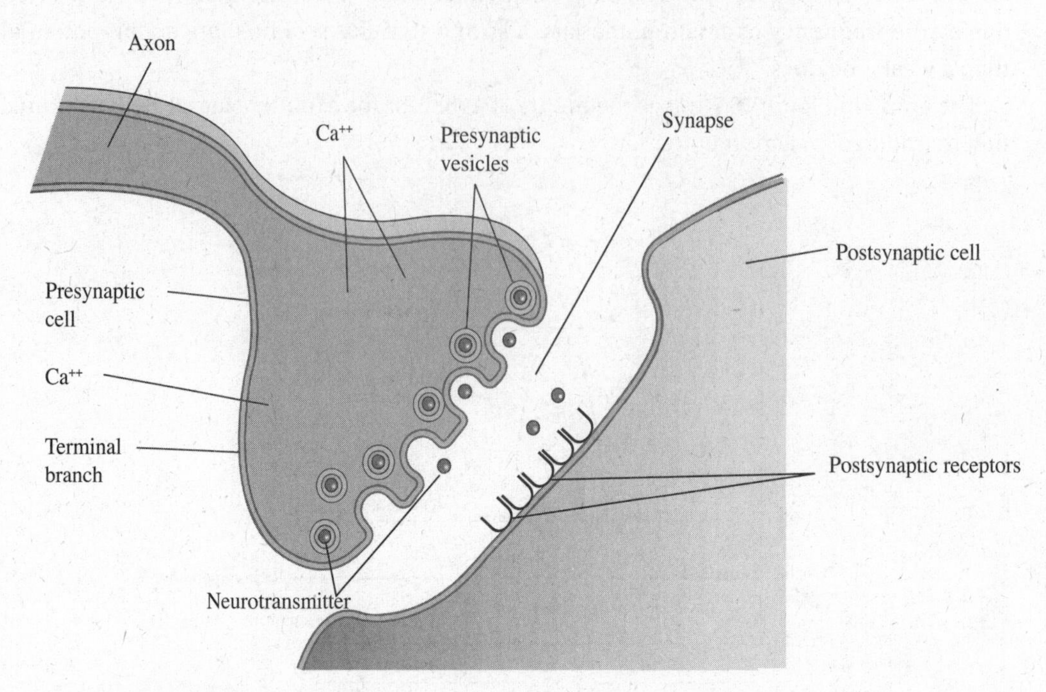

Figure 16.8 Terminal Branch of the Neuronal Axon and the Synapse

ORGANIZATION OF THE HUMAN BRAIN

Table 16.3 reviews the main parts of the brain. Clearly, each region receives information from other regions, integrates that information, and responds.

Table 16.3

The Brain	
Cerebrum	Controls learning, emotions, memory, and perception Divided into right and left hemispheres The left side of the brain receives information from and controls the right side of the body and vice versa
Cerebellum	Coordinates movement and balance; helps in learning and remembering motor skills Receives sensory information about the position of joints and muscles Monitors motor commands from the cerebrum and integrates the information as it carries out coordination
Brainstem—includes medulla oblongata	Controls several automatic homeostatic functions, including breathing, heart and blood vessel activity, swallowing, and digestion Receives and integrates several types of sensory information and sends it to specific regions of the forebrain Transfers information between the PNS and other parts of the brain

For the AP exam, focus on the basics of cell communication: how cells transmit, detect, and integrate signals. The structure and function of the human eye is a good example to use if you are asked to write about cell communication on a free-response question. See Figure 16.9.

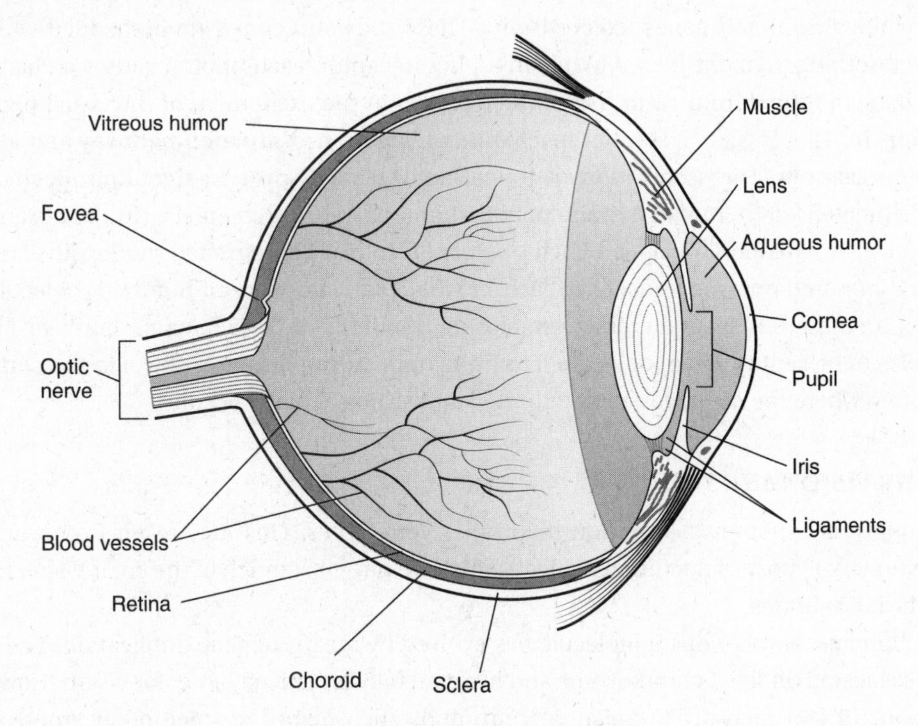

Figure 16.9 The Human Eye

VOCABULARY OF THE EYE

- **CONES:** Modified neurons; photoreceptors in the retina that distinguish different colors; embedded with visual pigments
- **CORNEA:** Tough, clear covering that protects the eye and allows light to pass through
- **HUMOR:** Fluid that maintains the shape of the eyeball
- **IRIS:** Colored part of the eye that controls the size of the pupil and how much light enters the eye
- **LENS:** Focuses light on the retina
- **OPSIN:** A membrane protein bound to a light-absorbing pigment molecule
- **PUPIL:** Small opening in the middle of the iris where light enters the eye
- **RETINA:** On the back of the eyeball; converts light to nerve impulses that are carried to the brain; consists of 5 layers of neurons
- **RETINAL:** Light-absorbing pigment in rods and cones
- **RHODOPSIN:** A visual pigment consisting of *retinal* and *opsin*; upon absorbing light, retinal changes shape and separates from opsin
- **RODS:** Modified neurons; photoreceptors in the retina that are extremely sensitive but distinguish only black and white; embedded with the visual pigment retinal

SIGNAL TRANSDUCTION PATHWAY AND AMPLIFICATION CASCADE IN THE EYE

STUDY TIP

Here's another example of a signal transduction pathway.

The sensory phase of vision begins when **photons** of light pass through the lens, which focuses the light onto the retina. The retina consists of several layers of neurons. When light strikes the retina, it is absorbed by photoreceptors in some of those neurons—rods (black and white vision) and cones (color vision). These rods and cones are embedded with the light-absorbing pigment *retinal.* While in a photoreceptor, each photon causes a change in the shape of retinal from *cis* to *trans,* which results in the excitement of the visual pigment **rhodopsin**. This triggers a familiar mechanism, a **signal transduction pathway** and **amplification cascade**. The stimulation of retinal activates a **G protein-signaling mechanism** that ultimately alters the membrane potential and closes Na^+ channels. (In the absence of light, the Na^+ channels are open.) Each single molecule of photoexcited rhodopsin activates several hundred enzyme molecules, each of which activates several hundred molecules of cGMP, a ubiquitous second messenger, closing hundreds of Na^+ channels, and causing an impulse to be sent to the optic nerve. From the optic nerve, impulses travel to the cortex of the brain where the messages are interpreted and seeing actually occurs.

COLOR VISION AND YOU

Mammals have poorer color vision than other vertebrates. This may result from our long evolutionary history of nocturnal living in which we have lost much of the color vision found in other vertebrates.

In humans, another opsin molecule has evolved by means of gene duplication. The gene for it is located on the X chromosome and has generally enhanced our color vision. However, in about 10% of men of European descent (and much higher in some other groups), the duplicated gene is nonfunctional. This results in red-green color blindness in those men.

CHAPTER SUMMARY

This chapter covered cell communication, specifically the fact that cells communicate by sending, transmitting, and receiving chemical signals.

- Similar cell communication mechanisms among cells of different organisms reflect a **shared evolutionary history**.
- **Negative feedback** is a common mechanism in the endocrine system to **maintain homeostasis**. For example, when the level of thyroxine in the blood is too low, the hypothalamus releases a **tropic hormone**, which stimulates the **anterior pituitary** to stimulate yet another gland, the **thyroid**, to release more **thyroxine**.
- Cells communicate using chemical signals within the **endocrine** and **nervous systems** to support the organism as a whole. **Endocrine glands** release hormones that travel long distances through the blood to reach all parts of the body. **Tropic hormones** send a signal to other glands to release another hormone.
- There are two types of hormones: **lipid** or **steroid hormones** and **protein hormones**.
- **Lipid hormones** can readily diffuse through the target cell membrane, binding to a receptor inside the cell to alter the expression of a gene and trigger a cellular response.
- **Protein hormones** must bind to a **cell surface receptor**, triggering a cellular response via a **secondary messenger** inside the cell.
- Neurons release **neurotransmitters** into a **synapse** to cause a local response in a neuron, gland, or muscle.

PRACTICE QUESTIONS

CHEMICAL SIGNALS

1. Which of the following hormones induces labor?
   (A) glucagon
   (B) adrenocorticotropic hormone
   (C) oxytocin
   (D) thyroxine

2. Which of the following hormones is released from the posterior pituitary?
   (A) glucagon
   (B) adrenocorticotropic hormone
   (C) oxytocin
   (D) thyroxine

3. Which of the following hormones stimulates the adrenal cortex?
   (A) glucagon
   (B) adrenocorticotropic hormone
   (C) oxytocin
   (D) thyroxine

4. Which of the following hormones controls the metabolic rate?
   (A) glucagon
   (B) adrenocorticotropic hormone
   (C) oxytocin
   (D) thyroxine

5. Which of the following hormones is produced in the pancreas?
   (A) glucagon
   (B) adrenocorticotropic hormone
   (C) oxytocin
   (D) thyroxine

6. Which of the following hormones causes blood sugar levels to increase?
   (A) glucagon
   (B) adrenocorticotropic hormone
   (C) oxytocin
   (D) thyroxine

7. The two cells that are pictured below have the same shape and structure; they only differ in size.

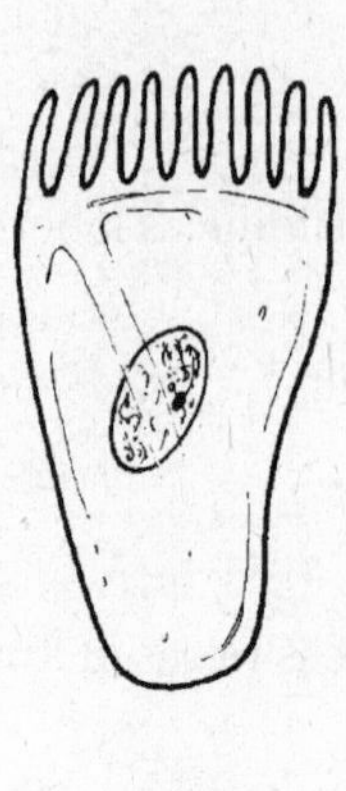

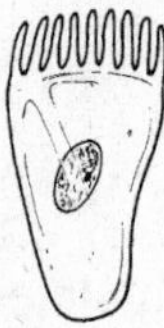

Which of the following statements is correct about these two cells?

(A) The ratio of the volume of the cytoplasm to the size of the nucleus is the same in both cells.
(B) The smaller cell can ingest and egest material faster because the ratio of the size of its plasma membrane to the volume of its cytoplasm is greater than that of the larger cell.
(C) The smaller cell can ingest and egest material faster because the ratio of the size of its plasma membrane to the volume of its cytoplasm is smaller than that of the larger cell.
(D) The nucleus in the larger cell controls that cell more effectively because the ratio of the size of the nucleus to the volume of the cytoplasm in the larger cell is so much greater than that of the smaller cell.

8. After eating a large Thanksgiving meal, which of the following is the most likely response to occur?

(A) Glycogen will be released from the liver.
(B) Glucagon will be released from the liver.
(C) Insulin will be released from the pancreas.
(D) The anterior pituitary will stimulate the pancreas and liver.

9. Homeostasis typically relies on negative feedback, not positive feedback. What is the correct reason for this fact?

(A) Negative feedback brings a process to completion.
(B) Negative feedback monitors shut off pathway activity when the activity passes a certain set point.
(C) Positive feedback works by signal transduction.
(D) Positive feedback relies on stimuli from only outside the body.

NERVOUS SYSTEM

10. Which of the following sequences describes the passage of a nerve impulse through a simple reflex arc in humans, as depicted in the figure below?

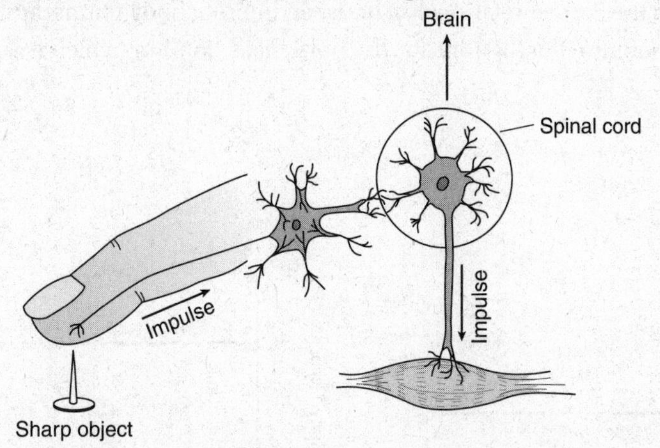

(A) receptor → effector → interneuron → motor neuron → sensory neuron
(B) receptor → sensory neuron → interneuron → effector → motor neuron
(C) sensory neuron → effector → motor neuron → interneuron → receptor
(D) receptor → sensory neuron → interneuron → motor neuron → effector

11. Feedback mechanisms maintain internal body temperatures within many animals. Faced with fluctuations in environmental temperatures, an animal manages its internal body temperature by regulating it or by conforming to the environment. Which graph below shows the correct relationship between internal body temperatures and environmental temperature fluctuations for the bottlenose dolphin (which is a mammal) and a fish?

(A)

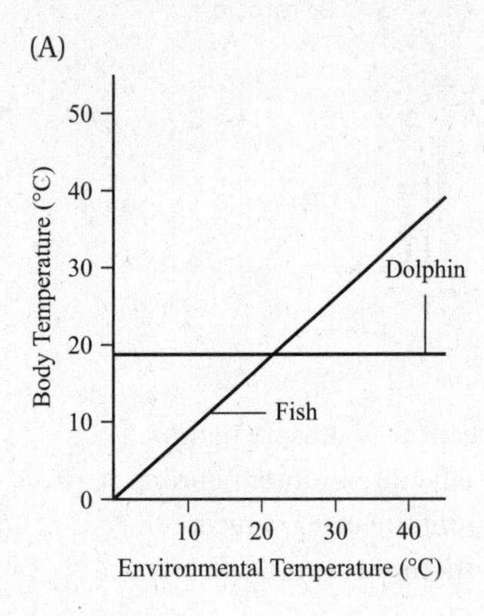

(B)

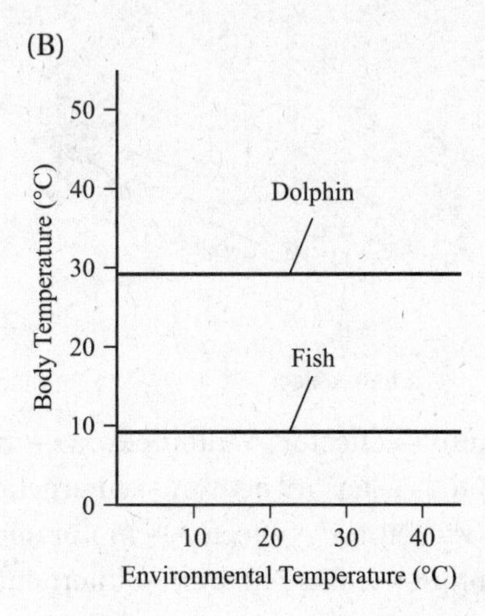

(C)

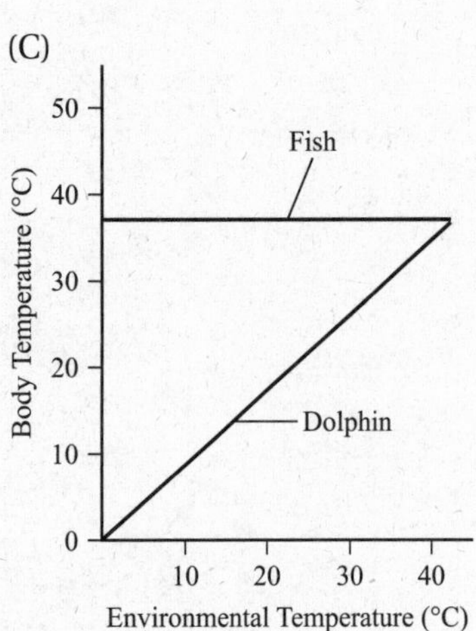

(D)

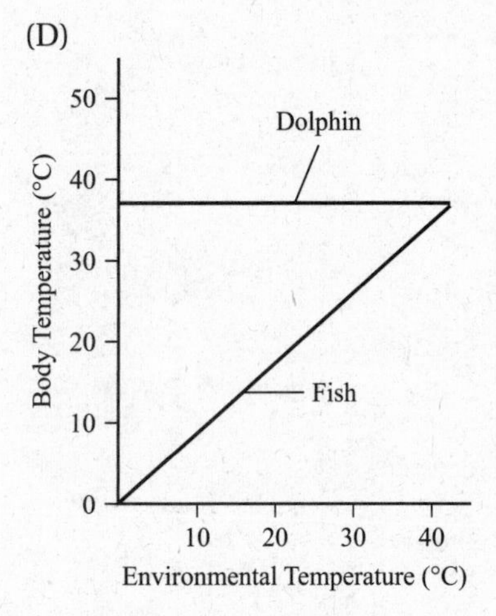

12. Below is a graph that shows the change in oxygen usage for a mouse at rest in an environment where the ambient temperature is changed.

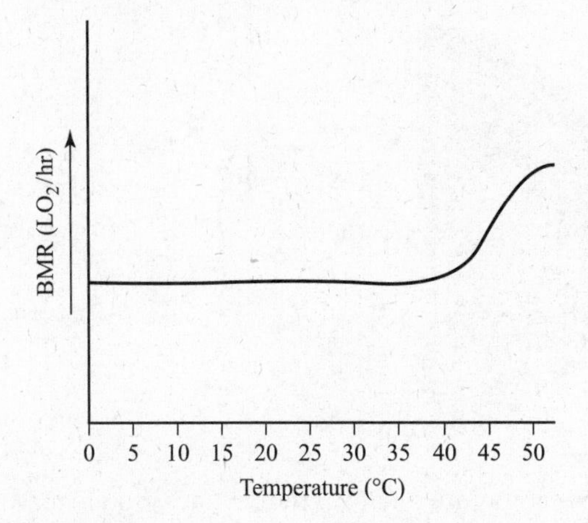

Oxygen usage is proportional to the rate of metabolism. BMR (basal metabolic rate) is the metabolic rate of a resting, fasting, nonstressed animal. Which of the following statements correctly describes what is shown in this graph, regarding the metabolism of a mammal, and provides the most likely reason for that occurrence?

(A) As the ambient temperature increases, a mammal's metabolic rate remains the same so that the animal can be active throughout the day.
(B) As the ambient temperature increases, the metabolic rate increases so that the animal can remain active throughout the day.
(C) As the ambient temperature increases well above normal body temperature, the animal must expend energy to cool itself down.
(D) As the ambient temperature decreases well below normal body temperature, the animal must expend energy to cool itself down.

13. This question explores energy budgets for two animals. The following pie chart shows the relative energy budget for the first animal: a 25 g female mouse.

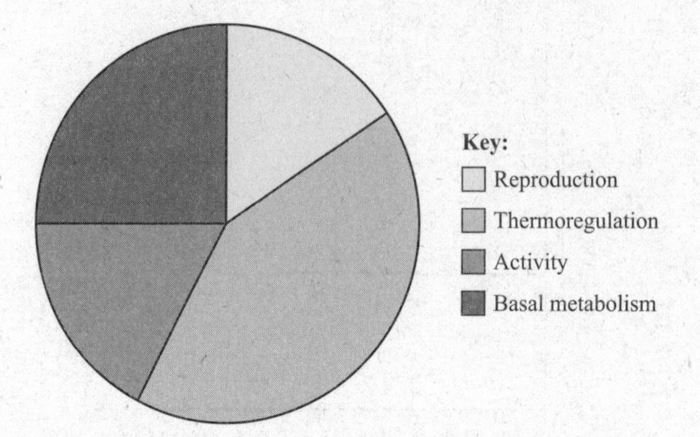

The mouse lives in a temperate environment, where food is readily available, but the animal's small size causes a rapid loss of body heat. (Basal metabolism is the minimum amount of energy needed to maintain all vital functions in an organism that is at complete rest.)

The second animal is the 4 kg male Adélie penguin that lives in the cold Antarctic environment. This penguin is well-insulated against the cold, but it must expend large amounts of energy swimming to catch food, incubating eggs laid by his partner, and bringing food to his chicks.

Based on all of this information, which of the following pie charts is the most accurate depiction of the relative energy budget of the 4 kg male Adélie penguin?

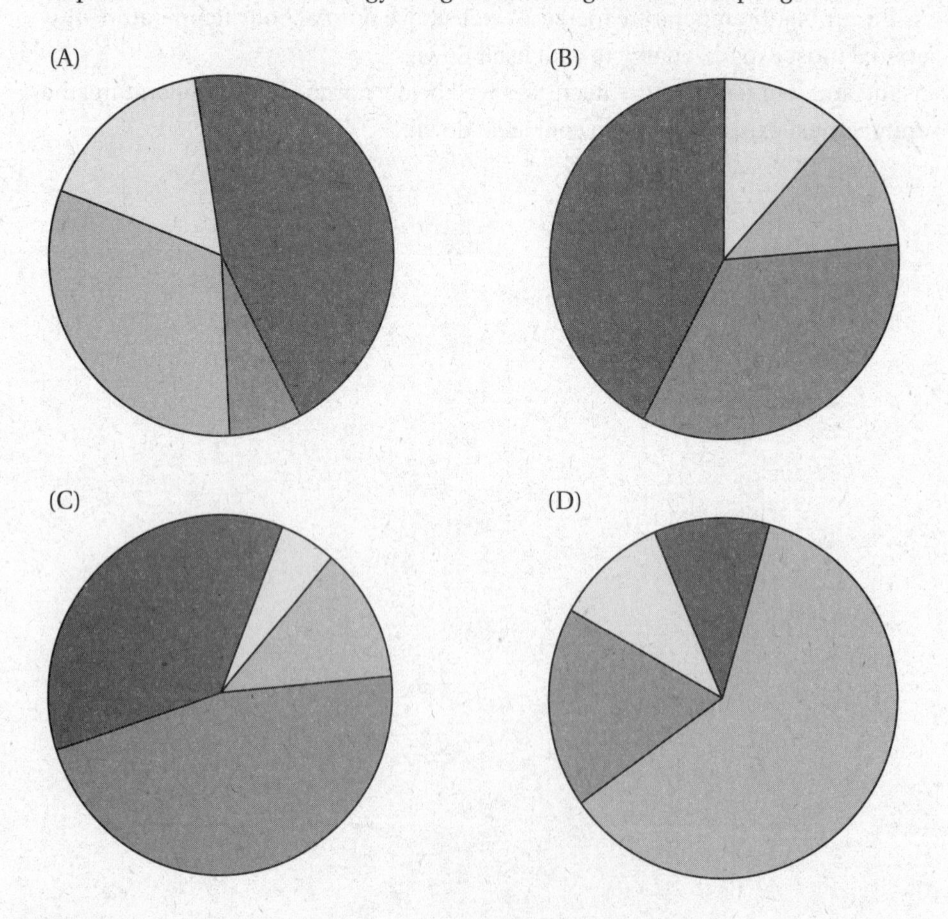

14. The threshold potential of a particular membrane measures –70 mV at time zero. After 10 minutes, it measures –90 mV. What is the best explanation for what has happened to the membrane?

    (A) It became depolarized.
    (B) The concentrations of Na^+ and K^+ became balanced.
    (C) The membrane became hyperpolarized.
    (D) The membrane became hypopolarized.

15. Changes in the membrane potential of a neuron occur because neuronal membranes have gated ion channels that open and close in response to various stimuli. For example, opening potassium ion gated channels in a resting neuron increases the membrane's permeability to potassium ions (K^+). The result is an increase in the net diffusion of K^+ out of the neuron, which results in a hyperpolarized (more polarized) membrane.

    Which of the following graphs correctly shows the resting membrane becoming hyperpolarized when a stimulus is applied?

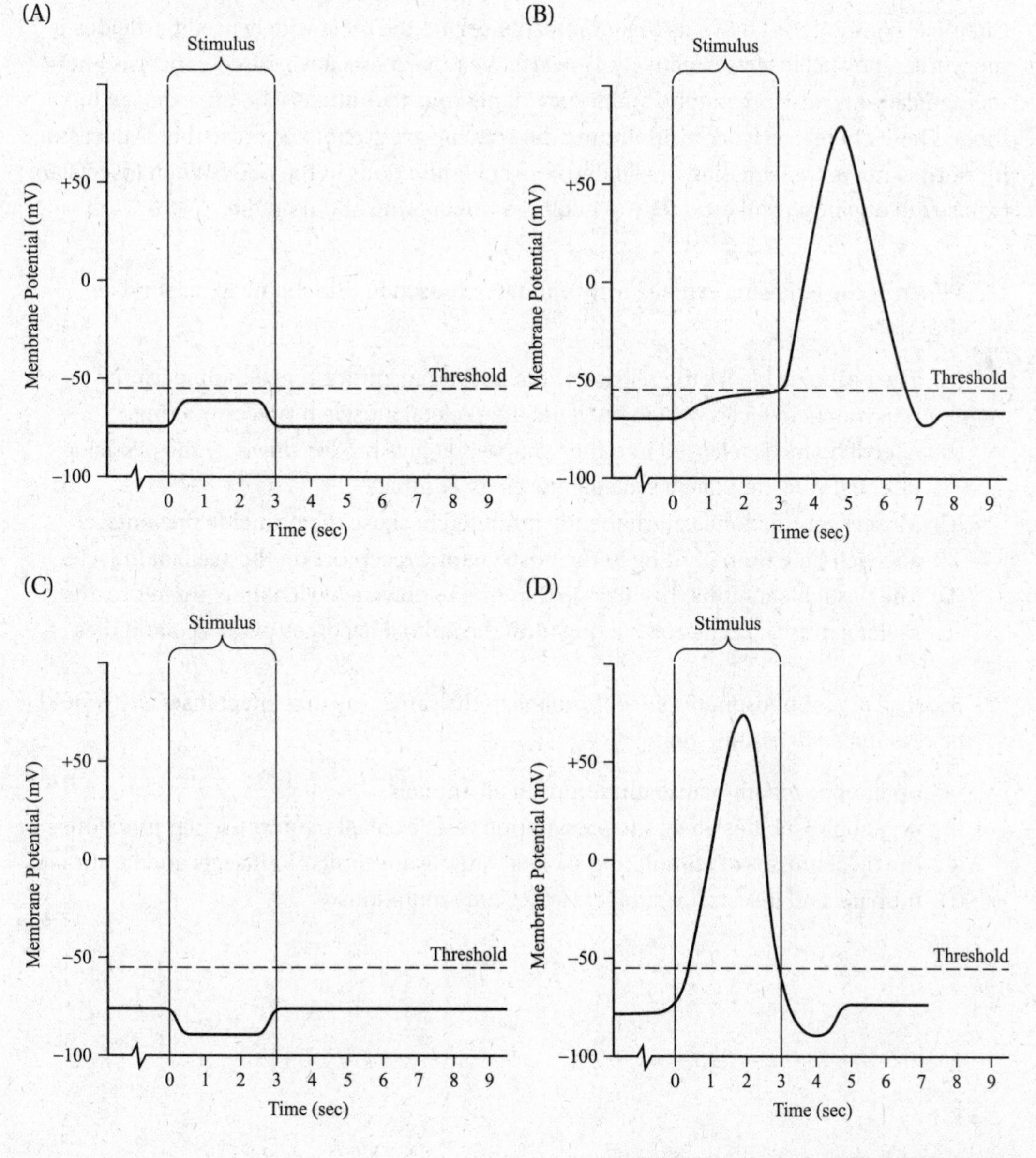

16. If the membrane potential were measured at –75 mV, what would be the value if the polarization across the membrane were increased?

    (A) –70 mV
    (B) –80 mV
    (C) zero
    (D) There would be no change because the membrane potential of a particular neuron can never change.

17. How does an axon respond to stimuli of different intensities?

    (A) by changing the amplitude of its action potentials
    (B) by changing the duration of its action potentials
    (C) by changing the frequency of its action potentials
    (D) The axon does not respond differently to stimuli of different intensities; an axon only responds in one way.

Questions 18–19

Chemical compounds known as organophosphates are the most widely used pesticides in the world. They kill insects effectively. However, health experts have raised concerns about their efficacy because organophosphates are dangerous to humans who are exposed to them. These chemicals work by inhibiting the enzyme acetylcholinesterase that is necessary for normal nerve transmission at all neuromuscular junctions in the body. When insects are exposed to organophosphates, they twitch for a brief period and then die.

18. Which of the following explains why animals exposed to this chemical respond as they do?

    (A) The pesticide blocks the release of the neurotransmitter acetylcholine from presynaptic vesicles, which prevents the skeletal muscle from contracting.
    (B) Acetylcholine is released into the synapse but gets broken down by the pesticide before it can stimulate a skeletal muscle to contract.
    (C) Muscle contraction is permanently inhibited because the pesticide prevents acetylcholine from binding to the postsynaptic receptors on the skeletal muscle.
    (D) The pesticide inhibits the enzyme that breaks down acetylcholine; therefore, the skeletal muscle keeps contracting until the animal becomes paralyzed and dies.

19. Because organophosphates affect humans in the same way that affect insects, it would be reasonable to assume that

    (A) organophosphates cause mutations in all animals
    (B) organophosphates affect the presynaptic vesicles of all neuromuscular junctions
    (C) the mechanisms of stimulating skeletal muscle are similar in insects and in humans
    (D) humans and insects are subject to the same mutations

Answers Explained

CHEMICAL SIGNALS

1. **(C)** Oxytocin induces labor and the production of milk by the mammary glands.

2. **(C)** Oxytocin is produced in the hypothalamus and stored and released from the posterior pituitary.

3. **(B)** Adrenocorticotropic hormone (ACTH) is a tropic hormone. A tropic hormone is one that stimulates another endocrine gland to release a substance. ACTH stimulates the adrenal cortex to release a glucocorticoid-like cortisol.

4. **(D)** The thyroid gland releases thyroxine, which controls the rate of metabolism.

5. **(A)** Glucagon is produced in the α cells of the pancreas and raises blood sugar by breaking down glycogen that is stored in the liver.

6. **(A)** Glucagon raises blood sugar levels, as explained in the answer to Question 5.

7. **(B)** The smaller cell is more efficient for two reasons. First, the ratio of the size of the nucleus in the smaller cell compared to the size of the cytoplasm is greater than that of the larger cell. Second, there is more surface area of membrane compared to the volume of the cytoplasm in the *smaller* cell. Choice A is incorrect because that statement is not true about both cells. Choice C is incorrect because it is the opposite of the correct answer. Choice D is incorrect because the ratio of the size of the nucleus to the volume of the cytoplasm in the larger cell is much smaller than that of the smaller cell.

8. **(C)** Insulin is pumped out of the pancreas to bring glucose blood levels back down to the homeostatic set point. Glycogen is the large molecule in which glucose is packed for storage. Glucagon is the hormone responsible for breaking down glycogen into glucose in the liver. Glucagon is not released from the liver. The anterior pituitary does not stimulate the pancreas and liver.

9. **(B)** Negative feedback works like a thermostat. When the temperature falls below the already established set point, the heater turns on. When the temperature rises to the set point, the heater turns off. Negative feedback maintains homeostasis. Positive feedback enhances an already-existing response. During labor, the pressure of the baby's head against sensors stimulates uterine contractions, which further stimulates the same sensors and so on. This loop stops only when the baby is born.

NERVOUS SYSTEM

10. **(D)** An effector is a muscle or a gland. The interneuron is located in the spinal cord.

11. **(D)** A dolphin is a mammal (a homeotherm or an endotherm) like humans. As a mammal, it maintains a constant body temperature close to ours—about 37°C. Fish do not regulate their body temperatures from inside. They do it by changing their behavior: they change their location and move to a warmer or cooler spot. Choice A is incorrect because a dolphin could not maintain a body temperature that low. Choices B and C are incorrect because they do not depict both the dolphin having a constant body temperature and the fish having a body temperature that varies with the environmental temperature.

12. **(C)** The mouse is a mammal, and as such, it maintains a constant body temperature. In this case, the temperature in the environment increased very high, and the animal had to expend energy to cool itself down. Similarly, when it gets cold, it has to expend extra energy to warm itself up.

13. **(C)** The stem of this question holds important information. The male penguin did not lay the eggs, so it does not expend a lot of energy on reproduction. It is well-insulated, so it does not have to spend a lot of energy making itself warm. It is a bird and all mammals and birds have a high metabolic rate because they must maintain a constant body temperature. That leaves activity. The question described an active lifestyle for the male penguin, so that must be a large piece of the pie chart. Of the four pie charts provided, the one for choice C best depicts the relative energy budget of the 4 kg male Adélie penguin.

14. **(C)** The gradient is steeper at –90 mV than it was at –70 mV. The membrane is now hyperpolarized. The threshold is higher. Depolarizing the membrane and passing an impulse are more difficult.

15. **(C)** The stem of the question asks you to identify a graph that shows the neuronal membrane becoming hyperpolarized (more polarized), not depolarized. When a membrane becomes hyperpolarized, the line on the graph moves downward, farther away from 0.0 mV, and it is harder to pass an impulse. Choice A is incorrect because when the neuron is stimulated, the value of the membrane potential drops from about –70 mV to almost –50 mV (meaning it becomes less polarized), and the line goes upward. A similar occurrence happens in choice B, which is why that choice is also incorrect. Choice D is incorrect because the stimulus causes an action potential: the neuron becomes *depolarized*, and you can see that an impulse passes.

16. **(B)** The membrane potential reads zero when there is no differential between one side of the membrane and the other. The greater the differential, the more negative the potential. Therefore, –80 mV is more polarized than –70 mV.

17. **(C)** A nerve response is all-or-none. There is no such thing as a small action potential. If the stimulus gets stronger, there will be an increase in the *frequency* of the action potentials.

18. **(D)** Neurotransmitter is normally released from presynaptic vesicles into the synapse, where it binds to postsynaptic vesicles on the skeletal muscle cell and causes the muscle to contract. Acetylcholinesterase is an enzyme that normally breaks down acetylcholine after the muscle contracts in order to prevent uncontrolled contracting, which would result in the depletion of ATP in the muscle and paralysis. This is exactly what is described in choice D.

19. **(C)** The mechanisms are similar because insects and humans descended from a common ancestor, albeit millions of years ago.

The Human Immune System

17

- → NONSPECIFIC INNATE DEFENSE MECHANISMS
- → ADAPTIVE IMMUNITY
- → TYPES OF IMMUNITY
- → BLOOD GROUPS AND TRANSFUSION
- → AIDS
- → POSITIVE FEEDBACK IN THE IMMUNE SYSTEM
- → OTHER TOPICS IN IMMUNITY

Remember, this chapter consists of illustrative examples. Although these topics are not required for the AP exam, studying them will broaden your understanding of Biology and provide you with examples to use as you write your answers to the free-response questions.

INTRODUCTION

We live in a sea of germs. They are in the air we breathe, the food we eat, and the water we drink and swim in. Our body has three cooperating lines of defense that are able to respond to new threats as quickly as they appear. This chapter covers this complex topic in detail. The material in this chapter is not required for the AP Biology exam as per the new College Board curriculum. That is why this chapter is placed in Part 3 of this book and why there are no classification references to the curriculum frameworks (such as ENE-1).

NONSPECIFIC INNATE DEFENSE MECHANISMS

First Line of Defense

The first line of **nonspecific defense** is a **barrier** that helps prevent **pathogens** (things that cause disease) from entering the body. Barriers include:

- Skin
- Mucous membranes, which release mucus that contains antimicrobial substances including **lysozymes**
- Cilia in the respiratory system to sweep out mucus (which contains trapped microbes)
- Stomach acid

Second Line of Defense

Microbes that get into the body encounter the second line of **nonspecific defense**, which is meant to limit the spread of invaders in advance of specific immune responses. This second line includes the following.

- **Inflammatory response includes:**
  - ✔ **Histamine** triggers **vasodilation** (enlargement of blood vessels), which increases blood supply to the area, bringing more **phagocytes**. It is secreted by **basophils** (a type of circulating white blood cell) and **mast cells**, found in the connective tissue. Histamine is also responsible for the symptoms of the common cold: sneezing, coughing, redness, itchy eyes, and runny nose. Each attempts to rid the body of pathogens.
  - ✔ **Prostaglandins** further promote blood flow to the area.
  - ✔ **Cytokines**, which are signaling molecules, are released by neutrophils and macrophages to promote blood flow to the site of injury or infection.
  - ✔ **Fever** is a response to certain pathogens. Substances released by activated macrophages cause the body's thermostat to reset to higher temperatures to increase metabolic activity and kill off invaders more rapidly.
- **Phagocytes** ingest invading fungal and bacterial microbes. Two types of phagocytes, called **neutrophils** and **macrophages**, migrate to an infected site in response to local chemical attractants. This response is called **chemotaxis**. **Neutrophils** engulf microbes and die within a few days. **Macrophages** ("giant eaters") extend **pseudopods** and engulf huge numbers of microbes over a long period of time. They digest the microbes with a combination of **lysozymes** and two toxic forms of oxygen: superoxide anion and nitric oxide.
- A **complement** is a group of proteins that leads to the lysis (bursting) of invading cells.
- **Interferons** block against cell-to-cell viral infections.
- **Natural killer (NK) cells** destroy virus-infected body cells (as well as cancerous cells). They attack the cell membrane, causing it to **lyse** (burst open) and die.

ADAPTIVE IMMUNITY

Third Line of Defense

STUDY TIP

Each B or T cell displays specificity for one particular epitope only.

The **adaptive** third line of defense relies on **B lymphocytes** and **T lymphocytes**, which arise from **stem cells** in bone marrow. Once mature, both cell types circulate in the blood, lymph, and *lymphatic tissue* (spleen, lymph nodes, tonsils, and adenoids). Both cell types recognize different specific **antigens**.*

The adaptive immune response is a specific response and involves three phases:

1. **RECOGNITION:** *Antigen receptors* on B and T lymphocytes recognize specific antigens or **epitopes** by binding to them. *Antigens* are any substance that elicits an immune response from B cells or T cells. *Epitopes* are an accessible piece of an antigen that elicits an immune response from a B or T cell. Each B cell or T cell displays specificity for one particular epitope. In order to recognize an antigen, it must be *presented* to a B or T cell by an *antigen-presenting cell.*
2. **ACTIVATION PHASE:** The binding of an antigen receptor activates B and T cells, causing them to undergo rapid cell division. The cells form populations of *effector cells* and *memory cells.*
3. **EFFECTOR PHASE:** After being activated, B cells produce a humoral response; they produce antibodies. T cells engage in a cell-mediated response. See Figure 17.3.

*Other blood cell types—eosinophils and basophils—play a lesser role in adaptive immunity and are beyond the scope of the AP course.

T Lymphocytes—T Cells

There are several types of T cells, but they all form in the bone marrow and mature in the thymus gland (from which they get their name). They fight pathogens in the **cell-mediated immune response**. In general, the activation process begins when T cell **antigen receptors** recognize and bind to antigens that are displayed on the surface of **antigen-presenting cells (APCs)** by a molecule called **MHC**, major histocompatibility complex. Examples of APCs are *macrophages, dendritic cells,* and sometimes *B cells.* Each T cell displays specificity for one particular **epitope**. Once activated, a T cell proliferates and forms a population of activated T *clones.* Some of the clones become **effector cells**, while the remainder become circulating **memory cells** that can rapidly respond to any exposure to the *same* antigen many years later. These memory cells are responsible for **immunological memory** and are the reason that a person does not generally get measles or chickenpox a second time.

REMEMBER

All mammalian immune systems include two types of specific responses: cell-mediated and humoral. This commonality reflects shared ancestry.

The two main types of T cells are **helper T cells (T_h)** and **cytotoxic T cells (T_c)**. Helper T cells are activated by an interaction with an APC. Cytotoxic T cells are activated by helper T cells. Once activated, **T_h cells** announce to the immune system that foreign antigens have entered the body. They trigger both humoral and cell-mediated immune responses. In addition to activating other helper T cells, T_h cells activate cytotoxic T cells (T_c), which kill infected cells, and also activate B cells, which produce antibodies. T_h cells activate these other cells by releasing **cytokines**, **interleukin-1 (Il-1)**, and **interleukin-2 (Il-2)**. Because T_h cells have CD4 accessory proteins on their cell surface, they are also referred to as **CD4 cells**. See Figure 17.3. The CD4 cells are specifically attacked by the human immunodeficiency virus (HIV).

Cytotoxic T cells, $T_{c,}$ attack and kill *body cells* that are infected with pathogens as well as those that transformed into cancer cells. They do it by a cell-mediated immune response. An activated cytotoxic T cell proliferates and differentiates into an **effector cell** and a **memory cell**. Activated T_c cells attack and kill infected cells by releasing **perforin** (a protein that forms pores in the target cell's membrane) and granzymes (enzymes that break down proteins) that cause the cell to lyse and die. The infecting microbes are released into the blood or tissue and are disposed of by circulating antibodies. Because T_c cells have CD8 accessory proteins on their surface, they are also referred to as **CD8 cells**. See Figure 17.1.

B Lymphocytes—B Cells

B cells mature in bone marrow, from which they get their name. A typical B lymphocyte or B cell has about 100,000 identical **antigen receptors** on its surface that it uses to recognize pathogens. Each person produces more than a million different B cells, each with a different antigen receptor. Each antigen receptor is a Y-shaped molecule consisting of 4 polypeptide chains: two identical heavy chains and two identical light chains. Each chain has constant and variable regions. Antibodies secreted by B cells are soluble forms of these antigen receptors. Figure 17.2 shows a typical B lymphocyte.

B cells provide adaptive immunity through the **humoral immune response**, meaning they produce **antibodies (immunoglobulins)**. When they become activated, B cells secrete about 2,000 antibodies per second over the cell's 4–5 day life span. A B cell becomes activated in several steps. First, an APC (**macrophage** or **dendritic cell**) presents an antigen or epitope on its cell surface using a **class II MHC** molecule. Next, a helper T cell that recognizes this **epitope-MHC molecule complex** is activated with the aid of **cytokines** secreted from the APC. See the right side of Figure 17.1. Once activated, the B cell undergoes multiple cell divisions, producing thousands of clones. Some of these clones become **effector cells**, also called **plasma cells**, that secrete antibodies. The remainder of the clones develop into long-term **memory cells** that can rapidly respond to any exposure to the *same* antigen many years later. These memory cells are responsible for **immunological memory** and are the reason that a person does not, for example, generally get measles a second time.

STUDY TIP

Examples of how cells signal each other are important to know for the AP exam.

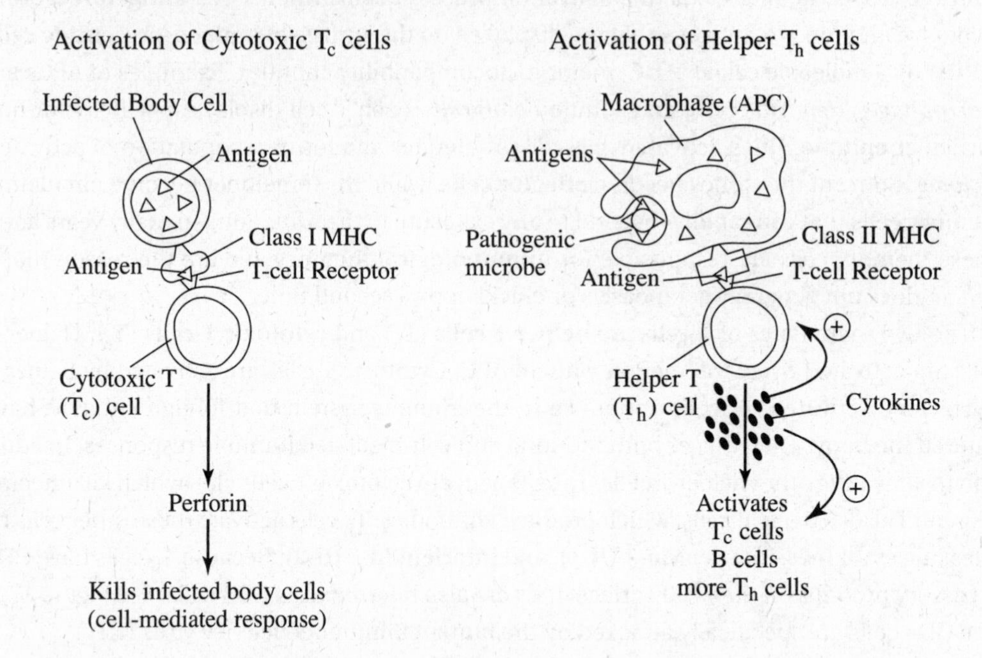

Figure 17.1 Activation of B and T cells

Regulatory T Cells and Self-Tolerance

Normally, the immune system exhibits *self-tolerance,* meaning it does not attack body cells. However, sometimes immature lymphocytes develop that have antigen receptors specific for the body's own cells. If allowed to escape, these cells would attack the body's cells, a situation characteristic of *autoimmune disease.* To avoid this, lymphocytes are tested for *self-reactivity* as they mature in the bone marrow. B and T cells that are identified as self-reactive are destroyed by **apoptosis**, programmed cell death. (Also see page 53.) Regulatory T cells, called T_{reg}, inhibit the activation of the immune system in response to *self-antigens.* T_{reg} secrete interleukin-10 (Il-10).

Major Histocompatibility Complex (MHC) Molecules

Major histocompatibility complex (MHC) molecules, also known as HLA (human leukocyte antigens), are a collection of cell surface markers that identify the cells as **self**. No two people, except identical twins, have the same MHC markers. There are two main classes of MHC markers: **class I** and **class II**.

- **Class I MHC molecules** are found on the surfaces of every nucleated body cell. They signal the immune system that they are "self," not foreign.
- **Class II MHC molecules** are found on specialized cells, including **macrophages**, **B cells**, dendritic cells, and **activated T cells**; see Figure 17.1.

APCs have both MHCI and MHCII on the cell surface.

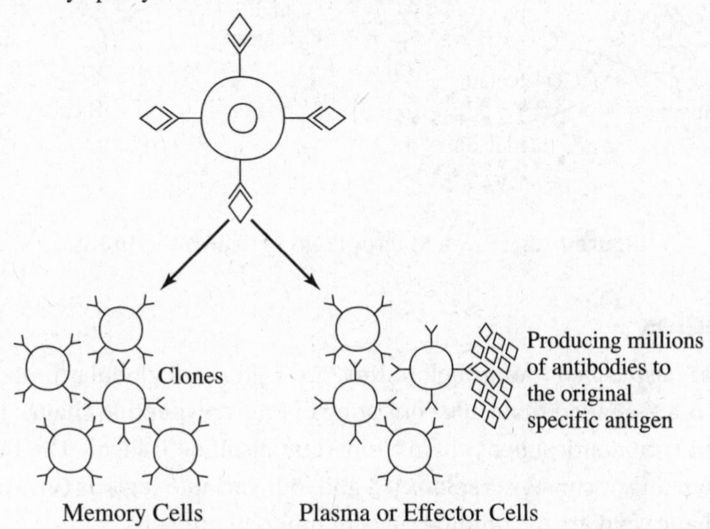

Figure 17.2 Clonal Selection

REMEMBER

Cells communicate with each other through direct contact with other cells or from a distance via chemical signaling.

APCs: Antigen-Presenting Cells

APCs do just what their name says. They present an antigen or a piece of an antigen, an *epitope*, to the immune system. In effect, they are saying, "Here is the enemy!" The process begins when an APC—a **macrophage**, **dendritic cell**, or **B cell**—takes in an antigen. Either the APC becomes infected with the antigen or engulfs it. Once inside the host, enzymes break apart the antigen into fragments and attach them to an MHC molecule in the cytoplasm. The MHC molecule, with the antigen fragment attached, then moves to the surface of the cell and displays it. Other cells of the immune system, such as T cells or other B cells, become activated if they can properly bind with the exposed epitope.

Clonal Selection

Clonal selection is a fundamental mechanism in the development of immunity. It is the means by which one particular lymphocyte that matches a specific antigen or epitope (also called **antigenic determinant**) is identified and activated. See Figure 17.2.

An antigen is presented to a steady stream of lymphocytes in lymph nodes by **APCs** (**antigen-presenting cells**) until a match is found. This triggers events that activate the lymphocyte. Once activated, the lymphocyte divides rapidly, forming a population of *clones* that develop into **effector** and **memory cells**. The effector cells are short-lived and begin battle immediately. They neutralize or destroy all of those identified antigens and any pathogen that produces them. The remaining clones develop into long-term **memory cells** that remain in the body for the rest of one's life and can rapidly respond to any future exposure to the *same* antigen, even many years later. Memory cells are responsible for what is called **immunological memory** and the fact that you generally don't get measles, for example, a second time.

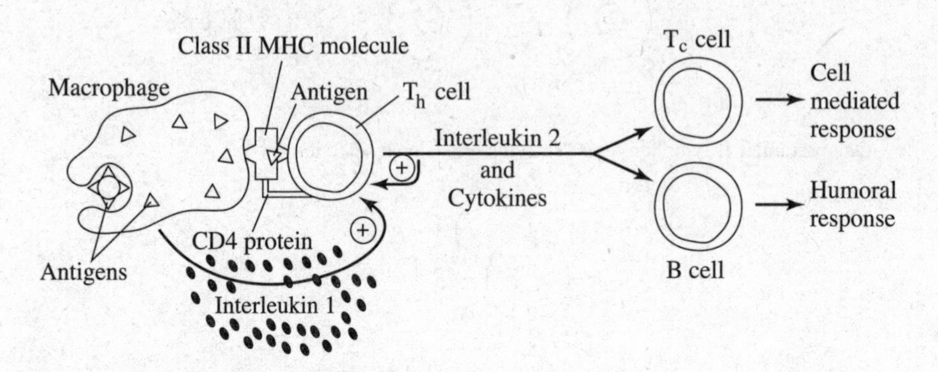

Figure 17.3 How a Macrophage Activates the Immune System

Antibodies

Antibodies, also called **immunoglobulins**, are a group of globular proteins. Each antibody molecule is a **Y-shaped molecule** consisting of four polypeptide chains: two identical heavy chains and two identical light chains joined by disulfide bridges. The molecule consists of four unchanging or **constant regions (C)** and four **variable regions (V)**. The tips of the Y have specific shapes and are the binding sites for different antigens.

It is now accepted that the blueprints for antibodies are made early in life, *prior* to any exposure to antigens. When you are exposed to an antigen, antibodies are chosen by **clonal selection** from a limitless variety as needed.

As a consequence, the variety in antibodies is *unlimited* and there is no viral disease for which humans cannot produce antibodies.

Figure 17.4 is a graph that shows immunological memory:

Day 1: First exposure to antigen A.

Day 15: Maximum production of antibodies against A; the individual will fight off the infection.

Day 28: There is another exposure (a second challenge) to antigen A.

Day 28 to Day 35: There is a secondary immune response that is *more rapid* and *more intense* than the primary immune response. This is due to immunological memory.

Day 28: First exposure to antigen B. This is another primary immune response. The fact that there was an earlier exposure to antigen A has *no bearing* on antibodies against antigen B because adapted immune reactions are very specific.

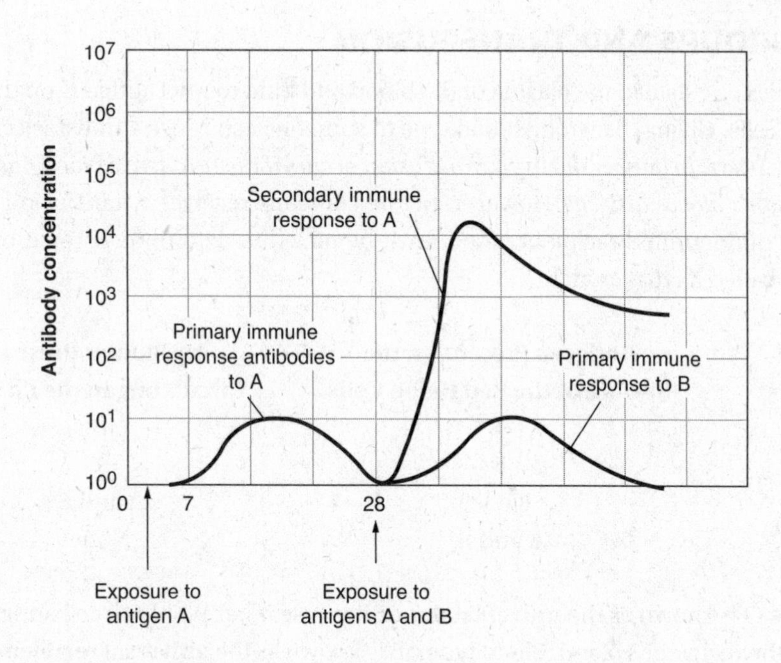

Figure 17.4 Specificity of Immune Response

STUDY TIP

Recognizing the same antigen during a second exposure to it results in a more rapid and enhanced immune response than the first exposure does.

TYPES OF IMMUNITY

Table 17.1 describes the two types of immunity.

Table 17.1

Types of Immunity	
Passive Immunity	**Active Immunity**
Temporary Antibodies are transferred to an individual from someone else. Examples are maternal antibodies that pass through the placenta to the developing fetus or through breast milk to the baby. Also, a person with a weak immune system often receives an injection of gamma globulin (IgG), which are antibodies culled from many people, to boost the weak immune system.	**Permanent** The individual makes his or her own antibodies after being ill and recovering or after being given an immunization or vaccine. A vaccine contains dead or live viruses or enough of the outer coat of a virus to stimulate a full immune response and to impart lifelong immunity.

BLOOD GROUPS AND TRANSFUSION

ABO antibodies circulate in the plasma of the blood and bind to ABO antigens on the surface of red blood cells. Giving the wrong blood type to someone can cause a transfusion reaction or even death. *Certain danger during a transfusion occurs if the recipient's blood has antibodies to the donor's blood antigens.* However, before someone receives a transfusion of blood, two samples of donor and recipient blood must be mixed to determine compatibility. This procedure is called a **cross-match**.

Blood Type	Antigens Present on the Surface of the Red Blood Cells	Antibodies Present Circulating in the Plasma
A	A	B
B	B	A
O	None	A and B
AB	A and B	None

Blood type O is known as the **universal donor** because it has no blood cell antigens to be clumped by the recipient's blood. Blood type AB is known as the **universal recipient** because there are no antibodies to clump the donor's blood.

Rh Factor

Rh factor is another antigen located on the surface of red blood cells. Most of the population—85 percent—have the antigen and are called Rh^+. Those without the antigen (15 percent) are Rh^-.

AIDS

AIDS stands for acquired immunodeficiency syndrome. People with AIDS are highly susceptible to opportunistic diseases, infections, and cancers that take advantage of a collapsed immune system. The virus that causes AIDS, **HIV** (**human immunodeficiency virus**), attacks cells that bear CD4 molecules on their surface, mainly helper T cells. HIV is a retrovirus. Once inside a cell, it reverse transcribes itself, using the enzyme **reverse transcriptase**, and integrates the newly formed DNA into the host cell genome. It remains in the nucleus as a provirus, directing the production of new viruses.

POSITIVE FEEDBACK IN THE IMMUNE SYSTEM

Positive feedback enhances an already existing process until some endpoint or maximum rate is reached. An example in the immune system can be seen in the activity of helper T cells. When a T_h cell becomes activated by a class II MHC molecule, it releases two cytokines: interleukin-I and interleukin-II. Il-2 stimulates B cells and other T cells into action. In this example, *interleukin-I enhances the activity of the already activated T_h cells, stimulating them more until they are activated to a maximum.* This is in contrast to negative feedback, which is a means to achieve stability.

OTHER TOPICS IN IMMUNITY

- Allergies are hypersensitive immune responses to certain substances called **allergens**. They involve the release of **histamine**, an anti-inflammatory agent, which causes blood vessels to dilate. A normal allergic reaction involves redness, a runny nose, and itchy eyes. Antihistamines can normally counteract these symptoms. However, sometimes an acute allergic response can result in a life-threatening response called **anaphylactic shock**, which can result in death within minutes.
- **Antibiotics** are medicines that kill bacteria or fungi. Whereas vaccines are given to prevent illnesses that are caused by viruses, antibiotics are administered after a person is sick.
- **Autoimmune diseases,** such as **multiple sclerosis**, **lupus**, **arthritis**, and **juvenile diabetes**, are caused by a terrible mistake of the immune system. The system cannot properly distinguish between **self** and **nonself**. It perceives certain structures in the body as nonself and attacks them. In the case of multiple sclerosis, the immune system attacks the **myelin sheath** surrounding certain neurons in the CNS.
- **Monoclonal antibodies** are antibodies that are produced by a single B cell that has been selected because it produces one specific antibody. Monoclonal antibodies are important in research and in the treatment and diagnosis of certain diseases.

Antigenic Variation

Just as animals have evolved the ability to fight off pathogens, pathogens have evolved mechanisms to protect themselves from animal immune systems. One way a pathogen escapes our immune system is by changing how it appears to its potential host. For example, the parasite that causes African trypanosomiasis ("sleeping sickness") periodically switches at random, among 1,000 different versions of the protein that it displays on its surface. This ability of a pathogen to alter its biochemical appearance is called **antigenic variation**.

This antigenic variability is also the reason that the flu (influenza) virus is a slightly different form every year or every few years. As the flu virus infects and replicates in one person after another, it undergoes frequent mutations. The accumulation of these mutations enables the virus to change its surface proteins, thus making it less vulnerable to our immune systems.

In addition, the human influenza virus can *exchange genes* with other viruses that have infected other animals, like pigs and chickens. This is called **horizontal gene transfer**. When this happens, the viral strain may become completely unrecognizable by our immune system. Viral transformation was the cause of the great pan-European flu epidemic of 1918–1919, which killed 20 million people. A virus's ability to jump from one species to another, from pigs to humans, was also the cause of the **H1N1 flu** pandemic of 2009. Luckily, in that case, scientists were able to make an effective H1N1 vaccine that averted an even greater disaster.

Overview of the Immune System

The immune system has four major characteristics:

1. **SPECIFICITY:** The entire system depends on the matching of antigens to antigen receptors and the matching of antigens to antibodies.
2. **DIVERSITY:** A wide variety of different cell types defend our bodies from pathogens.
3. **MEMORY:** B and T memory cells circulate for a lifetime, which results in a stronger and faster response to an antigen encountered previously.
4. **CAPACITY TO DISTINGUISH SELF FROM NONSELF:** Fortunately, most of the time, the immune system does not attack healthy body cells. This phenomenon is known as self-tolerance.

CHAPTER SUMMARY

This chapter covered the human immune system. The most important takeaways from this chapter include all of the examples of how cells communicate with each other.

- **Innate** (inborn) defense mechanisms are **nonspecific**. These include barriers to germs, such as skin, mucous membranes, cilia, stomach acid, inflammatory response, complement, and natural killer cells, among others. **Adaptive immunity** develops in response to exposure to antigens in one's lifetime. With the exception of antibodies passed from a mother to a nursing baby, it is not passed to the next generation. Adaptive immunity is based on **specificity**—the recognition of a particular antigen by its cell surface markers or receptors.
- **T cells** attack infected cells directly by **cell-mediated response**. **B cells** produce and release **antibodies** to neutralize antigens in **humoral response**. **Clonal selection** is a fundamental mechanism of the immune system whereby one particular lymphocyte that matches a specific antigen is identified and activated.
- There are two types of immunity: passive and active. **Passive immunity** arises when antibodies are passed from one individual to another, such as in mother's milk or an injection of gamma globulin. **Active immunity** results when a person makes his or her own antibodies after exposure to a particular antigen, such as after receiving an immunization or after being infected with a virus.
- Specificity of the immune response is of extreme importance. A **primary exposure** to an antigen evokes an immune response specific to that antigen *only.* A second exposure to the same antigen is more rapid and stronger than the first.
- Another important topic is regulation by **feedback mechanisms**. In this chapter, we reviewed an example of **positive feedback**. Helper T cells (T_h) become activated by an MHC molecule to release **interleukin** and **cytokines** which stimulates B cells and other T cells into action. These chemicals also enhance the activity of the already activated T_h cells until they reach their *maximum activity.* (This is different from negative feedback where the goal is *homeostasis.*)
- This chapter included many examples of the all-important topic of **cell communication**:
  Long-distance signaling: histamine, complement, interferons
  Local signaling: cytokines
  Cell-to-cell contact: activation of cytotoxic T cells, helper T cells, and clonal selection

PRACTICE QUESTIONS

1. What is the role of histamine in the immune response?

   (A) Histamine is an important part of the adaptive immune system.
   (B) Histamine contains the antimicrobial substance lysozyme, which destroys pathogens.
   (C) Histamine is a class of substances that kills germs by causing them to lyse or burst.
   (D) Histamine enhances blood flow to an area, bringing more germ-fighting cells to the area.

2. A person's humoral response to an antigen differs depending on whether or not the person has been exposed to that antigen before. Which of the following graphs correctly represents a humoral response when a person has been exposed to the same antigen more than once?

(A)

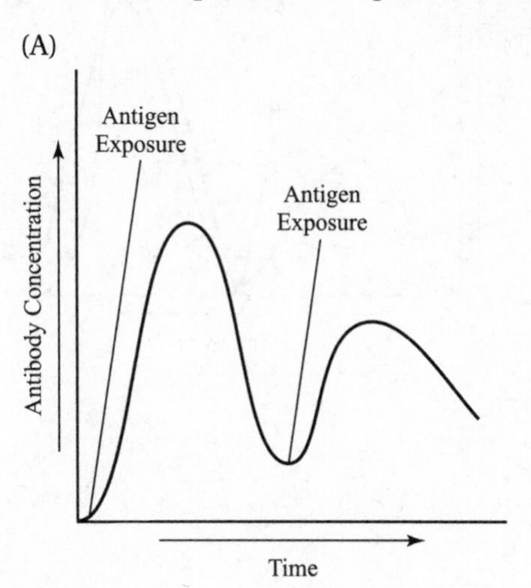

(B)

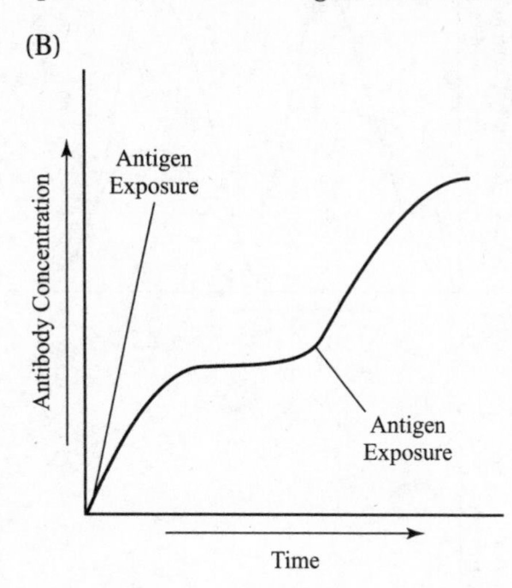

(C)

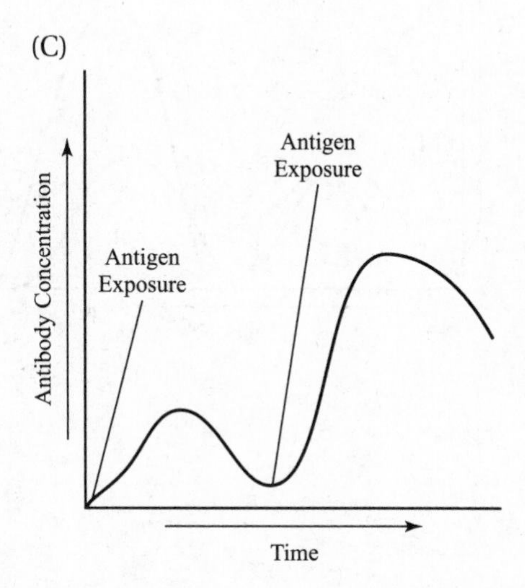

(D)

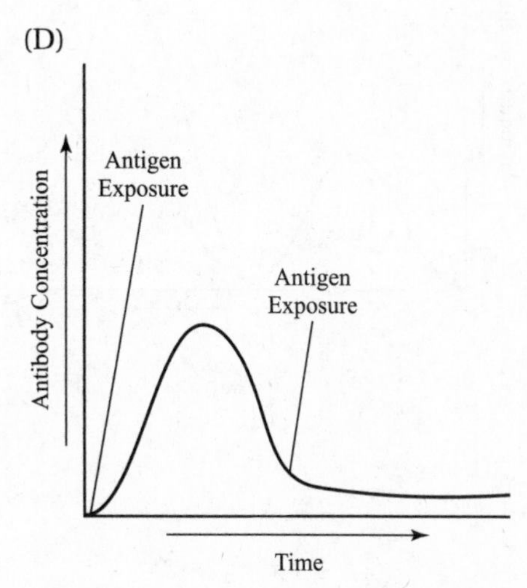

3. The humoral response to an antigen is specific. For example, antibody A will develop in response only to antigen A, but not to antigen B. Which of the following graphs correctly represents humoral responses to these three exposures to antigens?

I. First exposure to antigen A
II. Second exposure to antigen A
III. First exposure to antigen B

(A)

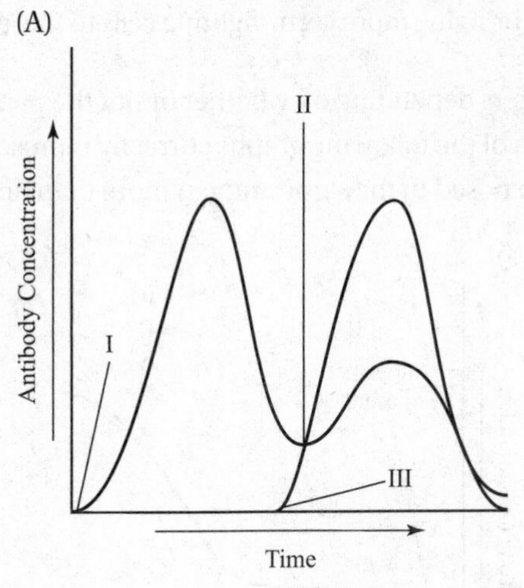

(B)

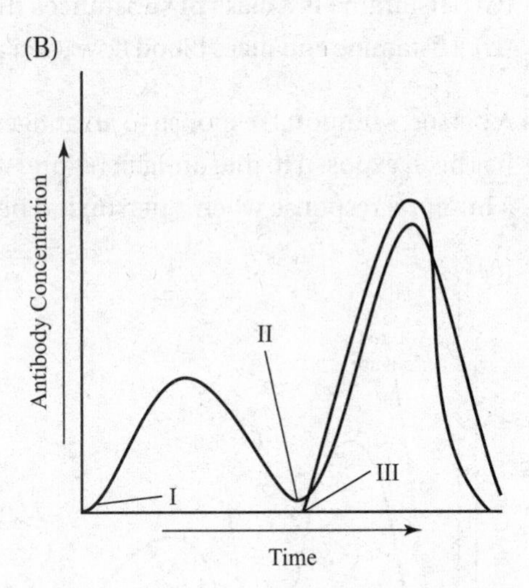

(C)

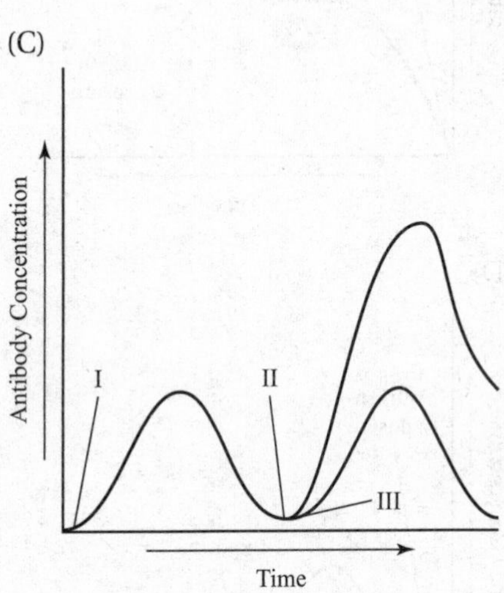

(D)

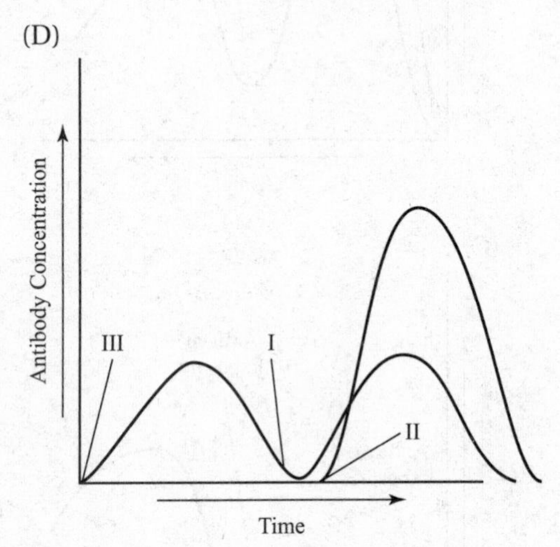

4. The following diagram shows an important step in fighting pathogens.

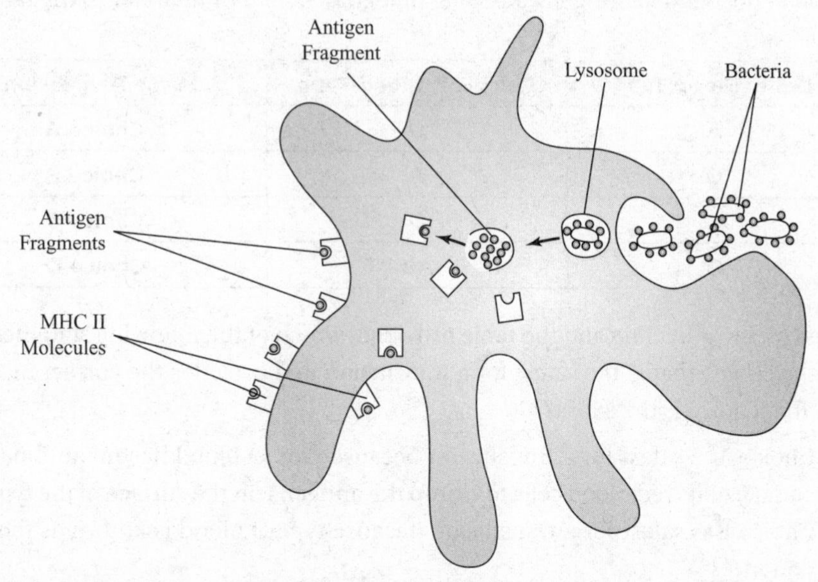

What type of cell is shown in this diagram, and what process is occurring?

(A) This is a T cell that is attacking a pathogen-infected cell.
(B) This is a B cell that is producing and releasing antibodies.
(C) This is a helper T cell (T_h) that is releasing cytokines and sounding the alarm that the body has been invaded.
(D) This is a macrophage that is engulfing bacteria and presenting them on its cell surface so that other immune cells can identify the invader.

5. Which of the following is true about immunity?

(A) T cells produce antibodies.
(B) Another name for antibody is antibiotic.
(C) Juvenile diabetes is an autoimmune disease.
(D) Macrophages are one type of B lymphocyte.

6. Certain danger during a blood transfusion occurs if the recipient's blood has antibodies to the donor's blood antigens. Keeping that guideline in mind, refer to the table below.

Donor Blood Type	Recipient Blood Type	Safe for Transfusion (?)
A	O	Choice A
O	A	Choice B
AB	O	Choice C
O	AB	Choice D

Based on the guideline and the table provided, which of the following statements identifies the choice that is the safest for a transfusion and provides the correct reasoning for why that choice is the safest?

(A) Choice A is safest for a transfusion because type O blood has no antibodies on the surface of its red blood cells to clump the antigens on the surface of the type A blood.
(B) Choice B is safest for a transfusion because type O blood is known as the universal donor.
(C) Choice C is safest for a transfusion because type O blood has no antigens on the surface of its red blood cells to clump the AB antibodies of the donor blood.
(D) Choice D is safest for a transfusion because type O blood has no antigens on the surface of its red blood cells, and type AB blood has no circulating antibodies.

7. Which of the following is an antigen-presenting cell (APC) that is also a part of innate immunity?

(A) B cell
(B) macrophage
(C) helper T cell
(D) cytotoxic T cell

8. Which of the following is responsible for the humoral immune response?

(A) B cell
(B) macrophage
(C) helper T cell
(D) cytotoxic T cell

9. Which of the following is correct about class I MHC molecules?

(A) They are only located on the surface of lymphocytes.
(B) They are only located on the surface of red blood cells.
(C) They are located on the surface of every body cell.
(D) They are only located on the surface of macrophages.

10. Which of the following is a lymphocyte that secretes cytokines to stimulate other lymphocytes?

(A) B cell
(B) macrophage
(C) helper T cell
(D) cytotoxic T cell

11. Which of the following is correct about the immune system?
    (A) Cytotoxic T cells (T_c) attack and kill body cells that have been infected with pathogens.
    (B) A fever is an example of the first line of defense.
    (C) Only the human immune system includes both cell-mediated and humoral specific responses.
    (D) A vaccination generally give permanent immunity against a wide range of pathogens.

12. Which of the following statements explains the significance of clonal selection in the immune system?
    (A) It is the way in which one particular lymphocyte (out of an array of lymphocytes), that matches a specific antigen or epitope, is identified and activated.
    (B) It is the way in which a macrophage or dendritic cell presents antigens to the immune system.
    (C) It is the way in which a cytotoxic T cell selects the correct body cells to attack.
    (D) It is the way in which a macrophage or dendritic cell selects which antigens to engulf and kill.

Question 13

Use the graph below to answer the following question.

A person is exposed twice to the same antigen on Days 1 and 28. The graph tracks the resulting antibody levels.

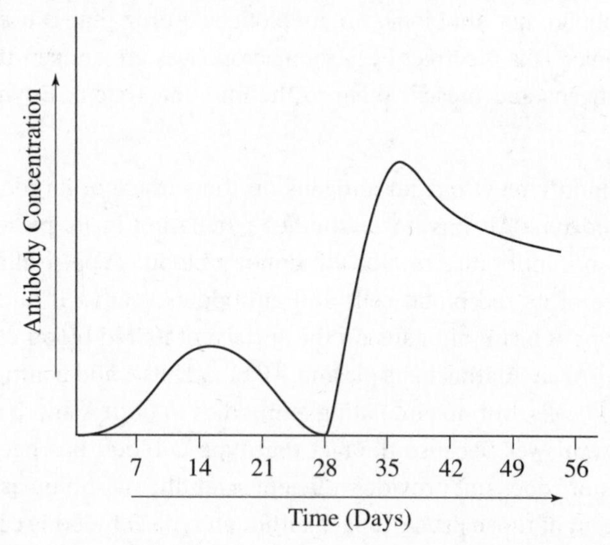

13. What is happening at Day 28?
    (A) The person feels much sicker after the second exposure to the antigen.
    (B) Antibody production increases more slowly for the second exposure, but reaches higher levels due to increased T cells.
    (C) Antibody production is faster and greater after the second exposure to the same antigen due to circulating memory B cells.
    (D) Once the body fights off one infection, it fights off all infections more quickly.

Answers Explained

1. **(D)** Choice D accurately describes the role of histamine in the immune response. Choice A is incorrect because histamine is a part of the innate immune system, not the adaptive one. Histamine triggers vasodilation and the inflammatory response. It does not contain any antimicrobial chemicals, and it does not kill any germs directly. Therefore, choices B and C are also incorrect.

2. **(C)** The second exposure to an antigen produces a more rapid and stronger reaction than the first exposure to that antigen. Circulating memory cells are credited with this more rapid response.

3. **(C)** I is the response to a first exposure to an antigen. II shows the second exposure to the same antigen. As stated in the answer explanation to Question 2, the second exposure to an antigen produces a more rapid and stronger reaction than the first exposure to that antigen. III shows exposure to a new antigen, and the immune response is the same as for any first exposure.

4. **(D)** This is an antigen-presenting cell (APC). The cell is clearly phagocytosing bacteria. It then digests the bacteria in a lysosome and places bits of it (the epitopes) in class II MHC molecules, which migrate to its cell surface. The purpose is to show the immune system what the "enemy" looks like biochemically.

5. **(C)** Juvenile diabetes results when a person's immune system destroys its own pancreatic cells, which normally produce insulin. Choice A is incorrect because antibodies are produced by B cells, not by T cells. Choice B is incorrect because another name for antibody is immunoglobulin, not antibiotic. An antibiotic is a drug that is used to treat bacterial infections. Choice D is incorrect because macrophages are antigen-presenting cells that gobble up antigens and present them to the immune system. B lymphocytes produce antibodies.

6. **(D)** Donor blood type O has no antigens on the surface of its red blood cells, while recipient blood type AB has no antibodies circulating in its plasma. Therefore, the recipient has no antibodies to clot the donor's blood. Type A blood has A antigens on the surface of its red blood cells and antibodies against B (anti-B) circulating in its plasma. Type B has B antigens on the surface of its red blood cells and antibodies against A (anti-A) circulating in its plasma. AB blood has A and B antigens on the surface of its red blood cells, but no circulating antibodies to both A and B antigens. Choice B is not the *best* answer because the fact that type O blood has been classified as the "universal donor" does not provide sufficient scientific reasoning as to why choice B is the safest option of those listed. In fact, although type O blood is called the "universal donor," in reality, it has antibodies circulating in its plasma. Thus, choice D is the only answer option that describes the safest choice for a transfusion and provides the correct reasoning for why that choice is safest.

7. **(B)** A macrophage is a phagocytic cell that gobbles up microbes as part of the innate immune system. As part of the adaptive immune system, the macrophage presents antigen pieces (epitopes) on its surface so that the immune system can recognize the invaders. Choice A, C, and D are incorrect because they are only a part of the adaptive immune system.

8. **(A)** The humoral immune response refers to antibodies. B cells produce antibodies by the millions, while T cells attack microbe-infected body cells directly.

9. **(C)** Class I MHC molecules are located on the surface of every body cell. They identify the type of cell it is (i.e., a liver cell or a kidney cell).

10. **(C)** Helper T cells sound the alarm that microbes have invaded a body. They stimulate other white blood cells by releasing cytokines. Although macrophages secrete cytokines, they are not lymphocytes.

11. **(A)** Choice A describes exactly what cytotoxic T cells do. A fever is an example of the second line of defense, which makes choice B incorrect. Choice C is a false statement. All mammals have both types of specific responses, which supports the important idea that all mammals share a common ancestry. Choice D is not correct because vaccines are specific. They only work against one type of antigen, or one that is very similar to it.

12. **(A)** Clonal selection is a fundamental mechanism in the development of immunity. An antigen is presented (by APCs) to a steady stream of lymphocytes in lymph nodes until a match is found. This triggers events that activate the lymphocyte. Once activated, the lymphocyte divides rapidly, forming an army of clones that develop into effector and memory cells.

13. **(C)** This is the second exposure to the antigen that stimulates the secondary immune response. Antibody production is much more rapid than during the first immune response because antibodies are being produced from circulating memory B cells. T cells do not produce antibodies.

Animal Reproduction and Development 18

- → ASEXUAL REPRODUCTION
- → SEXUAL REPRODUCTION
- → EMBRYONIC DEVELOPMENT
- → FACTORS THAT INFLUENCE EMBRYONIC DEVELOPMENT

INTRODUCTION

From an evolutionary standpoint, reproduction and passing one's genes to the next generation are the goals of life. Nutrition, transport, excretion, and the other life functions are the processes that enable an organism to live long enough to reproduce. This chapter primarily focuses on how animals reproduce and develop, although initially there is a brief review of asexual reproduction. A small section of this chapter covers spermatogenesis and oogenesis; the rest of meiosis was covered in Chapter 4.

The material in this chapter is not required for the AP Biology exam as per the new College Board curriculum. That is why this chapter is placed in Part 3 of this book and why there are no classification references to the curriculum frameworks (such as ENE-1). However, while reviewing this chapter, focus on the timing and coordination of embryonic development, embryonic induction, feedback mechanisms, apoptosis, and homeotic genes, as these topics will provide you with excellent examples to cite when writing your answers to the free-response questions on the exam.

Remember, this chapter consists of illustrative examples. Although these topics are not required for the AP exam, studying them will broaden your understanding of Biology and provide you with examples to use as you write your answers to the free-response questions.

ASEXUAL REPRODUCTION

Asexual reproduction produces offspring that are *genetically identical to the parent.* It has several advantages over sexual reproduction:

- It enables animals that are living in isolation to reproduce without a mate.
- It can create numerous offspring quickly.
- It does not require as much energy as sexual reproduction does, since sexual reproduction involves reproductive systems and hormonal cycles.
- Because offspring are **clones** of the parent (meaning the offspring are genetically identical to the parent), it is advantageous when the environment is stable.
- Reproduction occurs by mitotic cell division.

The following list shows the types of asexual reproduction in sample organisms:

- **Budding** involves the splitting off of new individuals from existing ones. (An example is hydra.)
- **Fragmentation** and **regeneration** occur when a single parent breaks into parts that regenerate into new individuals. (Examples are sponges, planaria, and starfish.)
- **Parthenogenesis** involves the development of an egg without fertilization. The resulting adult is haploid. (Examples are honeybees and whiptail lizards.)

SEXUAL REPRODUCTION

Sexual reproduction involves the fusion of two **haploid gametes (n)** to form a **diploid (2n)** cell called a **zygote**. The animal that develops from the gametes produces gametes by **meiosis**. The female gamete (the **egg** or **ova**) is large and **nonmotile**, while the male gamete (**sperm**) is much smaller and **motile**. Although there are some exceptions, for most animals, reproduction is entirely or primarily sexual.

Sexual reproduction is, energetically speaking, expensive. Animals must maintain many structures and systems related to producing offspring, including:

- Separate and elaborate male and female reproductive organs
- Endocrine glands and complex hormonal cycles
- Systems for timing reproductive cycles around environmental changes

Despite its complexity, sexual reproduction has one major advantage over asexual reproduction: **variation**. Each offspring is the product of both parents and may be better able to survive than either parent, especially in an environment that is changing.

The Human Male Reproductive System

- The **epididymis** is the tube in the testes where sperm gain motility.
- **Leydig cells** are clusters of cells, located between seminiferous tubules, that produce testosterone.
- The **prostate gland** is the large gland that secretes **semen** directly into the urethra.
- The **scrotum** is the sac outside the abdominal cavity that holds the testes. The cooler temperature there enables sperm to survive.
- **Seminal vesicles** secrete mucus, fructose sugar (which provides energy for the sperm), and the hormone **prostaglandin** (which stimulates uterine contractions) during sexual intercourse.
- **Seminiferous tubules** are the site of sperm formation in the testes.
- **Sertoli cells** provide nutrients for developing sperm.
- **Testes** (*testis*, singular) are the male gonads, where sperm are produced.
- The **urethra** is the tube that carries semen and urine.
- The **vas deferens** is the muscular duct that carries sperm during ejaculation from the epididymis to the urethra in the penis.
- See Figure 18.1.

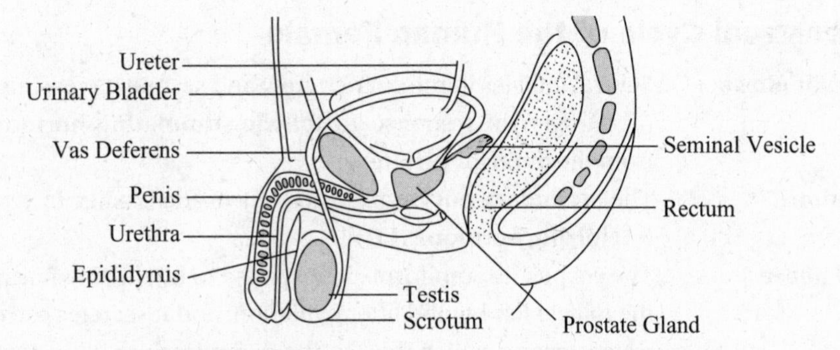

Figure 18.1 The Human Male Reproductive System

The Human Female Reproductive System

- The **ovaries** are where meiosis occurs.
- The **oviducts** or **fallopian tubes** are where fertilization occurs. After ovulation, the egg moves through the oviduct to the uterus.
- The **uterus** is where the **blastocyst** will implant and where the embryo will develop during the nine-month **gestation** if fertilization occurs.
- The **endometrium** is the lining of the uterus that thickens monthly in preparation for implantation of the blastocyst.
- The **vagina** is the birth canal. During labor and delivery, the baby passes through the **cervix** (the mouth of the uterus) and into the vagina.
- See Figure 18.2.

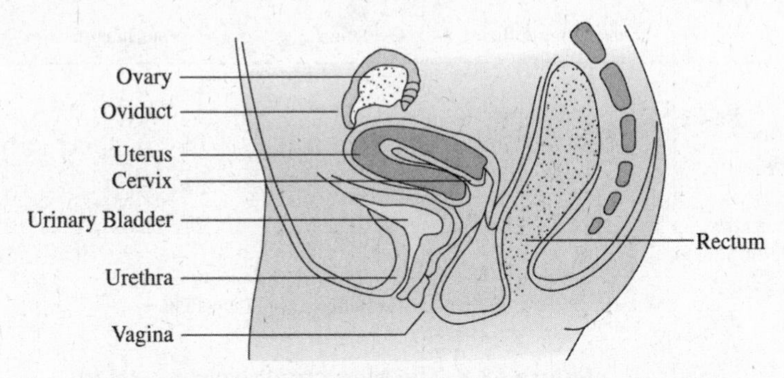

Figure 18.2 The Human Female Reproductive System

The Menstrual Cycle of the Human Female

Follicular phase	Several follicles in the ovaries grow and secrete increasing amounts of estrogen in response to **follicle-stimulating hormone (FSH)** from the **anterior pituitary**.
Ovulation	The **secondary oocyte** ruptures out of the ovaries in response to **luteinizing hormone (LH)**.
Luteal phase	The **corpus luteum** forms in response to luteinizing hormone. It is the follicle left behind after ovulation, and it secretes **estrogen** and **progesterone**, which thicken the **endometrium** of the **uterus**.
Menstruation	This is the monthly shedding of the lining of the uterus when implantation of an embryo does not occur.

Figure 18.3 illustrates the menstrual cycle, and Figure 18.4 illustrates the hormonal control of the menstrual cycle.

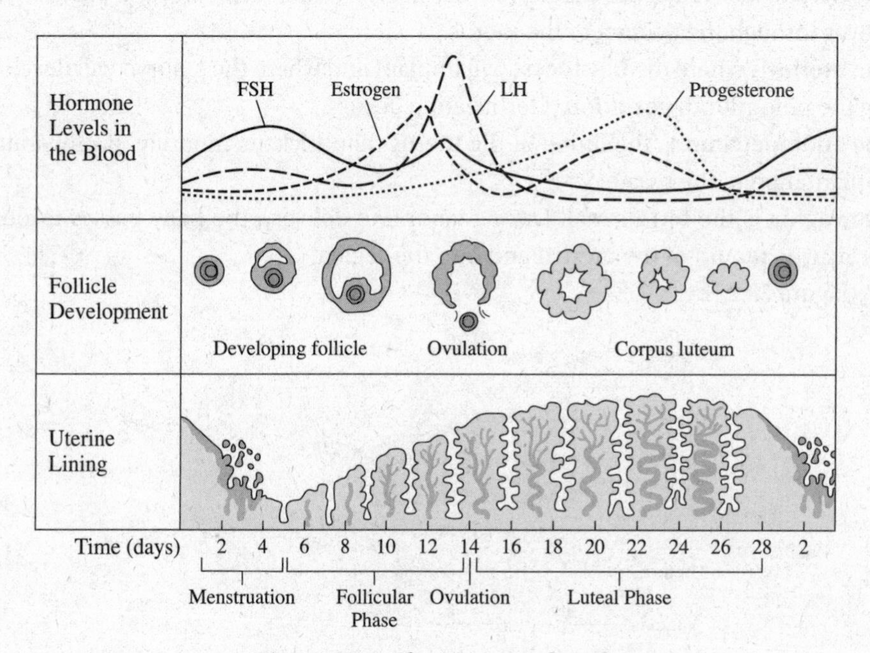

Figure 18.3 The Menstrual Cycle

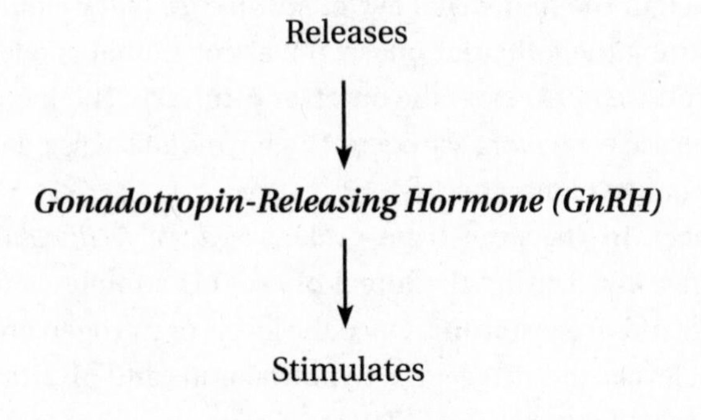

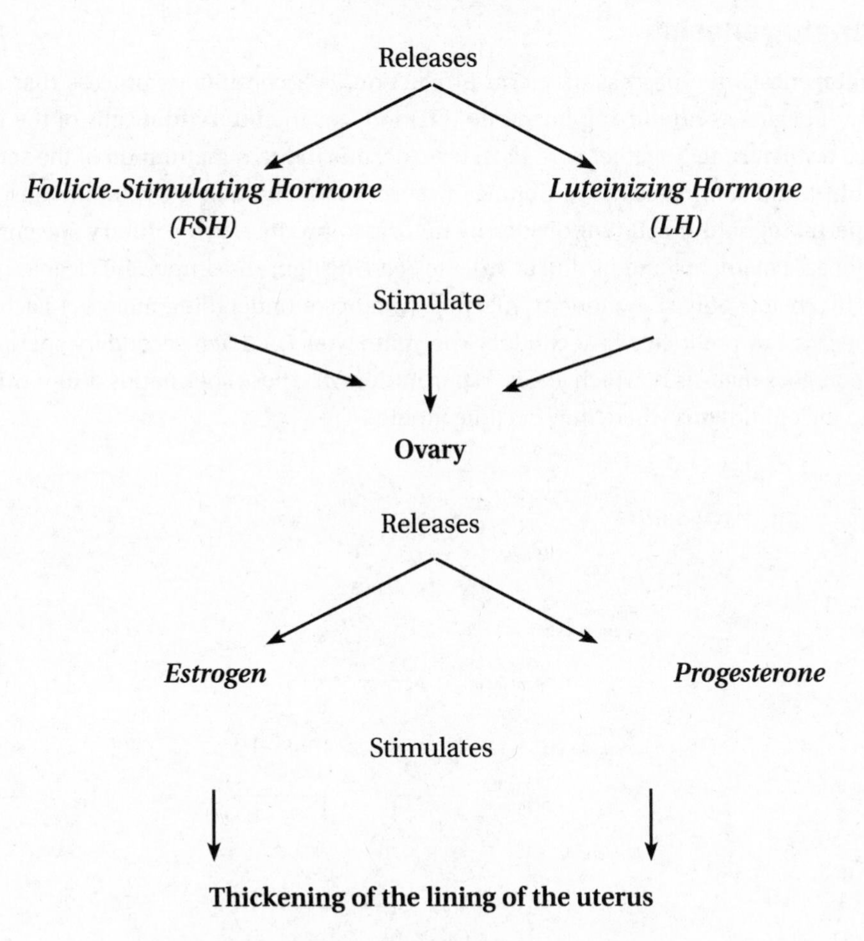

Figure 18.4 Hormonal Control of the Menstrual Cycle

STUDY TIP

Endocrine signaling molecules are specific and can travel long distances through blood to reach distant body parts.

Positive and Negative Feedback of the Menstrual Cycle

- **Positive feedback in the menstrual cycle**: *Positive feedback enhances a process until it is completed.* During the **follicular phase**, the estrogen that is released from the follicle stimulates the release of **LH** from the **anterior pituitary**. The increase in LH stimulates the follicle to release *even more* estrogen. The hormone levels continue to rise until the follicle matures and ovulation occurs.
- **Negative feedback in the menstrual cycle**: *Negative feedback stops a process once homeostasis is reached.* During the **luteal phase**, **LH** stimulates the **corpus luteum** to secrete **estrogen** and **progesterone**. Once the levels of estrogen and progesterone reach sufficiently high levels, they trigger the **hypothalamus** and **pituitary** to *shut off*, thereby **inhibiting** the secretion of LH and FSH.

REMEMBER

Organisms use feedback mechanisms to maintain their internal environment and respond to external environmental cues.

Spermatogenesis

Spermatogenesis, the process of sperm production, is a continuous process that starts at puberty. It begins as **luteinizing hormone** (**LH**) induces the **interstitial cells** of the testes to produce **testosterone**. Together with **FSH**, **testosterone** induces maturation of the **seminiferous tubules** and stimulates the beginning of sperm production. In the seminiferous tubules, each **spermatogonium cell** ($2n$) divides by mitosis to produce two **primary spermatocytes** ($2n$). (For ease of understanding and in order to keep the figure as simple and clear as possible, Figure 18.5 below only shows one primary spermatocyte undergoing meiosis.) Each undergoes meiosis I to produce two **secondary spermatocytes** (n). Each secondary spermatocyte then undergoes meiosis II, which yields 4 **spermatids** (n). These spermatids **differentiate** and move to the **epididymis** where they become motile.

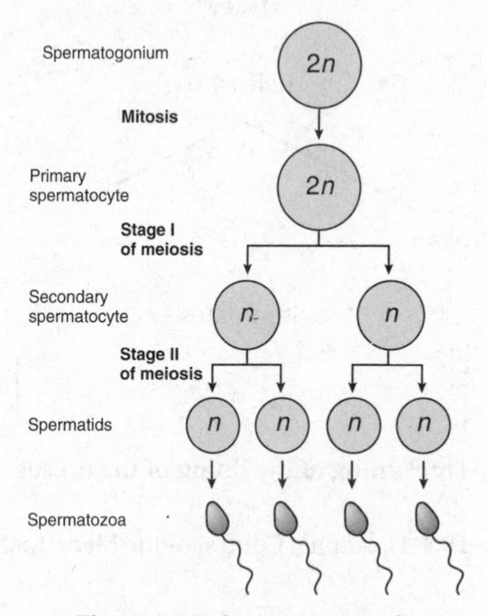

Figure 18.5 Spermatogenesis

Oogenesis

Oogenesis, the production of ova, begins prior to birth. Within the embryo, an **oogonium cell** ($2n$) undergoes *mitosis* to produce **primary oocytes** ($2n$). These remain quiescent within small follicles in the ovaries until puberty, when they become reactivated by hormones. **FSH** periodically stimulates the follicles to complete meiosis I, producing **secondary oocytes** (n), which are released at ovulation. Meiosis II then stops again and does not continue until **fertilization,** when a sperm penetrates the **secondary oocyte**. See Figure 18.6. Oogenesis differs from sperm formation in three ways. First, it is a **stop-start process**. It begins prior to birth and is completed after fertilization. Second, cytokinesis divides the cytoplasm of the cell unequally, producing one large cell and *two small polar bodies* which will disintegrate. Third, one primary oogonium cell produces only *one active egg cell.*

STUDY TIP

You may wish to review "Meiosis" in Chapter 4.

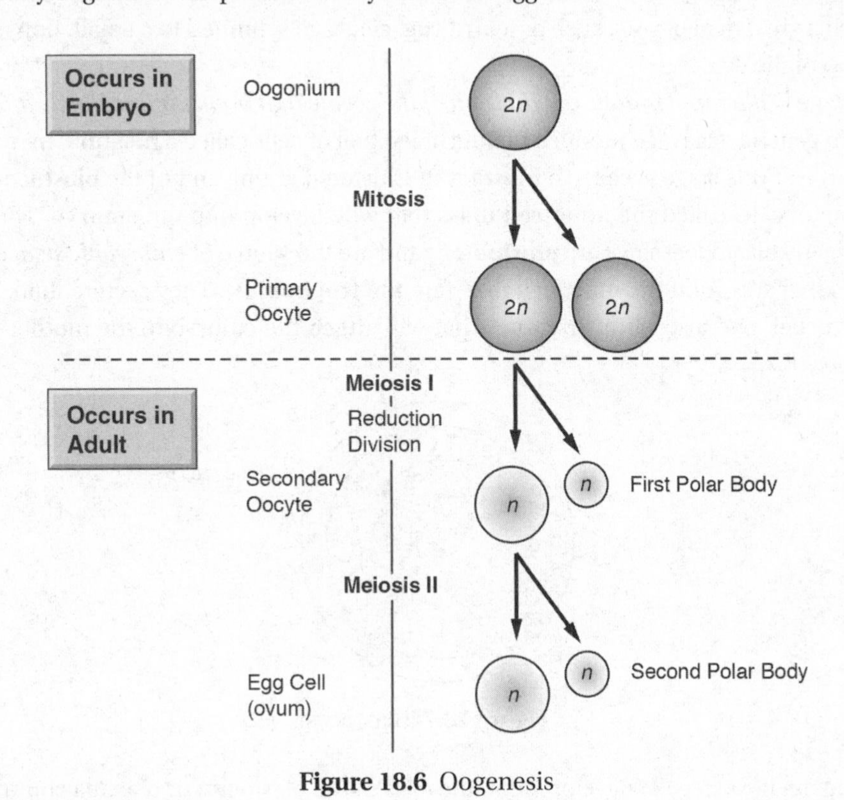

Figure 18.6 Oogenesis

Fertilization

Fertilization, the fusion of sperm and ovum nuclei, is a complex process. It begins with the **acrosome reaction**, when the head of the sperm, the **acrosome**, releases hydrolytic enzymes that penetrate the *jelly coat* of the egg. Specific molecules from the sperm bind with ***receptor molecules*** on the *vitelline membrane* before the sperm come in contact with the ovum's plasma membrane. This *specific recognition* ensures that the egg will be fertilized by only sperm from the same species. Once a sperm binds to receptors on the egg, the membrane is dramatically depolarized and no other sperm can penetrate the egg membrane. This change in the membrane is known as the *fast block to polyspermy.* This fast block lasts only a minute, just long enough to allow the *slow block to polyspermy.* The slow block converts the vitelline membrane into an impenetrable *fertilization envelope.*

Although the fusion of the sperm and egg triggers this activation, an unfertilized egg can be activated artificially by electrical stimulation or by injection with Ca^{++}. The development of an unfertilized egg is called **parthenogenesis**. The adult that results is haploid. **Drone honeybees** develop by natural parthenogenesis from unfertilized eggs and are haploid males.

EMBRYONIC DEVELOPMENT

Embryonic development is characterized by exquisite timing and coordination of many different events. Many genes are silenced by microRNAs (miRNA), while others are "turned on."

Programmed cell death (apoptosis) removes cells that would impede the proper development of others.

Embryonic development consists of three stages: **cleavage**, **gastrulation**, and **organogenesis**. The stages that are discussed below generally describe the development of all animal eggs but are typical of the **sea urchin** that has almost no yolk. In eggs with more yolk, such as those of the **frog**, cleavage is unequal, with very little cell division in the yolky region. In eggs with a great deal of yolk, such as a **bird egg**, cleavage is limited to a small, nonyolky disc at the top of the egg.

REMEMBER

Differentiation results from silencing specific genes.

Cleavage *is the rapid mitotic cell division of the zygote that occurs immediately after fertilization.* In general, cleavage produces a fluid-filled ball of cells called a **blastula**. In mammals, the embryo at this stage is called a **blastocyst**. Clustered at one end of the **blastocoel cavity** is a group of cells called the **inner cell mass** that will develop into the **embryo**. The cells of the very early blastocoel stage are **pluripotent** and are the source of *embryonic stem cell lines.* The cells that surround the inner cell mass are the **trophoblast**. They secrete fluid, creating the **blastocoel**, and also form structures that will attach the embryo to the mother's uterus. See Figure 18.7.

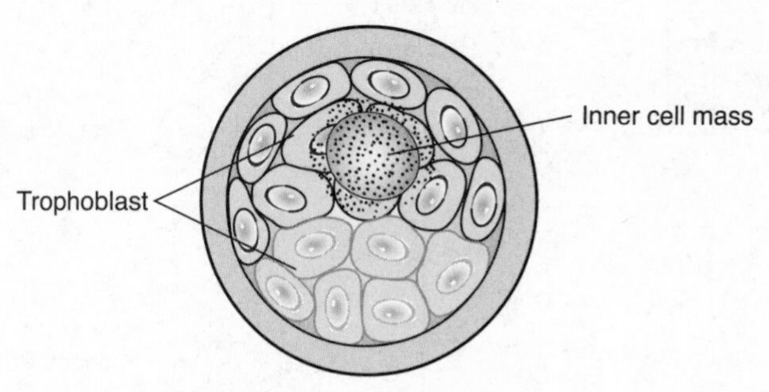

Figure 18.7 Blastocyst

During the next stage of development, the cells of the blastocyst or blastula communicate with each other and begin to differentiate. In many animals, the movement of the cells is so regular that it is possible to label a specific **blastomere** (individual cells of a blastocyst) and identify the tissue that results as embryonic development proceeds. Labeling these cells produces a fate map.

Gastrulation is a process that involves rearrangement of the blastula or blastocyst and begins with the formation of the **blastopore**, an opening into the blastula. In some animals, the blastopore becomes the mouth; in other animals (deuterostomes), the blastopore becomes the anus. Some of the cells on the surface of the embryo migrate into the blastopore to form a new cavity called the **archenteron** or primitive gut. As a result of this cell movement, gastrulation forms a **three-layered** embryo called a **gastrula**. These three differentiated layers, called **embryonic germ layers**, are the **ectoderm**, **endoderm**, and **mesoderm**. They will develop into all the parts of the adult animal.

- The **ectoderm** will become the **skin** and the **nervous system**.
- The **endoderm** will form the **viscera** including the lungs, liver, and digestive organs, and so on.
- The **mesoderm** will give rise to the **muscle**, **blood**, and **bones**. Some primitive animals (sponges and cnidarians) develop a noncellular layer, the **mesoglea**, instead of the mesoderm. See Figure 18.8.

Organogenesis is organ building. It is the process by which cells continue to **differentiate**, producing organs from the three embryonic germ layers. Three kinds of morphogenetic changes—folds, splits, and dense clustering called condensation—are the first evidence of organ building. Once all the organ systems have been developed, the embryo simply increases in size.

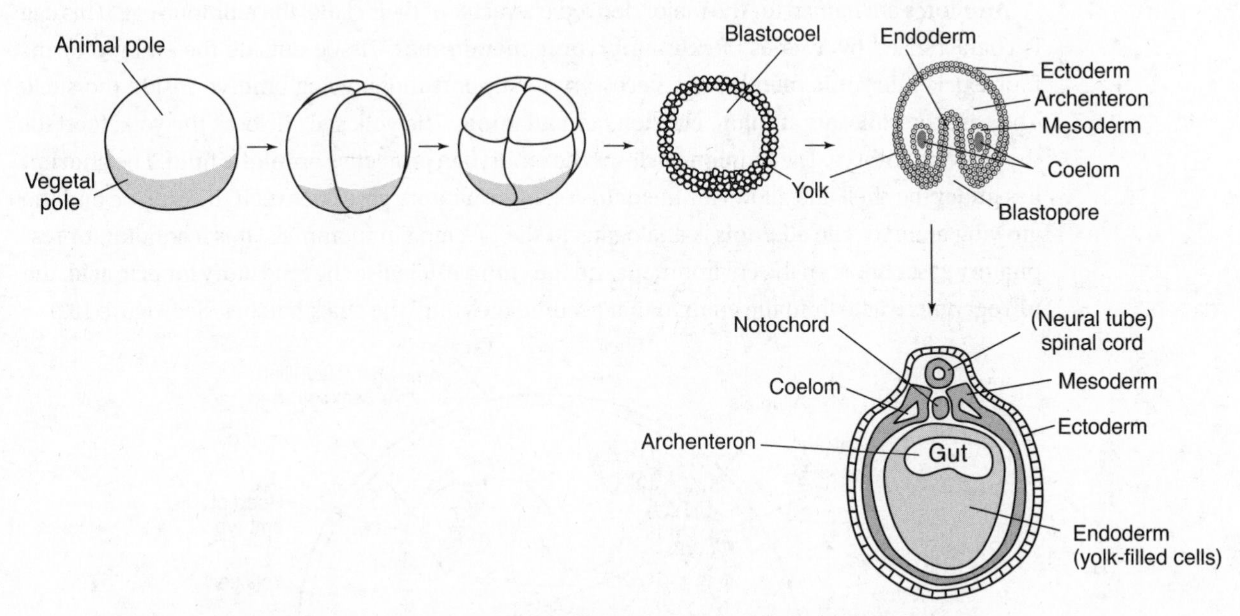

Figure 18.8 Cleavage, Gastrulation, and Differentiation in a Developing Embryo

The Frog Embryo

Fertilization: One third of the frog egg is yolk, which is massed in the lower portion of the egg. This yolky portion of the egg is called the **vegetal pole**. The top half of the egg is called the **animal pole** and has a **pigmented cap**. The animal-vegetal asymmetry dictates where the anterior-posterior axis forms in the embryo. Eggs are laid directly into water and fertilization is external. When the sperm penetrates the egg, the pigmented cap rotates toward the point of penetration and a **gray crescent** appears on the side opposite the point of entry of the sperm. The gray crescent is a marker of the future dorsal side and is critical to normal development of the growing embryo. (See "The Gray Crescent" on page 397.)

STUDY TIP

Timing and coordination of specific events are necessary for normal development, and are highly regulated.

Cleavage and gastrulation: Because of the presence of yolk, cleavage is uneven. The **blastopore** forms at the border of the gray crescent and the *vegetal pole.* Cells at the **dorsal lip** above (dorsal to) the blastopore begin to stream over the dorsal lip and into the blastopore in a process called *involution.* As these ectoderm cells stream inward by what is called *epibolic movement,* the blastocoel disappears and is replaced by another cavity called the **archenteron**. The region of mesoderm lining the archenteron that formed opposite the blastopore is called the **dorsal mesoderm**.

Organogenesis: In chordates, the organs that form first are the **notochord**, the skeletal rod that is characteristic of all chordate embryos, and the **neural tube**, which will become the **central nervous system**. The neural tube forms from the **dorsal ectoderm** just above the notochord. Both form by **embryonic induction** (see "Embryonic Induction" on page 397). After the blueprints of the organs are laid down, the embryo develops into a larval stage: the tadpole. Later, **metamorphosis** will transform the tadpole into a frog.

The Bird Embryo

Cleavage and gastrulation: The bird's egg has so much yolk that development of the embryo occurs in a flat disc or **blastodisc** that sits on top of the yolk. A **primitive streak** forms instead of a gray crescent. Cells migrate over the primitive streak and flow inward to form the **archenteron**. As cleavage and gastrulation occur, the yolk gets smaller.

Amniotes are named for the major derived character of their clade: the amniotic egg. This egg is characterized by a series of **extraembryonic membranes**: Tissue outside the embryo forms four **extraembryonic membranes** necessary to support the growing embryo inside the shell. They are the **yolk sac**, **amnion**, **chorion**, and **allantois**. The **yolk sac** encloses the yolk, food for the growing embryo. The **amnion** encloses the embryo in protective **amniotic fluid**. The **chorion** lies under the shell and allows for the diffusion of respiratory gases between the outside and the growing embryo. The **allantois** is analogous to the placenta in mammals. It is a conduit for respiratory gases between the environment and the embryo. It is also the repository for **uric acid**, the **nitrogenous waste** from the embryo that accumulates until the chick hatches. See Figure 18.9.

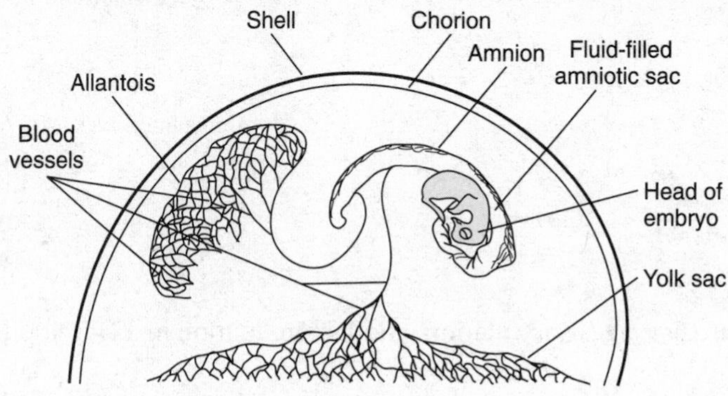

Figure 18.9 The Bird Embryo

FACTORS THAT INFLUENCE EMBRYONIC DEVELOPMENT

Cytoplasmic Determinants

NOTE

This topic is important because it's about cell communication.

When an eight-ball sea urchin embryo is separated into two halves, the future development of the two halves depends on the plane in which they are cut. If the dissection is longitudinal (A), producing embryos containing cells from both animal and vegetal poles, subsequent development is normal. If, however, the plane of dissection is horizontal (B), the result is four abnormally developing embryos. *This demonstrates that the cytoplasm surrounding the nucleus has profound effects on embryonic development.* The importance of the cytoplasm in the development of the embryo is known as **cytoplasmic determinants**; see Figure 18.10.

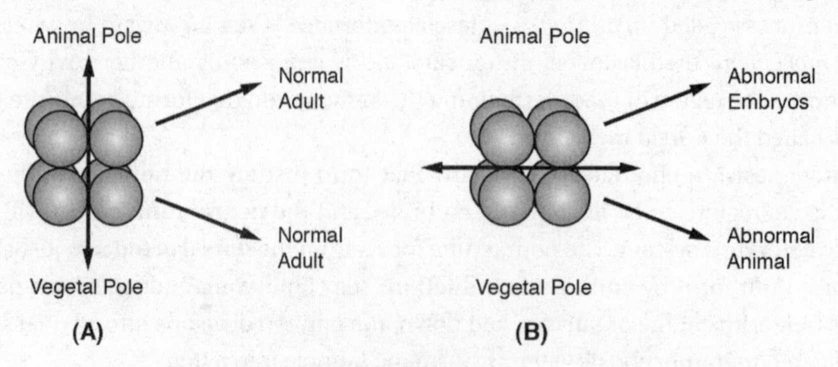

Figure 18.10 The Cytoplasm that Surrounds the Nucleus Affects Embryonic Development

The Gray Crescent

Hans Spemann, in his now-famous experiment, demonstrated the importance of the **cytoplasm** associated with the **gray crescent** in the normal development of an animal. He bisected embryos very early in their development in different ways. Only the cell that contained the gray crescent developed normally. In addition, these experiments provided more proof that the cytoplasm plays a major role in determining the course of embryonic development; see Figure 18.11.

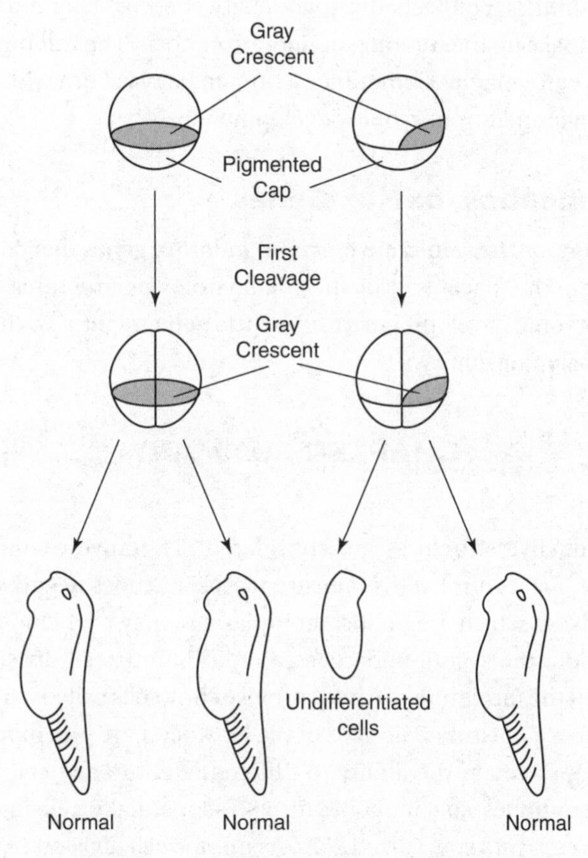

Figure 18.11 Distribution of the Gray Crescent Affects the Development of Embryos

Embryonic Induction

Embryonic induction is the ability of one group of embryonic cells to influence the development of another group of embryonic cells. Spemann proved that the **dorsal lip** of the blastopore normally initiates a chain of inductions that results in the formation of a **neural tube.** In the now-famous experiment, he grafted a piece of **dorsal lip** from one amphibian embryo onto the ventral side of a second amphibian embryo. What developed on the recipient was a complete secondary embryo attached at the site of the graft. *The dorsal lip induced the abdomen tissue above it to become neural tissue.* Because it plays a crucial role in development, Spemann named the dorsal lip the **primary embryonic organizer** or simply the organizer. After many years of study, scientists have identified that the protein β-catenin is a likely candidate for the transcription factor that triggers which cells become the organizers.

REMEMBER

Embryonic induction in development results in the correct timing of events.

Apoptosis

STUDY TIP

Programmed cell death plays an important role in normal development and differentiation.

Apoptosis is programmed cell death. (See also page 53.) We have seen that it is an integral part of the immune system. It is also crucial to the normal development of embryos. For instance, many more neurons are produced during development of the vertebrate nervous system compared to the number of neurons that exist in the adult. That is because neurons survive if they make connections with other neurons during development and self-destruct if they do not.

During early embryonic development, webbing, a remnant of our shared evolutionary history with other animals, connects the spaces between fingers and toes. As development proceeds, cells that make up the webbing undergo apoptosis. The webbing is eliminated, and the fingers and toes can separate. Sometimes a human baby is born with some residual webbing, the result of incomplete embryonic development.

Homeotic, Homeobox, or Hox Genes

Homeotic, **homeobox**, or **Hox genes** are **master regulatory genes** that control the expression of genes that regulate the placement of specific anatomical structures. They play a critical role in normal embryonic development. A **homeotic gene** might give the instruction "place legs here" in the developing embryo.

CHAPTER SUMMARY

- Human reproductive structures are considered illustrative examples for AP Biology. However, hormonal control of the menstrual cycle includes **negative** and **positive feedback mechanisms**, which are important topics and may be included on the AP exam.
- After fertilization, the zygote undergoes a rapid mitotic cell division called **cleavage**, which results in the formation of a hollow ball of tiny cells called a **blastocyst**, the mammalian version of a **blastula**. The cells of the blastocyst are **pluripotent** or **omnipotent**, meaning that they retain the ability to differentiate into any cell type. As embryonic development continues, specific genes turn off sequentially, causing the **differentiation** of omnipotent cells into specialized cells. As mitotic cell division continues, the process of **gastrulation** leads to the development of three embryonic layers: **ectoderm**, **mesoderm**, and **endoderm**. They will each develop into separate body structures: nerves and skin, blood and bones, and digestive organs, respectively.
- **Programmed cell death** (**apoptosis**) plays an important role in embryonic development. Certain cells must be removed in order for others to develop properly. An example is the reabsorption of a tadpole's tail during **morphogenesis**. The tail is resorbed, and legs develop. The topic of apoptosis is also important in the study of Biology because it connects with two other important topics: *cell signaling* and *gene regulation*.
- Classic experiments beginning in 1924 proved that the nucleus does not exert sole control over embryonic development. The cytoplasm also plays an important role. Hans Spemann established that the **dorsal lip** from an amphibian embryo is an **organizer** and can *induce* or influence the development of another group of embryonic cells with which it comes in contact. This process is known as **embryonic induction**.
- **Homeotic** or **Hox genes** are master regulatory genes that control the expression of genes that regulate the placement of specific anatomical structures. To demonstrate the role of homeotic genes, scientists manipulated them and developed a fruit fly with legs coming out of its head.

PRACTICE QUESTIONS

1. Which of the following gives rise to the lining of the digestive tract?
   (A) ectoderm
   (B) endoderm
   (C) mesoderm
   (D) blastopore

2. Which of the following gives rise to the brain and the eyes?
   (A) ectoderm
   (B) endoderm
   (C) mesoderm
   (D) blastopore

3. Which of the following gives rise to the blood?
   (A) ectoderm
   (B) endoderm
   (C) mesoderm
   (D) blastopore

4. Which of the following gives rise to the bones?
   (A) ectoderm
   (B) endoderm
   (C) mesoderm
   (D) blastopore

5. Refer to the following diagram to help you answer this question.

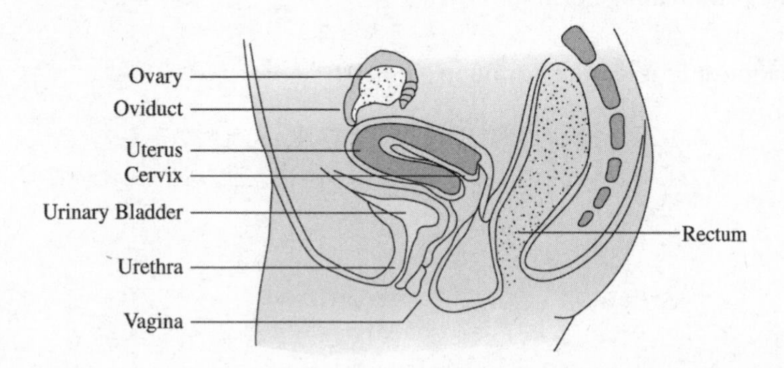

In human females, fertilization normally occurs in the ______ and implantation occurs in the ______.

   (A) ovary; uterus
   (B) fallopian tubes; uterus
   (C) ovary; oviduct
   (D) oviduct; vagina

6. Which of the following causes labor?
   (A) follicle-stimulating hormone (FSH)
   (B) oxytocin
   (C) gonadotropin-releasing hormone (GnRH)
   (D) estrogen

7. Which of the following is released by the hypothalamus and stimulates the anterior pituitary?
   (A) follicle-stimulating hormone (FSH)
   (B) oxytocin
   (C) gonadotropin-releasing hormone (GnRH)
   (D) estrogen

8. Which of the following stimulates the ovary to mature a secondary oocyte?
   (A) follicle-stimulating hormone (FSH)
   (B) oxytocin
   (C) gonadotropin-releasing hormone (GnRH)
   (D) estrogen

9. Which of the following is responsible for thickening the endometrial lining of the uterus?
   (A) follicle-stimulating hormone (FSH)
   (B) oxytocin
   (C) gonadotropin-releasing hormone (GnRH)
   (D) estrogen

10. Which of the following stimulates sperm production?
    (A) follicle-stimulating hormone (FSH)
    (B) oxytocin
    (C) gonadotropin-releasing hormone (GnRH)
    (D) estrogen

Questions 11–12

The following two questions refer to the information and table below.

Alfred Jost investigated the role of sex chromosomes, XY or XX, in sex determination in mammals. Here is Jost's procedure:

Jost surgically removed 50 rabbit embryos from 26 pregnant rabbits and excised the part of the genital structures that would become gonads (testes or ovaries). After the surgery, he placed the embryos back inside each mother's uterus, allowing them to develop to term. When the rabbit babies were born, he recorded their chromosomal sex as well as whether their genital structures developed as male or female. He also compared the results to a control group of rabbit babies that were born from pregnant rabbits that did not undergo any surgery.

Here are the results of his experiment:

	Appearance of Genitalia in Rabbit Babies	
Sex Chromosomes Present in Offspring	No Surgery (Control Group)	Embryonic Gonad Removed
XY	Male	Female
XX	Female	Female

11. Which of the following statements best explains what Jost's experiment demonstrated?
    (A) In the absence of a signal from male gonads, all embryos develop as male, regardless of whether they are genetically male (XY) or female (XX).
    (B) In the absence of a signal from male gonads, all embryos develop as female, regardless of whether they are genetically male (XY) or female (XX).
    (C) In the absence of a signal from female gonads, all embryos develop as female, regardless of whether they are genetically male (XY) or female (XX).
    (D) Determination of the sex of an embryo is controlled by genotype only: male (XY) or female (XX).

12. The results of Jost's experiment could be explained if some aspect of surgery, *other than gonad removal*, caused female genitalia to develop. If you were to repeat Jost's experiment, how would you improve the protocol to test this alternative hypothesis?
    (A) Remove only the gonads that will develop into testes from the embryos of the experimental group.
    (B) Repeat the experiment as is several times.
    (C) Carry out sham surgeries on the control group; remove the embryos from the mothers, but do not remove any genital structures.
    (D) Use more animals for both the experimental group and the control group.

13. Below is an illustration that shows the hormone levels in the blood, the follicle development, and the uterine lining during the menstrual cycle.

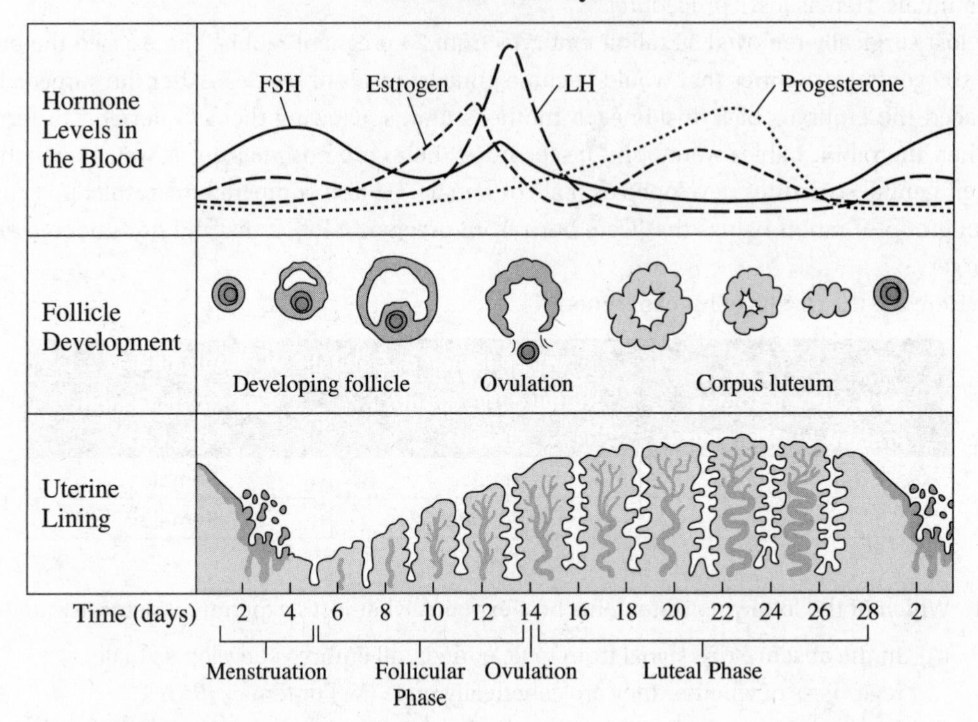

Which of the following correctly describes the role of luteinizing hormone (LH)?

(A) LH initiates the shedding of the uterine lining.
(B) LH stimulates the development of the follicle within the ovaries.
(C) LH initiates the ovulation of the secondary oocyte out of the ovaries.
(D) LH is responsible for the building up of the uterine lining.

14. Refer to the two diagrams below.

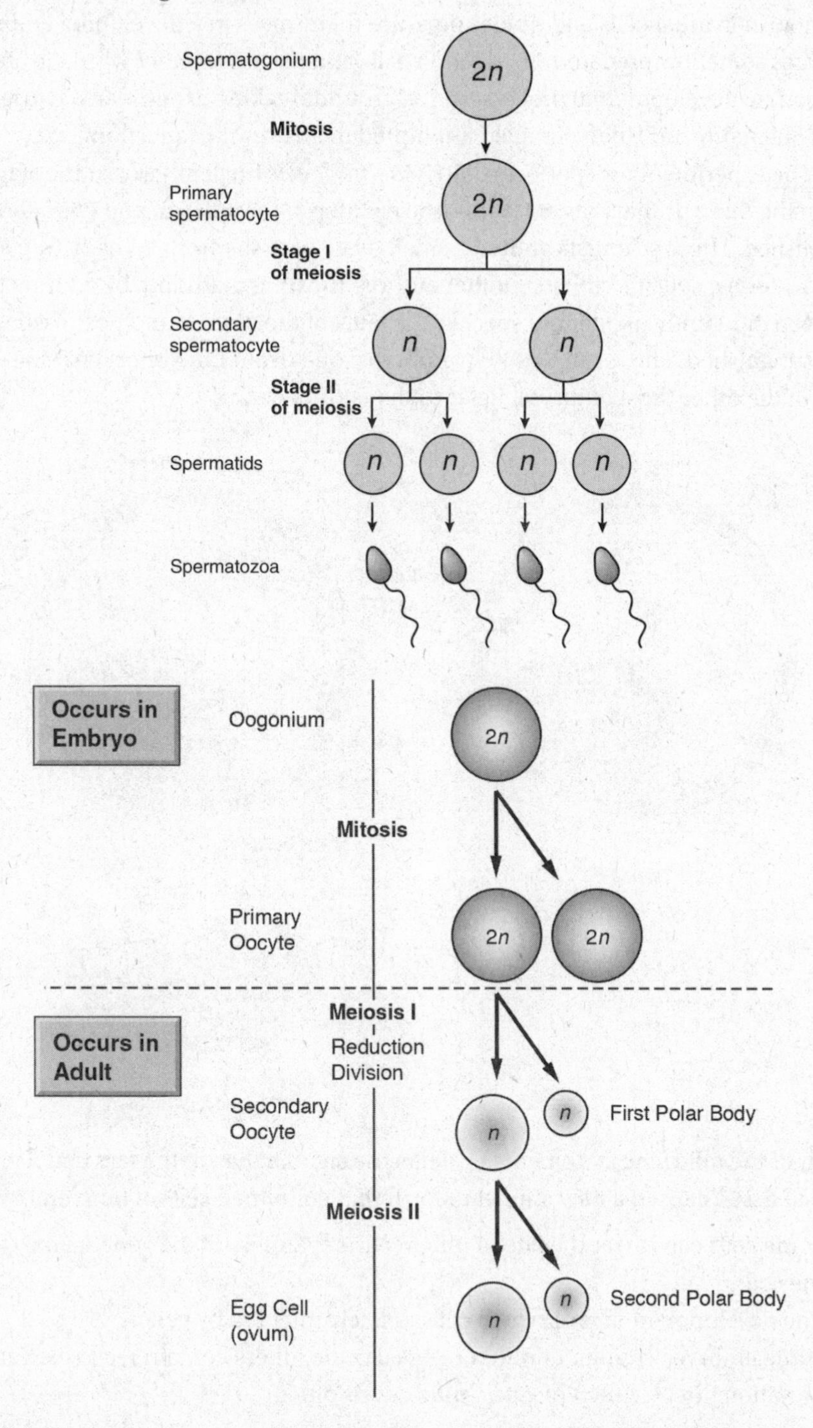

Which of the following is correct about meiosis and the production of gametes in human males and females?

(A) Egg production in females only involves mitotic cell division.
(B) In females, one primary oogonium cell produces four active egg cells.
(C) In females, one primary oogonium cell produces only one active egg cell.
(D) In males, sperm production involves three sequential meiotic cell divisions.

15. A important focus of developmental biology is the timing and mechanism of when and how individual cells and tissues organize themselves in a developing embryo. Are cell fates somehow predetermined, or do cells and tissues interact with one another to orchestrate developmental processes? The groundbreaking experiments carried out by Hans Spemann and Hilde Mangold attempted to answer this question.

    The experiment was performed in *Xenopus laevis* frog embryos at the stage during which the three primary germ layers—the ectoderm, mesoderm, and endoderm—are established. These scientists grafted a small piece of tissue from an early frog embryo (donor) onto a specific site on another embryo (host). In addition, by grafting tissue between differently pigmented species, the fates of the grafted and host tissues could be distinguished. The result was twin embryos: one (from the donor) growing out of the belly of the other (host embryo). Refer to the figure below.

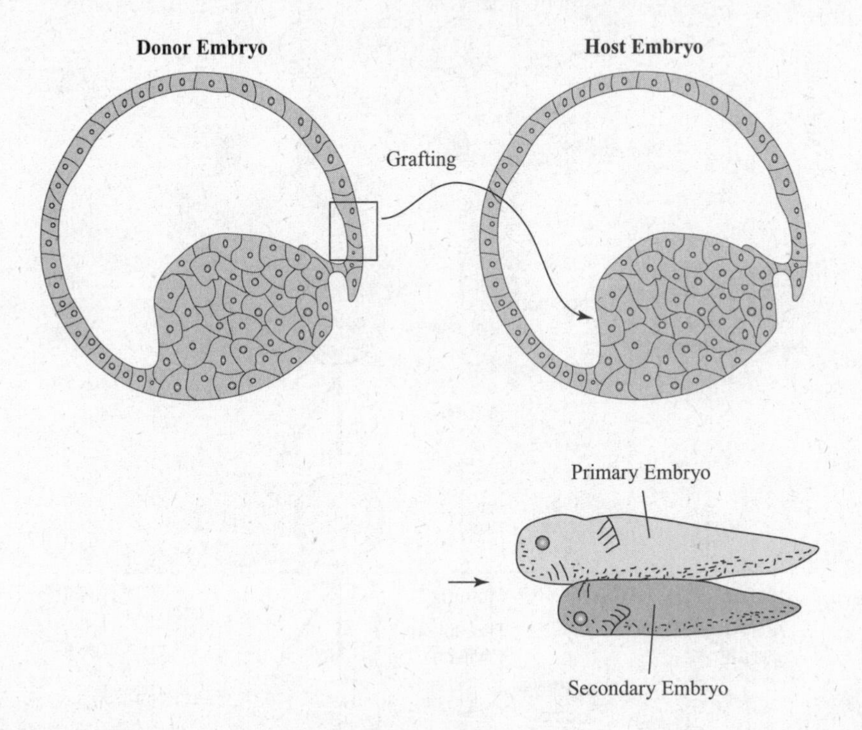

    Which of the following statements explains the significance of the fact that a grafted piece of tissue caused a new animal to form that contained cells of both embryos?

    (A) Some cells can direct the fate of other cells, regardless of the genetic makeup of the cells.
    (B) The development of embryonic cells is solely directed by genes.
    (C) Some embryos contain dominant genes, while others contain recessive genes.
    (D) Mutation directs much of embryonic development.

16. Hans Spemann also completed another set of experiments. Using microdissection tools, he bisected identical embryos very early in their development when a crescent of darkly pigmented cells became visible. The following diagram shows the two different ways he bisected two embryos and the results of the two surgeries.

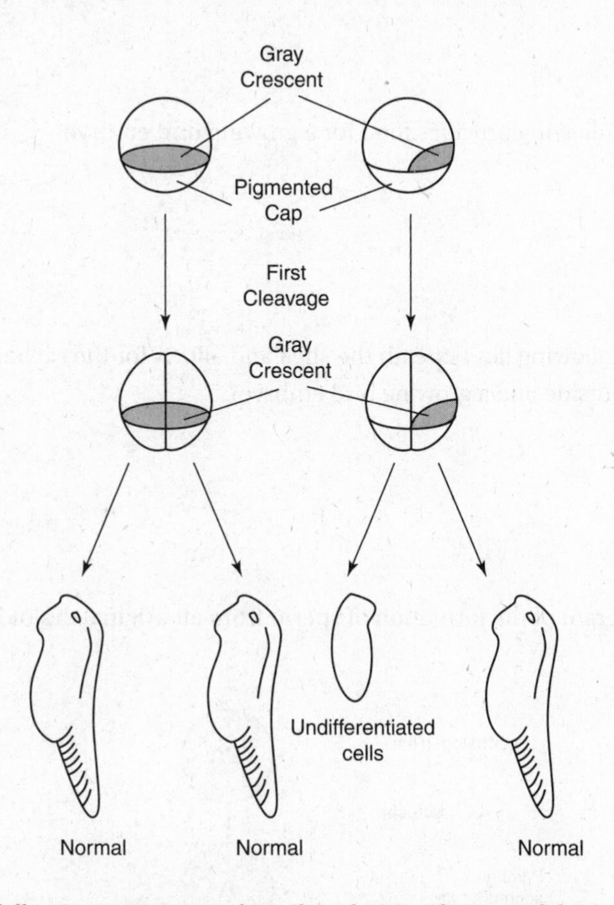

Which of the following statements describes the significance of the results of this experiment?

(A) Only embryos that contain a complete gray crescent can develop normally.
(B) The gray crescent is not necessary for normal embryonic development.
(C) Something in some cells of the embryo controls the fate of the development of the entire animal.
(D) The cells of the gray crescent contain special genes that are different from those in the rest of the embryo.

17. Which of the following is analogous to the placenta in mammals and is the repository for uric acid for bird embryos?

(A) yolk sac
(B) allantois
(C) amnion
(D) chorion

18. Which of the following provides food for a growing bird embryo?

(A) yolk sac
(B) allantois
(C) amnion
(D) chorion

19. Which of the following lies beneath the shell and allows for the exchange of O_2 and CO_2 between the outside and a growing bird embryo?

(A) yolk sac
(B) allantois
(C) amnion
(D) chorion

20. Below is a diagram of the formation of sperm from an original diploid spermatogonium cell.

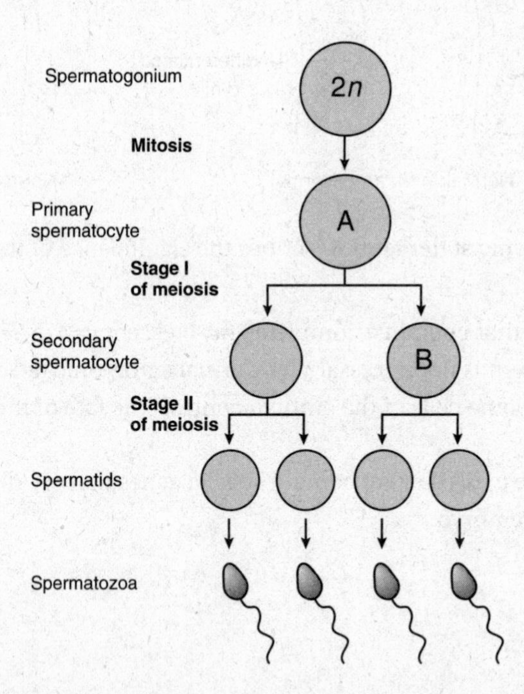

Which of the following statements is correct about the chromosome number of each cell?

(A) Cell A and cell B are both diploid.
(B) Cell A and cell B are both haploid.
(C) Cell A is haploid, and cell B is diploid.
(D) Cell A is diploid, and cell B is haploid.

21. Amniotes are named for the major derived character of their clade: the amniotic egg. This egg is characterized by a series of extraembryonic membranes that support the growing embryo inside the shell. These four extraembryonic membranes are the yolk sac, allantois, amnion, and chorion. Below is a diagram of a developing bird embryo.

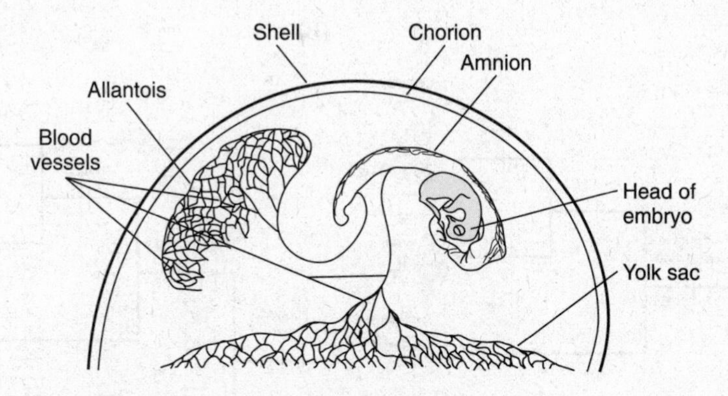

Which of the following is correct about the amnion?

(A) It lies under the shell and is the site of diffusion of respiratory gases between the outside and the growing embryo.
(B) It encloses the food for the growing embryo.
(C) It encloses the embryo in protective amniotic fluid.
(D) It is analogous to the placenta in humans and is a repository for nitrogenous waste.

22. The amniotic egg was an evolutionary innovation for terrestrial animals more than 300 million years ago. It allowed the embryo to develop on land without depending on an aquatic environment for reproduction. Based on the presence of an amnion, which of the following cladograms correctly shows the placement and relationship among four animals: bass, frog, turtle, and leopard?

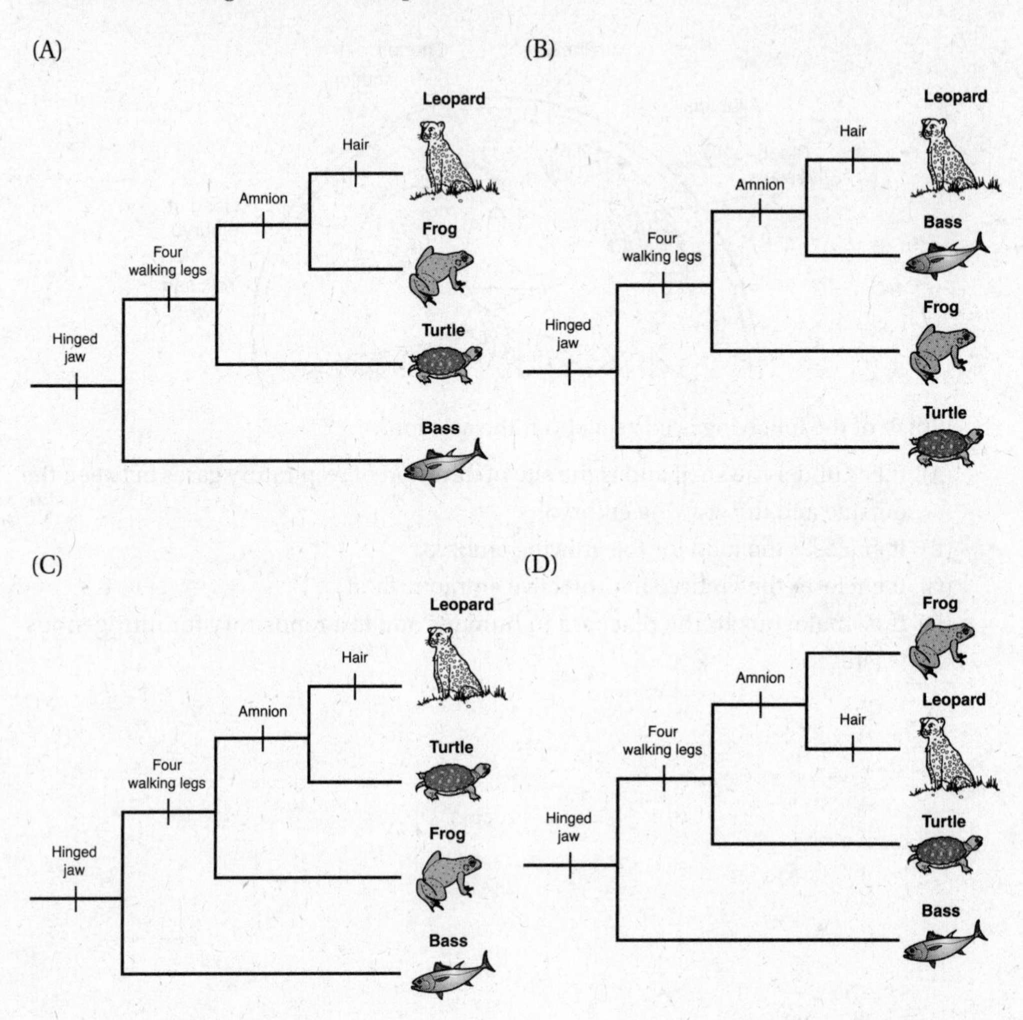

Answers Explained

1. **(B)** The endoderm gives rise to the viscera, the internal organs.
2. **(A)** The ectoderm gives rise to the skin and the nervous system. The eyes are part of the peripheral nervous system.
3. **(C)** The mesoderm gives rise to the blood, bones, and muscle.
4. **(C)** The mesoderm gives rise to the blood, bones, and muscle.
5. **(B)** Only choice B accurately describes the normal sites of fertilization and implantation in human females.
6. **(B)** Oxytocin is produced in the hypothalamus and released from the posterior pituitary.
7. **(C)** Gonadotropin-releasing hormone from the hypothalamus stimulates the anterior pituitary to release hormones, such as FSH and LH.
8. **(A)** FSH is released by the anterior pituitary and stimulates the follicle in the ovary to mature a secondary oocyte.
9. **(D)** Estrogen and progesterone are responsible for thickening the lining of the uterine wall in preparation for implantation of an embryo.
10. **(A)** FSH is active in males as well as females. In males, it stimulates sperm production; in females, it stimulates the maturation of a secondary oocyte in the ovary.
11. **(B)** In the experimental group, Jost removed both male and female tissue that would develop into gonads. All offspring were born female, regardless of the presence of a male genotype (XY). That also means that something in the gray crescent alters the expression of some genes.
12. **(C)** Notice that, in this experiment, Jost did not perform surgery on the control group. In an experiment, the control group and the experimental group must be identical in everything except for the one variable being tested. In this instance, that was not quite the case. Sham surgeries are often carried out in experiments on animals to fulfill that requirement.
13. **(C)** LH levels spike just before ovulation and as the uterine lining continues to thicken. Then, it just as suddenly decreases. Choice A is incorrect because the uterine lining is not shed until much after ovulation when LH levels are at their pre-spike levels. The follicle matures from day 1 to day 12 without the help of LH, so choice B is not correct. Choice D is incorrect because the level of LH decreases even as the uterine lining continues to build up.
14. **(C)** Choice (A) is not correct because in females, egg production begins with one mitotic cell division followed by two meiotic cell divisions. Choice (B) is not correct because one primary oogonium cell produces one active egg cell, not four active egg cells. Choice (D) is not correct because sperm production involves one mitotic cell division followed by two meiotic cell divisions.

15. **(A)** Clearly, the type of genes present is not the issue, since tissue from one individual (with one set of genes) was grafted onto another individual (with another set of genes). Therefore, choices (B), (C), and (D) are incorrect. It must be that a substance from the cytoplasm in the host embryo directed the grafted cells by altering the expression of some genes. This area from which Spemann and Mangold took the tissue (adjacent to the dorsal lip) was named the "organizer" because it directs neighboring cells to develop. It is a region of the cytoplasm (a "cytoplasmic determinant") that organized the neighboring cells.

16. **(C)** Remember that this embryo began as one fertilized egg that contained one nucleus plus the cytoplasm. Then, it divided into an embryo of cells, each with its own cytoplasm. What this experiment demonstrates is the importance of the cytoplasm to normal embryonic development. Some substance in the gray crescent cells determined the development of neighboring cells by altering the expression of some genes. Choice (A) is not correct because the embryos only needed half of the gray crescent to develop normally. Choice (B) is not correct because each embryo needed some of the gray crescent. All cells in an organism contain the same genes, so choice (D) is incorrect.

17. **(B)** The allantois is analogous to the placenta in mammals and also serves as the repository for uric acid, the nitrogenous waste from the embryo that accumulates until the chick hatches.

18. **(A)** The yolk sac contains the yolk, which provides food for the growing bird embryo until it hatches.

19. **(D)** The chorion is a thin membrane that lies just under the shell in a bird's egg. Oxygen and carbon dioxide diffuse across it, exchanging those gases between the embryo and the outside.

20. **(D)** The original cell, the spermatogonium, is diploid and produces the primary spermatocyte by mitosis. Therefore, the primary spermatocyte is also diploid. The secondary spermatocyte is produced by meiosis; thus, it is haploid.

21. **(C)** Choice (A) refers to the chorion. Choice (B) refers to the yolk sac. Choice (D) refers to the allantois.

22. **(C)** All of these animals have hinged jaws and descended from a common ancestor with a hinged jaw. The frog, turtle, and leopard all descended form a common ancestor with four walking legs. Only the turtle (a reptile) and the leopard descended from a common ancestor with an amniotic sac. Only the leopard has hair.

PART 4: INVESTIGATIVE LABS

The AP Biology exam will test both your thinking skills in science as well as your knowledge of Biology. The way that the exam accomplishes both of these goals is by asking you questions about **investigative labs**, which involve experimental designs, hypotheses, data analysis, and much more. Questions that cover investigative labs may require you to:

- Develop a hypothesis, design and evaluate an experiment, and identify proper experimental controls
- Analyze, evaluate, and graph data, using confidence intervals and/or error bars
- Predict the outcome(s) of an experiment
- Perform mathematical calculations using equations and formulas from the Reference Tables (which are provided on the exam)
- Develop and justify scientific theories based on evidence
- State the null hypothesis for an experiment

The 13 labs in this chapter may be somewhat different from the labs you've worked on in your class, but these labs are common versions of the labs that appear most often on the actual exam. Be aware that there are usually quite a few lab-based questions on the actual exam, so be sure to review this section of the book carefully! Reviewing these investigative labs and learning how they were set up, understanding how and why the data were collected in a specific way, and evaluating the conclusions that can be drawn from these experiments will better prepare you for these types of questions on test day.

19 Investigative Labs

- → GRAPHING
- → HISTOGRAMS
- → DEALING WITH DATA
- → DESIGNING AN EXPERIMENT
- → LAB #1: ARTIFICIAL SELECTION
- → LAB #2: MATHEMATICAL MODELING: HARDY-WEINBERG
- → LAB #3: BLAST—COMPARING DNA SEQUENCES
- → LAB #4: DIFFUSION AND OSMOSIS
- → LAB #5: PHOTOSYNTHESIS
- → LAB #6: CELL RESPIRATION
- → LAB #7: CELL DIVISION—MITOSIS AND MEIOSIS
- → LAB #8: BIOTECHNOLOGY—BACTERIAL TRANSFORMATION
- → LAB #9: BIOTECHNOLOGY—RESTRICTION ENZYME ANALYSIS OF DNA
- → LAB #10: ENERGY DYNAMICS—FOOD CHAIN
- → LAB #11: TRANSPIRATION
- → LAB #12: ANIMAL BEHAVIOR
- → LAB #13: ENZYME CATALYSIS
- → HOW TO CALCULATE THE RATE OF AN ENZYME-MEDIATED REACTION
- → CHI-SQUARE TEST AND NULL HYPOTHESIS

INTRODUCTION

The 13 AP investigative labs (referred to hereafter as "labs"), that are covered in this chapter, are not considered "cookbook science," where the teacher or lab manual dictates what to investigate and/or how to conduct experiments. According to the latest curriculum framework, students are supposed to design their own experiments, test their own hypotheses, analyze and evaluate their own data, and communicate their results to the class. Although the College Board does not dictate exactly what lab procedures you must use, they do require that certain labs be completed. For example, you are required to complete a lab on "Cell Respiration," but you are not required to complete the exact one that is described in this chapter. However, the 13 labs that are presented in this chapter do fulfill the College Board's requirements. It is important to take note of the fact that, while the latest curriculum framework does not include "Plants" (which are covered in Chapter 14 of this book), you must learn about a plant's structure and function in order to complete two of the required labs: one on "Photosynthesis" and one on "Transpiration."

Be sure to familiarize yourself with all of the labs that are discussed in this chapter because you will be asked to design and/or interpret lots of experiments, data, and graphs on the AP Biology exam.

GRAPHING

The purpose of showing data on a graph is to make the information clear and easily understood. Here are some guidelines and reminders for constructing graphs.

Label Your Graph

a. Title it.
b. Use the *x*-axis for the **independent variable**, the value that you control, such as time.
c. Use the *y*-axis for the **dependent variable**, the value that changes as a result of changes in the independent variable. For example, if in the course of an experiment, you take a measurement every minute and the chunk of potato gets heavier and heavier, then time is independent and the mass of the potato is dependent.
d. Include units when you label the *x*-axis and the *y*-axis (i.e., mL/sec).
e. There must be equal intervals along the *x*-axis. There must also be equal intervals along the *y*-axis.
f. Use // marks to show a break on the *x*- or *y*-axis if the values do not begin at zero.
g. Do not extend a graph line if you do not have data to do so. If the data do not start at zero or do not touch the *x*- or *y*-axis, then you may not extend the graph line to those points.
h. The graph line must not be extended beyond the last data point (extrapolation) unless you mean to show a prediction about what may happen. In that case, extend the solid line with a dashed line.
i. Be able to interpret standard error bars.

Plot the Data Points, Then Draw a Line of Best Fit

STUDY TIP

Can you draw a line of best fit? Learn how.

If you are instructed to connect the dots, do so. You may be asked to draw or interpret a **line of best fit** (or a curve of best fit). This may not be something you ever encountered in math class because the math teacher always provided you with values to graph that formed a perfectly straight line. In science, when you collect data, the numbers rarely fall into a straight line. However, if the line is not straight, you cannot make predictions by simply extending the line. Somehow, you must translate your rough data into a straight line. Here is how.

Plot your points as usual. Then, draw a straight line that best shows the trend (slope) and that takes into account the location of all the data points. If one data point really differs from all the others, you may ignore it. Also, when drawing a line of best fit, your line does not have to pass through any data points and it does not have to pass through zero. Figure 19.1 shows a scatter plot graph with data points plotted and a line of best fit drawn.

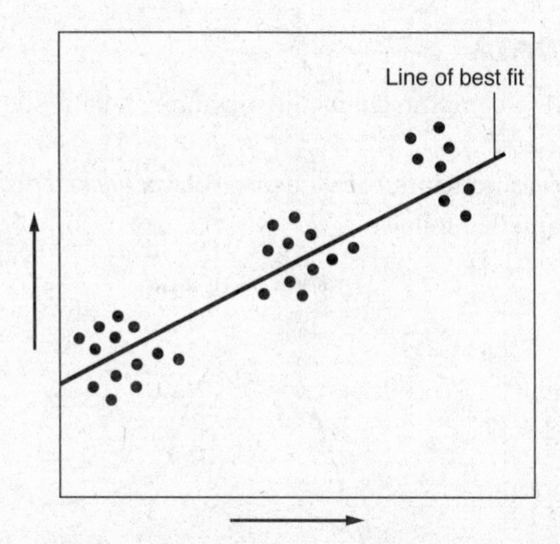

Figure 19.1 A Line of Best Fit on a Scatter Plot Graph

HISTOGRAMS

If you carried out an experiment and collected any type of data, a good first step is to organize the data on a **histogram**. To do this, take the entire range of values and divide it into a series of intervals. For example, let's pretend you measured the height of 100 trees in a forest. Your intervals could be 5' to 9', 10' to 14', 15' to 19', and 20' to 24'. Make sure that the intervals: (1) do not overlap, (2) are adjacent, and (3) are all the same size.

Examine the following intervals. Determine if they are acceptable or not and why.

1' to 4', 6' to 10', 13' to 20'

They are *not acceptable* intervals because they are not adjacent and there are unequal gaps between them. In addition, the intervals themselves are not of equal size.

Once you have decided on your intervals, place them on the *x*-axis. Next, count how many values fall into each interval and plot them on a graph, such as the histogram shown here:

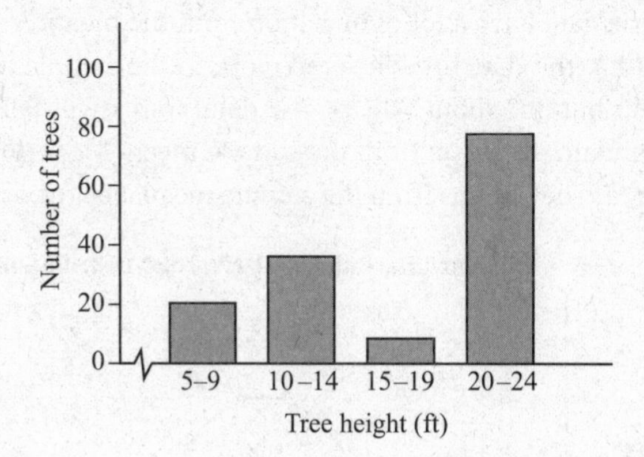

DEALING WITH DATA

A **normal distribution** is a common characteristic of much data, such as class grades or the height of students in a class.

If you take enough measurements, the data may show a *normal distribution* or *bell-shaped distribution.* See the graph that follows.

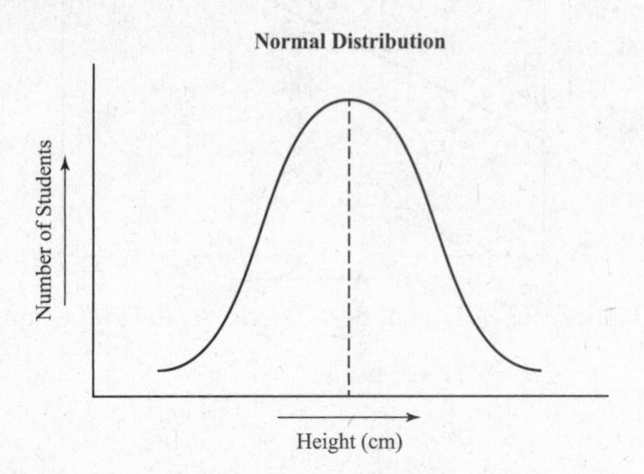

For a normal distribution, the appropriate descriptive statistics for the data set include the **mean**, **sample size**, **standard deviation**, and **standard error**.

The **mean** (or average) of the sample is the sum of the numbers in the sample divided by the total population in the sample. The mean summarizes the entire sample and might provide an estimate of the entire population's correct mean.

The **sample size** refers to how many members of the population are included in the study. Since it is not reasonable to count every tree in a forest, you take what you hope is a representative sample. You want to be confident that the sample size you chose accurately represents the entire population. Standard deviation and standard error will help you ensure that.

Standard deviation (SD) is a tool for measuring the *deviation* or *variability* from the mean in the sample population. In turn, this provides an estimate of the variation from the mean in the entire population from which the sample was taken. A large sample standard deviation indicates that the data have a lot of deviation from the mean. A small sample standard deviation indicates that the data are clustered close to the sample mean. On a graph that shows a normal distribution, about 68% of the data points will fall between +1 standard deviation and –1 standard deviation from the sample mean. More than 95% of the data will fall between ±2 standard deviations from the sample mean. See the graph that follows.

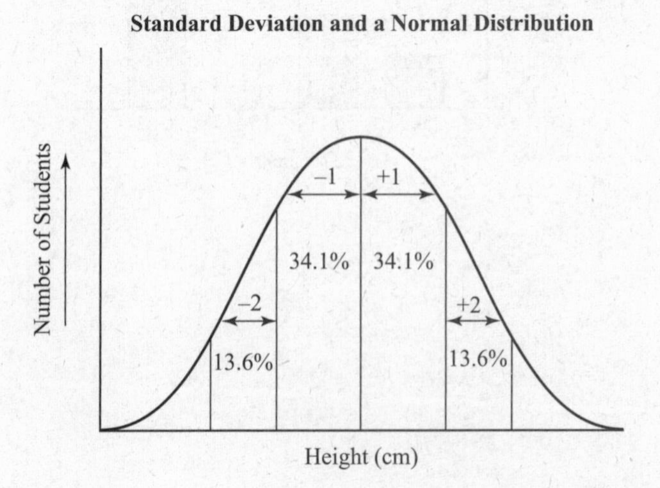

Standard error (SE) is a statistic that indicates the *reliability* of the mean, or how well the sample mean matches up to the true population mean. A small SE is an indication that the sample mean is a more accurate reflection of the actual population mean than a sample mean with a large SE. Taking a larger sample size will usually result in a smaller SE. If the sample mean is ±1 SE, the investigator can have an approximately 68% confidence that the range includes the true population mean. Moreover, a sample with a ±2 SE defines a range that has approximately a 95% confidence. In other words, if the sampling was repeated 20 times with the same sample size each time, the confidence limits would include the true population mean almost every time.

Error bars are a graphical representation of the variability of data. They are used on graphs to indicate how far from the reported value the correct value might be. Error bars can represent different things, such as standard deviation (SD), standard error (SE), or a confidence interval (e.g., a 95% interval). Whatever the error bars represent must be clearly stated on the graph.

Here is a bar graph of "Average Height of Girls and Boys in Middle School 184" that shows SE bars with a sample mean of ±1 SE. Actual values are not included.

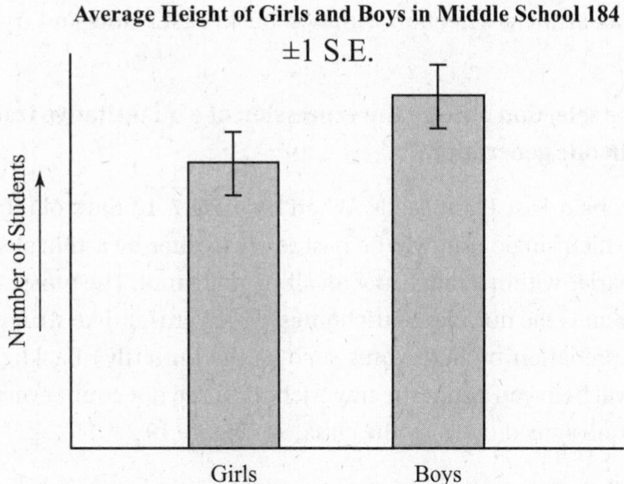

DESIGNING AN EXPERIMENT

A free-response question on the AP Biology exam may ask you to design an experiment. This makes many students nervous. However, if you follow these guidelines, you will have a good basis for devising a satisfactory experiment and writing a fine answer to that free-response question.

STUDY TIP

These guidelines for designing an experiment are very important.

1. **State a clear hypothesis**—what you expect to happen.
2. The **experiment must be reasonable to carry out and must work**. For example, having to set up 100 fish tanks is *not* reasonable.
3. A "controlled" experiment must **have a control**. The control must be exactly like the experimental except for the single factor you are testing. The control can be an organism left untreated.
4. Any experiment must have **only one variable**. For example, in an experiment in which you are testing the effect of various light intensities on the growth of plants, you must

place a **heat sink** between the light and the plant. Without something to absorb the heat from the light source, temperature would become a second variable.

5. **Have a large enough sample** to draw a reasonable conclusion. A sample of one organism is not acceptable. In Lab #1, "Artificial Selection," the sample size is 150.
6. **Experimental organisms must be as similar as possible**. They must all be of the same variety, size, and/or mass, whichever is appropriate. **You must state that fact.**
7. **State that the experiment must be repeated**. This reduces the possibility that an error occurred by chance, some random factor, or individual variations in the experimental organism.

LAB #1: ARTIFICIAL SELECTION

Introduction

In the classroom, students have few opportunities to study and measure natural selection in living multicellular organisms. Many labs use computer simulations instead. However, this lab is a good alternative for exploring selection and evolution in the classroom.

Objective

You will carry out an artificial selection with Wisconsin Fast Plants and try to answer the following question:

> **"Can extreme selection change the expression of a quantitative trait in a population in one generation?"**

Sprout some Wisconsin Fast Plant seeds. When they are 7–12 days old, examine them and decide as a class which single trait will be best to try to alter by artificial selection. The trait must be one that varies within a range, not an all-or-none trait. The most common trait studied in this experiment is the number of trichomes, which are hairlike structures on the leaves. (Trichomes deter predation by herbivores such as the butterfly.) Backlighting and using a magnifying glass will help you count the tiny trichomes. Do not count every hair. Instead, use a sampling technique agreed upon by the class. See Figure 19.2.

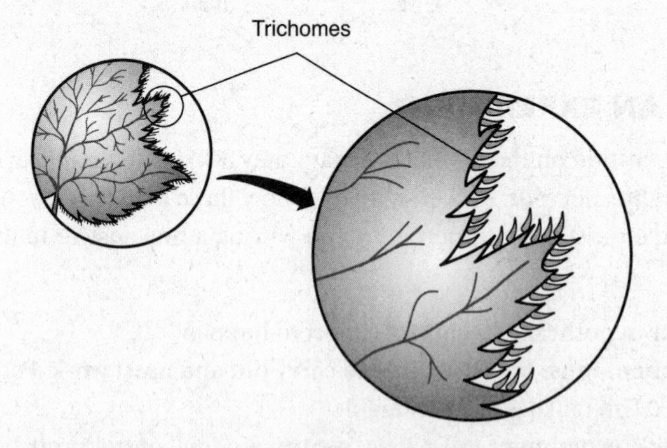

Figure 19.2 Trichomes on Wisconsin Fast Plant Leaves

An appropriate sample size for the experiment is 150 plants. Figure 19.3 is a histogram that shows the number of trichomes in the first plant population.

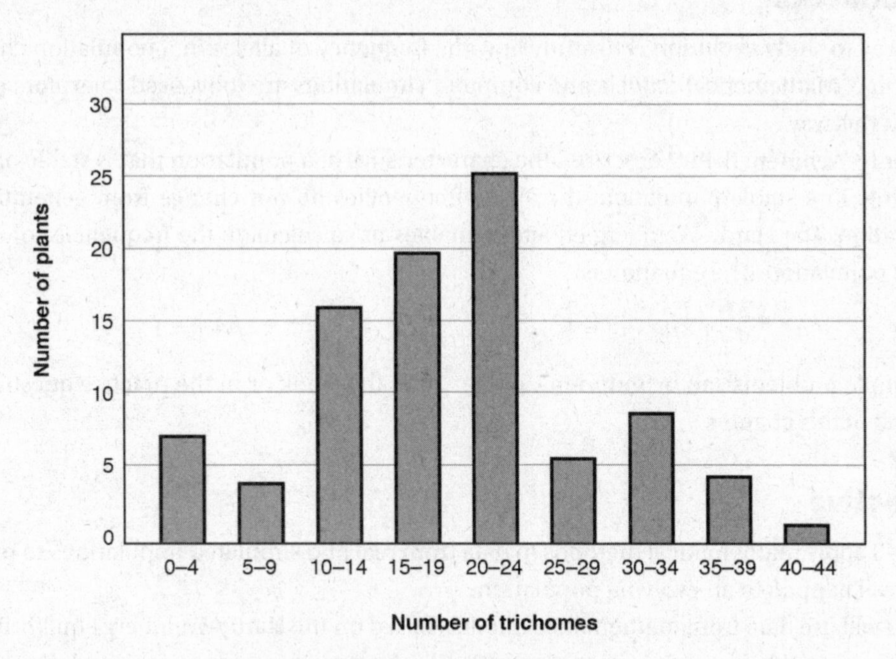

Figure 19.3 Trichome Numbers

Let's assume the class decides to breed the plants to increase the number of trichomes. Select the least hairy 10% of your original population of plants to use in your experiment. Perform one round of artificial selection by cross-pollinating the plants you have selected. After the F_1 plants germinate, mature, and flower, collect the seeds and grow them to produce another generation. As a control, pollinate an equal number of plants from the original population that were not part of your experimental group. Remember to keep the controls separate from the experimental plants. The histogram in Figure 19.4 shows the data from the F_2 experimental plants, which were selected for having fewer trichomes.

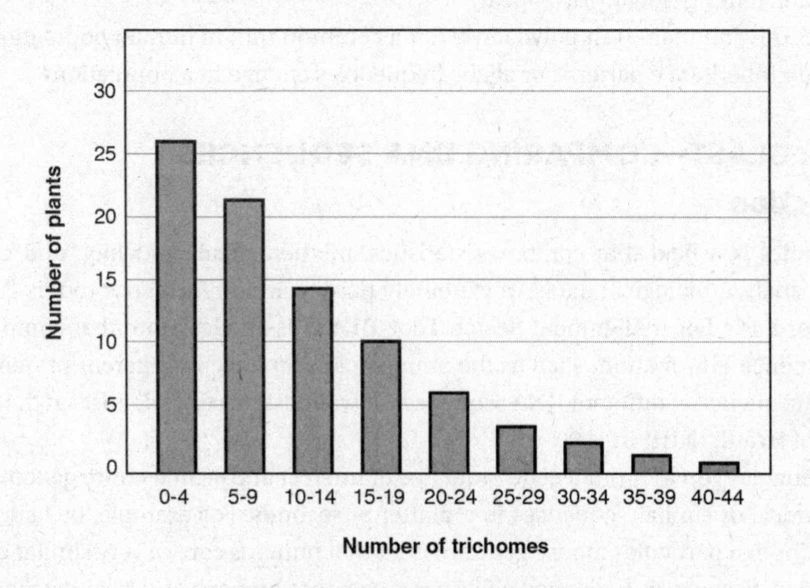

Figure 19.4 Trichome Numbers

Clearly, the average number of trichomes in the plant population decreased in one generation.

LAB #2: MATHEMATICAL MODELING: HARDY-WEINBERG

Introduction

One way to study evolution is to study how the frequency of alleles in a population changes over time. Mathematical models and computer simulations are tools used to explore evolution in this way.

Hardy-Weinberg theory describes the characteristics of a population that is stable or nonevolving. In a stable population, the allelic frequencies do not change from generation to generation. The Hardy-Weinberg equation enables us to calculate the frequencies of alleles in the population. The equation is:

$$p^2 + 2pq + q^2 = 1 \quad \text{or} \quad p + q = 1$$

Sample problems can be found in Chapter 11 of this book or in the practice questions at the end of this chapter.

Objective

You will apply mathematical methods to data from real and simulated populations to predict what will happen to an evolving population.

You will use data from mathematical models based on the Hardy-Weinberg equilibrium to analyze genetic drift and the effect of selection in the evolution of specific populations.

You will build a spreadsheet that models how a hypothetical gene pool changes from one generation to the next. Any modern spreadsheet will work, including Microsoft Excel, Google's online Google Docs, and Zoho's online spreadsheet. This spreadsheet will let you explore how different factors, such as selection, mutation, and migration, affect allelic frequencies in a population.

You will be able to answer questions such as:

- Why do recessive alleles like the one that causes cystic fibrosis stay in the human population?
- Why don't they gradually disappear?
- Why is the dominant trait polydactyly not a common trait in human populations?
- How do inheritance patterns or allelic frequencies change in a population?

LAB #3: BLAST—COMPARING DNA SEQUENCES

Introduction

Bioinformatics is a field that combines statistics, mathematical modeling, and computer science to analyze biological data. An extremely powerful bioinformatics tool is BLAST, an acronym for **B**asic **L**ocal **A**lignment **S**earch **T**ool. **BLAST** is an algorithm that compares biological sequence information, such as the amino acid sequence in different proteins or the nucleotide sequence of different DNA sequences. It was designed by scientists at the National Institutes of Health (NIH) in 1990.

BLAST enables you to input a gene sequence of interest and search entire genomic libraries for identical or similar sequences in a matter of seconds. For example, by using BLAST, you might input a particular mouse gene and find that humans carry a very similar gene. You might instead discover, as happened a few years ago, that humans and Neanderthals (*Homo neanderthalensis)* share some of the same DNA.

If you were to use BLAST to compare five DNA sequences from five different species, you would see something like the following on your computer screen:

Position:	1 2 3 4
Species *A*:	A**C**CGCT**A**CG**A**TTCGGC**T**AGCAT
Species *B*:	A**C**CGCT**G**CG**A**TTCGGC**C**AGCAT
Species *C*:	A**C**CGCT**A**CG**T**TTCGGC**T**AGCAT
Species *D*:	A**C**CGCT**G**CG**A**TTCGGC**T**AGCAT
Species *E*:	A**G**CGCT**G**CG**A**TTCGGC**T**AGCAT

Notice that species *A–D* all have a C in location 1 on the DNA, while species *E* has a mutation in that location.

Objective

In this lab, students will use BLAST to compare several genes and then use the information to construct a cladogram. A cladogram (or phylogenetic tree) is a visualization of the evolutionary relatedness of species. See Chapter 10 of this book for more information about cladograms. Figure 19.5 is a simple cladogram.

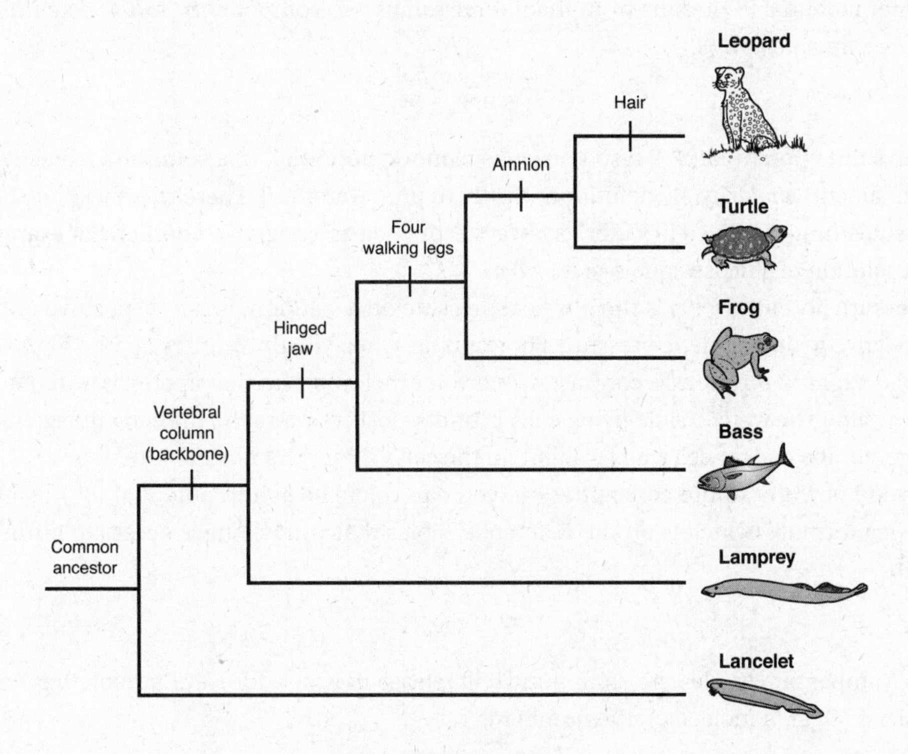

Figure 19.5 Cladogram

LAB #4: DIFFUSION AND OSMOSIS

Introduction

Water potential (Ψ)

- Is measured in units of pressure called megapascals (MPa)
- Measures the relative tendency for water to move from one place to another
- Is the result of the combined effects of solute concentration and pressure

Water potential	=	pressure potential	+	solute potential
Ψ	=	Ψ_p	+	Ψ_s

- **Water moves from high water potential to low water potential**

Water potential (Ψ) is a physical property that predicts the direction in which water will flow. Water flows from a region of high water potential to low water potential. Think of Ψ as water's potential energy. It is a powerful force and is measured in **megapascals** (MPa). One MPa equals approximately 10× atmospheric pressure. By definition, the Ψ of water in an open container under **standard conditions** of temperature and pressure (at sea level and at room temperature) is 0 MPa.

Water potential (Ψ) results from the interaction of two components: *solute potential* (Ψ_s) and *pressure potential* (Ψ_p).

$$\Psi = \Psi_s + \Psi_p$$

The **solute potential** (Ψ_s), also known as **osmotic potential**, of a solution is directly proportional to its *molarity*. By definition, the Ψ_s of pure water is 0. Therefore, the Ψ_s of a solution (something dissolved in water) is always expressed as a negative number. For example, a 0.5 M solution of glucose equals –1.22 MPa.

Pressure potential (Ψ_p) is the physical pressure on a solution. It can be positive or negative, relative to atmospheric pressure. For example, when you draw water up into a pipette, it is under *negative pressure*. In contrast, when you expel water from a pipette, it is under *positive pressure*. The water inside living cells is under positive pressure, pressing up against the cell membrane and/or cell wall to maintain the cell's shape and normal function.

If you know the solute concentration, you can calculate solute potential (Ψ_s) using the following formula (which is on the Reference Tables that you will have access to during the exam):

$$\Psi_s = -iCRT$$

i = Number of particles the compound will release in water: for NaCl (ionic), the number is 2; for sugar (a molecule), the number is 1.

C = Molar concentration determined from your experimental data. Place blocks of tissue (like potato cores) in solutions of different known molarities. Determine at which molarity the sample neither gains nor loses mass (crosses the *x*-axis). See page 311.

R = Pressure constant equals 0.0831 liter/MPa/mole K (which is on the Reference Tables that you will have access to during the exam). Note that one bar is approximately 1 atmosphere.

T = Temperature in degrees Kelvin equals 273 + °C of solution (which is on the Reference Tables that you will have access to during the exam).

This lab consists of three parts:

- Part 1: Make artificial cells to study the relationship of surface area to volume.
- Part 2: Create models of living cells.
- Part 3: Observe osmosis in living cells.

Part 1: Make artificial cells to study the relationship of surface area to volume.

OBJECTIVE

Fashion "cells" from agar or gelatin stained with phenolphthalein. Explore the rates of diffusion in cells of different sizes and shapes.

Calculate the ratio of surface area to volume in both small and large cells. Predict which cell(s) might eliminate waste and take in nutrients faster by diffusion—a small cell or a large cell. See Figure 19.6.

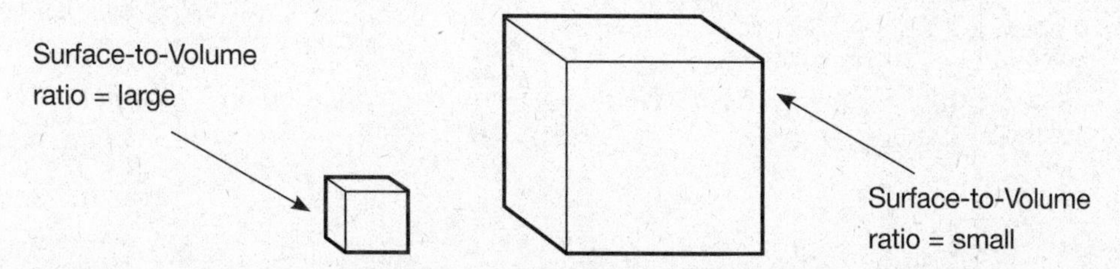

Figure 19.6 The Relationship Between Surface Area and Volume in Cells of Different Sizes

Part 2: Create models of living cells.

OBJECTIVE

Investigate the processes of diffusion, osmosis, and water potential in a model membrane system and in living cells. This section has three parts: A, B, and C.

PART 2A: OBSERVE DIFFUSION ACROSS A SEMIPERMEABLE MEMBRANE.

Fashion a length of semipermeable dialysis tubing into a bag, and fill it with two solutions: starch and glucose. Place the bag into a beaker that contains Lugol's iodine solution, and allow the system to stand for 30 minutes. The contents of the bag will turn blue-black because iodine molecules are small enough to diffuse into the bag and react with the starch. However, the iodine solution in the beaker will remain unchanged because the starch molecules are too large to diffuse out of the bag and mix with the iodine. See Figure 19.7.

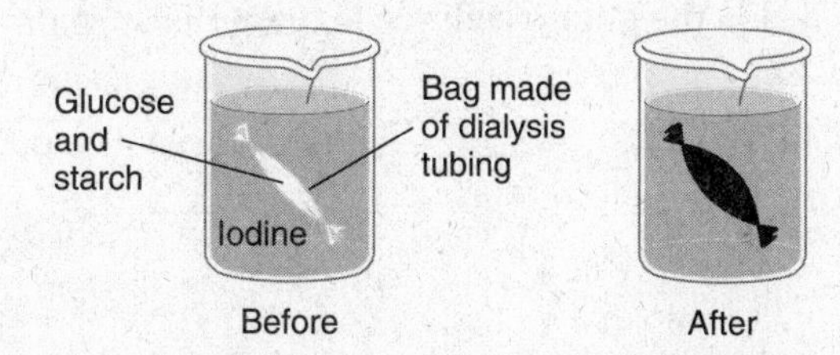

Figure 19.7 Diffusion Across a Semipermeable Membrane

PART 2B: CALCULATE THE MOLARITY OF AN UNKNOWN SOLUTION BY OBSERVING OSMOSIS.

Fashion 6 lengths of semipermeable dialysis tubing into 6 bags. Fill 5 of them with sucrose solutions of varying molarities: 0.2 M, 0.4 M, 0.6 M, 0.8 M, and 1.0 M. Fill 1 with distilled water. (The molecules of sucrose are too large to diffuse through the dialysis membrane.) Remove as much air from each bag as possible to allow room for expansion. Blot and weigh each bag. Place each bag into a beaker that contains distilled water. Allow them to sit for 30 minutes. Then remove the bags from the beakers. Blot and weigh the bags as before. Calculate the percent change in mass for each bag. Record and graph the data with the dependent variable (percent change in mass) on the y-axis and the independent variable (the various molarities) on the x-axis. The bag that contains distilled water will not change in mass because its contents are isotonic to the distilled water in the beaker; this bag is the control. *The mass of the other bags will increase. The bag with the lowest molarity will increase in mass the least, and the bag with the highest molarity will increase the most.* See Figure 19.8.

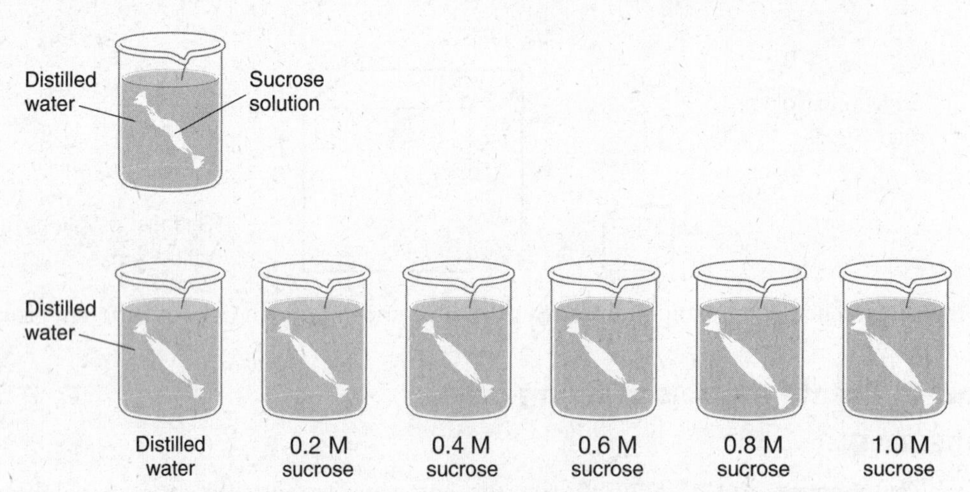

Figure 19.8 Calculating the Molarity of an Unknown Solution

PART 2C: DETERMINE THE MOLARITY OF A LIVING (POTATO) CELL.

Cut identical-size small cubes of potato and weigh them. Place each into a beaker covered with 300 mL of the following sucrose solutions: 0.2 M, 0.4 M, 0.6 M, 0.8 M, and 1.0 M and one with distilled water. Allow them to sit overnight, then weigh them again. Some pieces of potato will have gained mass and some will have lost mass. Calculate the percent change in mass for each. Plot a line of best fit on a graph, with the molarities of the solutions on the x-axis and the percent change in mass on the y-axis. The point of the **line of best fit** that crosses the x-axis is the molarity where there was no change in the mass of the potato and represents the molarity inside the potato cell; see Figures 19.9 and 19.10.

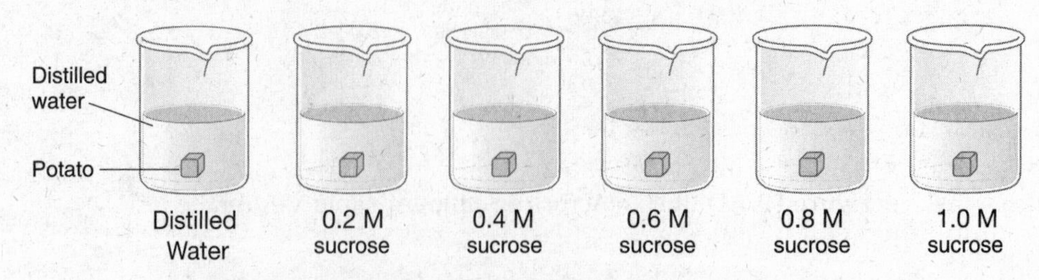

Figure 19.9 Determining the Molarity of a Living (Potato) Cell

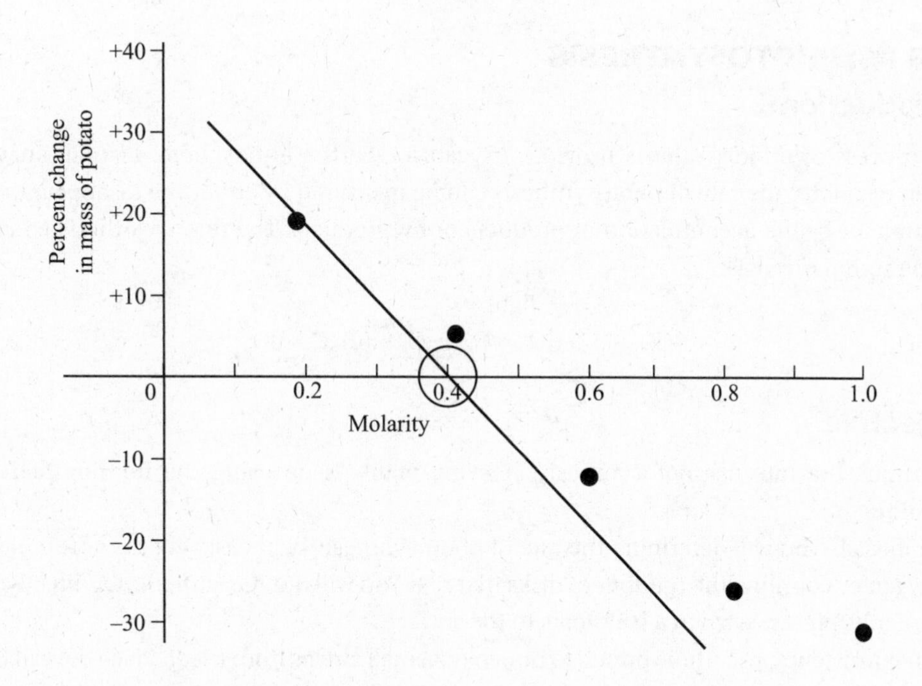

Figure 19.10 Determining the Molarity of the Potato Cell

Part 3: Observe osmosis in living cells.

OBJECTIVE

Observe plasmolysis in a living plant (elodea) cell.

Tear a thin piece of healthy elodea tissue, and prepare a wet mount. Locate a single layer of cells under the light microscope at 40× with clearly visible chloroplasts. While observing the slide, expose the tissue to 5% (hypertonic) sodium chloride solution. Observe changes in the cytoplasm of the cell as water suddenly diffuses out of the cell from high water potential to low water potential. The cytoplasm will shrink (plasmolysis), and the chloroplasts will condense into a small circle. If you wash away the salt water solution and rehydrate the cells with distilled water, you will see the cell fill up with water (turgor) and appear similar to the way it was initially. However, the cell is no longer alive; see Figure 19.11.

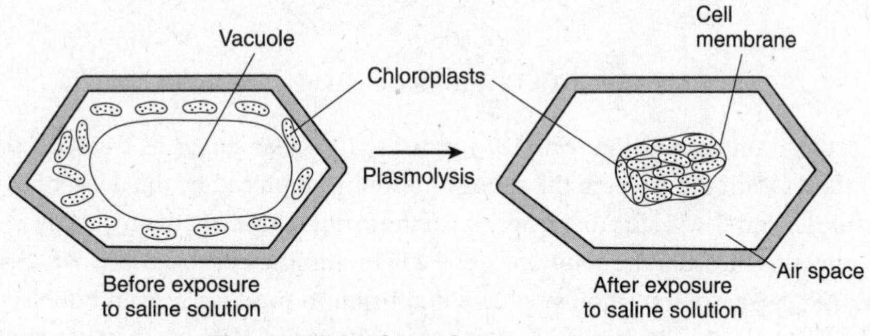

Figure 19.11 Plasmolysis in a Plant Cell

LAB #5: PHOTOSYNTHESIS

Introduction

The process of photosynthesis provides oxygen for Earth's atmosphere. Like all enzyme-driven reactions, the rate of photosynthesis can be measured by either the disappearance of substrate or by the accumulation of products or by-products. The photosynthesis equation can be summarized as:

$$6CO_2 + 12H_2O \xrightarrow{\text{light}} C_6H_{12}O_6 + 6H_2O + 6O_2$$

Objective

Determine the rate of photosynthesis in living plant tissue using the **floating leaf disk technique**.

In this lab, you will determine the rate of photosynthesis by measuring the accumulation of oxygen by counting the number of disks that rise. You will not measure the O_2 directly; you will infer its presence when a leaf floats to the surface.

For consistency, use a hole punch to punch out small circles from a leaf. Place the leaf disks into a syringe along with a baking soda solution (bicarbonate). Create a vacuum by pulling the syringe out about 1 cm while holding your thumb over the opening of the syringe. The vacuum removes air from the air spaces in the spongy mesophyll in the leaf and replaces it with the solution of baking soda. See Figure 19.12.

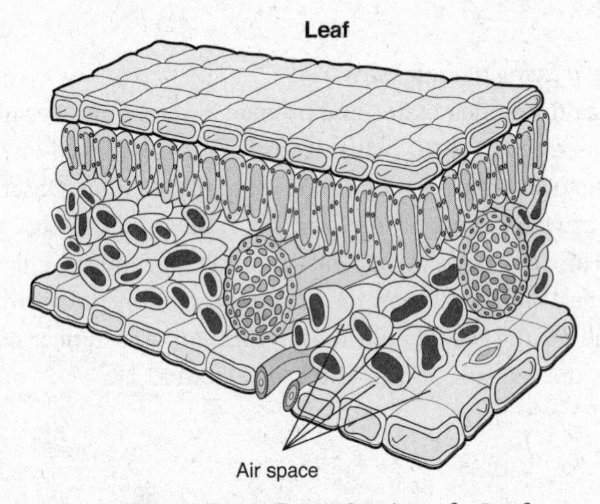

Figure 19.12 Cross-Section of a Leaf

The 1% bicarbonate solution ($NaHCO_3$) provides CO_2 that serves as a carbon source for photosynthesis. With the air from the spongy mesophyll replaced by the denser baking soda solution, the leaf disks will sink to the bottom of the syringe. Transfer them quickly to a beaker that also contains bicarbonate solution. Shine a light into the beaker so you can observe the leaf disks. As they carry out photosynthesis and begin to produce oxygen bubbles, the leaf disks will begin to rise in the bicarbonate solution. The speed at which they rise is a function of the rate of photosynthesis.

Begin collecting data as soon as you shine a light into the beaker. Stop timing when 50% of the disks have risen. See Figure 19.13.

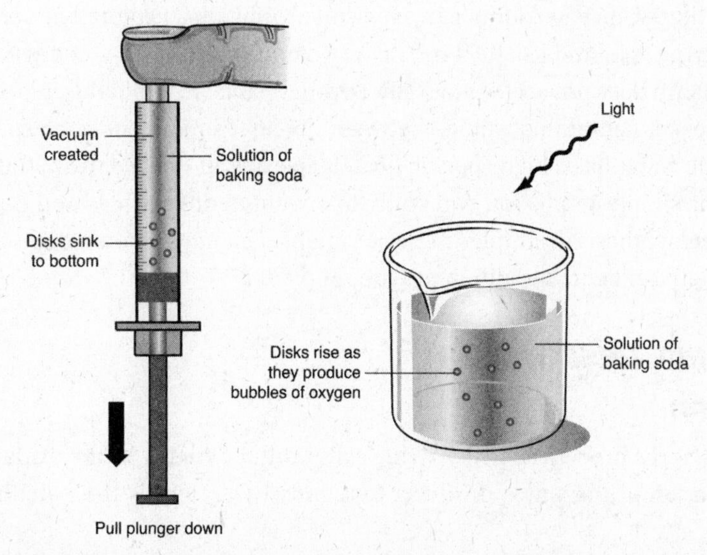

Figure 19.13 Disks Rise in the Solution as They Produce Oxygen

Figure 19.14 is a sample graph that shows the number of disks floating versus time.

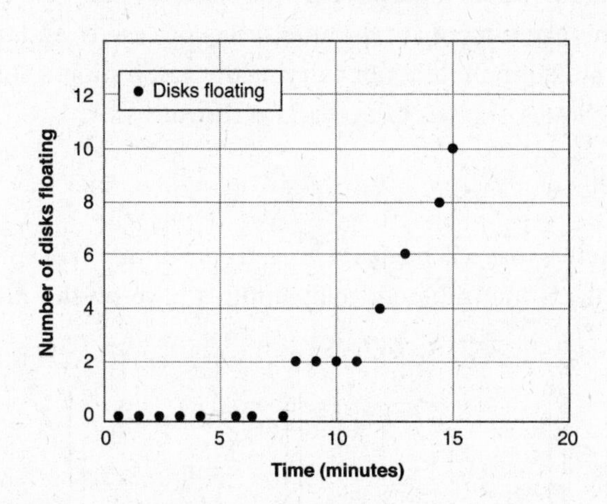

Figure 19.14 Graph of the Number of Disks Floating versus Time

Figure 19.15 is a graph that shows the rate of photosynthesis compared to light intensity based on this experiment.

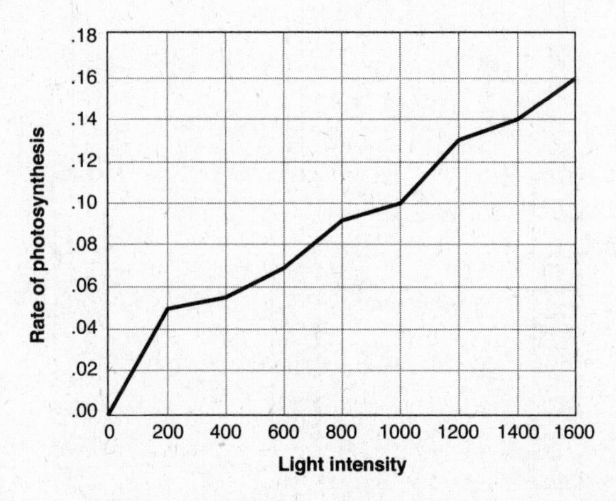

Figure 19.15 Increasing the Light Intensity Increases the Rate of Photosynthesis

Remember that when you conduct an experiment, you must explore one variable at a time. If you are studying light and its effect on the rate of photosynthesis, you must be careful that heat from the lamp does not accidentally affect your results. To avoid this, place a **heat sink**—a clear glass beaker containing water—between the light source and the experimental container. Since the water has a high specific heat, it absorbs all the heat from the light source.

By using this simple setup, you can study many different variables and how each affects the rate of photosynthesis. Examples of these variables include different light intensities, different wavelengths of light, and different types of plants.

LAB #6: CELL RESPIRATION

Introduction

Cell respiration is the process by which living cells produce energy by the oxidation of glucose with the formation of the waste products CO_2 and H_2O_2. Here is the equation for cellular respiration:

$$C_6H_{12}O_6 + 6O_2 \rightarrow 6CO_2 + 6H_2O + \frac{\text{686 kcal of energy}}{\text{mole of glucose oxidized}}$$

The rate of cell respiration can be measured by the volume of oxygen used. The cells used to study respiration in this experiment are sprouted sweet pea seeds, and the control consists of sweet pea seeds that are dormant and not carrying out respiration. (This apparatus can also accommodate invertebrates, such as insects and earthworms.)

Objective

Construct a **respirometers** that will measure oxygen consumption of sprouted or unsprouted seeds. Observe the effects that different temperatures have on the rate of respiration; see Figure 19.16.

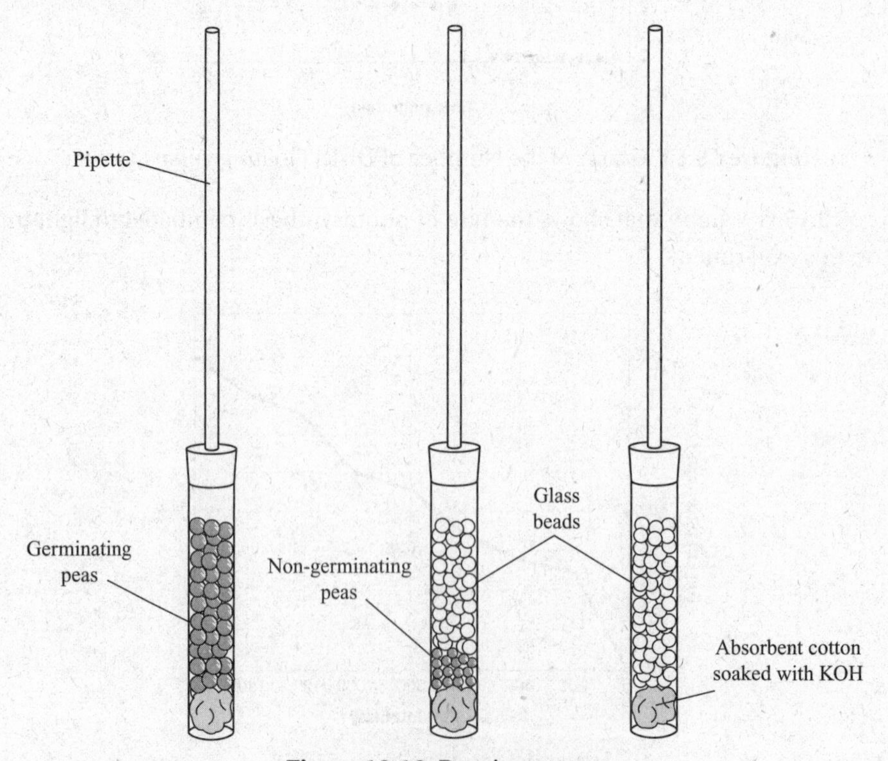

Figure 19.16 Respirometers

Since the sprouted seeds absorbed water and were larger than the unsprouted seeds, glass beads were added to the respirometer that held the unsprouted seeds to equalize the volume of empty space in both vessels. Set up six respirometers as shown in Table 19.1.

Table 19.1

Effects of Temperature on Respiration Rate		
Respirometer	**Temperature of Water**	**Contents**
1	25°C	Germinating seeds
2	25°C	Dry seeds + beads
3	25°C	Glass beads alone
4	0°C	Germinating seeds
5	0°C	Dry seeds + beads
6	0°C	Glass beads alone

Place a thin layer of cotton (soaked with KOH to absorb the CO_2 that is given off by respiration) at the bottom of each respirometer. This allows the change in volume inside the respirometer to measure *only* oxygen consumption.

Submerge the respirometers in water. As oxygen is used up inside the respirometer, water will flow into the respirometer through the pipette and will be a measure of oxygen consumption.

The results of the experiment are simply stated. The higher the temperature, the more oxygen is used and the faster the rate of respiration. Figure 19.17 shows a graph of the data.

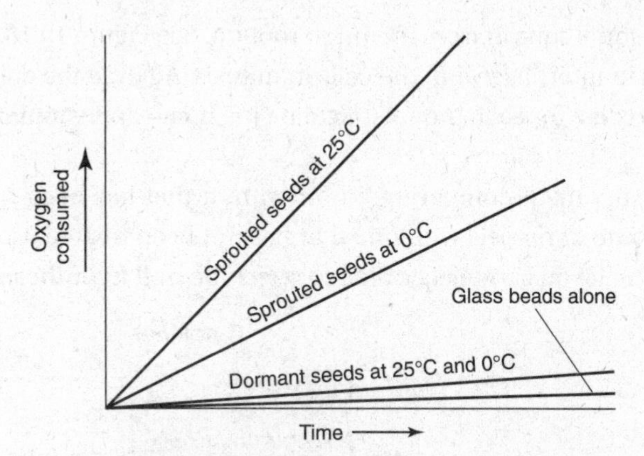

Figure 19.17 Oxygen Consumed in Respiring Seeds

LAB #7: CELL DIVISION—MITOSIS AND MEIOSIS

Introduction

See Chapter 4 for a review of mitosis and meiosis.

Objective

- Describe the events and regulation of the cell cycle.
- Explain how DNA is transmitted to the next generation by mitosis.
- Explain how DNA is transmitted to the next generation by meiosis followed by fertilization.
- Understand how meiosis and crossing-over lead to increased genetic diversity, which is necessary for evolution.

Part 1

Use models like pipe cleaners, beads, or "sockosomes" to learn about the stages of **mitosis**. Be able to answer the following questions:

- How does a chromosome replicate itself?
- Why do chromosomes condense before the nucleus divides?
- What happens if sister chromatids fail to separate properly?
- Can chemicals in the environment alter the process of mitosis in plants or animals? How would you design an experiment to prove your answer?

Part 2

Prepare a chromosome squash from an onion root tip. See Figure 19.18.

Count the cells in interphase and the cells in mitosis. Analyze the data.

Collect the class data for each group. Calculate the **mean** and **standard deviation** for each group.

Carry out an experiment comparing an onion root that has been treated with a chemical that alters the rate of mitosis with one that has not been treated. Complete a **chi-square analysis**. Determine if you can reject or fail to reject the **null hypothesis**.

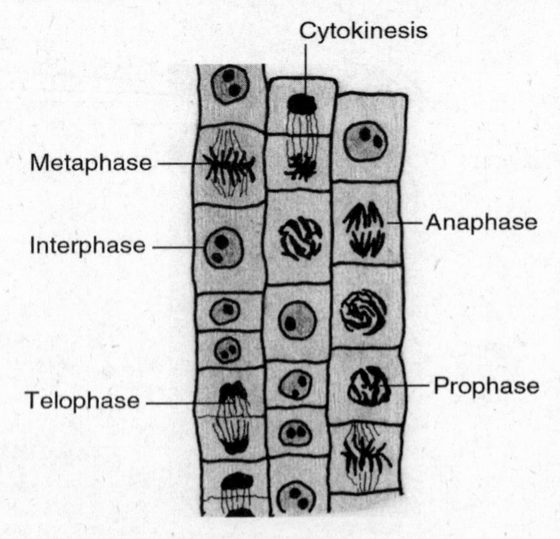

Figure 19.18 Onion Root Tip

Part 3

Become familiar with normal karyotypes and abnormal ones that are characteristic of cancer. Explore **HeLa cells** and **Philadelphia chromosomes**.

Part 4

Use models like pipe cleaners, beads, or "sockosomes" to learn about the stages of **meiosis**. Be able to answer the following questions:

- What is **crossing-over**?
- What is **independent assortment**?
- What happens if a pair of chromosomes fails to separate? How might this contribute to various genetic diseases?
- How are mitosis and meiosis fundamentally different?

Calculate the distance in map units between the gene for spore color and the centromere in the fungus *Sordaria fimicola.*

Sordaria fimicola is an **ascomycete** fungus that consists of fruiting bodies called **perithecia**, which hold many **asci**. Each ascus encapsulates eight **ascospores** or **spores**. When ascospores are mature, the ascus ruptures, releasing the ascospores. Each ascospore can develop into a new haploid fungus; see Figure 19.19.

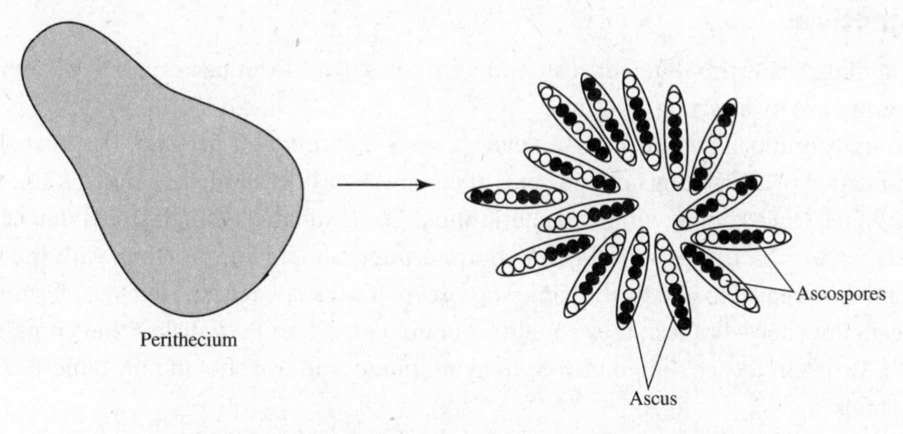

Figure 19.19 Fungus Fruiting Bodies with Spores

By observing the arrangements of ascospores from a cross between wild-type (black) and a mutant tan type *Sordaria fimicola* fungus, you can estimate the percentage of crossing-over that will occur between the centromere and the gene that controls the tan spore color. Count 50–100 hybrid asci, and calculate the distance in map units between the gene for spore color and the centromere using this formula,

Divide the number of crossover asci (2:2:2:2 or 2:4:2)
by the total number of asci × 100 and divide by 2

LAB #8: BIOTECHNOLOGY—BACTERIAL TRANSFORMATION

Introduction

Many years ago if a person was diabetic and needed insulin, the insulin was harvested from other mammals, such as pigs and cows, because their insulin was similar to human insulin. Today, human insulin is "manufactured" by bacteria in which scientists have inserted a gene that produces human insulin. This insulin is not similar to human insulin; it is human insulin. Several scientists took many years to figure out how to accomplish this. However, now it is done routinely.

Genes have been inserted in bacteria so the bacteria can "eat" oil and clean up major oil spills. Rice has been genetically modified to produce more protein. Cows have been genetically engineered to produce more milk. You may have seen the term GMO—genetically modified organism—on milk cartons or some other foods. The term means that the organism or food has been modified by altering its genes—by adding a foreign gene into its genome.

The science of inserting genes into a cell of an organism depends on small circles of exogenous DNA called **plasmids** that were first discovered in bacteria. Scientists insert a particular gene into a plasmid and then insert the plasmid into a bacterial cell. The host cell begins to produce the proteins encoded by the gene that was inserted into the plasmid. To force a bacterium to uptake a plasmid, the bacterium must be prepared in a special way. It must be made **competent.**

Objective

Cut lambda DNA with different restriction enzymes. Transform bacteria into an antibiotic-resistant form by inserting a plasmid.

To transform bacteria, *first make them competent to uptake a plasmid.* Use heat shock, a combination of alternating hot and cold, in the presence of calcium ions that disrupt the cell membrane. Once competent, the bacteria should be incubated with plasmids that carry the resistance to a particular antibiotic. A control sample should be run along with the experimental one, treated in exactly the same way except it does not get exposed to a plasmid. After the cells are allowed to rest, they should be poured onto four Petri dishes that contain Luria broth. Two Petri dishes should also contain antibiotic, and two should not. Table 19.2 shows this setup.

Table 19.2

Experimental Data			
Test Tube	**Plasmid**	**Petri Dish**	**Growth**
1	+	+ Antibiotic	+
2	+	No Antibiotic	+
3 (Control)	-	+ Antibiotic	-
4 (Control)	-	No Antibiotic	+

Figure 19.20 is an illustration of the Petri dishes that shows the results of this experiment.

- Plate 1. These bacteria were incubated with plasmid. There are 7 colonies growing here. Each colony consists of bacteria that have been **transformed** and are resistant to the antibiotic because they absorbed the plasmid.
- Plate 2. This plate contains a lawn of bacteria. Any bacteria would grow on this plate; there is no antibiotic to prevent it. However, this culture did receive plasmid. We can assume that this plate contains bacteria that are resistant to the antibiotic as well as those that are susceptible to it.
- Plate 3. There is no growth on this dish. These bacteria were not incubated with plasmid and the plate contains no growth because the dish contains antibiotic that killed the bacteria.
- Plate 4. The bacteria were not incubated with plasmid and the plate also contains a lawn of growth because there is no antibiotic to kill the bacteria.

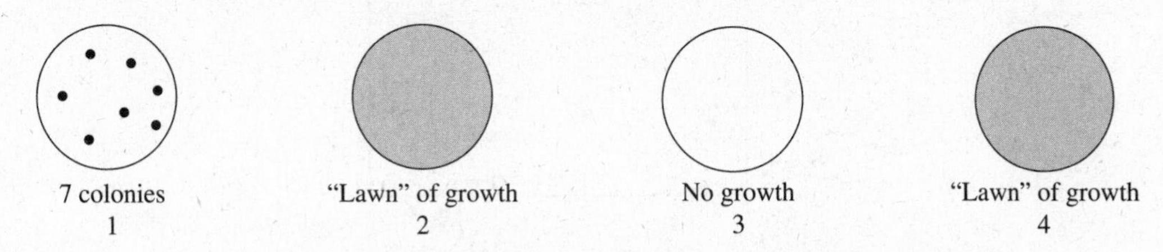

Figure 19.20 Plates that Show the Results of This Experiment

LAB #9: BIOTECHNOLOGY—RESTRICTION ENZYME ANALYSIS OF DNA

Introduction

In this experiment, lambda DNA is incubated with two different restriction enzymes and then run through an electrophoresis gel. Place the lambda DNA into three test tubes. Incubate one sample with *Eco*RI, incubate another sample with *HIND*III, and leave one untreated as a control. The two restriction enzymes will digest the DNA of two of the samples, cutting them into characteristic pieces called **restriction fragments**. The three samples should be placed into the wells on an agarose gel and drawn from the **cathode** to the **anode** by electric current. After staining, a pattern should result on the gel that reveals the **restriction fragments**. The control sample of DNA should remain uncut and should be seen as a large block of DNA near the well in lane 4. The two other samples were cut with restriction enzymes, and the restriction fragment or banding pattern should be visible. By measuring the distance of each fragment from the well (similar to determining the R_f value in paper chromatography) and comparing this distance against a standard, the size of the fragment of DNA can be determined; see Figure 19.21.

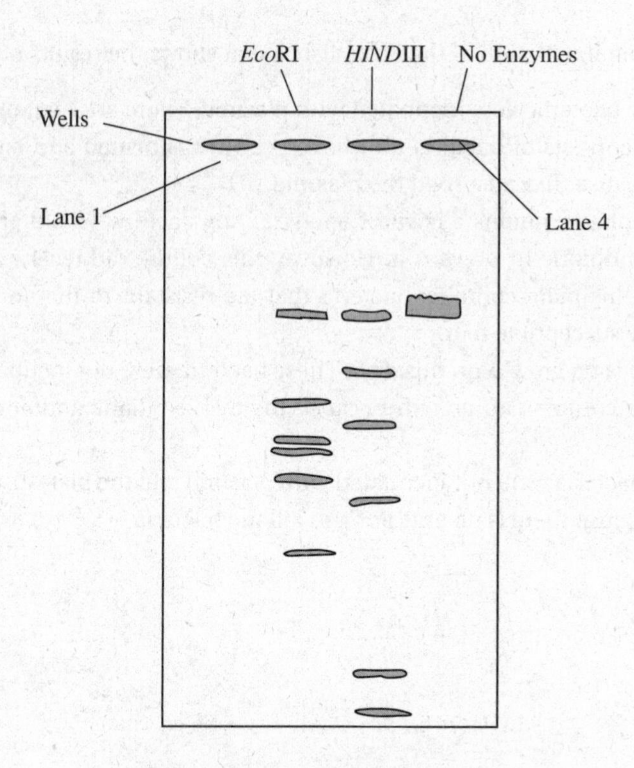

Figure 19.21 Electrophoresis Gel

Objective

Use restriction enzymes and gel electrophoresis to identify and form a biological profile for individuals.

LAB #10: ENERGY DYNAMICS—FOOD CHAIN

Introduction

Almost all life on this planet is powered, either directly or indirectly, by sunlight. Producers capture solar energy and convert it to biomass. The net amount of energy captured and stored by the producers in a system is the system's net primary productivity. In contrast, the gross primary productivity is a measure of the total energy captured. Different plants have different strategies of energy allocation that reflect their role in different ecosystems. For example, annual weedy plants allocate more of their biomass production to reproduction and seeds than do slower-growing perennials. The producers are consumed or decomposed, and their stored chemical energy powers the other trophic levels of the biotic community. Understanding energy dynamics in a biotic community will help you understand ecological interactions.

Objective

This lab has two parts. First, you will estimate the net primary productivity (NPP) of Wisconsin Fast Plants (the producers) growing under lights. Then, you will calculate the flow of energy from the Wisconsin Fast Plants to cabbage white butterfly larvae (the consumers that eat the plants).

Figure 19.22 models how energy flows through a plant.

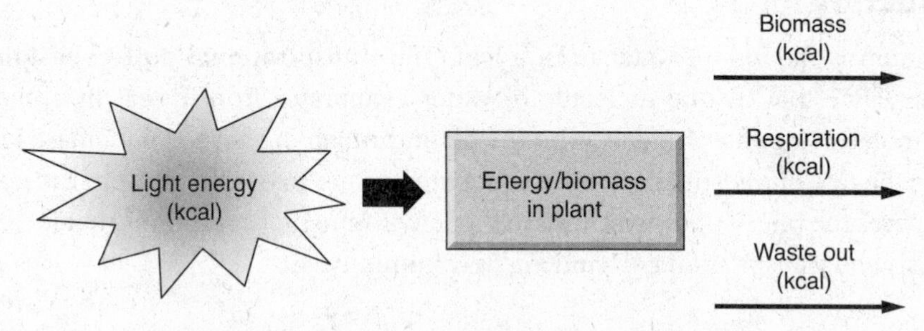

Figure 19.22 How Energy Flows Through a Plant

ESTIMATING NET PRIMARY PRODUCTIVITY (NPP) OF FAST PLANTS

Primary productivity is a rate. It is the energy captured by photosynthetic organisms in a given area per unit of time.

When light energy is converted to chemical energy in photosynthesis or transferred from one organism (a plant) to its consumer (insect), some energy will be lost as heat during the transfer.

In terrestrial ecosystems, productivity (or energy capture) is generally estimated by the change in biomass of plants produced over a specific time period. Measuring biomass or changes in biomass is simple. Take the mass of the organism(s) and record the mass over various time intervals.

Consider also that different organic compounds store different amounts of energy. Proteins and carbohydrates store about 4 kcal/g of dry weight. Fats store about 9 kcal/g of dry weight.

You must take this into consideration as you carry out your investigation.

Cabbage white butterfly larvae eat plants from the cabbage family. As with Wisconsin Fast Plants, measuring the energy flow into and out of these butterflies can be inferred from the biomass they gain and lose.

As butterfly larvae grow toward maturity, they pass through different developmental stages called **instars**. Use larvae that are already well along their 4th or 5th instar stage.

CONVERT BIOMASS MEASUREMENTS (GRAMS) TO ENERGY UNITS IN KILOCALORIES

For Wisconsin Fast Plants, assume that 1 gram (g) of dried biomass contains 4.35 kilocalories (kcal) of energy. Assume that the biomass of 4th-instar larvae is 40% of the wet mass.

Calculate the biomass of the larvae. For butterfly larvae, use an average value of 5.5 kcal/g of biomass to calculate the energy of each larva.

To determine the energy content in the larval frass (droppings), use 4.76 kcal of energy/g of frass. Calculate the frass lost per individual larva.

To determine the energy content of the brussels sprouts eaten by each larva, convert the wet mass of the sprout to dry mass and multiply by 4.35 kcal/g.

Use the estimated percentage of biomass (dry mass) in fresh Wisconsin Fast Plants that you previously calculated.

Graph your results.

LAB #11: TRANSPIRATION

Introduction

Transpiration is the loss of water from a leaf. The **transpirational pull–cohesion tension theory** explains that as one molecule of water evaporates from a leaf by transpiration, another molecule of water is pulled into the plant through the roots. An increase in sunlight increases the rate of transpiration (by causing more water to evaporate from the leaves) and also increases the rate of photosynthesis. Other environmental conditions that increase photosynthesis and transpiration are wind and low humidity.

Objective

In this lab, you will measure the rate of transpiration in a plant under various conditions in a controlled experiment.

Insert two-week-old bean seedlings (*Phaseolus vulgaris*) into a **potometer**, and measure the amount of water used by the plants under varying conditions of humidity (misting the leaves), light (a strong light bulb shining on the plant), and wind (a fan). Set up several potometers: one for each variable and one with only ambient conditions as the control; see Figure 19.23.

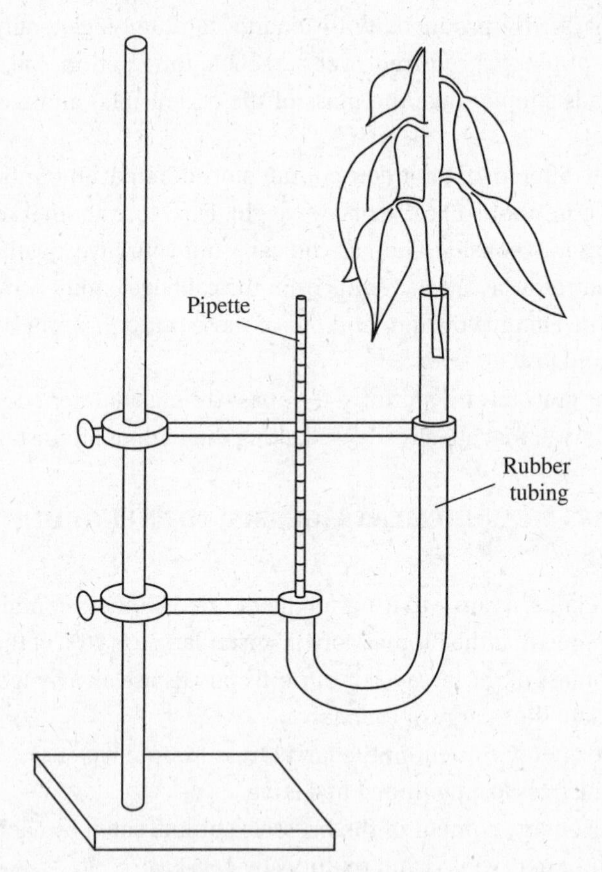

Figure 19.23 Potometer

Here is what you can expect to happen:

- Increasing the humidity, by placing a bag over or misting the leaves, will decrease transpiration.
- Increasing the light intensity will increase transpiration.
- Increasing the wind with a fan will lower the humidity around the leaves and thereby increase transpiration.

When you increase the light intensity, make sure that you also do not increase the temperature surrounding the plant, thus, introducing another variable. To accomplish this, place a **heat sink** (a large bowl or beaker of water) between the light and the plant to absorb the heat while allowing the light to pass through.

Since each plant does not have the same leaf surface area, your results will be meaningless unless you can demonstrate consistency. You must measure the surface area of the leaves on the plants and calculate the water loss in terms of square centimeters. To measure the surface area of the leaf, trace all the leaves from each plant onto graph paper with squares of known size. Construct a graph with "Total Water Loss (mL/cm^2)" on the *y*-axis and "Time (min)" on the *x*-axis, like the graph pictured below.

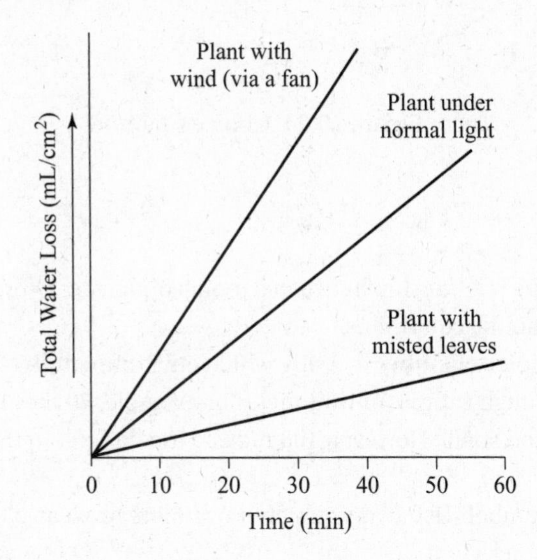

LAB #12: ANIMAL BEHAVIOR

Introduction

Drosophila melanogaster, the common fruit fly, has been studied by scientists since 1907. It lives throughout the world and feeds on rotting fruit. *Drosophila* is a model research organism. Students have been conducting genetic experiments with these flies for generations. However, most students today can conduct the classic genetic experiments using computer simulations instead of real fruit flies. In this experiment, you will use live fruit flies to study animal behavior.

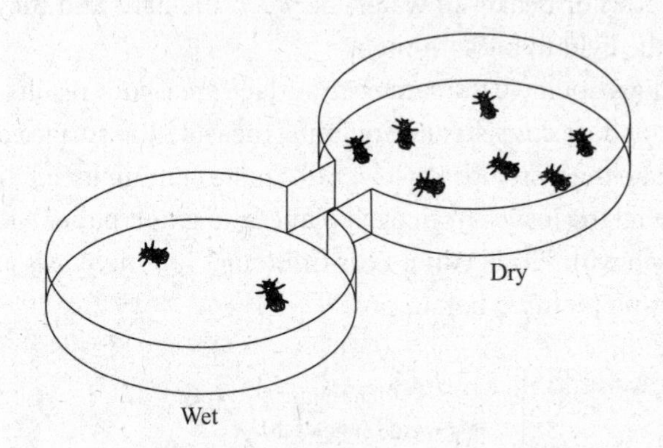

Figure 19.24 Choice Chamber

Objective

You will investigate the relationship between a model organism, *Drosophila*, and its response to different environmental conditions.

You will build a **choice chamber** to study which environment the flies prefer.

Choose a large enough sample of fruit flies. For example, 30 flies is a good sample size. A sample of 10 flies is too small. However, 100 may be too difficult to manage and might result in errors.

Build a choice chamber. Use two graduated cylinders or clear plastic water bottles with the necks cut off.

To test the hypothesis that fruit flies prefer specific lighting conditions, put them to sleep. Place the flies into the choice chamber, and seal the cover of the choice chamber with clear tape. Cover one side of the choice chamber to prevent light from entering it, and leave both chambers undisturbed. When the flies wake up, you can observe them. After what you determine is an appropriate amount of time, remove the cover and tape. Then place a piece of cardboard over each chamber to keep the flies in. Count the flies in each chamber.

To test the hypothesis that flies prefer one temperature over another, warm one half of the choice chamber. Before you put the flies into the chamber, make sure that you can control and measure the temperature of each chamber. Be careful that the temperature is not too warm or too cold. Carry out the experiment in the same way as before. After the same amount of time as before, carefully count the flies in each chamber to gather quantitative data.

You must always include a control in any experiment. In this case, your control is a chamber just like the experimental one except without altering the light or temperature. After all, you have to make sure that the flies are not moving from one part of the chamber to the other solely by chance.

LAB #13: ENZYME CATALYSIS

Introduction

In general, the rate at which enzymes catalyze reactions is altered by environmental factors such as pH, temperature, inhibitors, and salt concentration. In this lab, the enzyme used is one that is found in all aerobic cells: **peroxidase**. Its function in the cell is to decompose hydrogen peroxide, a by-product of cell respiration, into a less harmful substance. Here is the reaction:

$$2\,H_2O_2 \rightarrow 2\,H_2O + O_2\uparrow$$

In this lab, the enzyme peroxidase (which can be extracted from turnips) will be added to hydrogen peroxide (H_2O_2). As the H_2O_2 breaks down into water and oxygen, a color change will be observed. You will analyze this color change by either comparing the different resulting shades of brown to premade **reference tubes** or by taking an absolute reading of the light absorption from a **spectrophotometer**.

In a decomposition reaction of H_2O_2 by peroxidase, (as noted in the formula provided earlier), the easiest molecule to measure is the final product: oxygen. This is accomplished by using an indicator for the presence of oxygen. The compound **guaiacol** instantly binds to oxygen to form a compound that is brownish in color. *The greater the amount of oxygen gas that is present, the darker brown the solution will become.*

A series of dilutions of guaiacol will represent the relative changes that occurred during the reaction. You will use reference tubes (ranging from 1 to 10) to compare the relative amounts of oxygen produced.

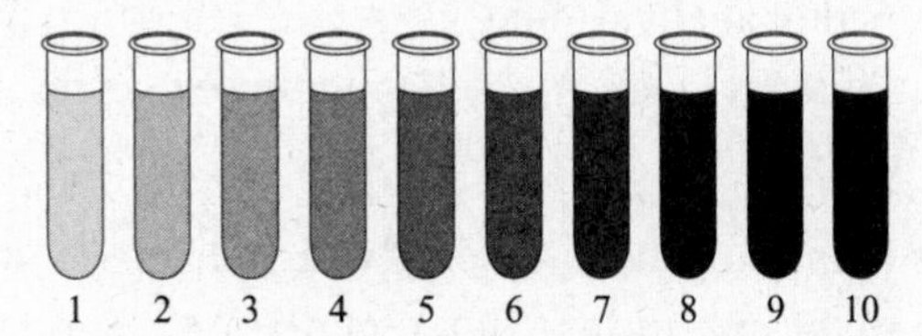

Objective

Mash up and liquefy a piece of turnip, and then filter it with cheesecloth and filter paper. Next, add the filtrate to distilled water to make your stock solution of peroxidase. To be consistent, use the same stock solution of peroxidase for every experiment.

Combine the H_2O_2 solution with the peroxidase enzyme at 4°C, and begin timing and observing the color changes. Take measurements at appropriate intervals (i.e., every 30 seconds). Repeat this procedure for different temperatures. Record your data.

A **spectrophotometer** has a detector that measures the transmittance or absorbance of light of a sample, depending upon how you set the dials. **Transmittance** refers to the amount of light that passes through the sample and strikes the detector. A perfectly transparent sample transmits 100% of the light. **Absorbance** is the opposite. It is a measurement of the light that is absorbed by the sample. In either case, the detector senses the light that is being transmitted through the sample and converts this information into a digital display. Be careful that, when looking at data from a spectrophotometer, you notice whether the data is for transmittance or for absorbance.

Table 19.3 represents sample data from a spectrophotometer. The higher the reading, the more light that is being absorbed by the sample and the darker brown the sample will be.

Table 19.3

Temperature (°C)	4	11	30	42	55	70	89	100
Absorption of Light % (According to the Spectrophotometer)	0.106	0.190	0.250	0.320	0.280	0.180	0.080	0.000

Below is a graph of the data from the spectrophotometer:

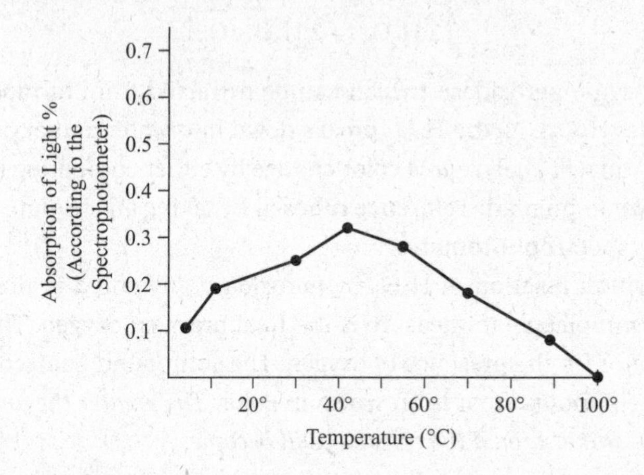

Initially, the enzyme activity increases slowly and reaches a maximum rate at approximately 42°C. Beyond that, the enzyme (protein) denatures as the heat continues to rise, until the enzyme does not function at all (at 100°C).

Questions that you may want to ask yourself (and these questions could lead to modifications of this experiment) include:

- What would happen to the enzyme function if you preheated the enzyme?
- What would happen to the enzyme at a different pH?
- What would happen to the enzyme function if you exposed the enzymes to acid before you began the experiment?
- Which has a greater effect on the rate of reaction: changing the concentration of enzymes or changing the concentration of the substrate?

HOW TO CALCULATE THE RATE OF AN ENZYME-MEDIATED REACTION

You can calculate the rate of an enzyme-mediated reaction by measuring either the disappearance of substrate over time or the appearance of product. Referring to the graph below, see how to calculate the rate of reaction for the first 20 seconds based on how much product is formed. Remember that rate $= \frac{\Delta y}{\Delta x}$.

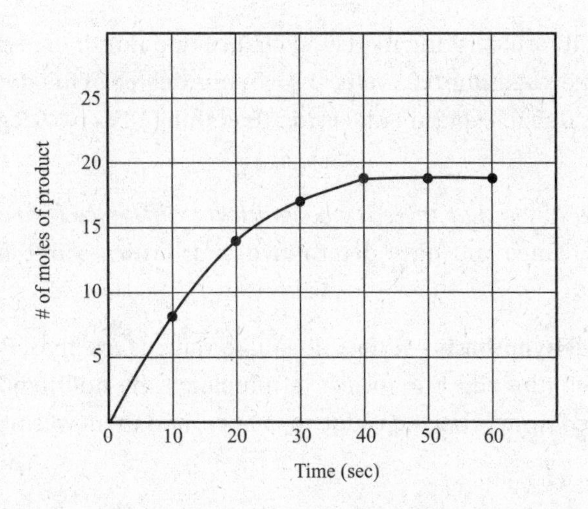

For this reaction, the rate for the first 20 seconds would be:

$$\frac{14 \text{ moles} - 0 \text{ moles}}{20 \text{ sec} - 0 \text{ sec}} = \frac{14}{20} = 0.70 \text{ moles/second of product}$$

Make sure you include the units label when writing your answer.

Now try a tricky question that fooled many students some years ago on the AP exam:

Using the same graph and reaction, calculate the rate of product formed from 40 to 50 seconds.

$$\frac{19 \text{ moles} - 19 \text{ moles}}{50 \text{ sec} - 40 \text{ sec}} = \frac{0}{10} = 0.0 \text{ moles per second}$$

The reason that no product formed is because the substrate was used up.

CHI-SQUARE TEST AND NULL HYPOTHESIS

A **chi-square** (χ^2) **test** compares your observed experimental data (*o*) to what you expected (*e*) and tells you if the two are statistically the same (due to chance) or statistically different (due to experimental error).

The equation from the reference table is: (chi-squared) χ^2 = (sum) $\Sigma = (o - e)^2 \div e$.

Degrees of freedom are equal to the number of distinct possible outcomes minus 1. If you expect two outcomes (i.e., black fur and white fur), that is one degree of freedom. If you expect black, white, and gray fur, there are three outcomes and two degrees of freedom. You need to determine degrees of freedom in order to use the chi-square table that follows.

CHI-SQUARE TABLE

	Degrees of Freedom							
p value	1	2	3	4	5	6	7	8
0.05 (95% confident)	3.84	5.99	7.81	9.49	11.07	12.59	14.07	15.51
0.01 (99% confident)	6.63	9.21	11.34	13.28	15.09	16.81	18.48	20.09

Calculate the critical value by using the chi-square equation. For one degree of freedom, if your chi-square value (*p*) is *less than* the critical value of 3.84, the difference in observed results from expected results is due merely to chance and not to experimental error. (Good thing.) If the χ^2 value exceeds the 0.05 value 3.84 for one degree of freedom, then the variation from the expected is due to experimental error. (Bad thing.) For two degrees of freedom, the critical value is 5.99.

The **null hypothesis** states *that there is no significant difference between the observed and expected frequencies.* Once you have determined your critical value from the Chi-Square Table, you can either:

Fail to reject the null hypothesis—If the chi-square value is *less* than the critical value, you *accept* or *fail to reject* (the odd but correct terminology) the null hypothesis, meaning the difference in observed from expected is due to chance, and there was no experimental error. Lucky you!

Reject the null hypothesis—If the chi-square value is *greater* than the critical value, you must reject the null hypothesis, meaning the difference between observed and expected is *not* due to chance, but is due to experimental error. The error may have been due to a technical error in carrying out the experiment or to some other factor that you failed to consider when devising the experiment.

CHAPTER SUMMARY

Twenty-five percent of instructional time in AP Biology is supposed to be devoted to hands-on laboratory work with an emphasis on inquiry-based investigations. That's a lot of time. Therefore, you can expect that lots of questions on the AP exam will focus on inquiry-related skills outlined by the 6 Science Practices. As you study these 13 labs, focus on the following skills (adapted from the College Board):

1. Describe and explain biological concepts, processes, and models.
2. Analyze drawings, graphs, and charts that show biological processes and concepts.
3. State the null hypothesis, or predict the results of an experiment, and identify testable questions based on an observation.
4. Construct a graph or chart from a data table.
5. Use mathematical calculations appropriately, and use and interpret confidence intervals and/or error bars.
6. Support a scientific claim with evidence from biological principles and concepts.

PRACTICE QUESTIONS

1. Below is an illustration of a beaker that contains a cube of plant tissue:

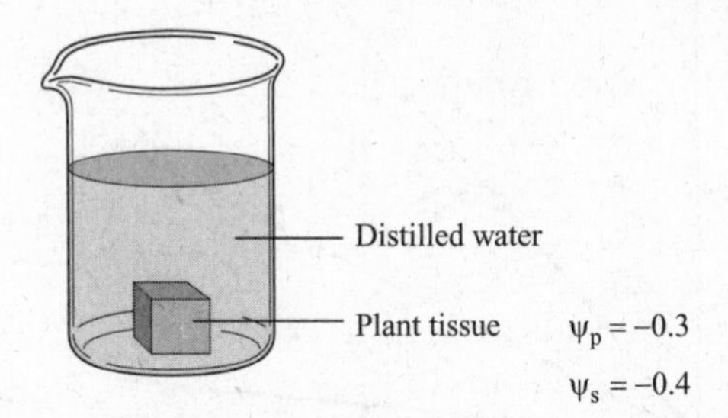

What is the water potential of the distilled water in the beaker, and what is the water potential of the plant tissue? Use this equation to help you solve this problem: $\Psi = \Psi_s + \Psi_p$

(A) The water potential of the distilled water in the beaker is 0. The water potential of the plant tissue is 0.
(B) The water potential of the distilled water in the beaker is –0.3. The water potential of the plant tissue is 0.
(C) The water potential of the distilled water in the beaker is 0. The water potential of the plant tissue is 0.3.
(D) The water potential of the distilled water in the beaker is 0. The water potential of the plant tissue is –0.1.

2. Which of the following correctly indicates the gradient of water potential from the highest to the lowest?

(A) root, soil, stem, leaf, air
(B) soil, root, stem, leaf, air
(C) air, leaf, stem, root, soil
(D) root, leaf, stem, soil, air

3. In a diffusion lab with 6 bags of dialysis tubing, why do you use the percent change in mass rather than simply change in mass?

(A) Doing so was arbitrary and applies to these instructions only; it really does not matter.
(B) Doing so is the convention.
(C) The percent change in mass is not as accurate.
(D) The bags did not all weigh exactly the same mass at the start.

4. Refer to the graph below.

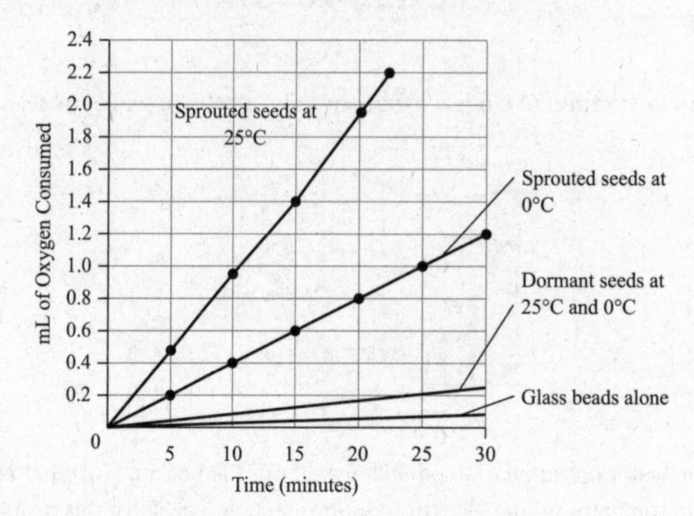

What is the rate of oxygen consumption in the germinating seeds at 0°C?

(A) 1.5 mL/min
(B) 0.04 mL/min
(C) 0.08 mL/min
(D) 25.0 mL/min

5. Which of the following statements is correct about cell division?

(A) Mitosis produces gametes.
(B) Mitosis produces daughter cells with half the haploid chromosome number.
(C) Meiotic cell division is responsible for the independent assortment of alleles.
(D) Most of the life of the cell is spent in meiotic cell division.

6. Below is a diagram that shows the “floating disk” method that you may have used to study photosynthesis.

Disks that are cut out of green plant leaves are soaked in a 1% solution of sodium bicarbonate ($NaHCO_3$) as a source of carbon dioxide. They are then transferred to a beaker that is also filled with sodium bicarbonate. They sink to the bottom of the beaker because the air spaces in the leaves have been replaced with $NaHCO_3$, making the leaves heavier. The experiment begins when a strong light is turned on. As the plant cells carry out photosynthesis, the disks slowly rise to the top of the solution in the beaker.

Which of the following is correct about this experiment?

(A) The plant cells produced bubbles of carbon dioxide and were carried upward by the light bubbles.
(B) You needed to test for the presence of oxygen in the bubbles with a glowing splint.
(C) The plant cells produced bubbles of oxygen, and thus the disks floated upward with the bubbles.
(D) The plant cells broke down sodium bicarbonate and lost mass.

7. If two bands on an electrophoresis gel have nearly the same base pair size, how might you best separate them?

(A) Make the gel from a more concentrated agarose solution; rerun the same DNA at a higher voltage.
(B) Make the gel from a more concentrated agarose solution; rerun the same DNA at a lower voltage.
(C) Make the gel from a less concentrated agarose solution; rerun the same DNA at a higher voltage.
(D) Make the gel from a less concentrated agarose solution; rerun the same DNA at a lower voltage.

8. You have two identical cultures of *E. coli* bacteria: culture A and culture B. You transform culture A, heat shocking it in the presence of a plasmid that is resistant to ampicillin. You transform culture B, heat shocking it in the presence of a plasmid that is resistant to tetracycline. You then prepare four agar Petri dishes, with and without antibiotic, as shown on the following table.

Plate I	With ampicillin only
Plate II	With tetracycline only
Plate III	With no antibiotics
Plate IV	With both antibiotics

What do you predict will be the results of plating the bacteria from culture A and culture B on all four agar Petri dishes, if both cultures of bacteria took up the plasmids they were exposed to?

(A) Bacteria B will grow on plates II and III only.
(B) Bacteria A will grow on plate I only.
(C) Bacteria B will grow on plate III only.
(D) Bacteria A will grow on plates I, II, and III.

9. The allele for the hair pattern called widow's peak is dominant over the allele for no widow's peak. In a population of 1,000 individuals, 841 show the dominant phenotype. How many individuals in this population would you expect to be hybrid for widow's peak?

(A) 16%
(B) 24%
(C) 48%
(D) 100%

10. In a certain population, the dominant phenotype of a certain condition occurs 91% of the time. What is the frequency of the dominant allele?

(A) 0.3
(B) 3%
(C) 0.7
(D) 7%

11. If a population is in Hardy-Weinberg equilibrium and the frequency of the recessive phenotype for a particular trait is 16% of the population, what will the frequency for the hybrid condition be in 500 years?

(A) 16%
(B) 24%
(C) 48%
(D) 64%

12. You need to take sample leaves from trees in a local forest as part of a broader study that you are carrying out on sunlight and the rate of photosynthesis. You are concerned about bias when taking samples. So you ask a friend, who has never taken a Biology class before, to take sample leaves for you. You believe that she will be able to do this task without bias, whereas you would be biased because it is your project. You instruct her to remove 100 healthy leaves randomly from the forest. Is this the most scientific way to take samples? If so, why? If not, why not?
    (A) Yes, it is. Your friend is unbiased, and the leaves will be taken randomly.
    (B) Yes, it is. The most important thing is that your friend does not know anything about the project and will take samples in an unbiased fashion.
    (C) No, it is not. Although the leaves will be chosen randomly, your friend is not really unbiased.
    (D) No, it is not. Sampling requires detailed and specific instructions from the investigator.

13. How many base pairs is the fragment of DNA that is circled in the following gel?

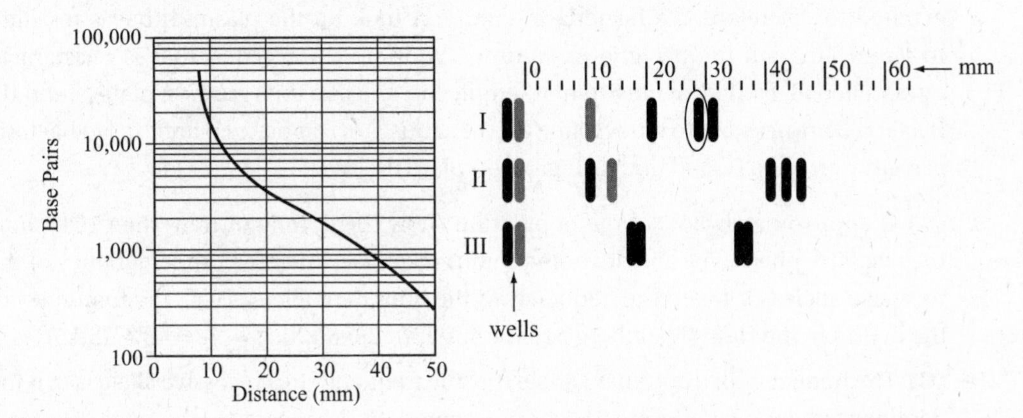

    (A) 1,100
    (B) 2,000
    (C) 2,800
    (D) 10,000

Answers Explained

1. **(D)** By definition, the water potential of an open container that is filled with distilled water is zero (0). Since the water potential equals the solute potential of the plant tissue (–0.4) plus the pressure potential (–0.3), the answer is –0.1. (SP 4)

2. **(B)** Water moves from the highest water potential to the lowest. In a plant, water moves from the soil to the roots to the stems and leaves and into the air spaces in the spongy mesophyll and out to the environment. (SP 4)

3. **(D)** Since the bags did not all weigh the same mass at the start, the only way to control for the variation is to show change in mass as a percentage. (SP 4)

4. **(B)** The rate is equal to the product divided by the time. Choose any point on the line for 0°C. For example, 0.4 divided by 10 equals 0.04. Also, 0.2 divided by 5 equals 0.04. This value (0.04) will be the resulting rate for any point on the graph. (SP 4)

5. **(C)** There are two cell divisions: meiosis, which is for the production of gametes only, and mitosis, which is for growth and repair. The independent assortment of alleles occurs during meiosis when homologous chromosomes line up on the metaphase plate and separate. Mitosis produces daughter cells that are identical to the parent cells. (SP 2)
6. **(C)** $NaHCO_3$ is the source of CO_2 for photosynthesis. As photosynthesis occurs, this CO_2 from the air spaces in the leaves is used up and replaced by bubbles of oxygen. This oxygen rises upward in the beaker, carrying the leaves with it. While it is correct that the test for the presence of oxygen is a glowing splint that bursts into flames, you did not test for the presence of oxygen in this experiment. You inferred oxygen's presence. The leaves did not lose mass because the CO_2 was replaced by O_2. (SP 3)
7. **(B)** If the gel is more concentrated, it will present more of an impediment and separate the two bands more effectively. Another method would be to lower the voltage and allow them to run very slowly and overnight. One must simply try both techniques to see which works better. (SP 6)
8. **(A)** The question told us to assume that each bacterium took up the plasmid it was exposed to. Therefore, the bacteria in culture A took up the plasmid that is resistant to ampicillin, and the bacteria in culture B took up the plasmid that is resistant to tetracycline. Bacteria A are resistant to ampicillin and can thus grow on plates I and III. Bacteria B are resistant to tetracycline and can thus grow on plates II and III. No bacteria can grow on plate IV, and both can grow on plate III. (SP 3)
9. **(C)** If approximately 84% of the population show the dominant trait, then 16% show the recessive phenotype and are homozygous recessive. Therefore, the frequency of the recessive allele is 0.4, and the frequency of the dominant allele is 0.6. The frequency of the hybrid in the Hardy-Weinberg equation is $2pq$: $2(0.6 \times 0.4) = .48 = 48\%$. (SP 5)
10. **(C)** The frequency of a dominant allele is p; the frequency of a recessive allele is q. If the dominant phenotype appears 91% of the time, then the recessive phenotype appears 9% of the time. If $q^2 = 9\%$ or 0.09, then $q = 0.3$. Therefore, $p = 0.7$. (SP 5)
11. **(C)** The population is in equilibrium; it will not change. If $q^2 = 0.16$, then $q = 0.4$ and $p = 0.6$. Therefore, the frequency of the hybrid is 48%. (SP 5)
12. **(D)** Experimental protocols must be detailed with no room for whim. The fact that the friend in question does not know any Biology is not relevant. She must be given specific instructions about which leaves to choose. The details must be clear and reasonable to follow. (SP 6)
13. **(B)** The DNA fragment is at 28 mm on the ruler. When you find 28 mm on the semi-log paper and move your finger up to touch the line, it reads at 2,000 base pairs. (SP 4)

PART 5: SCORING FOR SECTION II OF THE AP BIOLOGY EXAM

Once you've finished reviewing all of the subject area review chapters, the illustrative examples review chapters, and the chapter on investigative labs, you should be ready to tackle the **Section II free-response questions (FRQs)**. Section II of the AP Biology exam consists of two long free-response questions and four short free-response questions. These two question types have their own unique formats for this exam. The following chapter will provide an overview of both question types, and then present you with one practice long free-response question and one practice short free-response question, with sample answers for both and an analysis of how both sample answers would be graded on the actual exam.

For both question types, remember that the graders want to see an answer to the specific question that was asked, so be sure that your response directly answers all parts of the question. You must include correct, clear, and concise information *without contradicting yourself.* Learning how your answers will be graded, and studying the merits and flaws of the sample answers within the following chapter, will save you time (because you'll learn how to get straight to the point of each question), reduce your stress (because you will know exactly what the graders are looking for), and help improve your score on the actual exam!

How the College Board Grades Your Answers to the Free-Response Questions 20

→ **LONG FREE-RESPONSE QUESTION**
→ **SAMPLE ANSWER**
→ **ANALYSIS OF SAMPLE ANSWER**
→ **SHORT FREE-RESPONSE QUESTION**
→ **SAMPLE ANSWER**
→ **ANALYSIS OF SAMPLE ANSWER**

INTRODUCTION

If you want to write a high-scoring answer to a free-response question (FRQ), you have to know how it will be graded.

Section II of the AP Biology exam consists of two long free-response questions and four short free-response questions. The two long free-response questions take about 25 minutes each to answer, are usually worth 8 to 10 points each, focus on interpreting and evaluating experimental results (with and without graphing), and are broken down into four parts each. The four short free-response questions take about 8 to 10 minutes each, are usually worth 4 points each, focus on scientific investigations, conceptual analysis, analyzing models or visual representations, and analyzing data, and are also broken down into four parts each. For a complete list of all question types and the breakdown of each question type, refer to the "AP Biology Course and Exam Description" from the College Board: *https://apcentral.collegeboard.org/pdf/ap-biology-course-and-exam-description.pdf?course=ap-biology.*

Included within this chapter is a practice long free-response question (with a sample answer) followed by a practice short free-response question (also with a sample answer). Each time the sample answer includes relevant and correct terminology or a correct explanation of a concept, a point is given in the margin. Following each sample answer is an analysis of what the student would have or would not have received credit for. For each sample answer, you may want to first cover the numbers in the margin, read the answer, and then try to grade it yourself. Remember, you can only earn the maximum number of points that are stated for each part of the question.

LONG FREE-RESPONSE QUESTION

(Note: This question is worth 10 points total.)

(a) **Describe** and **explain** what an enzyme is and how it functions. (2 points)

(b) **Choose** one enzyme, and **identify** experimental design procedures (using different conditions of temperature or pH) that would demonstrate inhibition of an enzymatic reaction. (4 points)

(c) Below is a graph that shows the rate of two enzyme-mediated reactions:

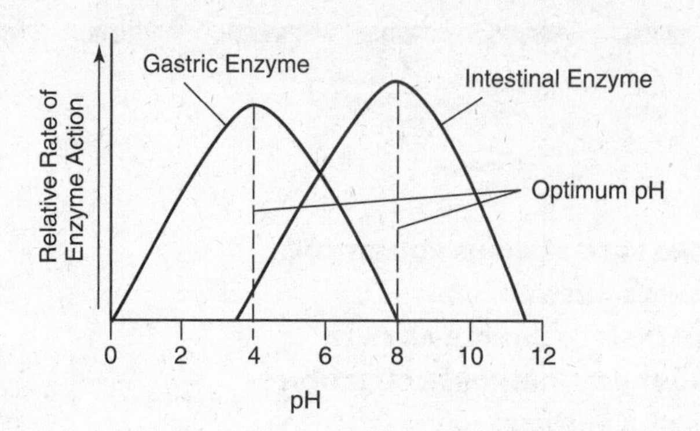

Analyze the results shown on this graph. (2 points)

(d) **Make** your own predictions about the results that you would expect for the experiment you designed in part (b). **Justify** your predictions. (2 points)

PRACTICE

How many points would you give this answer?

SAMPLE ANSWER

(a) An enzyme is a protein catalyst that reacts with every substrate and speeds up every type of reaction. An enzyme speeds up reactions by lowering the reaction's energy of activation (E_a), which is the amount of energy that is required to start the reaction. Enzymes are specific, meaning they only react in certain ways with certain substrates. This specificity is determined by the shape of the substrate and the shape (conformation) of the enzyme. The specific shape of an enzyme results from interactions among the R-groups of the amino acids that make up the enzyme. In order for the substrate to bond to an enzyme, the substrate slightly alters the shape of an enzyme, so that the fit is exact. The model for the specificity of enzyme function is called "lock and key."

1

1

1

(b) I would choose an enzyme that was recently extracted from the human liver. The easiest way to measure the amount of oxygen that is released is to use an indicator called guaiacol, which turns a brownish color when it binds to oxygen in a solution. One can determine how much oxygen is in a sample simply by observing how brown the solution is.

1

I would make a series of dilutions of guaiacol, plus different amounts of oxygen, that range from clear (little oxygen present) to dark brown (lots of oxygen present). I would then use a spectrophotometer to quantify how dark the color of the dilution is, which will help me then determine the amount of oxygen that is present in each tube. Then, I would use the guaiacol dilutions for comparison once I had results from the experiment.

1

The reaction that is the basis of this experiment is as follows:

Enzyme + Substrate → Product + Indicator → Brown color
Enzyme + Peroxide (H_2O_2) → Oxygen + Guaiacol → Brown color

To determine the rate of enzyme activity and the inhibition of an enzymatic reaction, I would carry out this experiment using the following different temperatures: 4°C, 15°C, 25°C, 38°C, 50°C, 70°C, and 100°C. 1

(c) The graph shows two enzyme-mediated reactions. One of these reactions shows an enzyme that was extracted from the stomach, an environment that is acidic. The other reaction shows an enzyme that was extracted from the intestine, an environment that is normally basic (alkaline).

The reaction with the gastric enzyme shows no reactivity at pH 0, with activity increasing in the acidic range until it reaches its maximum rate (velocity) of activity (V_{max}) at the optimum pH 4. Above pH 4, the reaction slows down, decreasing to zero by pH 8.

The reaction with the intestinal enzyme shows no reactivity at approximately pH 3.8, with increasing activity until it reaches its maximum rate (V_{max}) at its optimum pH 8. Above pH 8, the reaction slows down, eventually reaching zero by pH 11.8. 1

(d) In the experiment from part (b), I predict that the rate of the enzyme-controlled reaction would increase from zero to its maximum rate (at the optimal temperature [body temperature] of approximately 38°C), and then the reaction would be inhibited and the rate of the reaction would decrease until the reaction ceased totally. Refer to the following graph: 1

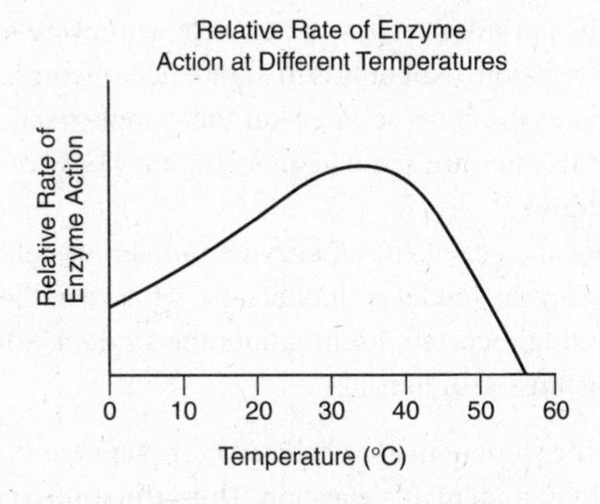

I made this prediction because each enzyme functions in the way that it does because of its unique shape or conformation. The rate of an enzyme reaction is optimal within a narrow range of temperatures. Outside that narrow range, the intramolecular bonds that hold an enzyme in its specific conformation break, causing the enzyme to lose its special shape. Enzymes denature at high temperatures, and, when they do, the reaction is inhibited. 1

ANALYSIS OF SAMPLE ANSWER

General Guidelines

The numbers in the margins of the sample answer represent points of merit in each part of the answer. Although the student may have presented many correct statements in each part of his or her responses, remember that *it is only possible to earn the maximum number of points that are stated for each part of the question.* For example, even if the student included 8 correct statements in his or her response to part (a), the student could only receive the maximum number of points for that part, which in this case is 2 points. If a term was used but not defined or explained correctly, no credit was awarded. If the student contradicted himself or herself within an answer to one part of the question, the student lost all points for that part of the question.

Out of 10 total points possible, this answer would have earned 6 points. The breakdown of these point distributions is as follows:

(a) The student received points for correctly stating the following:

- Enzymes lower a reaction's energy of activation (E_a), which is the amount of energy that is required to start the reaction. (1 point)
- Enzymes are specific because of their shape or conformation. (1 point)
- The specific shape of an enzyme results from interactions among the R-groups of the amino acids that make up the enzyme. (1 point)

The student also included some contradictory statements and/or incorrect statements and/or was missing information (for which he or she either lost points [only for the contradictory statements] or did not earn any points):

- An enzyme is a protein catalyst that reacts with every substrate and speeds up every type of reaction. (Not only is this statement incorrect, but it also **contradicts** the statement in the third sentence of the sample answer, which is correct and states that "Enzymes are specific, meaning they only react in certain ways with certain substrates.")
- The model for the specificity of enzyme function is called "lock and key." (The correct name for the model is "induced fit." However, the student would not lose points for stating incorrect information; the student would only lose points for contradicting himself or herself.)

In this case, the student made a contradictory statement, so he or she would have lost all points for part (a) of this question. **Thus, the student received 0 points out of a possible 2 points total for part (a).**

(b) The student received points for correctly stating the following:

- The easiest way to measure the amount of oxygen that is released is to use an indicator called guaiacol, which turns a brownish color when it binds to oxygen in a solution. (1 point)
- I would then use a spectrophotometer to quantify how dark the color of the dilution is, which will help me then determine the amount of oxygen that is present in each tube. (1 point)

- To determine the rate of enzyme activity and the inhibition of an enzymatic reaction, I would carry out this experiment using the following different temperatures: 4°C, 15°C, 25°C, 38°C, 50°C, 70°C, and 100°C. (1 point)

The student also included some incorrect statements and/or was missing information (for which he or she did not earn any points):

- The student did not name the enzyme that was used. (Peroxidase is an enzyme, found in all cells, that breaks down peroxide, which is a toxic by-product of cellular metabolism.)
- The student did not explain how the spectrophotometer measures color by the rate of absorbance or transmittance of light. (If the spectrophotometer is set to measure absorbency, then the darker brown the sample is, the more light will be absorbed by the sample and the more oxygen it will contain.)

Thus, the student received 3 points out of a possible 4 points total for part (b).

(c) The student received a point for correctly stating the following:

- The student correctly identified which enzymes are shown on the graph and correctly identified the arc of each enzyme reaction. (1 point)

The student also included some incorrect statements and/or was missing information (for which he or she did not earn any points):

- The student did not describe the graph perfectly. (The student did not specifically state the labels on the *x*-axis and the *y*-axis. Remember, when you draw a graph or describe a graph on the actual exam, your drawing or description must be complete, including the labels for all axes.)

Thus, the student received 1 point out of a possible 2 points total for part (c).

(d) The student received points for correctly stating the following:

- The student correctly made a prediction about the results that he or she would expect for the experiment designed in part (b). The prediction was that the enzyme-controlled reaction would be inhibited above approximately 38°C (body temperature) and that the decrease in the rate of the reaction would continue until the reaction ceased totally. (1 point)
- The student also correctly justified his or her prediction. The student correctly explained the cause of denaturation in an enzyme (that heat breaks intramolecular bonds that give the enzyme its specific shape and function). The student also stated that denaturation inhibits an enzyme-controlled reaction. (1 point)

Nothing was incorrect or missing from this response. **Thus, the student received 2 points out of a possible 2 points total for part (d).**

SHORT FREE-RESPONSE QUESTION

(Note: This question is worth 4 points total.)

In a population of royal flycatchers in Costa Rica, the allele for a large crown on the head (*C*) is dominant over the allele for a small crown on the head (*c*). It has been determined that, for this population, the percentage of royal flycatchers with small crowns has remained at 11% since 1850 (when the population was first studied).

(a) Describe one process that is responsible for the maintenance of the unchanged frequency of small-crowned alleles in this population of royal flycatchers. (1 point)

(b) Explain the standard by which scientists determine which individual is the fittest in any population. (1 point)

(c) Predict what would occur in this population if the royal flycatchers, that were born to a mating couple over several years, experienced a mutation and were all born with a very different colored crown. (1 point)

(d) Justify your prediction (from your answer to part (c)). (1 point)

PRACTICE

How many points would you give this answer?

SAMPLE ANSWER

(a) This population is in Hardy-Weinberg equilibrium. Thus, this population is isolated with no natural selection. 1

(b) According to Charles Darwin's theory of natural selection, only the fittest survive. The fittest individual is the individual that has the most offspring and passes more of its genes onto the next generation. 1

(c) If the mutation is advantageous, its frequency will increase in the population. If the mutation is disadvantageous, its frequency will decrease in the population. 1

(d) If the mutation is advantageous, the individual with the mutated trait will be more attractive and fitter than other individuals. That individual will then mate with more certainty and have more offspring. Thus, its offspring will carry the mutated trait and then pass it onto their offspring, increasing the frequency of the mutated trait in the population. Conversely, if the mutation is disadvantageous, the individual with the mutated trait will be less fit and will reproduce less. Thus, the frequency of the mutated trait will decrease in the population. 1

ANALYSIS OF SAMPLE ANSWER

General Guidelines

The numbers in the margins of the sample answer represent points of merit in each part of the answer.

Out of 4 total points possible, this answer would have earned all 4 points. The breakdown of these point distributions is as follows:

(a) The student received a point for correctly stating the following:

- This population is in Hardy-Weinberg equilibrium. Thus, this population is isolated with no natural selection. (1 point)

The student earned the full 1 point for this response. **Thus, the student received 1 point out of a possible 1 point total for part (a).**

Similarly, the student could have also earned the full 1 point for any of the following statements:

- This population is isolated with no mutation. (1 point)
- This population is isolated with no immigration from other populations of royal flycatchers. (1 point)
- This population is isolated with random mating. (1 point)

(b) The student received a point for correctly stating the following:

- The fittest individual is the individual that has the most offspring and passes more of its genes onto the next generation. (1 point)

The student earned the full 1 point for this response because it explained how scientists determine which individual is the fittest in any population (by determining which individual has the most offspring and passes more of its genes onto the next generation). **Thus, the student received 1 point out of a possible 1 point total for part (b).**

(c) The student received a point for correctly stating the following:

- If the mutation is advantageous, its frequency will increase in the population. If the mutation is disadvantageous, its frequency will decrease in the population. (1 point)

The student earned the full 1 point for this response because it predicted what would occur in this population if the mutation was advantageous and what would happen in this population if the mutation was disadvantageous. **Thus, the student received 1 point out of a possible 1 point total for part (c).**

(d) The student received a point for correctly stating the following:

- If the mutation is advantageous, the individual with the mutated trait will be more attractive and fitter than other individuals. That individual will then mate with more certainty and have more offspring. Thus, its offspring will carry the mutated trait and then pass it onto their offspring, increasing the frequency of the mutated trait in the population. Conversely, if the mutation is disadvantageous, the individual with the mutated trait will be less fit and will reproduce less. Thus, the frequency of the mutated trait will decrease in the population. (1 point)

The student earned the full 1 point for this response because it fully and clearly justified the prediction made in the answer to part (c) of this question. **Thus, the student received 1 point out of a possible 1 point total for part (d).**

PART 6: PRACTICE TESTS

The following section contains two full-length **practice tests** that mirror the format, content tested, and level of difficulty of the latest AP Biology exam. Immediately after each test, there is an answer key as well as detailed answer explanations for all questions. If you want to maximize your score, be sure to take at least one of these two tests weeks before the actual exam (do not wait until the day before the AP exam to take both of these tests). When taking these tests, if you can, try to mimic actual testing conditions: set aside the full three hours needed for the test, answer all questions in one sitting, and be sure to turn off your cell phone and music! However, if you're pressed for time and cannot spend the full three hours on each test, take one section of each test at a time (i.e., set aside 90 minutes and just complete Section I of Practice Test 1; then, when you have more time, set aside another 90 minutes and just complete Section II of Practice Test 1). Whichever method you choose, be sure to complete the entire practice test before reviewing the answer explanations. Review those answer explanations closely to see what topics you are solid on and what topics you may want to go back and review once more before test day.

Note that, on the actual exam, you will be provided with the official "AP Biology Equations and Formulas" sheet, which can be found on the College Board's AP Biology website. For convenience, you can find those equations and formulas in Appendix C of this book.

ANSWER SHEET
Practice Test 1

SECTION I

1. Ⓐ Ⓑ Ⓒ Ⓓ
2. Ⓐ Ⓑ Ⓒ Ⓓ
3. Ⓐ Ⓑ Ⓒ Ⓓ
4. Ⓐ Ⓑ Ⓒ Ⓓ
5. Ⓐ Ⓑ Ⓒ Ⓓ
6. Ⓐ Ⓑ Ⓒ Ⓓ
7. Ⓐ Ⓑ Ⓒ Ⓓ
8. Ⓐ Ⓑ Ⓒ Ⓓ
9. Ⓐ Ⓑ Ⓒ Ⓓ
10. Ⓐ Ⓑ Ⓒ Ⓓ
11. Ⓐ Ⓑ Ⓒ Ⓓ
12. Ⓐ Ⓑ Ⓒ Ⓓ
13. Ⓐ Ⓑ Ⓒ Ⓓ
14. Ⓐ Ⓑ Ⓒ Ⓓ
15. Ⓐ Ⓑ Ⓒ Ⓓ
16. Ⓐ Ⓑ Ⓒ Ⓓ
17. Ⓐ Ⓑ Ⓒ Ⓓ
18. Ⓐ Ⓑ Ⓒ Ⓓ
19. Ⓐ Ⓑ Ⓒ Ⓓ
20. Ⓐ Ⓑ Ⓒ Ⓓ
21. Ⓐ Ⓑ Ⓒ Ⓓ
22. Ⓐ Ⓑ Ⓒ Ⓓ
23. Ⓐ Ⓑ Ⓒ Ⓓ
24. Ⓐ Ⓑ Ⓒ Ⓓ
25. Ⓐ Ⓑ Ⓒ Ⓓ
26. Ⓐ Ⓑ Ⓒ Ⓓ
27. Ⓐ Ⓑ Ⓒ Ⓓ
28. Ⓐ Ⓑ Ⓒ Ⓓ
29. Ⓐ Ⓑ Ⓒ Ⓓ
30. Ⓐ Ⓑ Ⓒ Ⓓ
31. Ⓐ Ⓑ Ⓒ Ⓓ
32. Ⓐ Ⓑ Ⓒ Ⓓ
33. Ⓐ Ⓑ Ⓒ Ⓓ
34. Ⓐ Ⓑ Ⓒ Ⓓ
35. Ⓐ Ⓑ Ⓒ Ⓓ
36. Ⓐ Ⓑ Ⓒ Ⓓ
37. Ⓐ Ⓑ Ⓒ Ⓓ
38. Ⓐ Ⓑ Ⓒ Ⓓ
39. Ⓐ Ⓑ Ⓒ Ⓓ
40. Ⓐ Ⓑ Ⓒ Ⓓ
41. Ⓐ Ⓑ Ⓒ Ⓓ
42. Ⓐ Ⓑ Ⓒ Ⓓ
43. Ⓐ Ⓑ Ⓒ Ⓓ
44. Ⓐ Ⓑ Ⓒ Ⓓ
45. Ⓐ Ⓑ Ⓒ Ⓓ
46. Ⓐ Ⓑ Ⓒ Ⓓ
47. Ⓐ Ⓑ Ⓒ Ⓓ
48. Ⓐ Ⓑ Ⓒ Ⓓ
49. Ⓐ Ⓑ Ⓒ Ⓓ
50. Ⓐ Ⓑ Ⓒ Ⓓ
51. Ⓐ Ⓑ Ⓒ Ⓓ
52. Ⓐ Ⓑ Ⓒ Ⓓ
53. Ⓐ Ⓑ Ⓒ Ⓓ
54. Ⓐ Ⓑ Ⓒ Ⓓ
55. Ⓐ Ⓑ Ⓒ Ⓓ
56. Ⓐ Ⓑ Ⓒ Ⓓ
57. Ⓐ Ⓑ Ⓒ Ⓓ
58. Ⓐ Ⓑ Ⓒ Ⓓ
59. Ⓐ Ⓑ Ⓒ Ⓓ
60. Ⓐ Ⓑ Ⓒ Ⓓ

PRACTICE TEST 1

SECTION I

Time: 90 minutes

Directions: For each question, choose the best answer choice.

1. Below is a sketch of a molecule of the sex hormone, testosterone, that is derived from cholesterol.

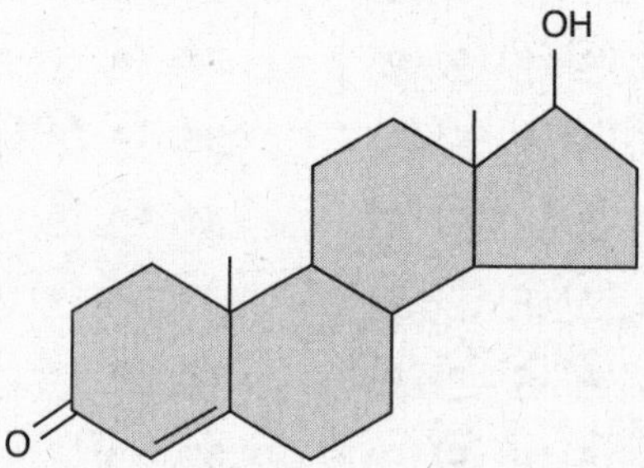

Which of the following statements best describes the action of this hormone on cells of the human gonads?

(A) The hormone acts as the first messenger when it binds to and activates the G protein-coupled receptor on the surface of cells in the testes. This activates the mobile G protein located inside the cell.

(B) The hormone enters cells in the testes by first binding with a membrane receptor, which causes a channel to open in the membrane, allowing the testosterone to flood into the cell.

(C) The hormone readily passes through the cell membrane and binds to a receptor in the cytoplasm. The hormone and receptor then enter the nucleus and act as a transcription factor that turns on one or more genes.

(D) The hormone binds with cAMP on the surface of the cell. Once attached to cAMP, the hormone enters the cell and initiates a signal transduction pathway.

2. Mice are mammals. As such, they are endotherms. They maintain internal heat metabolically. A lizard is an ectotherm and gains its body heat from the environment. Which of the following graphs most accurately depicts this situation?

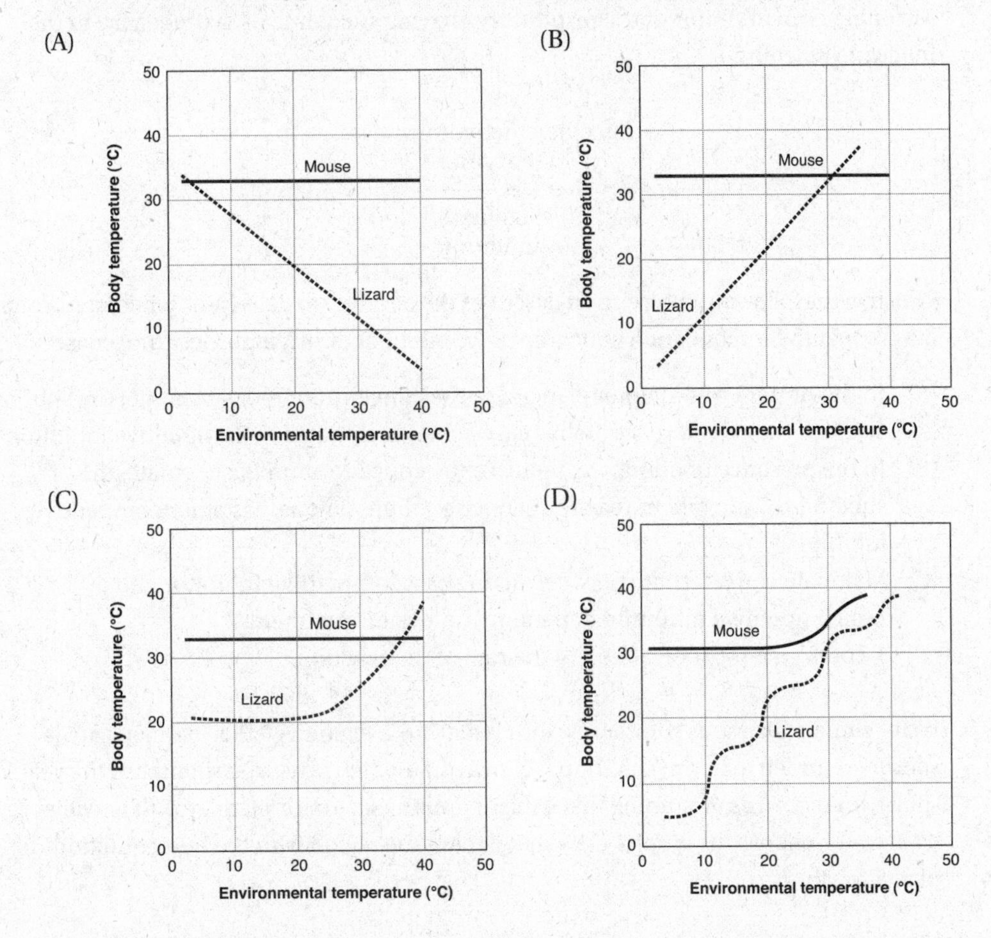

3. Which of following is the first result of depolarizing the presynaptic membrane of an axon? (Refer to the illustration below to help you answer this question.)

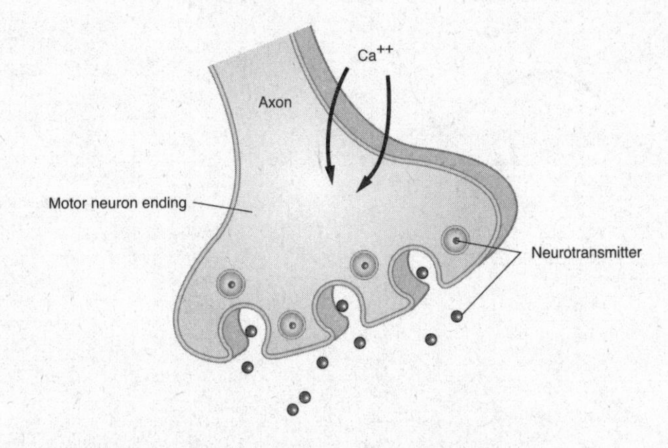

(A) The postsynaptic cell produces an action potential.
(B) Synaptic vesicles fuse with the presynaptic membrane.
(C) Voltage-gated calcium channels in the membrane open.
(D) Neurotransmitter is released into the synapse.

4. There are two types of enzyme inhibition: competitive and noncompetitive. Competitive inhibitors compete with the substrate for one active site, while noncompetitive inhibitors bind to another part of the enzyme and alter its shape. Malonate is an inhibitor of the respiratory enzyme succinate dehydrogenase in the following reaction:

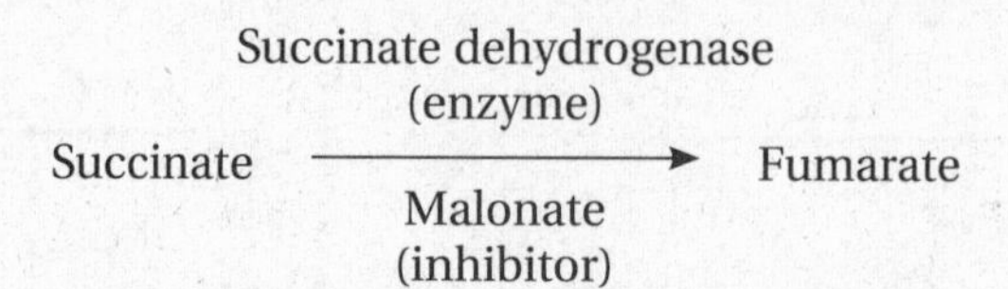

Which of the following choices best describes the best way to determine whether malonate is a competitive inhibitor or a noncompetitive inhibitor of succinate dehydrogenase?

(A) In the presence of malonate, increase the concentration of succinate, the substrate. If the rate of reaction increases, then malonate is a competitive inhibitor.
(B) In the presence of malonate, increase the concentration of succinate, the substrate. If the rate of reaction increases, then malonate is a noncompetitive inhibitor.
(C) Malonate can alternate between acting as a competitive inhibitor and a noncompetitive inhibitor, depending on what is required.
(D) A coenzyme must be affecting the rate of the reaction.

5. Water climate diagrams summarize the climate for a region. Average temperature is shown on the left axis, precipitation is shown on the right axis, and months of the year is shown on the *x*-axis. Assuming that a major limiting factor for plant growth is availability of water, choose the graph below that depicts the region with the best conditions for plant growth.

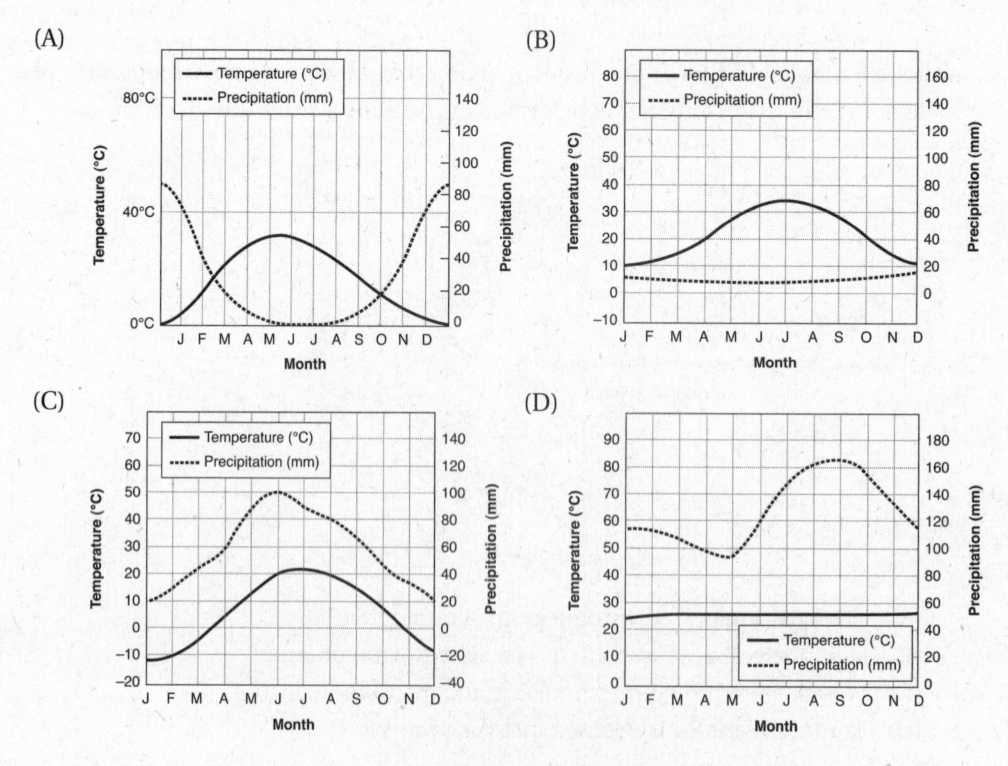

6. Below is a graph that shows the rate of photosynthesis of a green plant plotted against light intensity:

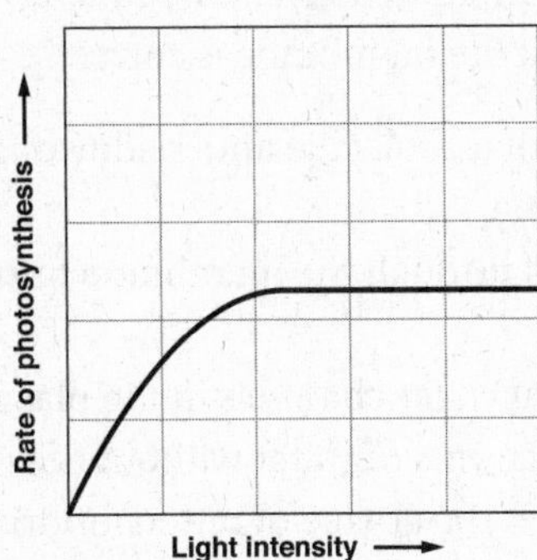

Which of the following statements accurately accounts for the shape of the graph?

(A) The rate of photosynthesis stops increasing because chlorophyll molecules begin to decompose due to increased heat, not increased light.
(B) The rate of photosynthesis slows as chlorophyll molecules, which are proteins, denature.
(C) Photosynthesis ceases when protons are no longer released by the photolysis of water.
(D) The rate of photosynthesis increases until the light-harvesting apparatus in the thylakoid membranes becomes saturated and cannot make use of additional light.

7. The lac operon is only found in prokaryotes and consists of structural genes and a promoter and operator. Which of the following statements explains why we study the lac operon?

(A) It represents a principal means by which gene transcription can be regulated.
(B) It explains how baby mammals utilize lactose from their mothers as they nurse.
(C) It illustrates how RNA is processed after it is transcribed.
(D) It illustrates possible control of the cell cycle and may lead to an understanding of cancer.

8. Below is the simplified equation for photosynthesis:

$$6CO_2 + 12H_2O \xrightarrow{\text{light}} C_6H_{12}O_6 + 6H_2O + 6O_2$$

Which of the following choices correctly traces the atom identified with an arrow?

(A)

$6CO_2 + 12H_2O \longrightarrow C_6H_{12}O_6 + 6H_2O + 6O_2$

(B)

$6CO_2 + 12H_2O \longrightarrow C_6H_{12}O_6 + 6H_2O + 6O_2$

(C)

$6CO_2 + 12H_2O \longrightarrow C_6H_{12}O_6 + 6H_2O + 6O_2$

(D)

$6CO_2 + 12H_2O \longrightarrow C_6H_{12}O_6 + 6H_2O + 6O_2$

9. The plasma membrane is selectively permeable and regulates what enters and leaves a cell. It consists of a phospholipid bilayer embedded with proteins. Which of the following statements about the membrane is correct?

    (A) Carbon dioxide is a polar molecule and readily diffuses through the hydrophilic layers of the membrane.
    (B) Starch readily diffuses through the membrane into the liver, where it is stored as glycogen.
    (C) Aquaporins are special water channels in the plasma membrane that facilitate the uptake of large amounts of water without the expenditure of energy.
    (D) Oxygen passes through the cristae of mitochondria mainly through ATP synthase channels.

10. Active transport involves the movement of a substance across a membrane against its concentration or electrochemical gradient. Active transport is mediated by specific transport proteins and requires the expenditure of energy. Which of the following statements describes an example of active transport?

    (A) Glucose is transported across some membranes by carrier proteins down a concentration gradient.
    (B) Freshwater protists, such as ameba and paramecia, have contractile vacuoles that pump out excess water.
    (C) Countercurrent exchange in the capillaries of fish gills enables fish to absorb large amounts of oxygen from the surrounding water where oxygen levels are low.
    (D) A red blood cell will lyse (burst) if placed into a hypotonic solution because large amounts of water flood into the cell.

11. Below is an illustration of an animal cell:

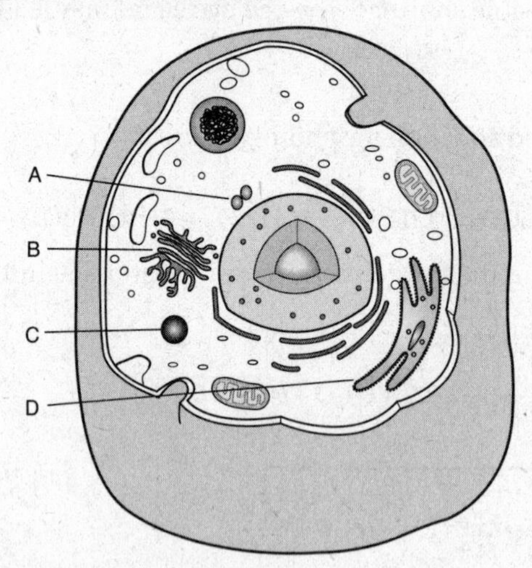

Which of the following statements is correct?

    (A) Structure A detoxifies poisons in the cell.
    (B) Structure B packages proteins for export.
    (C) Structure C synthesizes RNA.
    (D) Structure D consists of cytoskeleton.

12. Chronically high levels of glucocorticoids can result in obesity, muscle weakness, and depression. This looks like several diseases but actually is only one: Cushing syndrome. Excessive activity of either the pituitary gland or the adrenal gland can cause this disease. To determine which gland has abnormal activity in a particular patient, doctors use the drug dexamethasone, a synthetic glucocorticoid that blocks ACTH (adrenocorticotropic hormone) release.

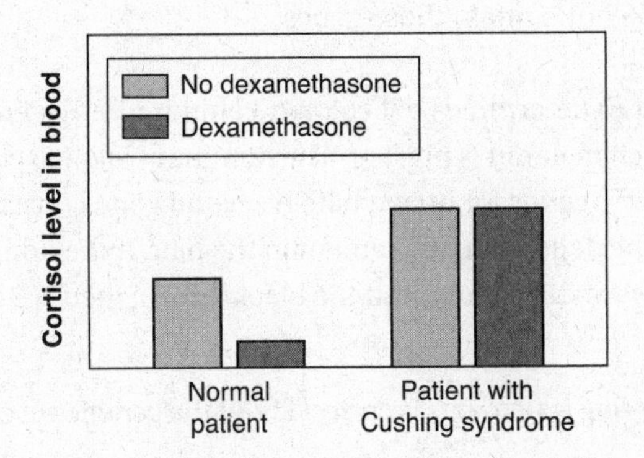

Based on the graph, which gland is affected in the patient with Cushing syndrome and what is the reasoning behind your answer?

(A) The pituitary gland is affected in the patient with Cushing syndrome because, although ACTH is blocked, the pituitary is still sending a signal to the adrenal glands.
(B) The pituitary gland is affected in the patient with Cushing syndrome because blocking ACTH has no effect on cortisol levels.
(C) The adrenal gland is affected in the patient with Cushing syndrome because the pituitary is prevented from stimulating the adrenal glands, and yet cortisol levels are still high.
(D) The adrenal gland is affected in the patient with Cushing syndrome because the pituitary is sending a signal to the adrenal glands and the adrenal glands have stopped producing cortisol.

13. Below is the final reaction in the citric acid cycle. It shows the regeneration of oxaloacetate.

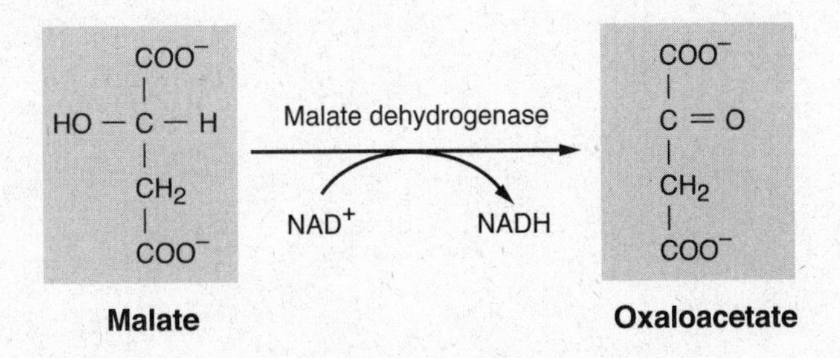

After studying this reaction, determine which of the following statements is correct.

(A) The enzyme malate dehydrogenase is allosteric.
(B) This reaction is exergonic; the released energy is absorbed by NAD^+.
(C) This reaction is a reduction reaction.
(D) NAD^+ is oxidized into NADH.

14. Which of the following statements about the DNA in one of your liver cells is correct?

    (A) Your liver cells contain the same DNA as your white blood cells.
    (B) Most of the DNA in your liver cells will be transcribed and translated.
    (C) Most of the DNA in your liver cells will be transcribed in the nucleus and translated in the cytoplasm.
    (D) DNA in your liver cells contains genes that are unique to storing glycogen, while other cells do not contain those genes.

15. Two genes (*B* and *E*) determine coat color in Labrador Retrievers. Alleles *B* and *b* code for how much melanin is present. The dominant allele *B* codes for black hair; the recessive allele *b* codes for brown hair. A second gene consisting of two alleles, *E* and *e*, codes for the deposition of pigment in the hair. If the dominant allele (*E*) is not present, regardless of the genotype at the black/brown locus (*B* or *b*) the animal's fur will be yellow.

    Which of the following statements is correct about the genetics of coat color in Labrador Retrievers?

    (A) *BBEe* will be brown.
    (B) *BbEe* will be brown.
    (C) *Bbee* will be brown.
    (D) *BBee* will be yellow.

16. Within a cell, the amount of protein that is synthesized using a given mRNA molecule depends in part on which of the following?

    (A) DNA methylation suppresses the expression of genes.
    (B) Transcription factors mediate the binding of RNA polymerase and the initiation of transcription, which will determine how much protein is manufactured.
    (C) The speed with which mRNA is degraded will determine how much protein is synthesized.
    (D) The location and number of ribosomes in a cell will solely determine how much protein is synthesized.

17. Oxygen is carried in the blood by the respiratory pigment hemoglobin, which can combine loosely with four oxygen molecules, forming the molecule oxyhemoglobin. To function properly, hemoglobin must bind to oxygen in the lungs and drop it off at body cells. The more easily the hemoglobin binds to oxygen in the lungs, the more difficult the oxygen is to unload at the body cells. Below is a graph that shows two different saturation-dissociation curves for one type of hemoglobin at two different pH levels:

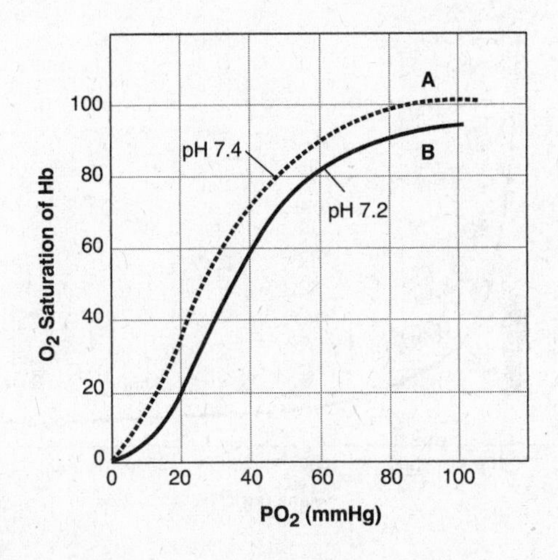

Based on your knowledge of Biology and the information provided, which of the following statements about the hemoglobin curves is correct?

(A) Hemoglobin B has a greater affinity for oxygen and therefore binds more easily to oxygen in the lungs.
(B) Hemoglobin A is characteristic of a mammal that evolved at sea level where oxygen levels are high.
(C) In a more acidic environment, hemoglobin drops off oxygen more easily at body cells.
(D) Hemoglobin A is found in mammals with a higher metabolism.

18. Banana plants, which are triploid, are seedless and therefore sterile. What is the most likely explanation for why banana plants are sterile?

(A) Because they are triploid, bananas cannot generate gametes because homologous pairs cannot line up during meiosis.
(B) Because they are triploid, there are no male or female banana plants.
(C) Because they are triploid, bananas cannot pair up homologues during mitosis.
(D) Because they are triploid, bananas cannot carry out crossing-over.

19. Animals maintain a minimum metabolic rate for basic functions such as breathing, heart rate, and maintaining body temperature. The minimum metabolic rate for an animal at rest is the basal metabolic rate (BMR). The BMR is affected by many factors, including whether an animal is an ectotherm or endotherm; its age and sex; and its size and body mass. Below is a graph that shows the relationship of BMR per kilogram of body mass to body size for a group of mammals:

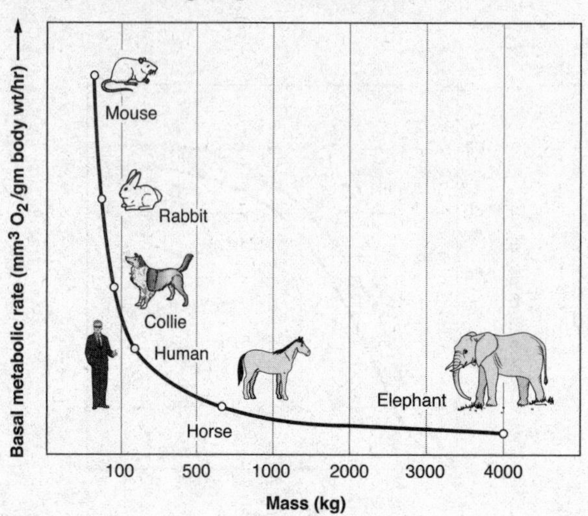

Which of the following statements correctly describes the relationship between BMR and body mass?

(A) The relationship between BMR and body mass is a direct one. The larger the body mass, the higher the BMR is and the greater the breathing rate is.
(B) The relationship between BMR and body mass is a direct one. The larger the body mass, the higher the BMR is and the lower the breathing rate is.
(C) The relationship between BMR and body mass is inversely proportional. The larger the body mass, the lower the BMR is and the lower the breathing rate is.
(D) The larger the animal, the faster the heart rate and breathing rate both are.

20. An important characteristic of the human immune system is that it is specific. If you are exposed to Antigen A, you will develop antibodies against A, but not against any other antigen. However, exposure to Antigen A does not alter how your body will respond to an exposure to another antigen for the first time.

Below is a graph that depicts a first exposure to Antigen A on Day 3 with a subsequent, primary immune response. A second exposure to Antigen A on Day 30 results in a secondary immune response due to the presence of circulating memory cells that release antibodies against Antigen A.

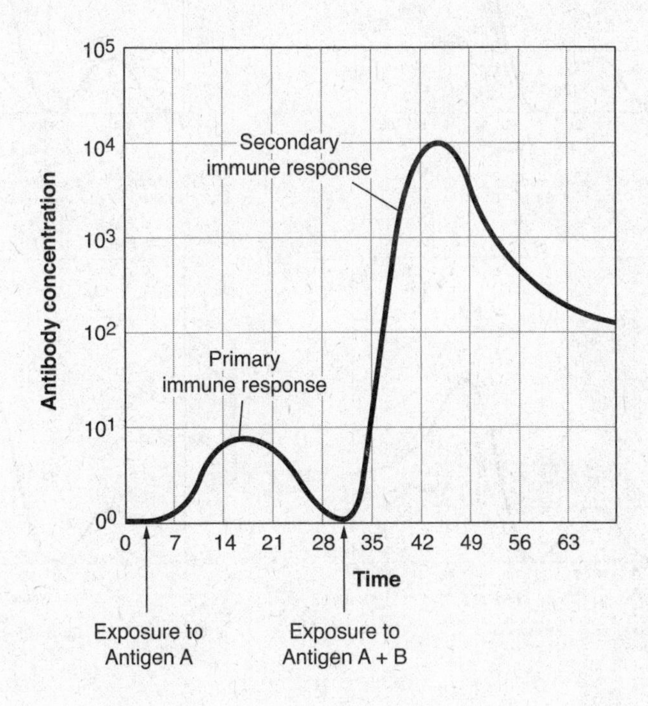

(continues on the next page)

Given the information provided, if there is also a first exposure to Antigen B on Day 30, which of the following graphs accurately depicts the correct immune response to Antigen B?

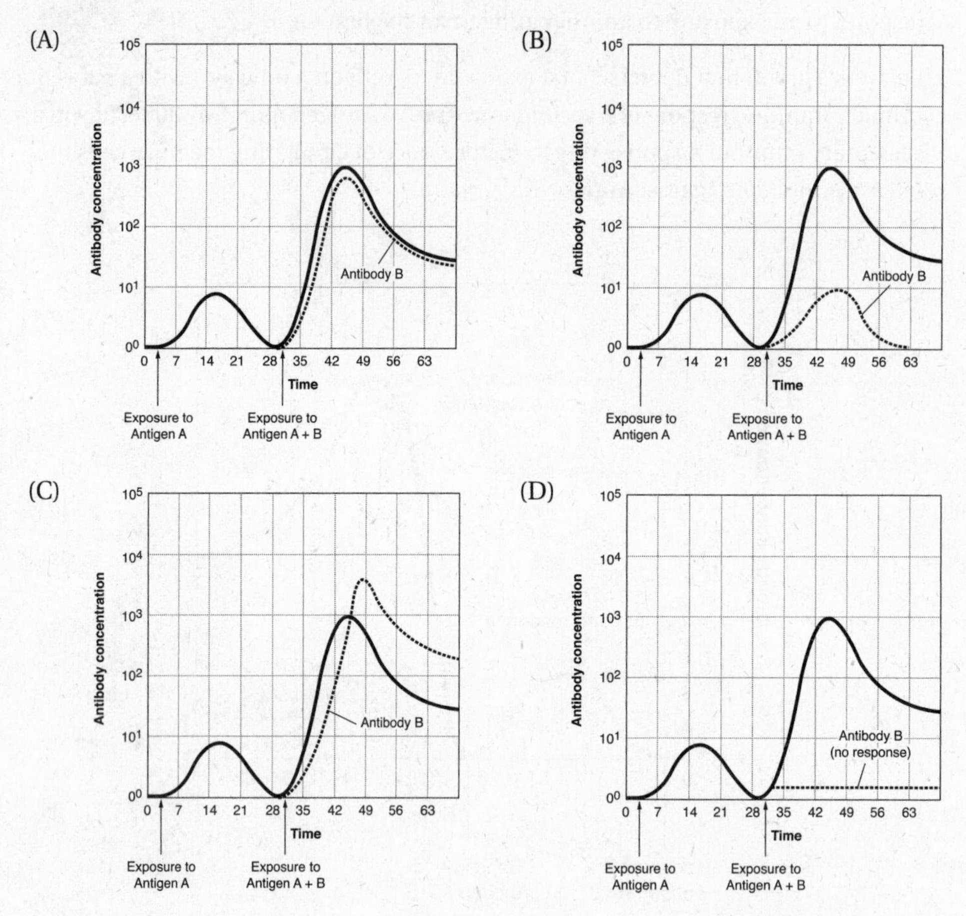

(A) Graph A. The primary response to Antigen B is almost as fast and large as the secondary immune response to Antigen A because the entire immune system was activated by the first exposure to Antigen A.
(B) Graph B. Immune responses are specific. The fact that there was an earlier exposure to Antigen A has no bearing on the response to Antigen B on Day 30.
(C) Graph C. The response to Antigen B on Day 30 is larger than the secondary immune response to Antigen A because the immune system has already been activated and all new responses are heightened.
(D) Graph D. There is almost no immune response to Antigen B because the immune system is fully engaged in a secondary response to Antigen A.

21. Some athletes take a form of steroids (known as androgenic-anabolic steroids or just anabolic steroids) to increase their muscle mass and strength. The main anabolic steroid hormone that is produced by the body is testosterone. Androgenic-anabolic steroids have approved medical uses; however, improving athletic performance is not one of them. Taking these drugs does not come without risks.

Below is a diagram of the human male endocrine system related to sperm production.

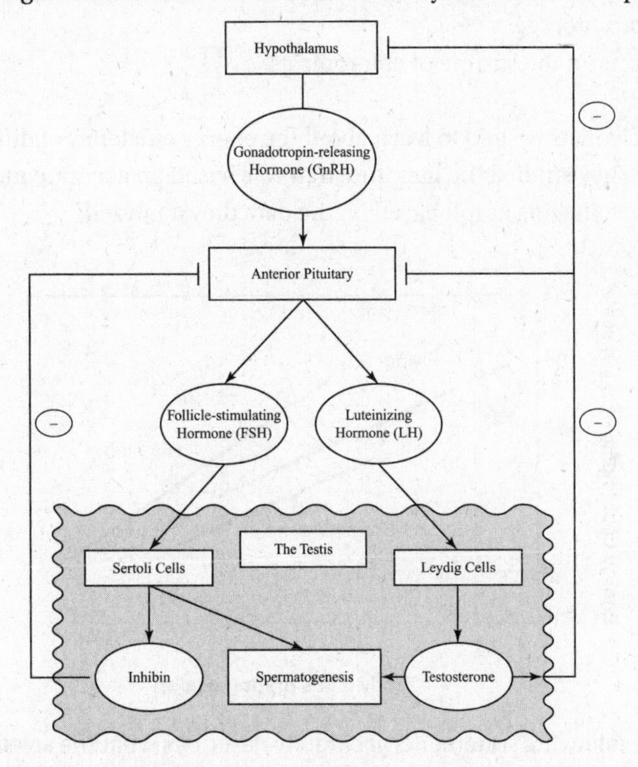

What might be a consequence for a man who regularly uses performance-enhancing drugs, such as androgenic-anabolic steroids?

(A) increased FSH production
(B) stimulation of the anterior pituitary
(C) an increase in GnRH (gonadotropin-releasing hormone)
(D) a reduction in spermatogenesis

22. Before the human genome was sequenced by the Human Genome Project, scientists expected that they would find about 100,000 genes. In fact, they discovered that humans have only about 24,000 genes. How can we exhibit so many different traits from so few genes?

(A) Modification of histone proteins usually increases the function of genes.
(B) Epigenetics enables one gene to produce many different traits.
(C) This is proof that pseudogenes and introns are expressed.
(D) A single gene can produce more than one trait because of alternative splicing.

23. Which of the following statements about the light-dependent reactions of photosynthesis is correct?

(A) They provide the carbon that becomes incorporated into sugar.
(B) They produce PGA, which is converted to glucose by carbon fixation in the light-independent reactions.
(C) Water is split apart, providing hydrogen ions and electrons to NADP for temporary storage.
(D) They occur in the stroma of chloroplasts.

24. A group of scientists wished to learn about the energy efficiency of different modes of locomotion. They studied the literature that was based on accurate measurements and produced the following graph based on the data they analyzed:

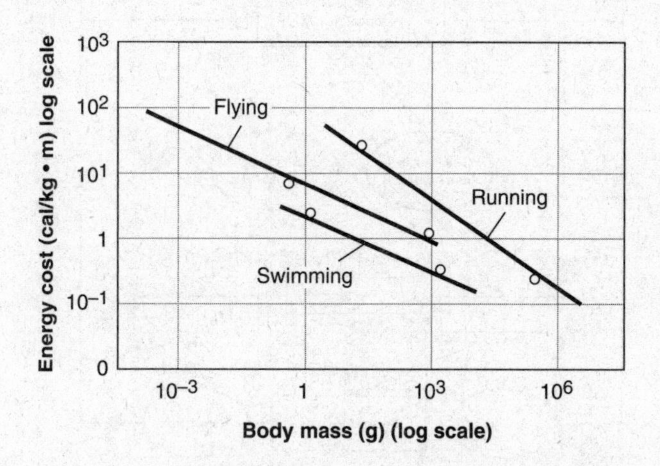

Which of the following statements accurately describes what the scientists discovered?

(A) An animal with a large body mass expends more energy per kilogram of body mass than a small animal does, regardless of the type of locomotion.
(B) Swimming involves overcoming drag as well as gravity.
(C) Running is the least energy efficient mode of locomotion.
(D) The best parameter to measure in this type of experiment is CO_2 consumption.

25. Which aspect of cell structure best reveals the unity of all life?

(A) All cells are surrounded by a plasma membrane.
(B) All cells have at least one nucleus.
(C) All cells carry out cellular respiration in mitochondria.
(D) The surface-to-volume ratio of all cells is the same.

26. Refer to the codon table below for this question.

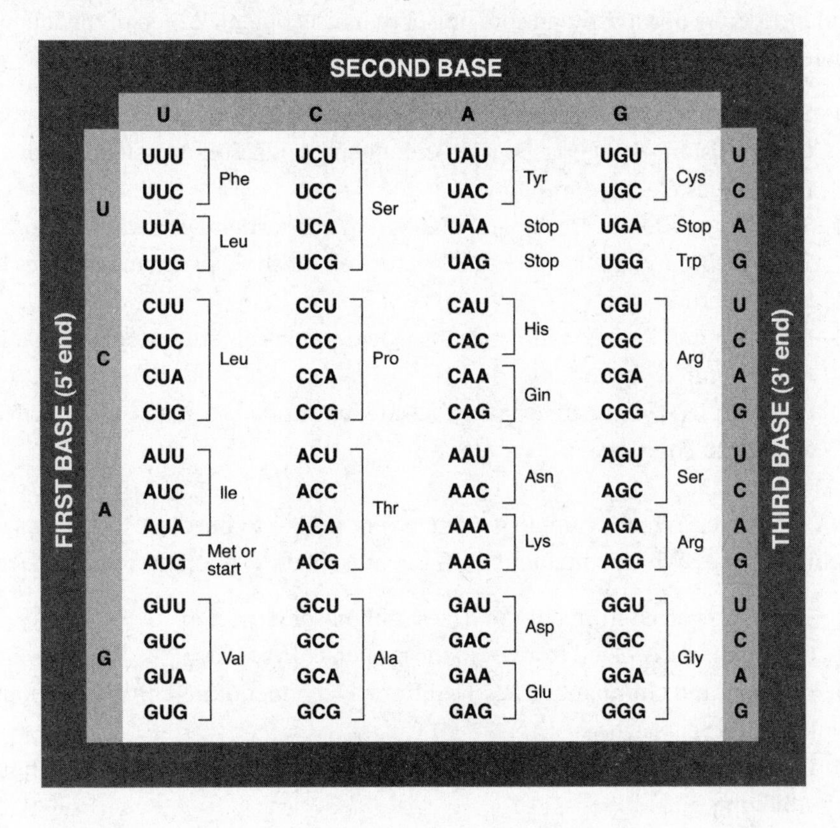

SECOND BASE

FIRST BASE (5′ end)	U		C		A		G		THIRD BASE (3′ end)
U	UUU	Phe	UCU	Ser	UAU	Tyr	UGU	Cys	U
	UUC		UCC		UAC		UGC		C
	UUA	Leu	UCA		UAA	Stop	UGA	Stop	A
	UUG		UCG		UAG	Stop	UGG	Trp	G
C	CUU	Leu	CCU	Pro	CAU	His	CGU	Arg	U
	CUC		CCC		CAC		CGC		C
	CUA		CCA		CAA	Gin	CGA		A
	CUG		CCG		CAG		CGG		G
A	AUU	Ile	ACU	Thr	AAU	Asn	AGU	Ser	U
	AUC		ACC		AAC		AGC		C
	AUA		ACA		AAA	Lys	AGA	Arg	A
	AUG	Met or start	ACG		AAG		AGG		G
G	GUU	Val	GCU	Ala	GAU	Asp	GGU	Gly	U
	GUC		GCC		GAC		GGC		C
	GUA		GCA		GAA	Glu	GGA		A
	GUG		GCG		GAG		GGG		G

Here is a small stretch of mRNA that would be translated at the ribosome:

...AUG CUG **A**AA UCAGGG...

Suppose a spontaneous mutation altered the boldface A and changed it to a U. What effect, if any, would this have on the protein formed?

(A) Because of redundancy in the code, there would be no change in the protein formed.
(B) The amino acid sequence formed from this stretch of DNA would be Met–Leu–Lys–Ser–Gly.
(C) The polypeptide would not form because translation would stop at UAA.
(D) Translation would continue and a polypeptide would form because AUG codes for start as well as for methionine.

27. Stretching out from the equator is a wide belt of tropical rainforests that are being cut down to provide exotic woods for export to the U.S. and to make land available to graze beef cattle. Which of the following statements is a negative consequence of clear-cutting the tropical rainforests?

(A) Indigenous populations will have access to modern conveniences.
(B) There will be less precipitation in those clear-cut areas.
(C) U.S. markets will have access to less expensive beef.
(D) Deforestation will allow more sunlight to penetrate areas that were kept dark by dense vegetation.

28. Mosquitoes that are resistant to the pesticide DDT first appeared in India in 1959 (within 15 years of widespread spraying of that insecticide). Which of the following statements best explains how the resistant mosquitoes arose?

(A) Some mosquitoes experienced a mutation after being exposed to DDT that made them resistant to the insecticide. Then their population expanded because these mosquitoes had no competition.
(B) Some mosquitoes were already resistant to DDT when DDT was first sprayed. Then their population expanded because all the susceptible mosquitoes had been exterminated.
(C) DDT is generally a very effective insecticide. One can only conclude that it was manufactured improperly.
(D) Although DDT is effective against a wide range of insects, it is not effective against mosquitoes.

29. DNA sequences in many human genes are very similar to the corresponding sequences in chimpanzees. Which statement gives the most likely explanation for this fact?

(A) Humans evolved from chimpanzees millions of years ago.
(B) Chimpanzees evolved from humans millions of years ago.
(C) Humans and chimpanzees evolved from a recent common ancestor about 6 million years ago.
(D) Humans and chimpanzees evolved from a distant common ancestor about 4 billion years ago.

Questions 30 and 31

Two ecologists, Peter and Rosemary Grant, spent thirty years observing, tagging, and measuring finches (a type of bird) in the Galápagos Islands. They made their observations on Daphne Major—one of the most desolate of the Galápagos Islands. It is an uninhabited volcanic cone where only low to the ground cacti and shrubs grow. During 1977, there was a severe drought and seeds of all kinds became scarce. The small, soft seeds were quickly eaten by the birds, leaving mainly large, tough seeds that the finches normally ignore. The year after the drought, the Grants discovered that the average width of the finches' beaks had increased.

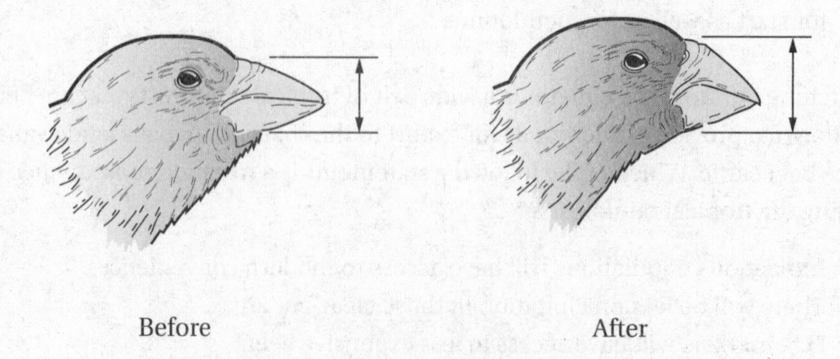

30. Which of the following statements best explains the increase in beak size?

(A) Finches with bigger and stronger beaks were able to attack and kill the finches with smaller beaks.
(B) Finches with bigger beaks were larger animals with stronger wings that could fly to other islands and gather a wide variety of seeds.
(C) During the drought, the finches' beaks grew larger to accommodate the need to eat tougher seeds.
(D) Finches with larger beaks were able to eat the tougher seeds and were healthier and reproduced more offspring that inherited the trait for wider beaks.

31. Which of the following statements best explains the mechanism behind the change in beak size?

(A) A new allele appeared in the finch population as a result of a mutation.
(B) A change in the frequency of a gene was due to selective pressure from the environment.
(C) A new trait appeared in the population because of recombination of alleles.
(D) A new trait appeared in the population because of genetic drift.

Questions 32 and 33

Answer the following two questions based on this pedigree for the biochemical disorder known as alkaptonuria. Affected individuals are unable to break down a substance called alkapton, which colors the urine black and stains body tissues. Other than that, there are no significant risks or medical consequences that result from this disorder. In this pedigree, males are shown as squares, and females are shown as circles. Affected individuals are shown in black. If there is a carrier condition, it is not displayed.

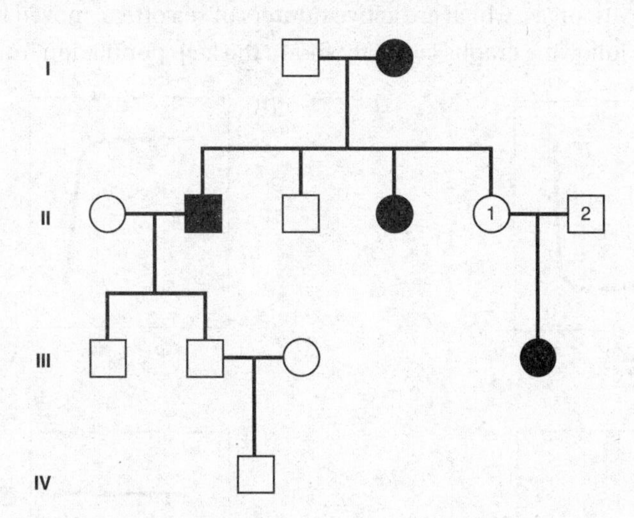

32. Which of the following best states the pattern of inheritance shown in the pedigree?

(A) The trait is sex-linked dominant.
(B) The trait is sex-linked recessive.
(C) The trait is autosomal dominant.
(D) The trait is autosomal recessive.

33. Which of the following statements is supported by the information given in the pedigree?

(A) The P generation mother is X-X.
(B) The P generation father is X-Y.
(C) If parents 1 and 2 in row II have another child, the chance that the child would be afflicted with alkaptonuria is 25%.
(D) The genotype of woman 1 is X-X.

---

34. Sea otters in the North Pacific are a keystone species. That means they are not abundant in a community. However, they do exert major control over other species in the community. Below is a food chain in which the sea otter is a keystone species:

Kelp → Sea urchin → Sea otter → Orca

Below are two graphs that show the populations of sea otters and sea urchins from 1970 to 2000:

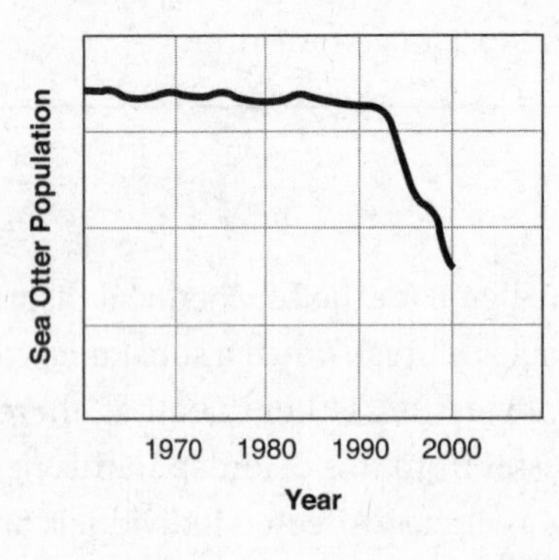

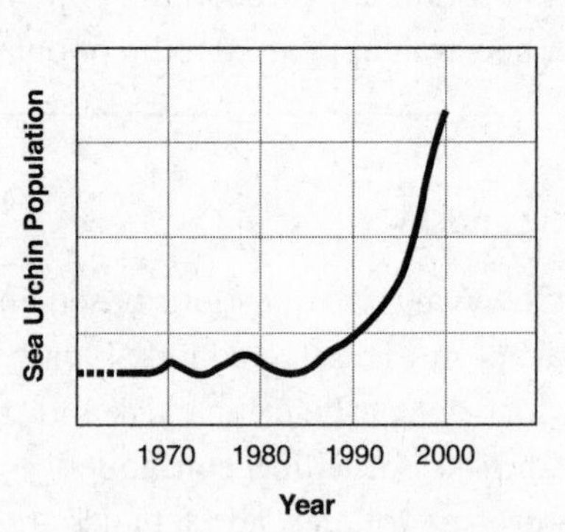

In the late 1990s, orcas, which are active hunters of sea otters, moved into the area. Which of the following graphs correctly shows the kelp population from 1970 to 2000?

(A)

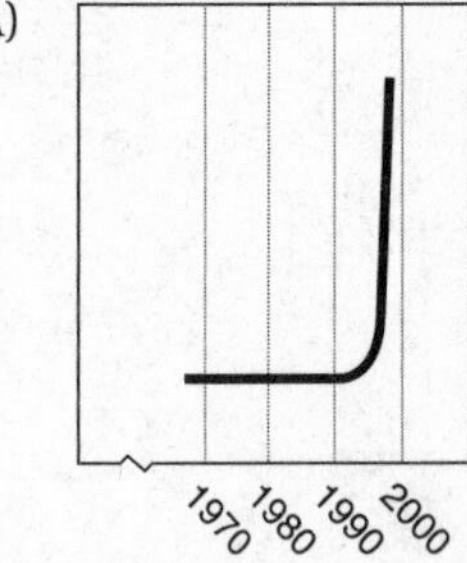

(B)

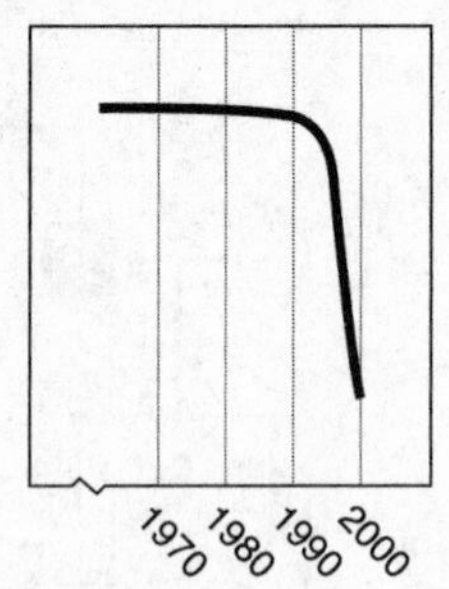

(C)

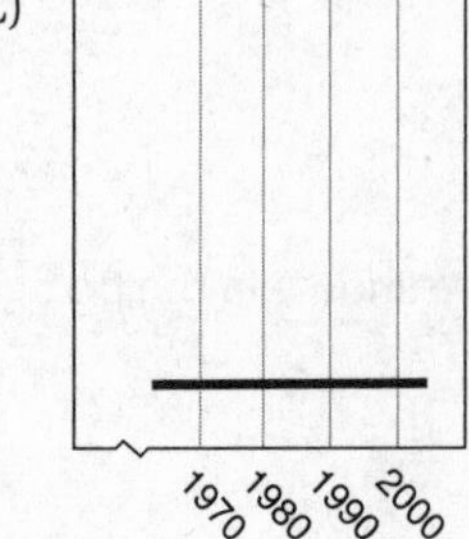

(D)

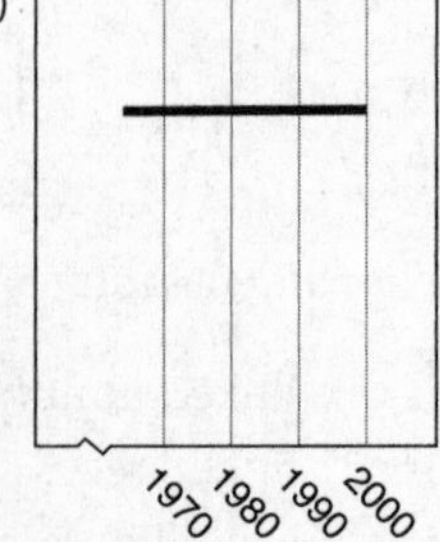

Questions 35–38

Four vials were set up to investigate bacterial transformation. Vials 1 and 2 each contained *E. coli* bacteria that had been made competent and had then been mixed with a plasmid containing the gene for ampicillin (pAMP) resistance. Vials 3 and 4 both contained *E. coli* that had also been made competent but had not been mixed with a plasmid. Each vial of bacteria was poured onto a nutrient agar plate. Vials 1 and 3 were poured onto plates that contained the antibiotic ampicillin. Vials 2 and 4 were poured onto plates that did not contain ampicillin.

The figure below shows what the nutrient agar plates looked like. The shading represents extensive growth, and the dots represent individual bacterial colonies.

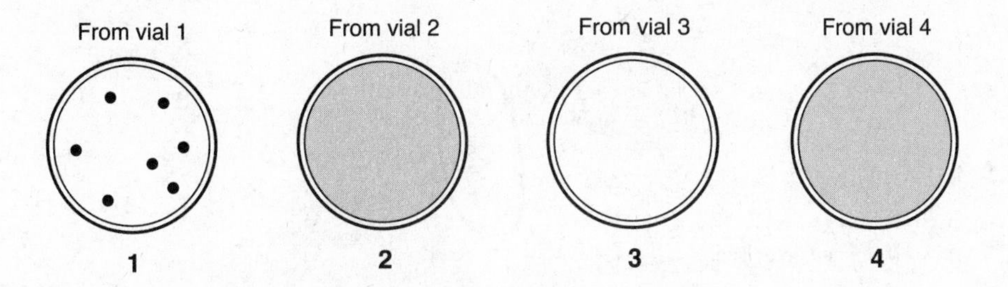

35. Plates that have only ampicillin-resistant bacteria growing on them include which of the following?

    (A) Plate 1
    (B) Plate 2
    (C) Plate 3
    (D) Plates 1 and 3

36. Which of the following statements explains why there was no growth on Plate 3?

    (A) The heat shock method, which was used to make the *E. coli* competent, killed the bacteria.
    (B) Those particular *E. coli* bacteria were inhibited from growing by the nutrient in the agar.
    (C) Those bacteria were not transformed.
    (D) The bacteria died because they have a short life span.

37. Which of the following statements best explains why there were fewer colonies on Plate 1 than on Plate 2?

    (A) The bacteria on Plate 2 did not transform.
    (B) There was no antibiotic in the agar on Plate 2 that would have restricted the growth of the bacteria.
    (C) The transformation of bacteria on Plate 2 was more successful.
    (D) The bacteria on Plates 1 and 2 were not taken from the same culture.

38. In a variant of this experiment, the plasmid contained GFP (green fluorescent protein) in addition to ampicillin resistance. Which of the following plates would have the highest percentage of bacteria that would fluoresce?

(A) Plate 1
(B) Plate 2
(C) Plate 3
(D) Plate 4

---

39. Below is an illustration of a neuromuscular junction in a patient with an autoimmune disease. Acetylcholine (ACh) is the stimulatory neurotransmitter.

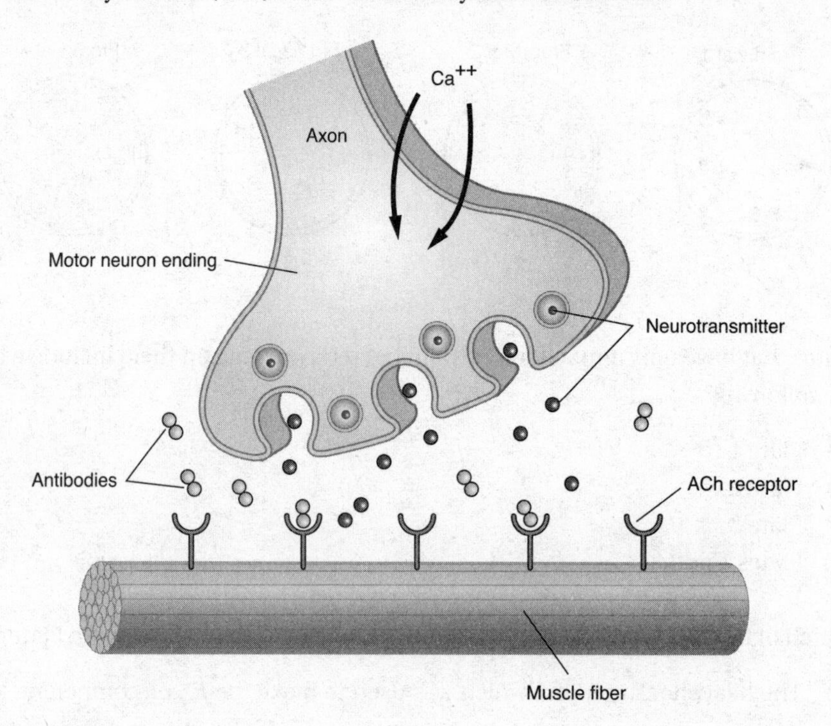

Which of the following predicts what will happen in the continued presence of the antibodies?

(A) Ca^{++} ions will flood into the motor neuron ending, increasing the release of more ACh.
(B) The amount of neurotransmitter being released will decrease.
(C) The number of action potentials in the motor neuron will decrease.
(D) The antibodies will destroy the postsynaptic receptors, and the muscle response will diminish.

40. In all living things, function dictates form and vice versa. Since cells have different functions, they look different. Below are illustrations of four cells.

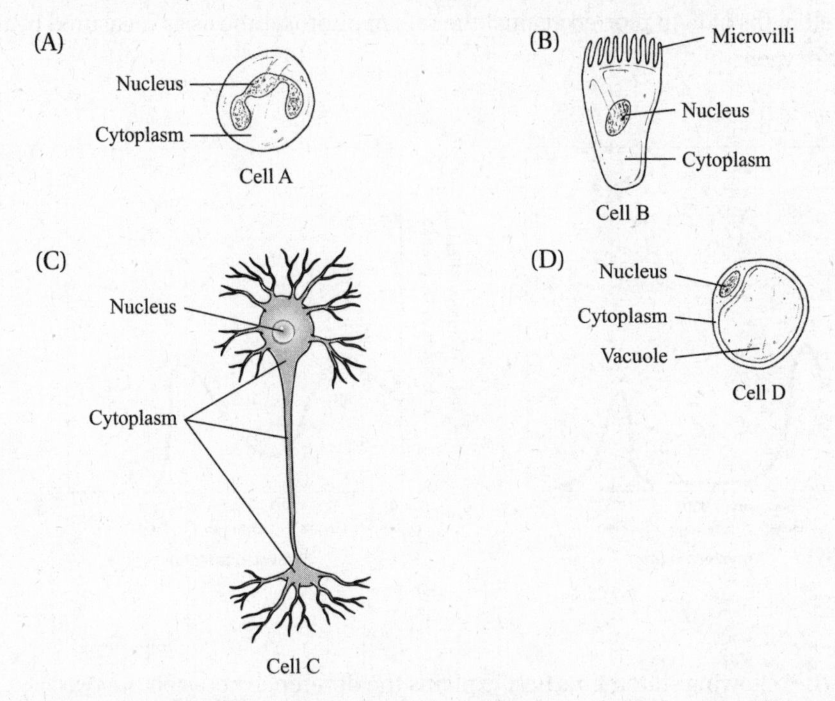

Which of the following statements identifies the cell that is most likely a storage cell and provides the correct evidence for that answer?

(A) Cell A is a storage cell because storing particles requires extra energy, and this cell has a large nucleus.
(B) Cell B is a storage cell because it has projections that can grab onto particles and bring them into the cell to be stored.
(C) Cell C is a storage cell because it has many projections from the cell.
(D) Cell D is a storage cell because it has a large vacuole.

41. In 1953, Stanley Miller, while working under the guidance of Harold Urey at the University of Chicago, carried out a series of experiments with substances that mimicked those of early Earth. His experimental setup yielded a variety of amino acids that are found in organisms alive on Earth today. The purpose of these experiments best supports which of the following hypotheses?

(A) The basic building blocks of life originated in outer space and came to Earth on comets or meteorites.
(B) The molecules necessary for life to develop were located in deep-sea vents.
(C) The molecules necessary for life to develop could have formed under the conditions of early Earth.
(D) The molecules necessary for life on Earth were self-replicating proteins, just like the ones produced in Miller's experiments.

42. The graph on the left (*A*) shows an absorption spectrum for chlorophyll *a* extracted from a plant. The graph on the right (*B*) shows an action spectrum from a living plant, with wavelengths of light plotted against the rate of photosynthesis as measured by the release of oxygen.

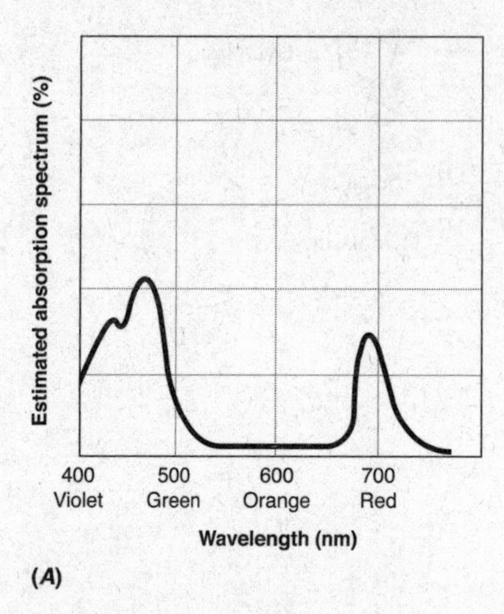

(*A*)

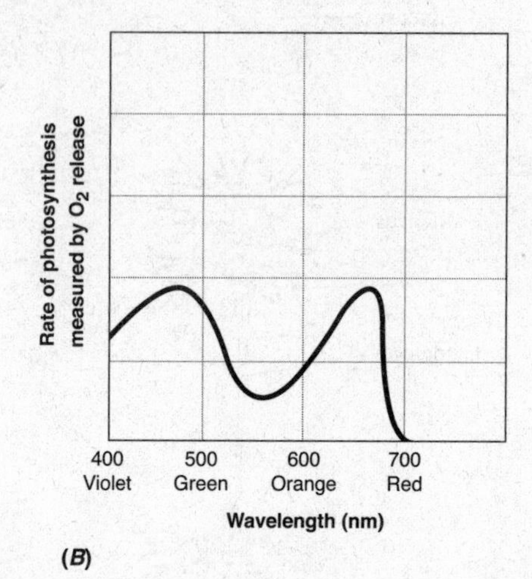

(*B*)

Which of the following statements best explains the difference between the two spectra?

(A) Graph *A* plots the absorption spectrum of a red plant; Graph *B* plots an absorption spectrum of a green plant.
(B) The chlorophyll from Graph *A* cannot carry out the light-dependent reactions, but the chlorophyll from Graph *B* can.
(C) The data from Graph *A* characterize several photosynthetic pigments that reflect almost no light; the data from Graph *B* characterize chlorophyll *a*, which reflects only green light.
(D) Graph *A* shows an absorption spectrum for an unusual type of chlorophyll *a*.

Questions 43 and 44

Cystic fibrosis is the most common inherited disease in the U.S. among people of European descent. Four percent are carriers of the recessive cystic fibrosis allele. The most common mutation in individuals with cystic fibrosis is a mutation in the CFTR protein that functions in the transport of chloride ions between certain cells and extracellular fluid. These chloride transport channels are defective or absent in the plasma membranes of people with the disorder. The result is an abnormally high concentration of extracellular chloride that causes the mucus that coats certain cells to become thicker and stickier than normal. Mucus builds up in the pancreas, lungs, digestive tract, and other organs. This buildup of mucus leads to multiple effects, including poor absorption of nutrients from the intestines, chronic bronchitis, and recurrent bacterial infections.

43. Scientific work has been carried out to measure where the relative amounts of CFTR protein are localized in the affected cells.

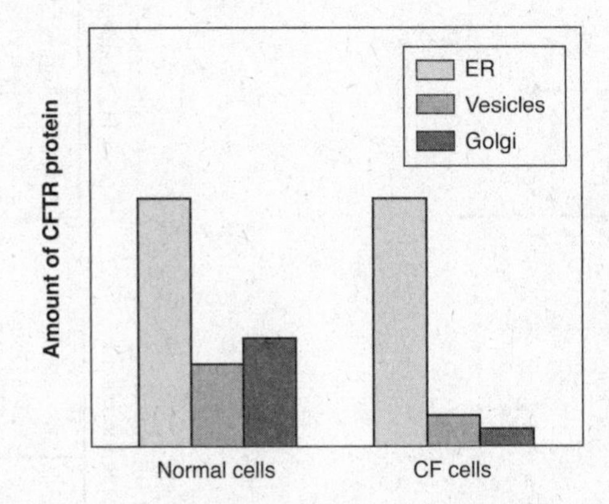

Based on the information and graph above, which of the following statements about CFTR protein is correct?

(A) Transcription is not occurring.
(B) Translation is not occurring.
(C) CFTR protein does not fold properly after it is synthesized.
(D) CFTR protein is not being packaged in the cytoplasm.

44. From the description of cystic fibrosis above, which of the following statements is correct?

(A) The disease is similar to cancer in that it has both an environmental and a genetic cause.
(B) Cystic fibrosis is an example of a genetic disease caused by polygenic inheritance because several genes must be mutated in order for the disease to occur.
(C) Cystic fibrosis is an example of a disease in which one mutated gene causes multiple effects.
(D) A person who has one cystic fibrosis allele will have the disease.

45. The graph below shows the pH difference across the mitochondrial cristae membrane over time as a normal cell is respiring.

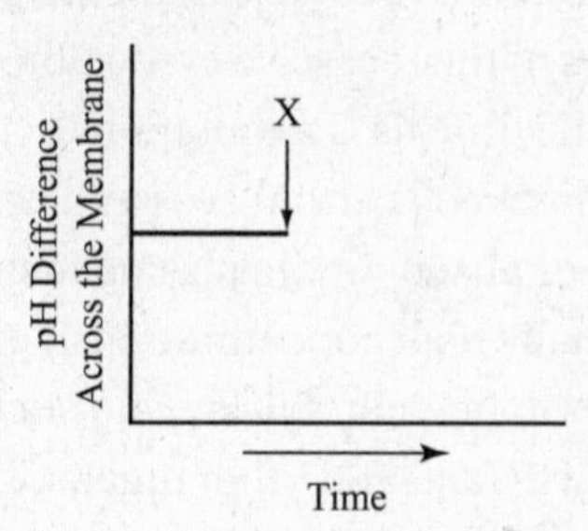

At time "X," a metabolic poison is added to the system, which immediately and completely shuts down all functions of the mitochondrial ATP synthase molecule. Which of the following graphs correctly demonstrates how the rest of the line in the graph would appear?

(A)

X
pH Difference
Across the Membrane
Time

(B)

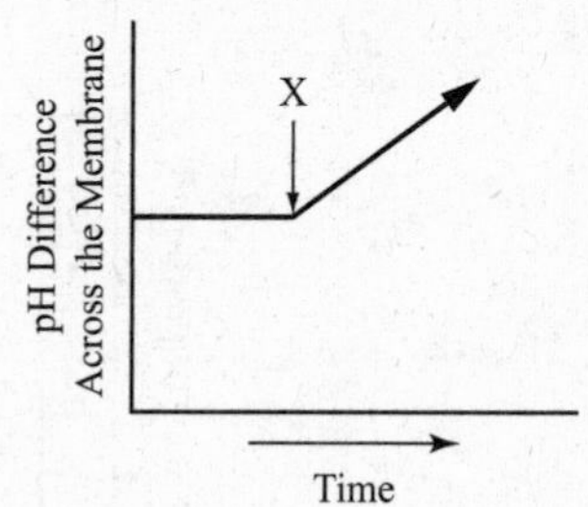

(C)

X
pH Difference
Across the Membrane
Time

(D)

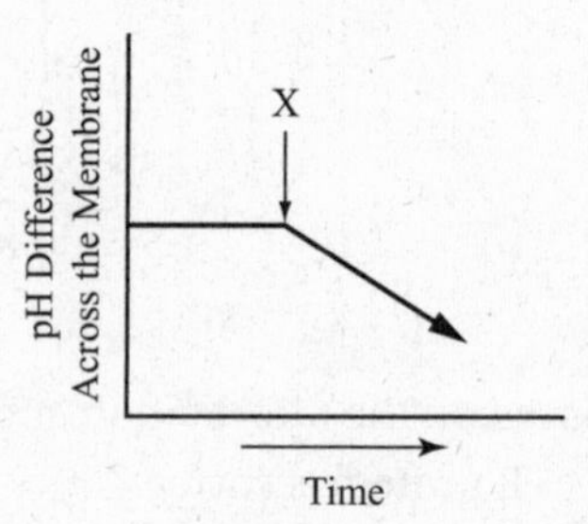

46. Based on the phylogenetic tree below, which of the following statements is correct?

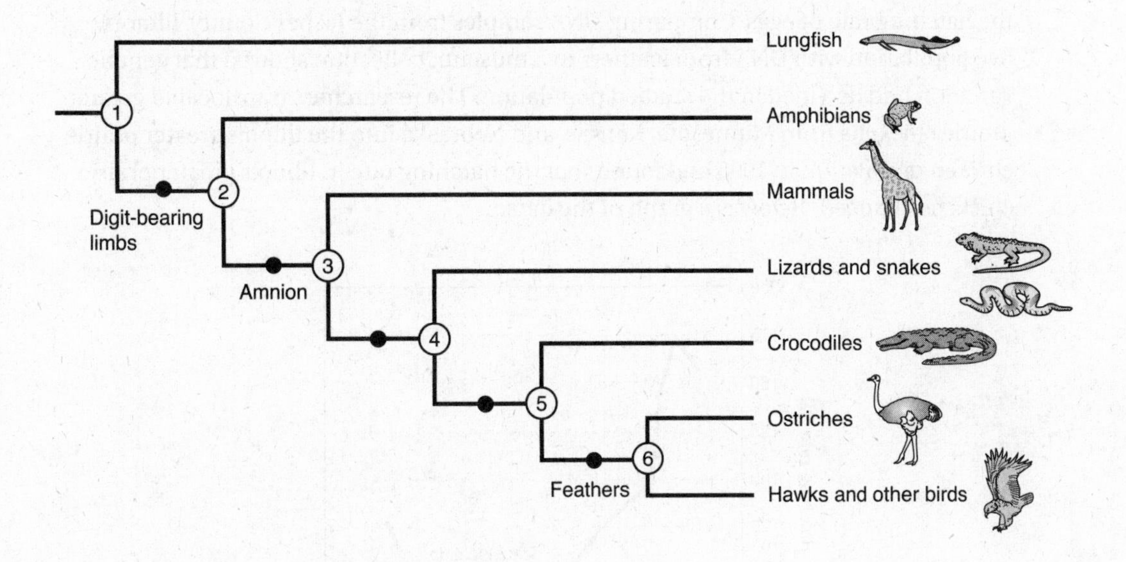

(A) All tetrapods have amnions.
(B) Crocodiles and birds share a recent common ancestor.
(C) Amphibians and lungfish are more closely related than amphibians and mammals.
(D) Amphibians lay amniotic eggs.

47. On Andros Island in the Bahamas, populations of mosquitofish, *Gambusia hubbsi*, colonized a series of ponds. These ponds are no longer connected. However, the environments are very similar except that some ponds contain predatory fish, while others do not. In high predation ponds, selection has favored the evolution of a body shape that enables rapid bursts of speed. In low predation ponds, another body type is favored—one that is well-suited to swim for long periods of time.

When scientists brought together sample mosquitofish from the two types of ponds, they found that the females mated only with males that had the same body type as their own. Which of the following statements best describes what happened to the mosquitofish as evidenced by the mating choice in the female fish?

(A) Reproductive isolation caused geographic isolation.
(B) Reproductive isolation was not complete.
(C) Geographic isolation brought about reproductive isolation.
(D) A mutation caused a sudden change in many fish that brought about reproductive isolation.

48. From 1970 to 2003, researchers who were studying the greater prairie chicken observed that a population collapse mirrored a reduction in fertility as measured by the hatching rate of eggs. Comparing DNA samples from the Jasper County, Illinois, live population with DNA from feathers in a museum collection showed that genetic variation had declined in the studied population. The researchers translocated greater prairie chickens from Minnesota, Kansas, and Nebraska into the Illinois greater prairie chicken population in 1993 and found that the hatching rate in Illinois greater prairie chickens changed. Below is a graph of the data:

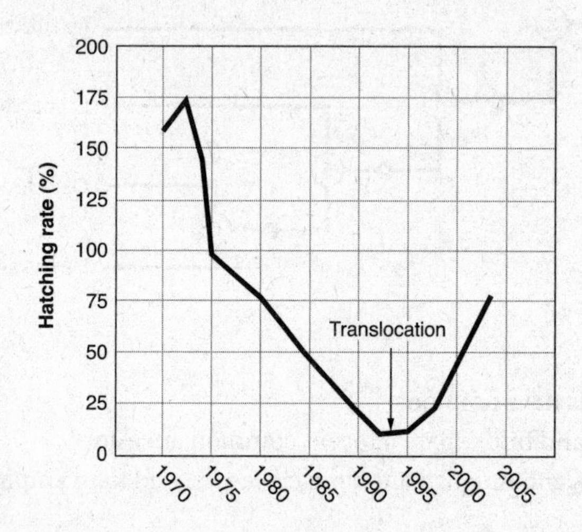

Which of the following statements most accurately explains what happened to the studied population of greater prairie chickens after translocation and why?

(A) The hatching rate increased because genetic variation declined.
(B) The hatching rate increased because genetic variation increased.
(C) The hatching rate decreased because genetic variation decreased.
(D) The hatching rate decreased because the translocated animals were invasive and grew to dominate the population.

49. Which of the following statements highlights the evolutionary significance of the fact that the wing of a bat and the pectoral fin of a whale have the same internal bone structure?

(A) A bat and a whale share a common origin, and their internal bone structures reflect a shared ancestry.
(B) These structures are analogous and do NOT reflect a common ancestry.
(C) The internal structure is similar because, eons ago, the bat and the whale evolved in similar environments.
(D) The anatomies of both animals evolved because they were needed for the specific environment the animals lived in.

50. In a study of dusky salamanders, *Desmognathus ochrophaeus*, scientists brought individuals from two different populations into the laboratory and tested their ability to mate and produce viable, fertile offspring. Below is the graph of the data:

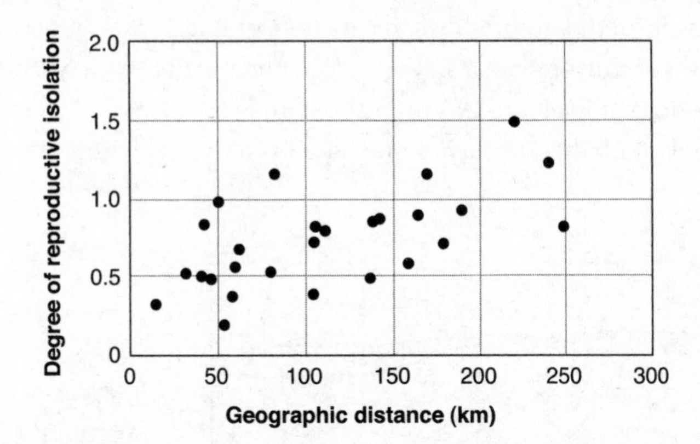

The degree of reproductive isolation is represented by an index ranging from 0 (no isolation) to 2 (complete isolation).

Which of the following statements best describes the evolutionary history of dusky salamanders?

(A) Mutations alone caused the two populations to diverge.
(B) Both mutations and genetic drift caused the two populations to diverge.
(C) Both mutations and genetic drift caused the two populations to become separate species after they were separated by a great distance.
(D) Reproductive isolation between the two populations increases as the distance between them increases.

51. Refer to the figure below, which shows a normal body cell from a female and a normal body cell from a male. Which of the following statements best describes what a Barr body is and its significance?

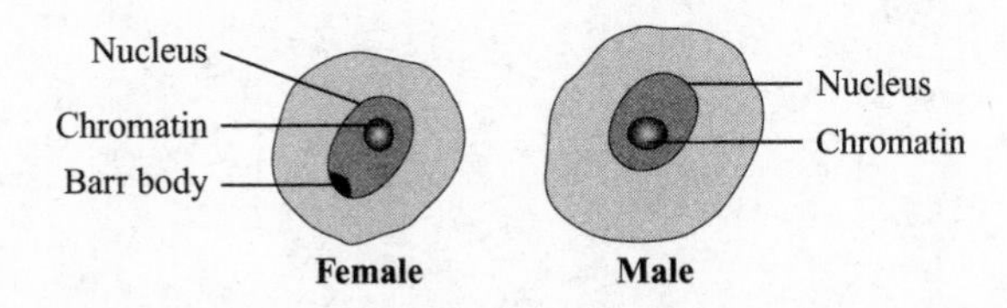

(A) It is an inactivated Y chromosome and results in a man being sterile.
(B) It is an inactivated Y chromosome, and the person who has it appears female.
(C) It is an inactivated X chromosome and results in females with half their cells having one X inactivated and the other half of their cells having the other X inactivated.
(D) It is an inactivated X chromosome and results in females who are sterile.

52. Scientists carried out a series of experiments to study innate immunity in fruit flies. They began with a mutant fly strain in which a pathogen is recognized, but the signaling that would normally trigger an innate response is blocked. As a result, the flies did not make any antimicrobial peptides (AMPs). The scientists then genetically engineered some of the mutant flies to express significant amounts of a single AMP, either defensin or drosomycin. They then infected the flies with a fungus, *Neurospora crassa*. The 6-day survival rate was monitored and recorded.

Below is a graph that displays some data from the experiment:

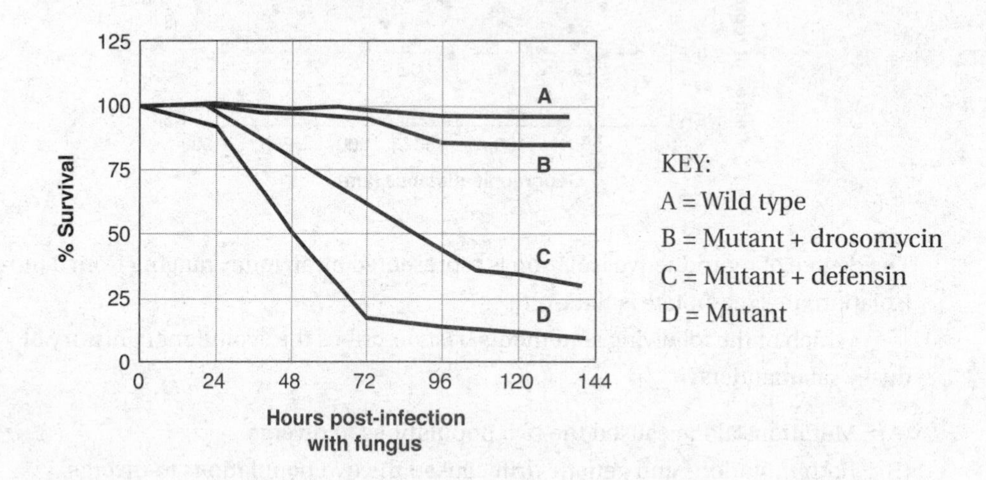

Which of the following statements is supported by the data?

(A) Each AMP provided minimal immunity against the fungal infection.
(B) A mutant with both AMPs provided the best immunity.
(C) Drosomycin provided strong immunity against the fungal infection.
(D) There was no control in this experiment.

Questions 53 and 54

An experiment was carried out with guppies, which are brightly colored, popular, aquarium fish. Three hundred guppies were added to 12 large pools. Cichlids, a voracious predator, were added to 4 of the pools. Killifish, which rarely eat guppies, were added to 4 other pools. No other fish were added to the last 4 pools. After about 16 months, a time period that represents 10 generations for guppies, all the guppies were analyzed for size and coloration. Below are the data that were collected:

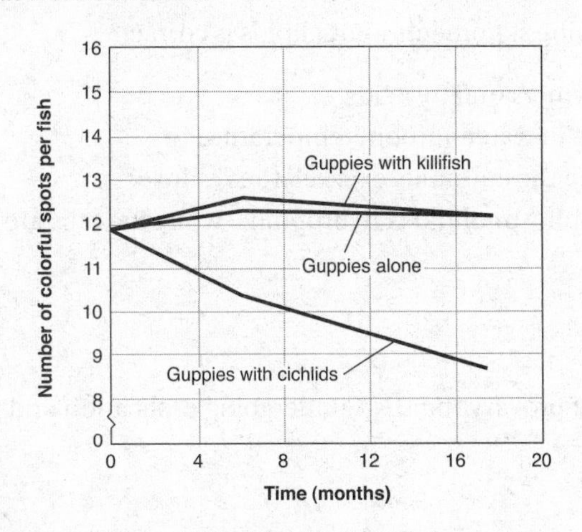

53. Which of the following statements is supported by the data?

    (A) Cichlids were an agent of selection, eating the more visible, brightly colored guppies.
    (B) Killifish were an agent of selection, eating the more visible, brightly colored guppies.
    (C) Guppies, as a group, experienced a change in coloration, from brightly colored to dull in order to survive.
    (D) Individual guppies experienced a mutation that caused a change in coloration and enabled them to avoid being eaten.

54. Which of the following statements best expresses the point of this experiment?

    (A) Mutations can be brought about by the environment.
    (B) Mutations occur randomly.
    (C) Agents of selection are not always readily apparent.
    (D) Evolution does not always require millions of years.

---

55. In the Tularosa Basin of New Mexico are black lava formations surrounded by light-colored sandy desert. Pocket mice inhabit both areas. Dark-colored mice inhabit the lava formations, while light-colored mice inhabit the desert. Which of the following statements is correct about this scenario?

    (A) The two varieties of mice descended from a recent common ancestor.
    (B) The mouse population was originally one population that diverged into two species because of mutations.
    (C) Selection favors some phenotypes over others.
    (D) Originally the mice were all dark colored. As the lava decomposed into sand, the color of some mice changed because that color was favored in that environment.

56. What is the best explanation for the fact that tuna, sharks, and dolphins all have a similar streamlined appearance?

(A) They all share a recent common ancestor.
(B) They all sustained the same set of mutations.
(C) They acquired a streamlined body type to survive in their particular environment.
(D) The streamlined body has a selective advantage in that environment.

57. Which of the following statements about lipids is correct?

(A) Lipids are polymers of fatty acids.
(B) Lipids dissolve in water at room temperature.
(C) Lipids make up the chitinous exoskeletons of insects.
(D) The head of a phospholipid is hydrophilic, while the tails are hydrophobic.

Questions 58 and 59

Below is an illustration of prokaryotic DNA undergoing replication and transcription simultaneously.

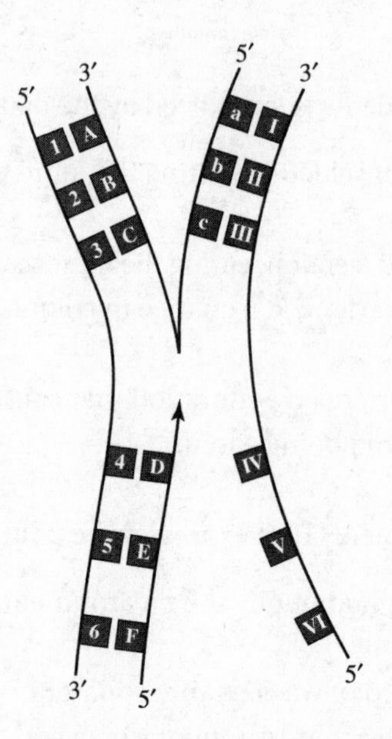

58. If 1 is thymine, what is a?

(A) adenine
(B) thymine
(C) cytosine
(D) uracil

59. If 4 is adenine, what is D?

(A) adenine
(B) thymine
(C) cytosine
(D) uracil

60. Yeast cells carry out both aerobic and anaerobic respiration. A yeast cell that is fed on glucose is moved from an aerobic environment to an anaerobic environment. Which of the following statements is correct and gives the correct reason why it is correct?

(A) The cell would die because it cannot make ATP.
(B) The cell would need to consume glucose at a much greater rate because aerobic respiration is much more efficient as compared with anaerobic respiration.
(C) The cell would need to consume another food source other than glucose because it will not be able to make adequate ATP with only glucose.
(D) The cell would begin to divide rapidly because larger cells require more oxygen and glucose than smaller ones.

SECTION II

Time: 90 minutes

Directions: Questions 1 and 2 are long free-response questions that require about 25 minutes each to answer and are worth 8 to 10 points each. Questions 3, 4, 5, and 6 are short free-response questions that require about 8 to 10 minutes each to answer and are worth 4 points each. You are advised to spend the first 10 minutes of this section as a "reading period" for planning your responses; however, you may begin writing your responses right away. On the actual exam, you must write your responses in the space provided for each question because *only material written in the space provided will be scored.* Your responses must be written out in paragraph form. Outlines, bulleted lists, and/or diagrams alone are not acceptable.

1. An animal's behaviors (such as getting food, finding a mate, escaping predators, and caring for its young) are essential for its survival. The common fruit fly, *Drosophila melanogaster*, is a model organism that is often used to study animal behavior because it is inexpensive, it is easy to grow in a laboratory, and it has a short life span.

   (a) An easy behavior to study is what choice an animal makes or what preference it has. **Explain** the significance of "animal preference" or "choice" in terms of the animal's survival. (1 point)

   (b) **Draw** a "choice chamber" to investigate whether the fruit flies prefer one environment over another. **Explain** how this apparatus would allow for data collection related to preference. **State** the null hypothesis for this investigation. (4 points)

   (c) **Draw** and **label** a histogram to show the results for your experiment. Make sure that you include a title, and make sure that the *x*-axis and the *y*-axis are appropriately labeled. Then, **analyze** your data. (3 points)

   (d) **State** whether you would reject the null hypothesis or fail to reject the null hypothesis, given the data. **Explain** your reasoning for your statistical decision. (2 points)

2. Cellular respiration is the process by which living cells extract energy from the oxidation of glucose, which results in the formation of the waste products CO_2 and H_2O. The rate of cellular respiration can be determined by measuring the volume of oxygen used or the volume of carbon dioxide released. The overall equation for the complete oxidation of glucose is:

$$C_6H_{12}O_6 + 6O_2 \rightarrow 6CO_2O + 6H_2O + 36\ ATP$$

   (a) **Devise** an experiment that will measure the rate of cellular respiration in a living organism. **Create** TWO treatments under different conditions. **Draw** and **label** your apparatus, and **explain** how it works. (2 points)

(b) **Graph** the predicted data from your experiment on the grid lines below. Be sure that your graph has an appropriately labeled *x*-axis and *y*-axis. (4 points)

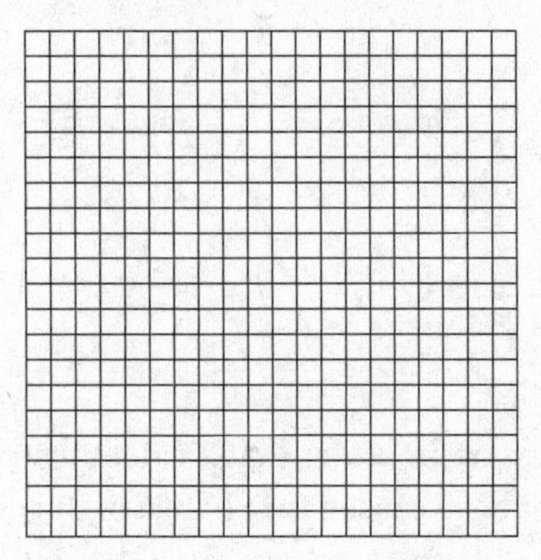

(c) **Analyze** your data by comparing the two treatment groups. (1 point)

(d) **Describe** a different experiment using EITHER a different apparatus OR a different organism to study. **Predict** the results of that different experiment, and **justify** your predictions. (3 points)

3. Some plants have trichomes on their leaves. Trichomes are hair-like structures that deter predation by herbivores. The number of trichomes on the leaves varies with the plant; some can have hundreds, and some have only a few dozen.

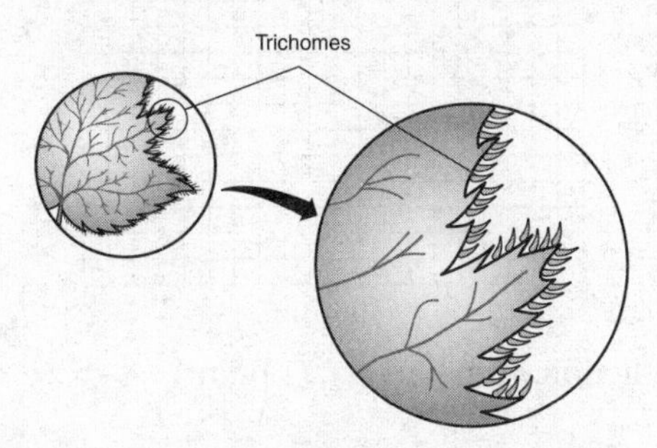

(a) **Describe** the process of artificial selection. (1 point)

(b) **Identify** the experimental procedures that you would use to demonstrate the process of artificial selection, focusing on the number of trichomes on a plant species (1 point)

(c) **Predict** the results you would expect from the experiment described in part (b). (1 point)

(d) **Justify** your predictions from part (c). (1 point)

4. Natural selection can alter the frequency of inherited traits in a population, depending on which traits are favored. Below is a graph that shows the evolution of a population over time.

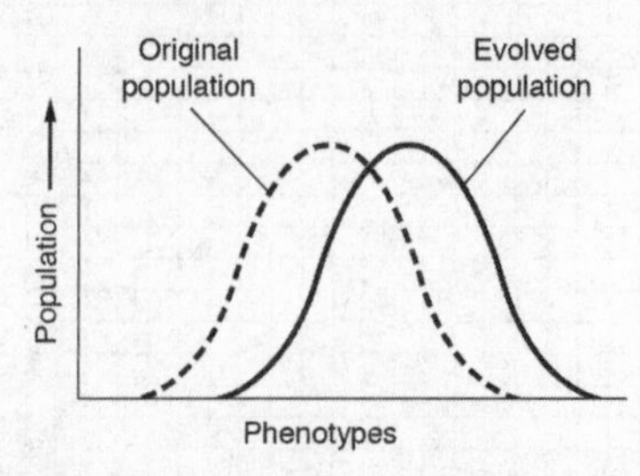

(a) **Describe** the process or processes that are demonstrated in this graph. (1 point)
(b) **Explain** what might have caused the change in the population, as shown on this graph. (1 point)
(c) On the grid below, **draw** an appropriately labeled graph that describes a different population before-and-after a change. **Predict** the change in the environment that could have caused the change in this population. (1 point)

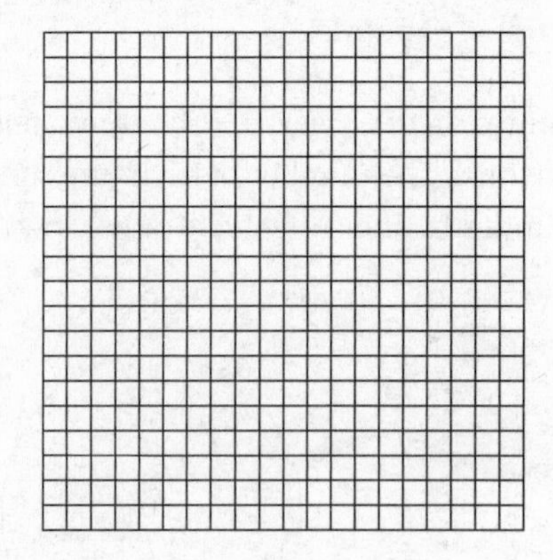

(d) **Justify** your prediction from part (c). (1 point)

5. Refer to the model below.

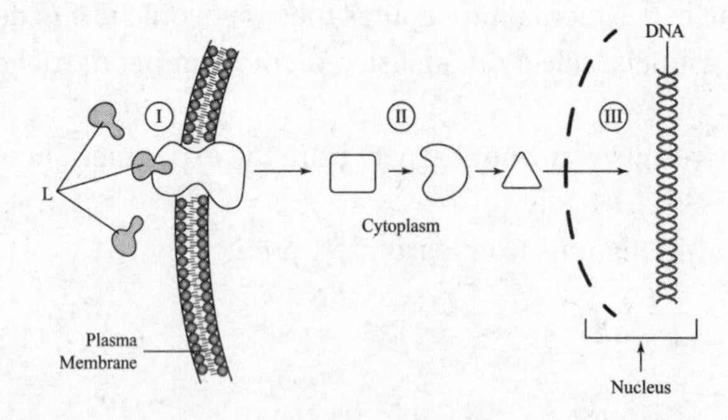

(a) **Describe** the process that is shown in this model, and **explain** what its characteristics are. (1 point)
(b) **Explain** the relationships between the different molecules shown in this model. (1 point)
(c) **Represent** the relationships of the molecules shown in this model. (1 point)
(d) **Explain** how the biological process that is represented visually in this model relates to a broader biological process. (1 point)

6. Apoptosis is genetically programmed cell death. Defects in the regulation of apoptosis contribute to many diseases, including cancers where cells grow out of control. The molecule miR-215 is a signaling protein that seems to be related to apoptosis in some cells. Below are two graphs: A and B:

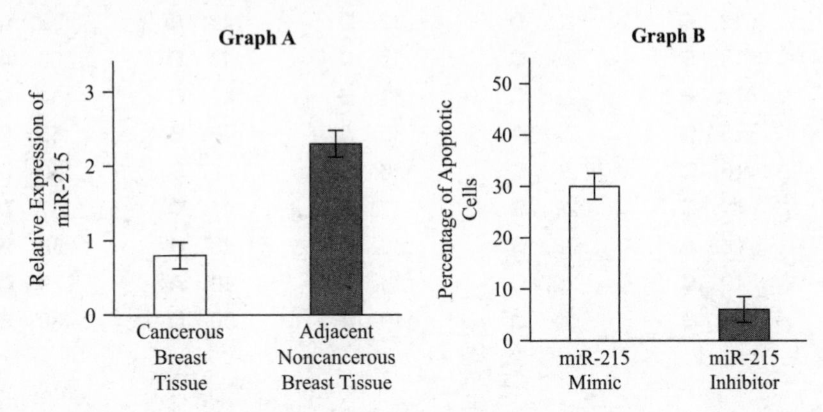

Graph A shows the relative expression of miR-215 in cancerous breast tissue and adjacent noncancerous breast tissue. Graph B shows the results of an experiment, demonstrating the relationship between apoptotic cells and an miR-215 mimic and an miR-215 inhibitor.

(a) **Describe** the data shown in Graph A, in terms of miR-215 and the presence or absence of cancer. (1 point)

(b) **Describe** the data shown in Graph B, in terms of miR-215 and the percentage of apoptotic cells in the experiment. (1 point)

(c) Do these data support a researcher's plan to design a drug that mimics miR-215 in order to slow the growth of cancerous cells in breast tissue? **Explain** your reasoning for your answer. (1 point)

(d) **Explain** how the results shown in these graphs support a relationship between cell death, expression of miR-215, and cancer. (1 point)

ANSWER KEY

Section I

1. **C**	11. **B**	21. **D**	31. **B**	41. **C**	51. **C**
2. **B**	12. **C**	22. **D**	32. **D**	42. **B**	52. **C**
3. **C**	13. **B**	23. **C**	33. **C**	43. **D**	53. **A**
4. **A**	14. **A**	24. **C**	34. **B**	44. **C**	54. **D**
5. **D**	15. **D**	25. **A**	35. **A**	45. **B**	55. **C**
6. **D**	16. **C**	26. **C**	36. **C**	46. **B**	56. **D**
7. **A**	17. **C**	27. **B**	37. **B**	47. **C**	57. **D**
8. **B**	18. **A**	28. **B**	38. **A**	48. **B**	58. **B**
9. **C**	19. **C**	29. **C**	39. **D**	49. **A**	59. **D**
10. **B**	20. **B**	30. **D**	40. **D**	50. **D**	60. **B**

Section II

See the Answers Explained section.

ANSWERS EXPLAINED

Section I

1. **(C)** The sex hormone testosterone is lipid soluble (hydrophobic) and dissolves directly through the cell membrane. Choices A and B correctly describe the action of a hydrophilic signal. Cyclic AMP is a secondary messenger found in the cytoplasm, not on the surface of a cell.

2. **(B)** The body temperature of an ectotherm increases with the ambient temperature. A homeotherm or an endotherm can raise and lower its body temperature to maintain a steady body temperature.

3. **(C)** This question focuses on the order of activity at the synapse. An electrical impulse travels down the axon, causing voltage-gated calcium channels in the presynaptic membrane to open. Calcium ions flood into the terminal branch. This causes presynaptic vesicles to fuse with the membrane and release their contents, neurotransmitter, into the synapse.

4. **(A)** If malonate is a competitive inhibitor, it competes with the enzyme (succinate dehydrogenase) for succinate. If malonate is a noncompetitive inhibitor, increasing the concentration of succinate will not increase the rate of reaction because the enzyme is being inhibited by malonate.

5. **(D)** This region has the most consistently warm temperature and the highest precipitation.

6. **(D)** The graph flattens out but doesn't decrease. Therefore, photosynthesis is still going on. That indicates that choices B and C cannot be correct. Choice A is not correct because chlorophyll molecules do not decompose due to heat or light; they are meant to absorb light and tolerate heat.

7. **(A)** Remember that operons are only found in prokaryotes, not in higher organisms (like mammals). However, higher animals may utilize a similar mechanism. For that reason, you must eliminate choice B. Operons are a means of controlling the transcription of genes, whereas choice C focuses on what happens after transcription. Operons have nothing to do with the cell cycle, which rules out choice D.

8. **(B)** Oxygen (O_2) in the air comes from molecules of water (H_20) that plants absorb at their roots. The process of hydrolysis releases oxygen from water into the air.

9. **(C)** Carbon dioxide is a balanced and nonpolar molecule that can readily diffuse through the membrane. Starch is too large to diffuse through a membrane. Only protons pass through ATP synthase channels in mitochondrial membranes.

10. **(B)** Whenever a pump is involved, the mechanism is active transport. Glucose is always transported by either simple diffusion or facilitated diffusion. However, neither process requires the expenditure of energy. Choices C and D refer to passive transport.

11. **(B)** Structure B is the Golgi apparatus, which packages and secretes proteins made by the ribosomes on the ER. The centrioles, shown as structure A, consist of microtubules. Structure C is a lysosome. Structure D consists of ribosomes on the ER.

12. **(C)** The patient is still producing too much cortisol even though the pituitary is not sending a signal to the adrenal glands. ACTH, which would normally stimulate the adrenal to release cortisol, is blocked by the dexamethasone. If the adrenals were healthy, they should not be producing anything. Unfortunately, they are. Compare the patient with the normal patient.

13. **(B)** Energy is released, and NAD^+ binds to the hydrogen released by malate.

14. **(A)** All cells in your body contain the same genes; the difference is which ones are turned on and which are turned off. Also, less than 3% of a human's DNA is transcribed and translated.

15. **(D)** The trait for deposition of pigment (*E*) dominates the trait for the production of coat color (*B* or *b*). If there is no dominant trait for deposition, the animal will be yellow or golden. An animal with the genotypes *BBEe* or *BbEe* will be black. An animal with the genotypes *bbEE* or *bbEe* will be brown. However, the genotype *BBee* will result in a yellow dog. This is an example of non-Mendelian inheritance.

16. **(C)** Choices A and B refer to transcription of DNA. However, the question asks about protein manufactured from an RNA molecule. Choice D is incorrect because several other processes could block protein production before mRNA gets to the ribosomes.

17. **(C)** The curve to the right shows a lower affinity for oxygen. CO_2 dissolved in water produces carbonic acid. What this graph demonstrates is that the blood near respiring cells, which release carbon dioxide, make the surrounding area more acidic. This environment causes hemoglobin to release its oxygen to the cells where it is needed.

18. **(A)** Choice B is incorrect because no evidence is provided in the question about bananas not having separate sexes. Choice C is incorrect because meiosis, not mitosis, produces gametes. Choice D is incorrect because, while it is true that there can be no crossing-over if there is no synapsis (when homologous pairs line up to exchange DNA), gametes could still form. However, there would not be any variation in the gametes.

19. **(C)** The energy to maintain each gram of body mass is inversely related to body size. This graph describes a relationship between X and Y that is inversely proportional. (When the relationship is directly proportional, the line goes diagonally from the lower left to upper right.) Each gram of mouse requires 20 times as many calories as a gram of elephant even though the whole elephant uses more calories than a whole mouse. The higher the metabolic rate, the higher the oxygen and food requirement per unit of body mass.

20. **(B)** This may seem tricky, but choice B is correct. Immune responses are specific. Each challenge with a new antigen results only in a primary immune response for that antigen.

21. **(D)** The minus signs inside the circles on the diagram stand for inhibition. Notice that the presence of testosterone directly inhibits the anterior pituitary and the hypothalamus. The Sertoli cells release a substance (inhibin) that also inhibits the anterior pituitary. Counter to what one might think, a man who takes testosterone may inhibit his sperm production. This loop is an example of negative feedback.

22. **(D)** Modification of histone proteins, as in methylation, silences genes. Epigenetics refers to the expression of a trait due to non-DNA factors. Pseudogenes and introns are not expressed.

23. **(C)** All the other choices refer to the light-independent reactions.

24. **(C)** According to the data, an animal with a large body mass expends less energy per kilogram of body mass, regardless of the type of locomotion. You can conclude that because the lines for "Flying," "Swimming," and "Running" all slope downward and to the right. Swimming is the most energy efficient mode of locomotion, whereas running is the least energy efficient mode of locomotion. You can conclude this because the "Swimming" line is placed lowest on the graph, whereas the "Running" line is above the other lines on the graph. Choice A is incorrect because it is a false statement given the data. Choice B is incorrect because the swimmer is supported by the water. Choice D is incorrect because the best parameters to measure in this type of experiment are O_2 consumption, or the amount of CO_2 that is given off, not CO_2 consumption.

25. **(A)** Choice B is incorrect because prokaryotes do not have internal membranes, and they do not have a defined nucleus. Instead, they have DNA in a nucleoid region. Choice C is incorrect because all cells carry out cellular respiration, but they do not necessarily do it in mitochondria. Choice D is incorrect because the surface-to-volume ratio of cells varies with the cell's size. Smaller cells have a greater surface-to-volume ratio.

26. **(C)** UAA codes for a stop sequence. Translation would cease at that point in the process.

27. **(B)** Rainforests are responsible for the cycle of rain, absorption of water by roots, transpiration from leaves, cloud formation, and rain again. Without trees to absorb and evaporate water, the region will dry up and perhaps become a desert under a hot sun.

28. **(B)** Resistance to a certain chemical arises because exposure to the chemical kills all the insects except the ones that are already resistant to it. The resistant population is selected for in this way and reproduces a new population in which all the individuals are resistant.

29. **(C)** Humans and chimps share an ancestor that walked upright about 6–7 million years ago.

30. **(D)** This is classic Darwinian evolutionary theory—the survival of the best adapted.

31. **(B)** Choices A, C, and D all have the same problem. A new trait may have appeared by a single mutation, by recombination of alleles, or by genetic drift. However, that does not explain how the birds, on average, came to have larger beaks.

32. **(D)** Look at individuals 1 and 2. Neither has the condition, but they have a daughter who has the condition. Sex-linked dominant and autosomal dominant can be eliminated because neither parent has the condition. Sex-linked recessive can also be eliminated because for the father to pass the trait to his daughter, he would have to have the condition. For the daughter to have the condition, she would have had to inherit two affected X chromosomes—one from each parent.

33. **(C)** Since the trait is inherited as autosomal recessive, parents 1 and 2 are each hybrid. The chance that any child of theirs will have the condition is 25%.

34. **(B)** When orcas moved into the area, the population of sea otters declined. This caused the population of sea urchins to increase and, ultimately, the kelp population to decline.

35. **(A)** The few colonies growing on Plate 1 consist of bacteria that have taken up the plasmid and are resistant to ampicillin. Plate 2 consists of both antibiotic-resistant bacteria as well as nonresistant bacteria. There is no antibiotic in the agar on Plate 2 to distinguish the two bacteria.

36. **(C)** The bacteria on Plate 3 were not transformed. That means they did not uptake a plasmid and were not resistant to the antibiotic. There is antibiotic in the agar on that plate, which killed the nontransformed bacteria.

37. **(B)** The few colonies growing on Plate 1 consist of bacteria that have taken up the plasmid and are resistant to ampicillin. Plate 2 consists of both antibiotic-resistant bacteria as well as nonresistant bacteria. There is no antibiotic in the agar on Plate 2 to distinguish the two bacteria.

38. **(A)** There would be fluorescent bacteria on both Plates 1 and 2. However, 100% of the bacteria on Plate 1 would fluoresce while only a percentage of the bacteria on Plate 2 would do so.

39. **(D)** In an autoimmune disease, the immune system mistakenly attacks its own body structures. The antibodies attack the postsynaptic muscle receptors, not the neuron function.

40. **(D)** Cell D happens to be a fat-storing cell from an animal. It has the obvious trait of a large vacuole. Cell A is a white blood cell. Cell B is an intestinal cell with microvilli projections. Cell C is a neuron.

41. **(C)** Some scientists believe that some complex molecules necessary for life on Earth might have come from outer space. That might be true. However, the purpose of Miller's experiments was to demonstrate that the molecules necessary for life to develop could have formed under the conditions of early Earth. He did not succeed in demonstrating that the first molecules were self-replicating.

42. **(B)** Graph *A* shows the absorption spectrum for chlorophyll that was extracted from a living plant. Once separated from the grana and stroma, chlorophyll by itself cannot carry out photosynthesis.

43. **(D)** Ribosomes produce protein for export. Once synthesized, the protein is packaged into vesicles in the Golgi for secretion. The graph shows that the amount of CFTR is high where synthesis occurs, in the ER, and low in the vesicles and Golgi. So, the CFTR protein was synthesized in the ribosomes but not packaged in the Golgi. Choice A is incorrect because transcription must have already occurred since the graph shows that CFTR protein already exists. Therefore, CFTR is transcribed and translated, but not packaged in the Golgi. Thus, choice B is also incorrect. Choice C is incorrect because there is no information in the stem of the question and/or on the graph that discusses "folding" of the CFTR protein.

44. **(C)** Choice A is incorrect because cystic fibrosis is solely an inherited disease. No environmental component causes this disease. It is autosomal recessive. In order to have the disease, a person must have inherited two mutated genes, one from each parent. Therefore, choice D is incorrect. Choice B is incorrect because polygenic inheritance involves the inheritance of several genes. Examples are height, skin, or hair color—any trait in which a tremendous variety occurs in a population. In cystic fibrosis, one mutated gene causes multiple medical problems, such as chronic bronchitis, recurrent bacterial infections, and more.

45. **(B)** This question refers to the mitochondrial cristae membrane. The energy for the production of ATP is supplied by the potential energy of the proton gradient that accumulates in the outer compartment. As protons flow down this gradient from the outer compartment through the ATP synthase channel to the matrix, ADP gets phosphorylated and ATP is produced. If the ATP synthase channel were to be shut down (by some metabolic poison), protons would continue to accumulate in the outer compartment. Remember, that as protons (H+) accumulate, the pH value decreases. The y-axis of the graph is "pH Difference Across the Membrane." As the protons accumulate, the difference in the pH between outside and inside the membrane increases. If the y-axis had been simply "pH," the answer would have been choice D.

46. **(B)** This image is a phylogenetic tree, not a cladogram, because the horizontal lines vary in length and represent time. The lungfish group (longest line) is the oldest while hawks and other birds (shortest line) are the most recently evolved animals. Choice A is incorrect because, although all of the animals on this phylogenetic tree are tetrapods, two of the animals (the lungfish and the amphibians) do not have amnions. Choice C is incorrect because amphibians and mammals both have digit-bearing limbs, but lungfish do not. Choice D is incorrect because amphibians do not lay amniotic eggs; mammals, lizards and snakes, crocodiles, ostriches, and hawks and other birds do lay amniotic eggs.

47. **(C)** These fish clearly exhibited geographic isolation because the two populations lived in different ponds with different selective factors. The pressure for selection came from the predatory fish. After years of being separated, the two fish populations could no longer mate. Therefore, geographic isolation brought about reproductive isolation.

48. **(B)** Prior to translocation, the greater prairie chickens had become inbred, diversity had decreased, and fertility had declined. You can easily eliminate choice C and choice D because they refer to a decreasing hatching rate after translocation, which is counter to what can be seen on the graph. The hatching rate increased because genetic variation increased as a result of the introduction of a new population into the area. Since genetic variation must have increased, not decreased, due to the introduction of a new population, rule out choice A as well.

49. **(A)** These internal structures are *homologous*, not analogous, and they reflect a shared ancestry. Choice B is incorrect because it is the opposite of choice A, which is the correct answer. Choice C is incorrect because it describes structures that are analogous, like the streamlined shape of the dolphin (a mammal) and tuna (a fish). Analogous structures, by definition, do not reveal a common ancestry. Choice D is incorrect because body parts and systems do not evolve because they are "needed."

50. **(D)** The question does not reveal anything about mutations or genetic drift. The point here is that, in this case, divergence is a function of geographic distance.

51. **(C)** Every cell in the mammalian females has two X chromosomes, but one of them will be deactivated in every cell. Which one of the two becomes inactive occurs randomly and independently in each embryonic cell. So, females are considered to be mosaics because they contain two types of cells, in terms of what genes a cell contains. This does not apply to men because men only have one X chromosome in each cell.

52. **(C)** The survival of mutant flies + drosomycin was greater than that of mutant flies + defensin. Choice A is incorrect because clearly the AMPs provided some immunity. You see that drosomycin provided strong immunity (line B on the graph) and the defensing provided weaker immunity (line C on the graph) but both still had a survival rate than the mutant alone (line D on the graph). Choice B is incorrect because there was no mutant tested that had both AMPs. Choice D is incorrect because the wild flies (line A on the graph) are the control.

53. **(A)** Killifish do not eat guppies, but cichlids do. So cichlids were the agent of selection. No organism or population changes in order to survive. If individuals are not adapted, they die. No individual changes during its lifetime. Rather, the frequency of a particular trait in a population may change.

54. **(D)** The evolution in this population—the change in frequency of the trait for brightly colored spots—occurred rapidly, within about 16 months. Evolution does not always require millions of years. The rate of change is a function of the pressure from the environment to change.

55. **(C)** The question does not give you enough background information to determine anything more than that some traits are favored more than others in a particular environment.

56. **(D)** Once again, the environment provides the direction for evolutionary change. Tuna and sharks are fish; dolphins are mammals. Fish and mammals do not share a recent common ancestor.

57. **(D)** Choice D is a correct description of a phospholipid. Choice A is not correct because lipids are not polymers since polymers are chains of repeating *identical* units. Lipids are molecules that consist of two different substances: 1 glycerol plus 3 fatty acids. Choice B is not correct because lipids do not dissolve in water. Choice C is not correct because the exoskeletons of insects are made of chitin (a polysaccharide), not lipids.

58. **(B)** First, orient yourself in the drawing. The process at the top of the illustration is replication; the bottom is transcription. So if 1 is thymine, then its complement, I, is adenine. Therefore, if I is adenine, then its base pair is thymine.

59. **(D)** The bottom of the diagram represents transcription. Therefore, if 4 is adenine, then D is its complement in RNA, which is uracil.

60. **(B)** Anaerobic respiration is much less efficient at making ATP than is aerobic respiration. Glycolysis produces only 2 net ATPs from each molecule of glucose. The yeast can survive on glucose alone.

Section II

1. (a) The choices that an animal makes about where to make a nest or which berries to eat can be the difference between life and death for that animal. Making poor choices can leave an animal vulnerable to predators or submissive to competing animals in its own population. A weakened animal is also less likely to mate, produce offspring, and pass its genes onto the next generation.

   (b) Refer to the choice chamber shown below:

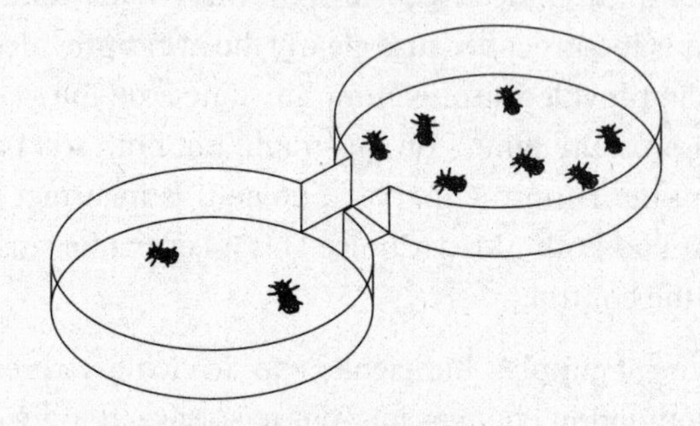

Choice Chamber

(Note that you could also make a choice chamber by taping together 2 two-liter soda bottles and placing the experimental flies inside before sealing them.)

For this experiment, I will place 30 fruit flies into the chambers and then seal them closed with tape. Next, I will choose one variable (related to preference) to test (such as a preference for the presence or absence of light). Then, I will cover one side of the choice chamber with paper towels and leave the other side exposed to light. After allowing some time to pass (for the flies to get used to this environment), I will make observations for 30 minutes. Then, I will record the number of flies that remain in each chamber at the end of the experiment.

The null hypothesis is that fruit flies demonstrate no preference for either chamber (the presence or absence of light).

(c) Below is a histogram that shows the results for this experiment.

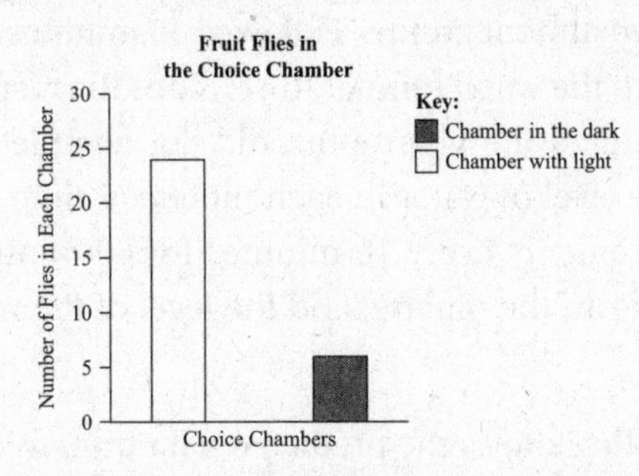

After the 30 minutes that were allotted for this experiment, 24 fruit flies preferred the chamber with light while 6 fruit flies preferred the chamber in the dark.

(d) Given the predicted data, I would reject the null hypothesis because clearly the fruit flies demonstrate a preference for the presence of light. They choose the chamber with light over the chamber in the dark: 4 to 1. The data show that, by the end of the experiment, 24 fruit flies preferred the chamber with light, while only 6 fruit flies preferred the chamber in the dark.

2. (a) **Treatment Group 1:** I constructed three respirometers and labeled them, as shown in the figure below.

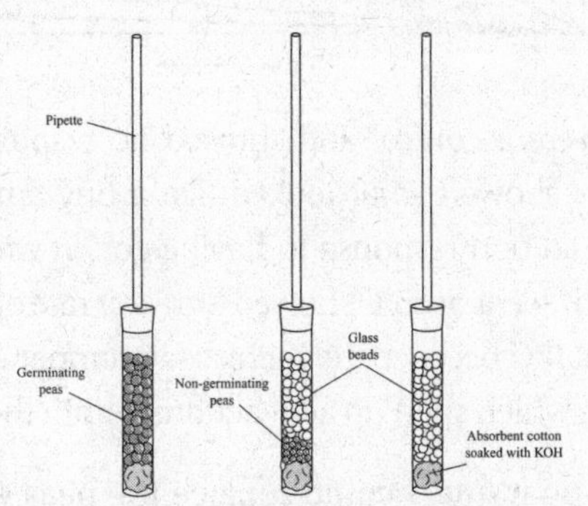

The first respirometer contained germinating (respiring) peas. The second respirometer contained non-germinating peas. Since the volume of the non-germinating peas is smaller than the volume of the germinating peas, I added glass beads to the second respirometer to make the total volume in the tube with the non-germinating peas the same as the total volume in the tube with the germinating peas. In addition, because I introduced glass beads into the experiment, I also included a third respirometer that was only filled with glass beads to control for the possibility that they somehow use air. The respirometer with the germinating peas was the experimental, whereas the apparatuses with the non-germinating peas and the glass beads only were the controls. (Note that I placed absorbent cotton that was soaked with KOH into the bottom of all three respirometers to bind with the CO_2 that is released during cellular respiration. CO_2 combines with KOH to form a precipitate that will prevent the CO_2 from interfering with the results.) Finally, I submerged the three respirometers completely into a bucket of water that was 0°C, keeping them upright.

Treatment Group 2: I set up another three respirometers (exactly like the ones used in Treatment Group 1), but I placed them in a water bath that was 25°C, keeping them upright.

Finally, for both treatment groups, I allowed 15 minutes for the respirometer tubes to equilibrate with the water temperature. Since the respirometers were filled with air, the water entered the pipette but did not completely fill it. I then observed and recorded the level of water in each pipette at time zero. I made observations and took measurements every 10 minutes for 60 minutes. As the peas respired, they drew in air from the pipette, and the level of the water in the pipette moved downward.

(b) Below is a graph that shows the predicted data from this experiment.

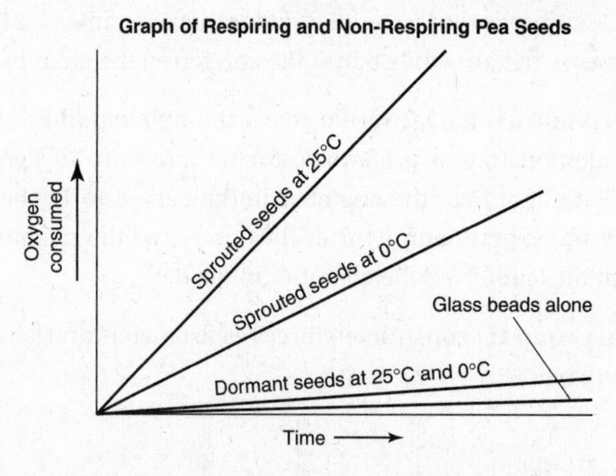

(c) The glass beads were a control and showed no respiration. The non-respiring (dry or dormant) seeds showed what looked like a tiny amount of respiration, but that was probably just an odd response to having gotten wet during the experiment. The respiring (sprouted) seeds at 25°C showed a greater rate of respiration than the respiring (sprouted) seeds at 0°C because the increase in temperature caused an increased rate of enzyme activity, which sped up all reactions within the respiring seeds.

(d) Using the same apparatus, I could replace the peas with fruit flies. Fruit flies are small, and the container is large enough to provide them with adequate oxygen. I would place one respirometer tube in 0°C of water and the other respirometer tube in water that is 25°C. The flies in the respirometer tube that is 25°C would have a faster respiration rate than the flies in the respirometer tube that is 0°C because the increased temperature would increase the enzyme activity and respiration rates. The results will be the same as they were with the peas because fruit flies respire just as peas do and just as every other living thing does.

3. (a) Artificial selection is the intentional breeding of plants or animals. Plants and animals that have desirable traits are bred with other individuals with the same desired traits. The result is an individual with one or more exceptional traits. Racehorses are bred for speed. Tomatoes and strawberries are bred for their size and color. Dogs are bred for appearance or personality.

(b) For this experiment, I would attempt to breed plants that have the largest number of trichomes possible. I would cross-pollinate plants that both have the greatest number of trichomes of all the plants available for this experiment. Then, I would gather the seeds they produce and cross-pollinate the plants that grow from those seeds. I would repeat this procedure a few times with each new generation of seeds produced.

(c) Each generation will produce plants with more and more trichomes.

(d) The phenotype of the plant (having lots of trichomes) is evidence that the plant carries the genes for having many trichomes. By selectively breeding plants that have many trichomes, this experiment is selecting for plants that carry the genes for having many trichomes.

4. (a) This graph shows the process of directional selection. This is when a population with one array of traits changes over time. For example, the population could have changed from having one color that predominates to having a different color that predominates. Another example could be that the original population was shorter and now the evolved population is taller. No single individual changed; *it was the percent of alleles in the population* that changed. Notice that both populations are shown in a bell curve distribution.

(b) This change in the population could have been a result of changes in the environment. For example, in pre-industrial England, white moths of one species were camouflaged and were selected for, while dark moths were visible, more often eaten by predators, and selected against. However, in post-industrial England, dark moths were camouflaged and were selected for, while light moths were more visible and eaten by predators, making them selected against. The change in the population was brought about by a change in the environment: industrialization caused excessive soot pollution that cover everything and affected which traits were selected for and which traits were selected against.

(c) Refer to the graph below, which demonstrates disruptive or diversifying selection in response to a change in the environment.

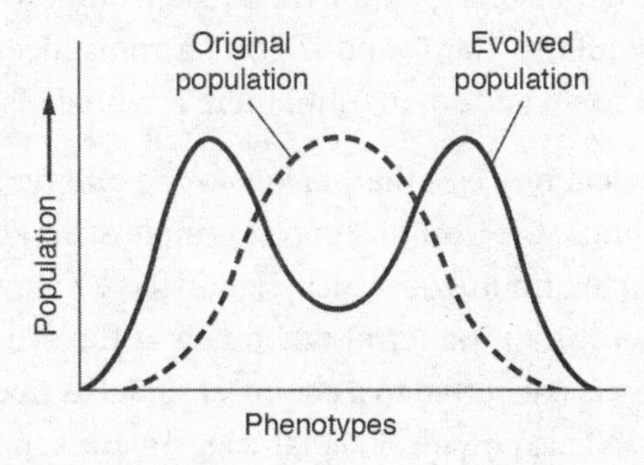

One possible change in the environment that could have caused the change in this population can be explained this way: Imagine that a lava flow suddenly covered an expansive plain, substantially darkening it, while leaving the adjacent highlands unaffected and light-colored. What was once a large, light, homogeneous area became divided into two differently colored regions. The animals that once moved freely across the plains now had to contend with two different environments, and the two populations became separated by this change in the environment. Thus, they became two distinct populations over time.

(d) In this scenario, as shown on the graph in part (c), one original bell-shaped population (that included a wide variety of individuals) evolved into two distinct populations. Dark-colored animals were safer in the darker regions and tended to remain in that environment. Light-colored animals were safer in the lighter regions and tended to remain in that environment. Over time, pressure from the environment caused extreme types to have the advantage and to be selected for at the expense of individuals with intermediate traits.

5. (a) This model demonstrates how a signal that is sent to a cell triggers a response from the cell. This process consists of three parts: reception, transduction, and response. It is a major way in which cells communicate. It exists in cells across every kingdom and controls a myriad of cellular processes, like the initiation of apoptosis or the release of hormones. One version of this process is the signal transduction pathway.

(b) This complex process relies on the interaction of many molecules that must bind to each other in specific ways. The three shapes at "L" in the model (called ligands) represent hundreds of molecules that send a signal. At the area labeled "I" on the model, you see that only one specially shaped ligand can bind to the receptor on the membrane surface. This binding triggers a chain reaction in the transducers, or relay molecules, at the area labeled "II" on the model. Finally, the last relay molecule enters the nucleus and initiates a response that is carried out when a gene is transcribed. In the model shown, the ligand is probably a polypeptide or protein because it cannot dissolve through the cell membrane, which consists of a lipid bilayer. Only lipid signaling molecules have the characteristics to do that. If the ligand were a lipid, it could dissolve directly through the cell membrane and bind with a receptor inside the cell in the cytoplasm.

(c) The response to the signal in the cytoplasm (at the area labeled "II" on the model) is like dominoes falling one after the other. Each domino knocks over the next one in line to fall. Similarly, the ligand triggers a transducer, which triggers another transducer (and so on and so on) until there is a final response.

(d) A broader biological process, that relates to the one represented in the model, is "how cells communicate." Another good example of how cells communicate can be seen in neurons. An impulse (electrical signal) passes along an axon to the terminal branch of the neuron to the terminal branch at the synapse. At the synapse, the electrical impulse is converted to a chemical signal as neurotransmitter is released into the synapse. This chemical signal triggers a response in another neuron, a gland, or a muscle. This neuronal model is similar to the model pictured in this question in that it depends on the interaction of many different molecules. It also depends on the presence of receptors. It differs, however, because there is an electrical impulse in that cell communication model, whereas there is none in the model pictured in the question.

6. (a) Graph A shows that the expression of miR-215 is relatively low in cancerous breast tissue (about 0.8) but is relatively high in adjacent noncancerous breast tissue (about 2.3). In other words, when the expression of miR-215 is present in high concentrations, there is less cancer. Therefore, the presence of miR-215 seems to correlate with healthy tissue.

(b) Graph B shows that apoptosis is high in the presence of a substance that mimics miR-215 (about 30%) and that apoptosis is low in the presence of a substance that inhibits miR-215 (about 6%).

(c) Sometimes, normal cells become cancerous—something causes them to keep dividing, grow out of control, and form a tumor (a clump of cancerous tissue). In most cases in our bodies, thankfully, these cancerous cells are directed to die by apoptosis before they become a large tumor. According to Graph B, apoptosis is enhanced in tissue with an miR-215 mimic. Furthermore, when miR-215 is inhibited, so is apoptosis. Therefore, an miR-215 mimic might slow the growth of a tumor by causing precancerous or cancerous cells to die by apoptosis, as the researcher predicts.

(d) The data on Graph A show that high levels of miR-215 are found in normal tissue but are lacking in cancerous breast tissue. This suggests that miR-215 protects against breast cancer. (Note that this question only refers to breast cancer, not to other causes of cancer.) Furthermore, the data on Graph B suggest a mechanism for how the expression of miR-215 protects against cancer: it triggers apoptosis (cell death), which removes cancerous cells from tissue.

ANSWER SHEET
Practice Test 2

SECTION I

1. Ⓐ Ⓑ Ⓒ Ⓓ	16. Ⓐ Ⓑ Ⓒ Ⓓ	31. Ⓐ Ⓑ Ⓒ Ⓓ	46. Ⓐ Ⓑ Ⓒ Ⓓ
2. Ⓐ Ⓑ Ⓒ Ⓓ	17. Ⓐ Ⓑ Ⓒ Ⓓ	32. Ⓐ Ⓑ Ⓒ Ⓓ	47. Ⓐ Ⓑ Ⓒ Ⓓ
3. Ⓐ Ⓑ Ⓒ Ⓓ	18. Ⓐ Ⓑ Ⓒ Ⓓ	33. Ⓐ Ⓑ Ⓒ Ⓓ	48. Ⓐ Ⓑ Ⓒ Ⓓ
4. Ⓐ Ⓑ Ⓒ Ⓓ	19. Ⓐ Ⓑ Ⓒ Ⓓ	34. Ⓐ Ⓑ Ⓒ Ⓓ	49. Ⓐ Ⓑ Ⓒ Ⓓ
5. Ⓐ Ⓑ Ⓒ Ⓓ	20. Ⓐ Ⓑ Ⓒ Ⓓ	35. Ⓐ Ⓑ Ⓒ Ⓓ	50. Ⓐ Ⓑ Ⓒ Ⓓ
6. Ⓐ Ⓑ Ⓒ Ⓓ	21. Ⓐ Ⓑ Ⓒ Ⓓ	36. Ⓐ Ⓑ Ⓒ Ⓓ	51. Ⓐ Ⓑ Ⓒ Ⓓ
7. Ⓐ Ⓑ Ⓒ Ⓓ	22. Ⓐ Ⓑ Ⓒ Ⓓ	37. Ⓐ Ⓑ Ⓒ Ⓓ	52. Ⓐ Ⓑ Ⓒ Ⓓ
8. Ⓐ Ⓑ Ⓒ Ⓓ	23. Ⓐ Ⓑ Ⓒ Ⓓ	38. Ⓐ Ⓑ Ⓒ Ⓓ	53. Ⓐ Ⓑ Ⓒ Ⓓ
9. Ⓐ Ⓑ Ⓒ Ⓓ	24. Ⓐ Ⓑ Ⓒ Ⓓ	39. Ⓐ Ⓑ Ⓒ Ⓓ	54. Ⓐ Ⓑ Ⓒ Ⓓ
10. Ⓐ Ⓑ Ⓒ Ⓓ	25. Ⓐ Ⓑ Ⓒ Ⓓ	40. Ⓐ Ⓑ Ⓒ Ⓓ	55. Ⓐ Ⓑ Ⓒ Ⓓ
11. Ⓐ Ⓑ Ⓒ Ⓓ	26. Ⓐ Ⓑ Ⓒ Ⓓ	41. Ⓐ Ⓑ Ⓒ Ⓓ	56. Ⓐ Ⓑ Ⓒ Ⓓ
12. Ⓐ Ⓑ Ⓒ Ⓓ	27. Ⓐ Ⓑ Ⓒ Ⓓ	42. Ⓐ Ⓑ Ⓒ Ⓓ	57. Ⓐ Ⓑ Ⓒ Ⓓ
13. Ⓐ Ⓑ Ⓒ Ⓓ	28. Ⓐ Ⓑ Ⓒ Ⓓ	43. Ⓐ Ⓑ Ⓒ Ⓓ	58. Ⓐ Ⓑ Ⓒ Ⓓ
14. Ⓐ Ⓑ Ⓒ Ⓓ	29. Ⓐ Ⓑ Ⓒ Ⓓ	44. Ⓐ Ⓑ Ⓒ Ⓓ	59. Ⓐ Ⓑ Ⓒ Ⓓ
15. Ⓐ Ⓑ Ⓒ Ⓓ	30. Ⓐ Ⓑ Ⓒ Ⓓ	45. Ⓐ Ⓑ Ⓒ Ⓓ	60. Ⓐ Ⓑ Ⓒ Ⓓ

PRACTICE TEST 2

SECTION I

Time: 90 minutes

Directions: For each question, choose the best answer choice.

1. Oxygen is carried in the blood by the respiratory pigment hemoglobin, which can combine loosely with four oxygen molecules to form the molecule oxyhemoglobin. To function properly, hemoglobin must bind to oxygen in the lungs and drop the oxygen off at the body cells. The more easily the hemoglobin binds to oxygen in the lungs, the more difficult it will be for the hemoglobin to unload the oxygen at the body cells. Below is a graph that shows two different saturation-dissociation curves for one type of hemoglobin at two different pH levels:

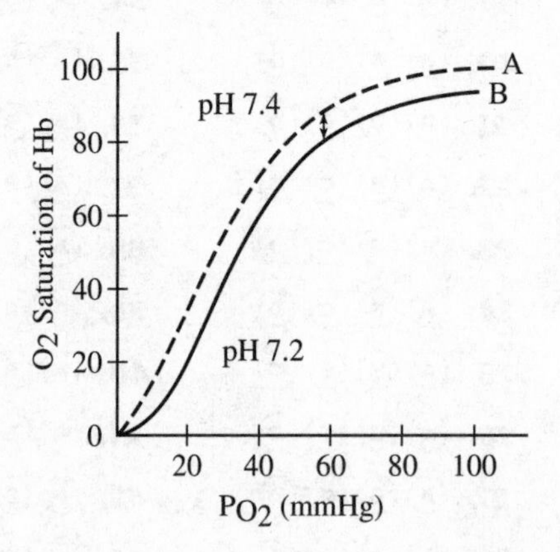

Based on your knowledge of Biology and the information in this graph, which of the following statements about the hemoglobin curves is correct?

(A) Hemoglobin B has a greater affinity for oxygen and will therefore bind more easily to oxygen in the lungs.
(B) Hemoglobin B is characteristic of a mammal that evolved at a high elevation where oxygen is rare.
(C) When CO_2 levels in the blood are high, as shown in hemoglobin B, hemoglobin releases oxygen more readily.
(D) Hemoglobin A is the type found in mammals with a higher metabolism. Hemoglobin B is characteristic of mammals with a lower metabolism.

PRACTICE TEST 2

2. Refer to the figure below.

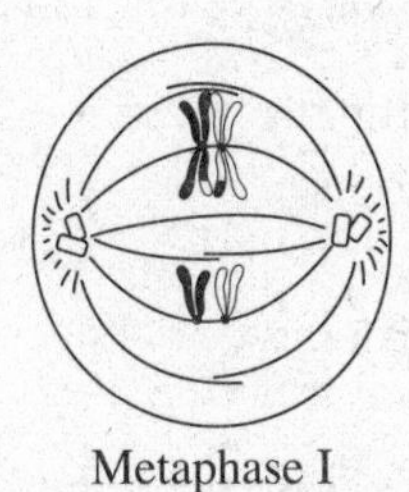

Metaphase I

How homologous pairs of chromosomes line up on the metaphase plate in metaphase I is random. This independent orientation of chromosomes results in which of the following?

(A) an increase in the number of gametes
(B) an increase in the number of possible combinations of alleles
(C) an increase in the number of offspring
(D) a decrease in the variation of the gametes

3. Energy is harvested during cellular respiration in stages. Which of the following correctly states which stage of cellular respiration harvests the most energy and provides the correct explanation for why that occurs?

(A) The most energy is released during the Krebs cycle because it is here that pyruvate is completely broken down into CO_2.
(B) The most energy is released during the Krebs cycle because, in addition to the production of ATP, both $FADH_2$ and NADH are produced. Each of those molecules will release 2 ATPs and 3 ATPs, respectively.
(C) The most energy is released during oxidative phosphorylation because, in addition to the phosphorylation of ADP into ATP, all the potential energy held in NADH and $FADH_2$ is transferred to ATP.
(D) The most energy is released during oxidative phosphorylation because H_2O is completely broken down into H^+ and O_2.

4. Cellular respiration in eukaryotes involves a series of coordinated enzyme-catalyzed reactions that harvest free energy from simple carbohydrates. Which of the following statements about cellular respiration is correct?

(A) Electron transport chain reactions are limited to the mitochondria in eukaryotic cells and prokaryotic plasma membranes.
(B) Energy is produced by oxidative phosphorylation during the citric acid cycle.
(C) Electrons that are delivered by NADH and $FADH_2$ to the electron transport chain are passed to a series of electron acceptors as they move toward the terminal electron acceptor: oxygen.
(D) Pyruvate is a waste product of cellular respiration; it is released along with CO_2 and water vapor.

5. Which of the following statements is correct about a man who has hemophilia and whose wife does not have the disease nor does she have any relatives with the disease?

   (A) All his daughters will have the disease.
   (B) All his sons will have the disease.
   (C) All his sons will be carriers.
   (D) All his daughters will be carriers.

6. Which of the following choices identifies the phase of meiosis when recombination occurs and provides the correct evidence for that claim?

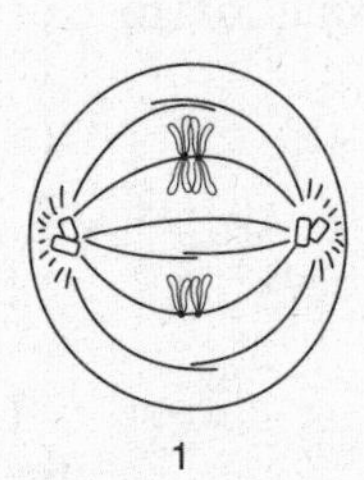

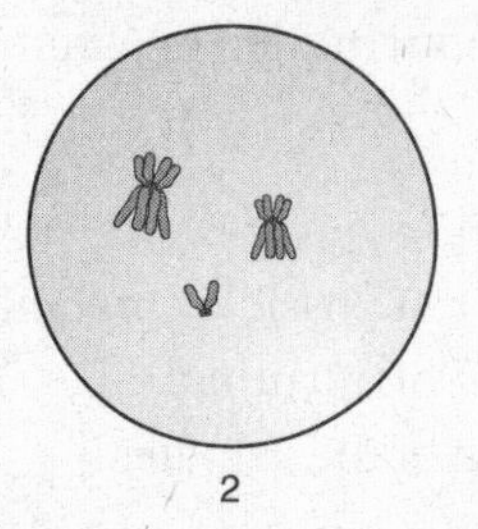

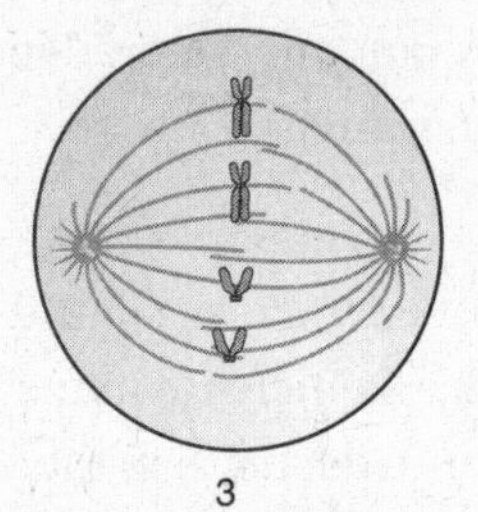

   (A) Picture 1, because sister chromosomes are lined up on the metaphase plate
   (B) Picture 1, because homologous pairs are lined up on the metaphase plate
   (C) Picture 2, because homologues are paired up in synapsis
   (D) Picture 3, because homologous pairs are exchanging genetic material

7. The table below shows the results of a breeding experiment that examined the inheritance of flower color (red or white) and pod shape (smooth or wrinkled). True-breeding parents were crossed to produce F_1 offspring, which were then testcrossed with homozygous recessive individuals.

Results from Crosses with Pea Plants

Parental Crosses	**Phenotypes of F_1 Offspring**	**Phenotypes of Testcross Offspring (Number of Individuals)**	
Red × White	Red	Red (455)	White (471)
Smooth × Wrinkled	Smooth	Smooth (501)	Wrinkled (498)
	Testcross		
Red/Smooth × White/Wrinkled	Red/Smooth (385) Red/Wrinkled (379) White/Smooth (375) White/Wrinkled (382)		

The ratio of phenotypes in the offspring from the testcross with F1 plants that had red flowers and smooth pods suggests that the genes for flower color and seed pod shape are located

   (A) on a mitochondrial chromosome
   (B) close together on the same chromosome
   (C) on different chromosomes
   (D) on the X chromosome

8. Hydrangea flowers have one gene for flower color. Plants of the same genetic variety have flowers that range in color from blue to pink with the color varying due to the type of soil in which they were grown. Which of the following statements best explains this phenomenon?

   (A) The alleles for flower color show incomplete dominance where neither trait is dominant; expression of the genes shows a blending of traits.
   (B) The alleles for flower color are codominant; both traits show dependence on the environment.
   (C) In this case, the environment alters the expression of a trait.
   (D) The genes for flower color show polygenic inheritance.

Questions 9 and 10

A female fruit fly hybrid for both a gray body (*Gg*) and normal wings (*Nn*) is crossed with a male with a black body (*gg*) and vestigial wings (*nn*): *GgNn* × *ggnn*.

The F_1 results of the cross are shown in the chart below:

A	*B*	*C*	*D*
Gray/Normal	Black/Vestigial	Gray/Vestigial	Black/Normal
969	941	190	184

9. Which of the following statements best explains the results?

   (A) The two alleles for body color and wing structure are located on separate chromosomes and assorted independently.
   (B) The two alleles for body color and wing structure are located on separate chromosomes and assorted independently, but there was some crossing-over.
   (C) The two alleles are linked and located on the same chromosome (immediately next to one another) and were inherited together.
   (D) The two alleles are linked and located on the same chromosome but are far apart and experienced some crossing-over.

10. Which of the following illustrations depicts the most likely location of the alleles for body color and wing structure?

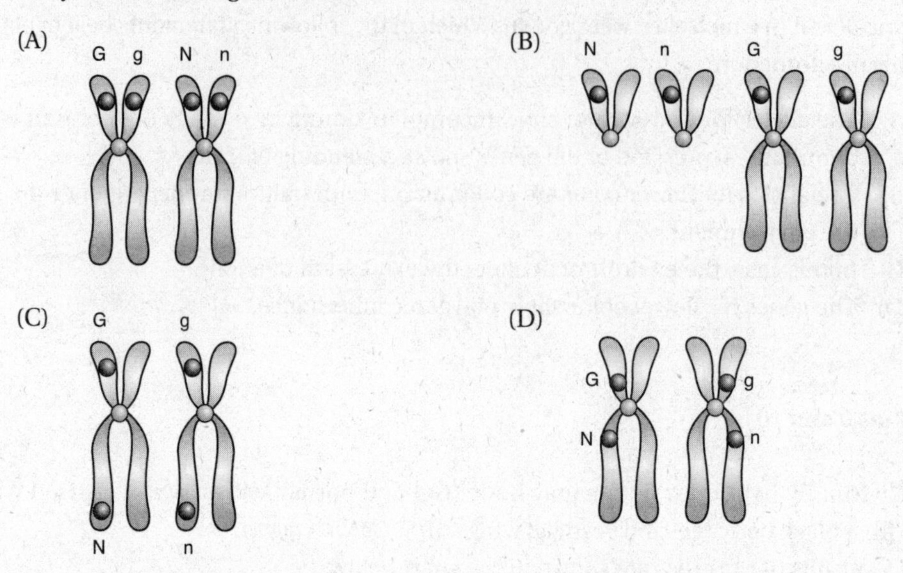

---

11. Which of the following is an example of a coupling of an exergonic reaction with an endergonic reaction?

    (A) Unicellular organisms that live in freshwater, such as ameba, must pump out excess water using their contractile vacuoles.
    (B) The enzyme lactase binds with lactose to produce molecules of glucose and galactose.
    (C) Electrons that escape from chlorophyll *a* are replaced by those released by the hydrolysis of water.
    (D) The flow of electrons down an electron transport chain in mitochondria powers the pumping of protons against a gradient into the outer compartment.

12. Which of the following statements about positive feedback is correct?

    (A) When the level of thyroxine in your bloodstream is too low, the anterior pituitary releases TSH to stimulate the thyroid to release more thyroxine.
    (B) It is a complex mechanism that is only found in more advanced mammals, like humans.
    (C) Glucagon causes the release of glucose from the liver when blood sugar levels fall below a certain point.
    (D) It is a mechanism that enhances an already existing response.

13. What is the most likely hypothesis for why the most species diversity is found in the tropics rather than in the temperate regions of the world?

    (A) The food chains are longer in the tropics because of all the available sunlight, and thus the tropics can support more life.
    (B) The tropics have the highest temperatures, which cause a faster rate of mutation and speciation.
    (C) The tropics were the first region to separate away from Pangaea.
    (D) The tropics have the most direct sunlight and the most rain.

14. The frequency of a particular allele in a population of 1,000 birds in Hardy-Weinberg equilibrium is 0.3. If the population remains in equilibrium, what would be the expected frequency of that allele after 500 years?

    (A) It will increase if the allele is favorable, or it will decrease if it is unfavorable for individuals in that population.
    (B) Not enough information is provided to determine if the frequency of the allele will remain the same or change.
    (C) The frequency of the allele will remain at 0.3 because the population is in Hardy-Weinberg equilibrium.
    (D) The frequency of the allele will remain the same; this is an example of the bottleneck effect.

Questions 15 and 16

You carry out an experiment to study transpiration in plants using a two-week-old bean plant. You set up a potometer by cutting the stem of a plant and securing it into clear flexible tubing that has been tightly connected to a calibrated pipette. Water fills the potometer from the plant stem to the tip of the pipette.

You measure water loss from your potometer at 10-minute intervals and plot your data on the graph below as Line *B*:

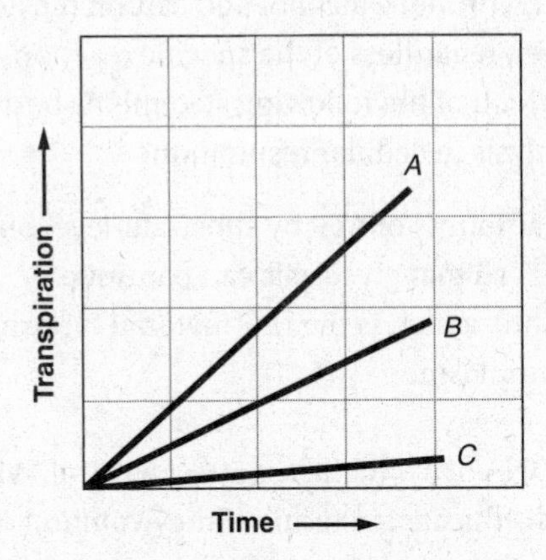

15. You then explore what happens if you expose the plant to different conditions. Which of the following statements is correct based on proposed changes to the potometer?

    (A) Increasing the green wavelengths of light will increase the rate of transpiration and account for Line *A* on the graph.
    (B) Placing a plastic bag over the plant will increase the rate of transpiration and account for Line *A*.
    (C) Placing a fan near the plant will decrease the rate of transpiration and account for Line *C*.
    (D) Painting the surface of most of the leaves with clear nail polish will decrease the rate of transpiration and account for Line *C*.

16. Which of the following is correct about transpiration in plants?

(A) Water moves upward from the roots to the shoots in the xylem by active transport from an area of high osmotic potential to an area of low osmotic potential.
(B) Plants lose water through stomates in their leaves.
(C) The movement of water up a tree is limited to about 20 meters.
(D) The higher the humidity, the faster the rate of transpiration.

---

17. When the first tiny prokaryotic cell took up residence inside a larger prokaryotic cell, it heralded the advent of the eukaryotic cell and led to an explosion of new life on Earth. Since then, most cells on Earth have internal organelles. Which of the following best summarizes an advantage of having internal membranes and organelles?

(A) DNA can reproduce more efficiently.
(B) Even though prokaryotes do not have mitochondria, they contain structures that carry out the same function.
(C) Organelles separate specific reactions in the cell and increase metabolic efficiency.
(D) Compartmentalization enables prokaryotes to reproduce more quickly.

18. Many different fermentation pathways occur in different organisms in nature. For example, skeletal muscle cells convert pyruvate into lactic acid when no oxygen is present. Yeast cells can produce alcohol and carbon dioxide under the same circumstances. However, regardless of the specific reactions, the purpose of glycolysis is an important one. Which of the following statements best describes the most important role of glycolysis in cellular respiration?

(A) It produces large amounts of ATP by substrate-level phosphorylation.
(B) It reoxidizes NADH so that glycolysis can continue.
(C) It produces pyruvate, which is the raw material for oxidative phosphorylation.
(D) It occurs in the cytoplasm.

19. Human DNA and the DNA of a banana are 50% identical. Which of the following statements best explains this fact in terms of the evolution of life on Earth?

(A) Humans evolved from an organism that resembled a banana more than a billion years ago.
(B) All organisms on Earth evolved from an ancestral eukaryotic cell about 1.5 billion years ago.
(C) Humans and bananas evolved from a recent common ancestor within the last 6 million years.
(D) Bananas and humans have experienced thousands of mutations that have made us appear so very different from bananas.

20. Which of the following statements is correct about Charles Darwin's theory of evolution by natural selection?

(A) New species arise suddenly after long periods of no change.
(B) Mutations are the main source of all variation in a population.
(C) Every population contains tremendous variation.
(D) Traits that are not used by an organism or species will disappear.

21. Which of the following statements is supported by the information presented in the graph below, which shows a comparison of two regions in Oregon, one deforested and the other undisturbed? Note that both regions are in close proximity to the same stream.

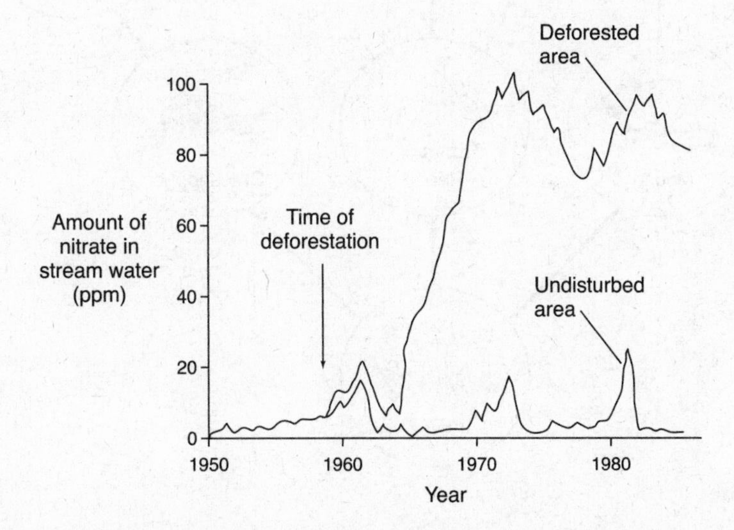

(A) Roots of plants are efficient at absorbing nitrates from the soil.
(B) Variation in the amount of nitrate in the stream water is due to periods of intense rain.
(C) Replanting trees after an area has been clear-cut can prevent nitrate runoff.
(D) The presence of trees in an area causes an increase of nitrates in the soil.

PRACTICE TEST 2

22. Below is a food web for a habitat that is threatened by developers who will remove three-fourths of the grass in the area on which the mice and rabbits feed:

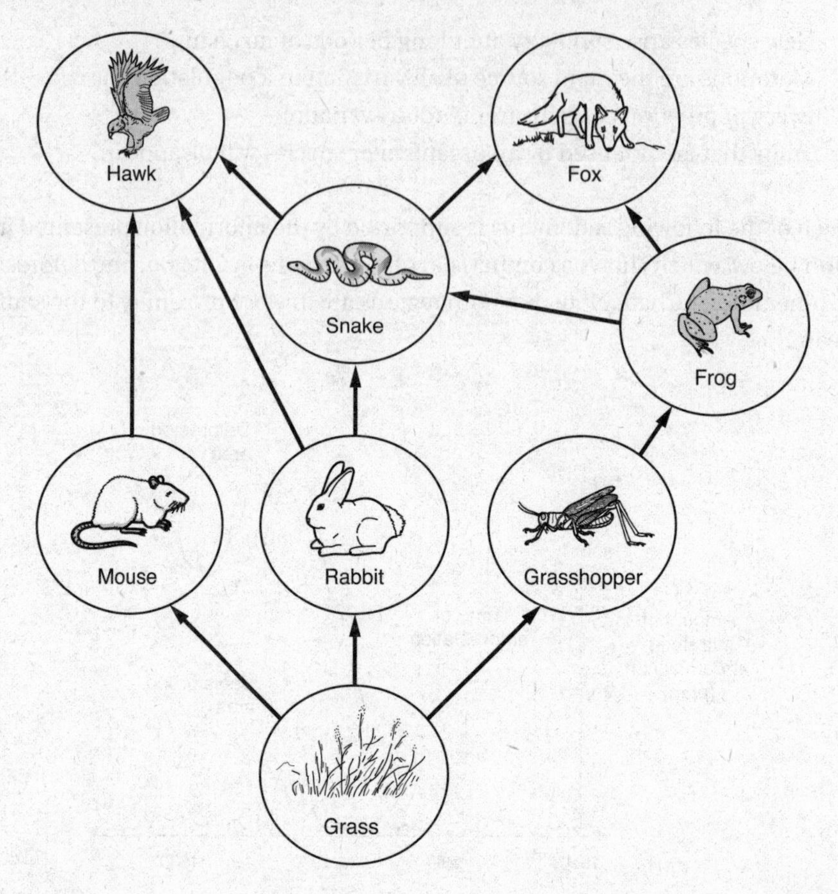

Which of the following statements describes what will most likely happen to the wildlife in the area?

(A) The hawk will begin to eat frogs instead of snakes and mice.

(B) Since three trophic levels are shown in the food web, 75% of the organisms in this food web will die.

(C) Based on the number of trophic levels, about 25% of the mice and rabbits will die.

(D) The hawk population will decrease.

23. Which of the following statements about apoptosis is correct?

(A) Apoptosis is a special type of cell division that requires numerous types of cell signaling.
(B) The fact that apoptosis is widespread across several kingdoms is evidence that it evolved early in the evolution of eukaryotes.
(C) Apoptosis plays a crucial role in the movement of chromosomes prior to gamete formation.
(D) Apoptosis plays an important role in the translation of mRNA at the ribosomes.

24. Below are illustrations of four amino acids:

Serine (A)	Lysine (B)	Glutamate (C)	Glycine (D)
COO	COO^-	COO^-	COO^-
H_3N-C-H	H_3N-C-H	H_3N-C-H	H_3N-C-H
CH_2OH	CH_2	CH_2	H
	CH_2	CH_2	
	CH_2	COO^-	
	NH_3^+		

Molecule A is serine and has polar side chains. Molecule B is lysine, which is basic and positively charged. Molecule C, glutamate, is acidic and negatively charged. Molecule D is glycine and has nonpolar side chains.

Which of the following statements about these amino acids is correct?

(A) Molecule A can readily dissolve through a plasma membrane.
(B) Only Molecules B and C can readily dissolve through a plasma membrane.
(C) Molecules A, B, and C can all readily dissolve through a plasma membrane.
(D) Only Molecule D can readily dissolve through a plasma membrane.

25. Which of the following statements describes the structural level of a protein that is least affected by hydrogen bonding?

(A) Primary structure depends on the sequence of amino acids.
(B) Tertiary structure has a shape that is dependent on the interactions of side chains of amino acids.
(C) Quaternary structure results from the aggregation of more than one polypeptide unit.
(D) An α–helix is an example of a secondary structure of a polypeptide.

26. The *BCL-2* gene codes for a protein that normally inhibits apoptosis. In some cases, the *BCL-2* gene is mutated and becomes activated inappropriately. Cells with this mutated, overexpressed gene fail to undergo apoptosis; they continue to divide and form cancerous tumors. The mutated gene causes many cancers, including lymphoma and breast, colon, and prostate cancers. Oblimersen is a drug now in clinical trials that is an antisense RNA. It binds to *BCL-2* mRNA and inactivates it. Which of the following illustrations accurately demonstrates the action of the antisense drug Oblimersen?

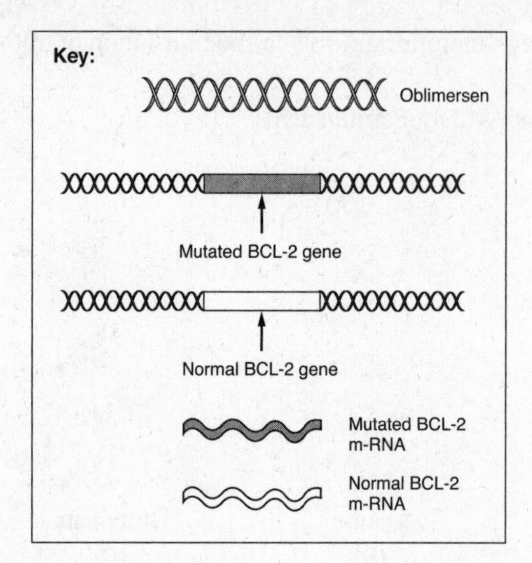

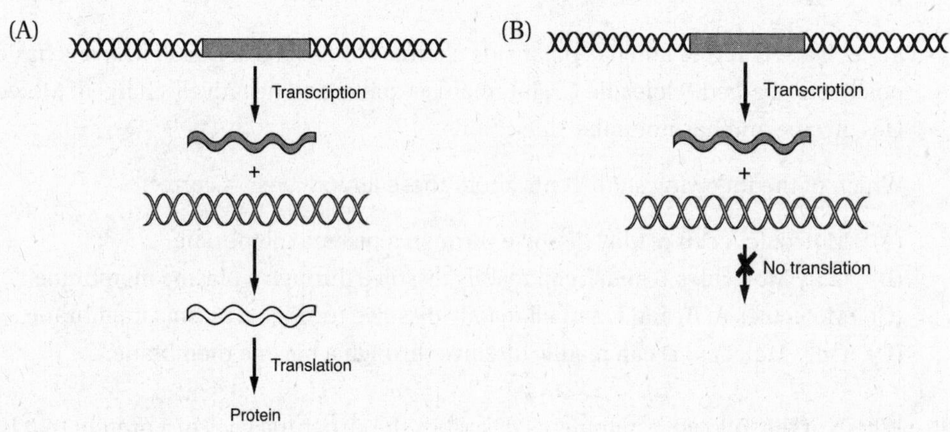

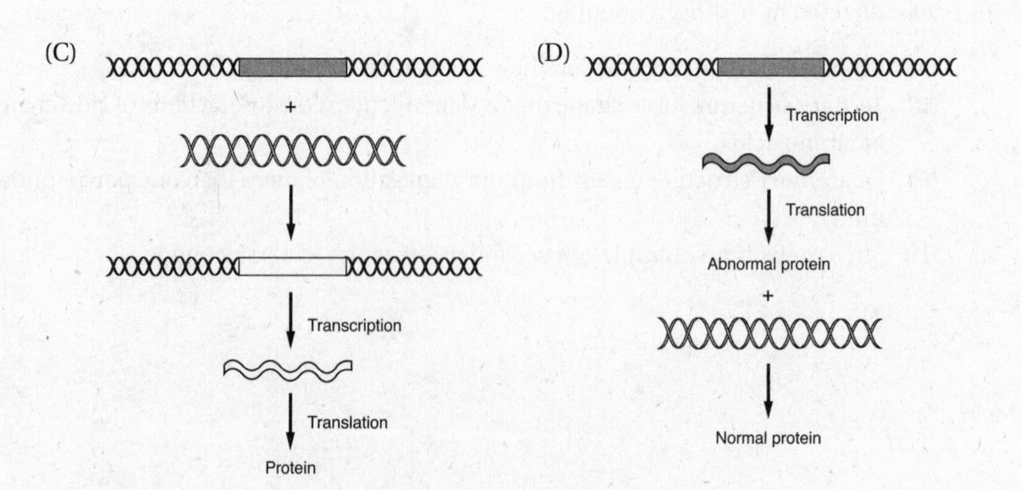

27. Figure 1 shows the growth of the filamentous green algae *Spirogyra* in a flask of sterilized pond water that contains nitrate. If phosphate is added to the flask, the growth curve changes to that shown in Figure 2.

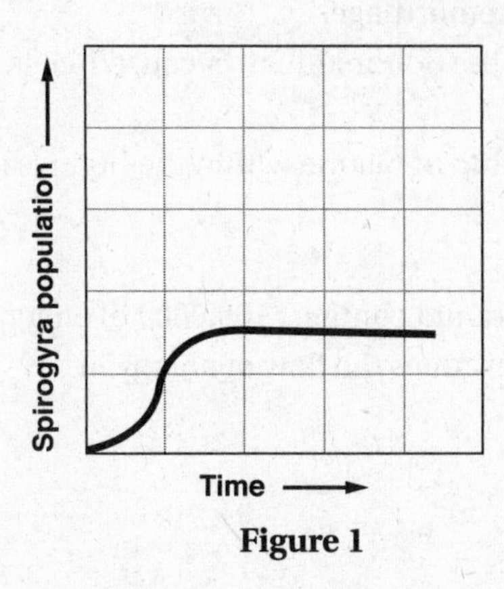

Figure 1

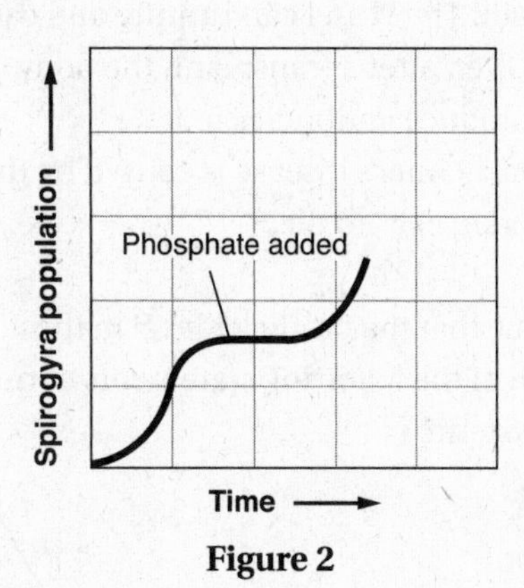

Figure 2

Which of the following graphs is the best prediction of the algae growth if more nitrate (NO_3) is added along with the phosphate?

(A)

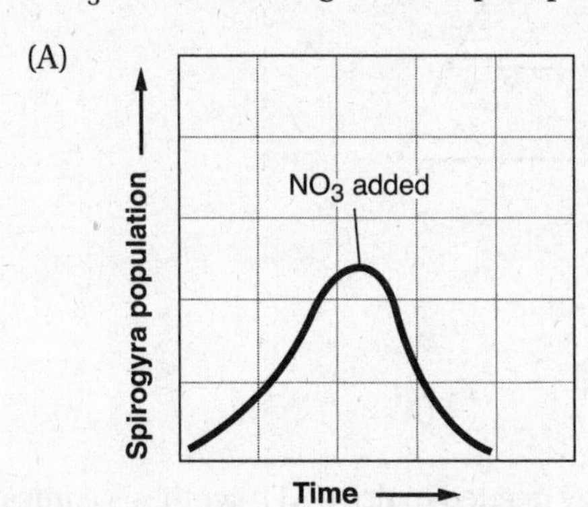

(B)

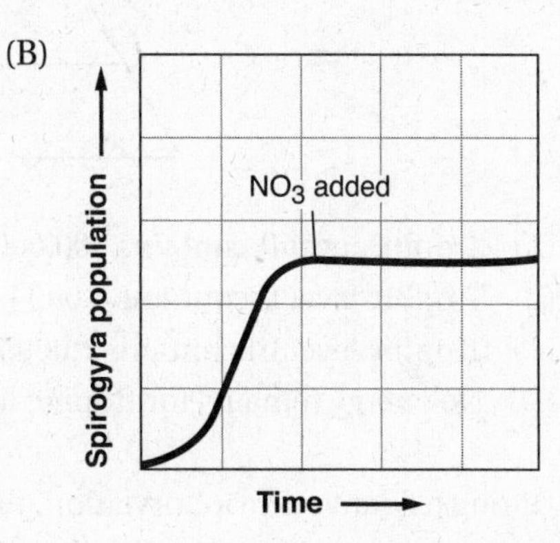

(C)

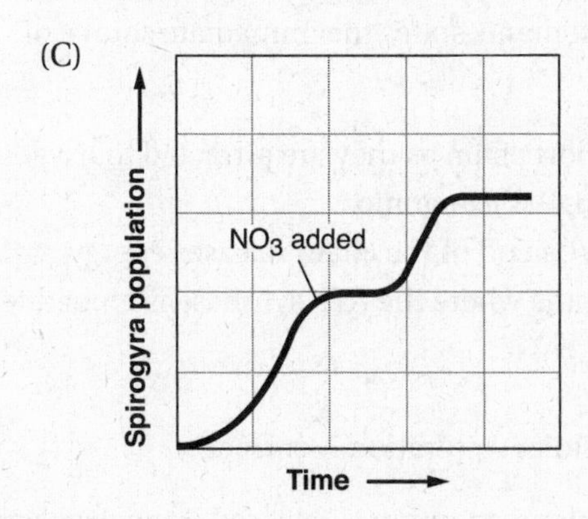

(D)

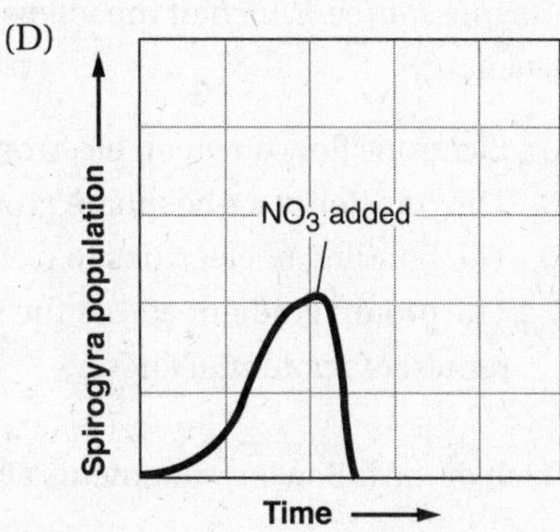

28. Which of the following statements correctly characterizes an autoimmune disease?

(A) T cells attack healthy body cells.
(B) An aneurysm is a weakness in the wall of a blood vessel that causes it to balloon out. The vessel can rupture and cause a hemorrhage.
(C) Often after a transplant, the body rejects the donor kidney because T cells produce antibodies to it.
(D) Alzheimer's disease is caused by the buildup of plaque within the brain, causing brain cells to die.

29. Assume that the producer level in this food pyramid contains 100,000 J of energy. Which of the following statements correctly describes the flow of energy in this system?

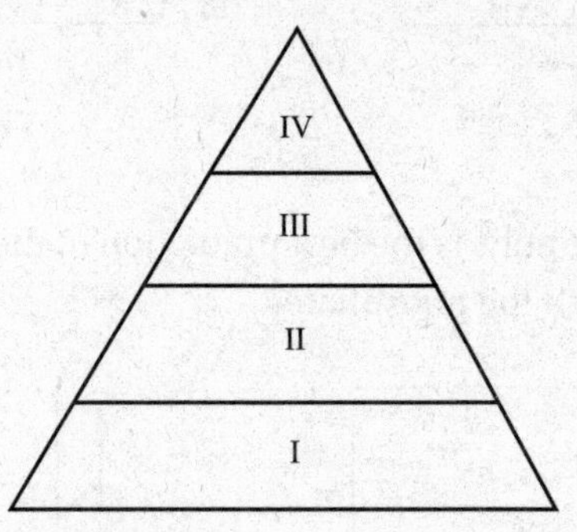

(A) Trophic level II contains 100,000 J.
(B) Trophic level II contains 1,000 J.
(C) Trophic level III contains 1,000 J.
(D) No energy remains for trophic level IV.

30. During oxidative phosphorylation, the energy needed to drive ATP synthesis comes from one source. Which of the following statements states the immediate source of that energy?

(A) Electrons flow down an electron transport chain as they are attracted to oxygen.
(B) The transfer of a phosphate group to ADP is exergonic.
(C) The bonding of electrons to oxygen at the end of the chain releases energy.
(D) The proton gradient across the membrane where the ATP synthase is embedded represents potential energy.

31. Which of the following statements about cellular respiration is correct?

(A) Most CO_2 that is produced during cellular respiration is released from glycolysis.
(B) Protons are pumped through the ATP synthase by active transport.
(C) The final electron acceptor of the electron transport chain is NAD^+.
(D) ATP is formed because an endergonic reaction is coupled with an exergonic reaction.

32. The light reactions of photosynthesis supply the Calvin cycle with which of the following?

(A) The light reactions provide oxygen for the light-independent reactions.
(B) ATP and NADPH provide the power and raw materials for the Calvin cycle.
(C) Water entering the plant through the roots provides hydrogen directly to the Calvin cycle.
(D) CO_2 released by the light-dependent reactions provides the raw material for the Calvin cycle.

33. Below is an illustration that shows cyclic photophosphorylation in the grana of plant cells:

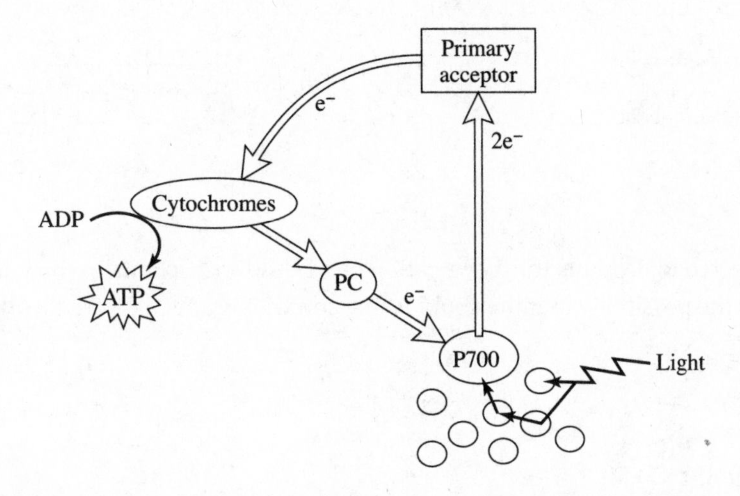

Which of the following statements about this process is correct and includes the right reasoning for why that statement is correct?

(A) It is similar to oxidative phosphorylation in cellular respiration because its function is to generate ATP.
(B) It is similar to glycolysis because it involves an electron transport chain.
(C) It is the opposite of cellular respiration because it releases oxygen rather than utilizes it.
(D) It is the opposite of the citric acid cycle because ATP is utilized, not produced.

34. Unlike large populations, small populations are vulnerable to various processes that draw these populations down an extinction vortex toward becoming an increasingly smaller population, until no individuals survive. Which of the following statements correctly identifies the factors that endanger a population?

(A) Inbreeding and loss of genetic variation threaten a population.
(B) Migration of new individuals into the population threatens a population.
(C) Mutation reduces the health of a population.
(D) Breeding with individuals from a different population may cause the extinction of the first population due to a decrease in diversity.

35. The following human pedigree shows three generations of deafness in one family. Squares represent males, circles represent females, and shaded symbols represent individuals affected with a disorder.

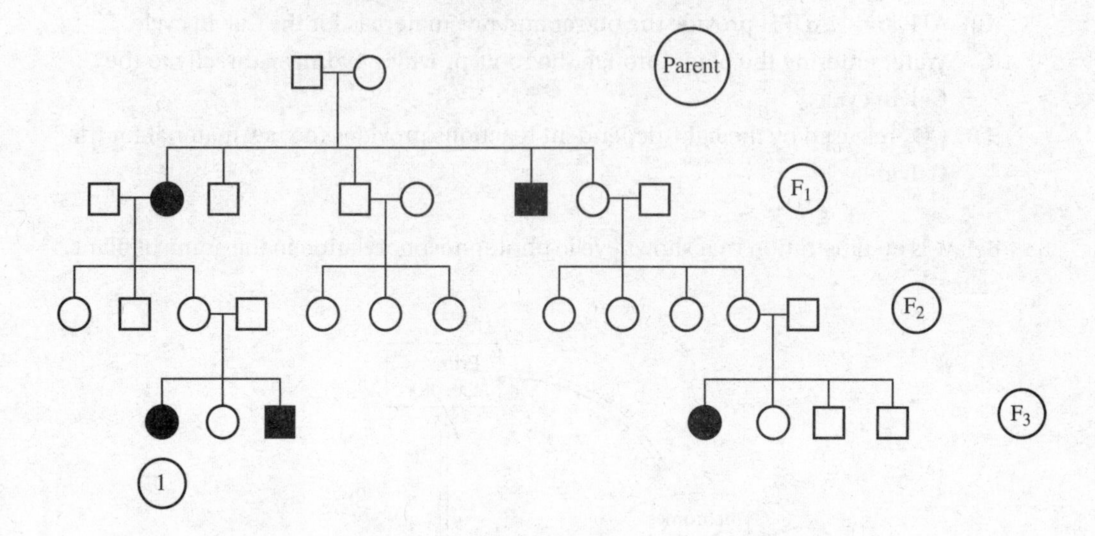

The affected male in the third generation has a child with a female who is a carrier. What is the possibility that the couple's first son will be affected with the disorder?

(A) 25%
(B) 50%
(C) 75%
(D) 100%

36. The disaccharide lactose is available to *E. coli* in the human colon if the host drinks milk. Digestion of lactose into glucose and galactose begins with the hydrolysis of lactose by the enzyme β-galactosidase, which is encoded by gene Z in the *lac* operon. Which of these diagrams correctly depicts the *lac* operon when lactose is being utilized?

(A)

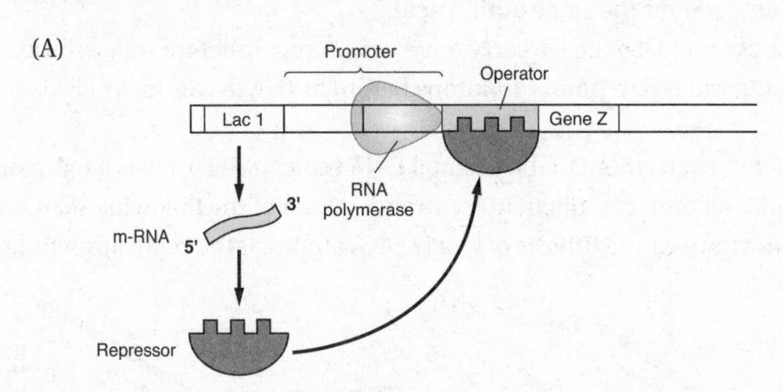

(B)

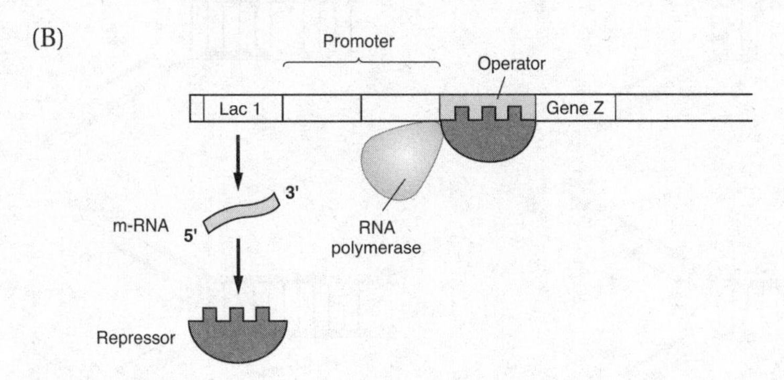

(C)

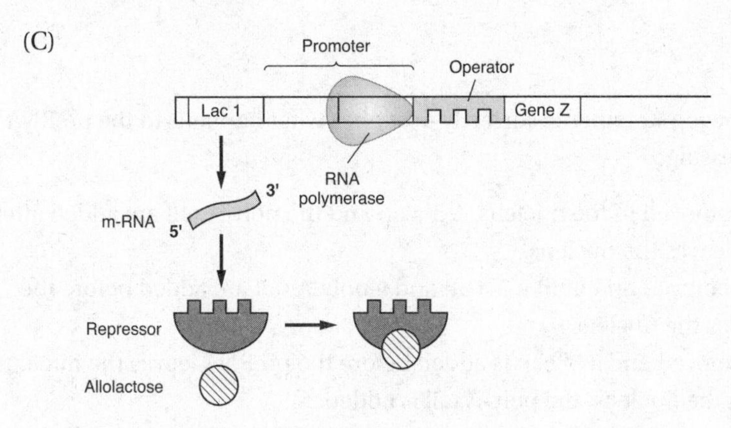

(D)

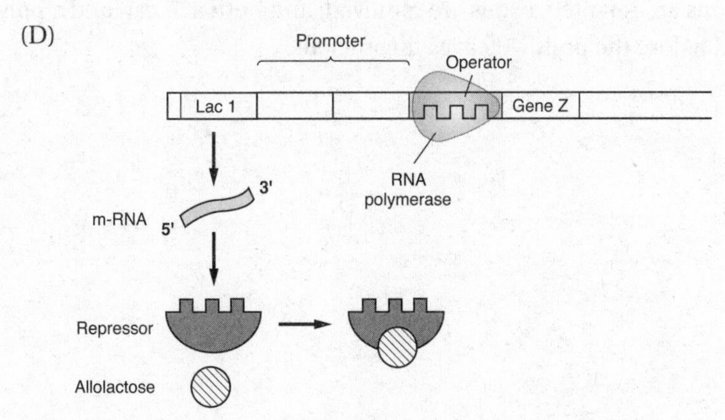

PRACTICE TEST 2

37. Which of the following statements explains how a point mutation can have no negative impact in the protein formed from a sequence in DNA?

(A) The first codon of a gene is always AUG, a leader sequence; a mutation in that sequence will have no effect on the protein produced.
(B) Several codons code for the same amino acid.
(C) DNA polymerase and DNA ligase carry out excision repair before transcription.
(D) RNA processing will repair point mutations before mRNA leaves the nucleus.

38. When DNA replicates, each strand of the original DNA molecule is used as a template for the synthesis of a second, complementary strand. Which of the following sketches most accurately illustrates the synthesis of a new DNA strand at the replication fork?

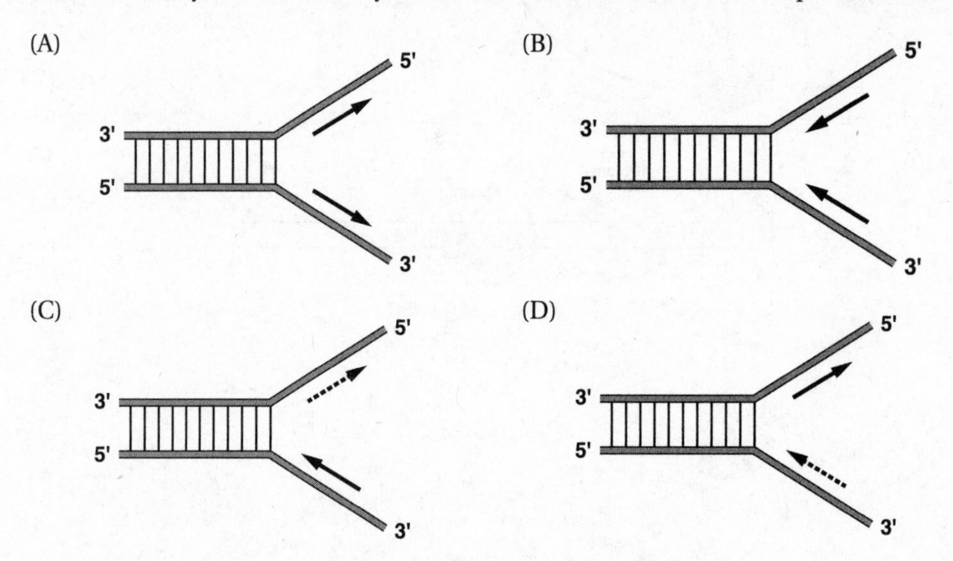

39. Which of the following statements correctly describes what happens to the preRNA during RNA processing?

(A) Introns are removed in the nucleus. A 5′ cap and the poly-A tail are added after the preRNA leaves the nucleus.
(B) Introns are removed and both a 5′ cap and a poly-A tail are added before the preRNA leaves the nucleus.
(C) Exons are removed and a 5′ cap is added before the preRNA leaves the nucleus. After leaving the nucleus, the poly-A tail is added.
(D) Point mutations are repaired, exons are removed, and both a 5′ cap and a poly-A tail are added before the preRNA leaves the nucleus.

40. Which of the following statements about the role of the restriction enzyme EcoRI is correct?

(A) It glues back together strands of DNA after replication.
(B) It recognizes and cuts specific short DNA sequences, like GAATTC.
(C) It assists in the transcription of DNA into mRNA in eukaryotic cells.
(D) It assists in copying a specific sequence of DNA during replication.

41. Small nuclear ribonucleoproteins (snRNPs) are RNA-protein complexes in eukaryotic cells. They are abundant and highly conserved across species. Which of the following statements accurately describes the important role snRNPs carry out in a cell?

(A) snRNPs splice out introns from mRNA after mRNA gets to the cytoplasm.
(B) snRNPs assist in the translation of polypeptides at the ribosomes by removing introns.
(C) snRNPs replace exons with introns in post-transcriptional RNA during RNA processing.
(D) snRNPs remove intervening sequences from post-transcriptional RNA during RNA processing.

42. The pH of two lakes is measured. Lake *A* has a pH of 8.0; Lake *B* has a pH of 6.0. Which of the following statements about these lakes is correct?

(A) Lake *B* is alkaline.
(B) Lake *B* is 100 times more alkaline than Lake *A*.
(C) The hydrogen ion concentration of Lake *B* is 10^4 M.
(D) The pH of both lakes will decrease in response to acid rain.

Questions 43 and 44

A scientist wanted to understand the role of the gonads and their hormones in the development of the sex of an individual. He carried out an experiment on 2 groups of pregnant rabbits. One group was the control group, on which he performed a sham surgery. The other group was the experimental group, on which he performed surgery on the embryos at a stage before sex differences were observable. He surgically removed the portion of each embryo that would develop into ovaries or testes. When the babies were born, he made note of their chromosome sex and the sexual differentiation of the genitals. His data is presented in the table below:

	Appearance of Genitals	
Sex Chromosome of Individual	**Sham Surgery**	**Embryonic Gonad Removed**
XY	Male	Female
XX	Female	Female

43. Which of the following statements is a correct conclusion to draw from this experiment?

    (A) In rabbits, female development requires a hormonal signal from the female gonad.
    (B) In rabbits, male development requires a hormonal signal from the male gonad.
    (C) In mammals, the male is dominant.
    (D) In mammals, all embryos develop male gonads without signals to do otherwise.

44. Which of the following predictions is most likely to occur if the scientist were to replace the surgically removed gonad with a crystal of testosterone (that is the appropriate dose for the animal's size)?

    (A) Only the chromosomal male embryo would develop testes.
    (B) Only the chromosomal female embryo would develop testes.
    (C) Both chromosomal male and female embryos would develop male gonads.
    (D) Neither chromosomal male nor female embryos would develop male gonads.

---

45. Which of the following statements about meiosis is correct?

    (A) The daughter cells are genetically identical to the parent cells.
    (B) Homologues pair during prophase II.
    (C) DNA replication occurs before meiosis I and meiosis II.
    (D) The number of chromosomes is reduced.

46. Which of the following is correct about organisms that are the first to colonize a habitat after a volcanic eruption and lava flow covered the area with half a meter of lava?

    (A) They are the fiercest competitors in the area.
    (B) They maintain the habitat as it is for their own kind.
    (C) They change the habitat in a way that makes it more habitable for other organisms.
    (D) They are small invertebrates.

47. In the microscopic world of a pond, paramecia are ferocious predators that prey on smaller protists. In a classic experiment, two species of paramecia were grown separately in cultures. The species in culture A was *P. caudatum*. The species in culture B was *P. bursaria*. Then, the two species were combined in one culture dish (culture C).

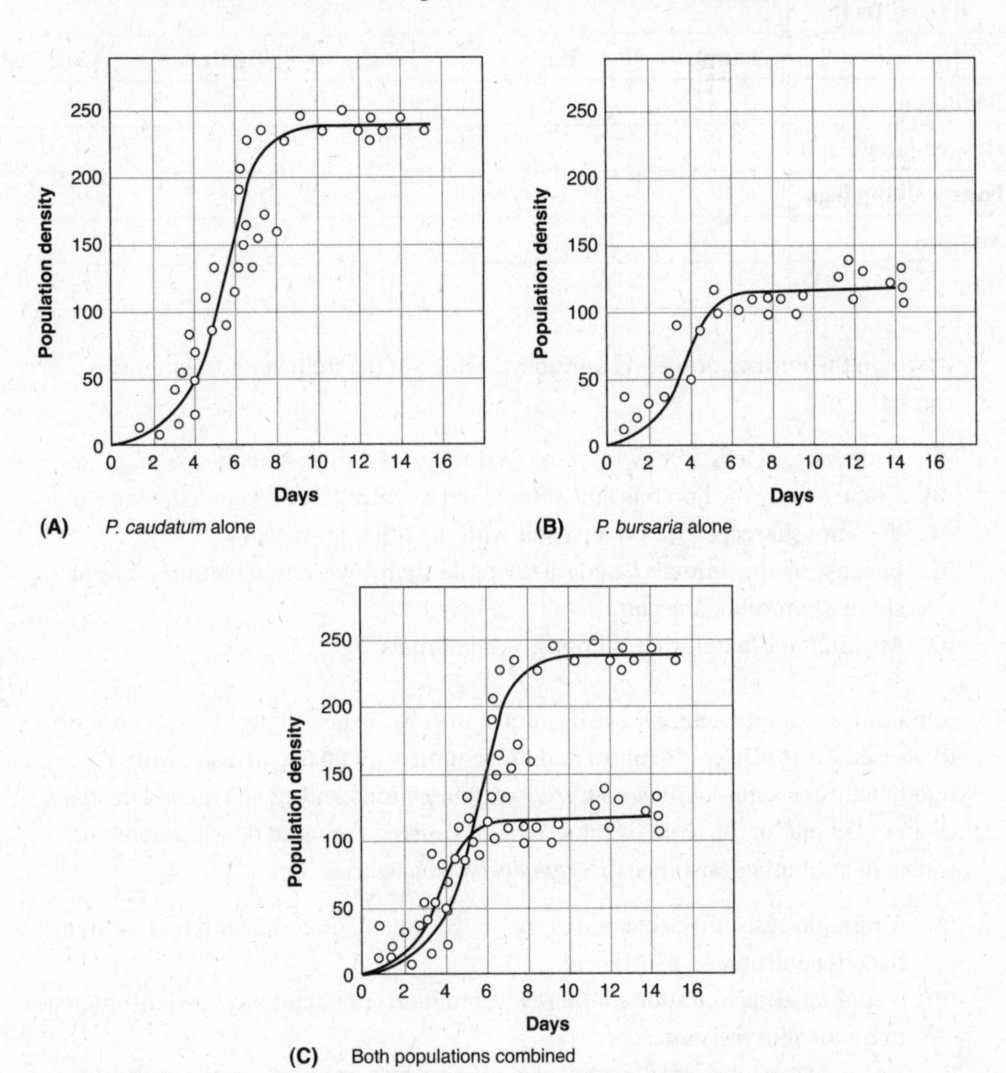

Which of the following is the most likely explanation for the growth pattern of the two populations combined in culture C?

(A) *P. caudatum* is driving *P. bursaria* to extinction because *P. caudatum* is the fitter species.
(B) *P. caudatum* and *P. bursaria* share a niche.
(C) *P. caudatum* and *P. bursaria* occupy different niches.
(D) *P. caudatum* is feeding on *P. bursaria* but only to a limited degree.

48. A trait that is shared by many organisms is called a shared ancestral trait. A new trait that appears for the first time in an organism is called a shared derived trait. Below is a character table that shows various animals that have a trait (+) or lack a trait (–).

Derived Traits					
	Lamprey	Bass	Frog	Turtle	Lion
Backbone	+	+	+	+	+
Hinged jaws	–	+	+	+	+
Four walking legs	–	–	+	+	+
Amnion	–	–	–	+	+
Hair	–	–	–	–	+

Based on the information and table above, which of the following statements is correct?

(A) Four walking legs first evolved in the turtle. It is a shared derived trait.
(B) Because only the lion has hair, it must have evolved in a separate lineage and does not share a common ancestor with the other animals.
(C) Because all the animals listed on the table share a vertebral column, they all share a common ancestor.
(D) An amnion is a trait that is unique to mammals.

49. Antibiotic-resistant bacteria have become a worldwide health problem. According to the Centers for Disease Control and Prevention, over 80,000 invasive MRSA (methicillin-resistant *Staphylococcus aureus*) infections and 12,000 related deaths occur every year in the United States. What is causing the rapid development and spread of antibiotic resistance in *Staphylococcus* bacteria?

(A) Antibiotic-resistant bacteria can exchange genetic material with nonresistant bacteria and spread resistance.
(B) A spontaneous mutation in the DNA of millions of bacteria caused this increase in methicillin resistance.
(C) Normal *Staphylococcus* bacteria can outcompete antibiotic-resistant bacteria.
(D) Recombination of homologous chromosomes resulted in a new trait in a single bacterium, which spread throughout the world.

50. Which of the following statements about ancient Earth is correct?

(A) Earth was formed almost 2 billion years ago.
(B) It is agreed upon by all scientists that all life on Earth began at the bottom of the oceans at hydrothermal vents.
(C) There have been 5 major extinctions that almost wiped out all life on Earth.
(D) The presence of free oxygen facilitated the rapid formation of organic molecules.

51. Intact chloroplasts are isolated from dark-green leaves by low-speed centrifugation and placed into six tubes that contain cold buffer. A blue dye, DPIP, which turns clear when reduced, is also added to all the tubes. The amount of decolorization is a function of how much reduction and, therefore, how much photosynthesis occurs. Each tube is exposed to different wavelengths of light. A measurement of the amount of decolorization is made, and the data are plotted on a graph. Although the wavelengths of light vary, the light intensity in each tube is the same.

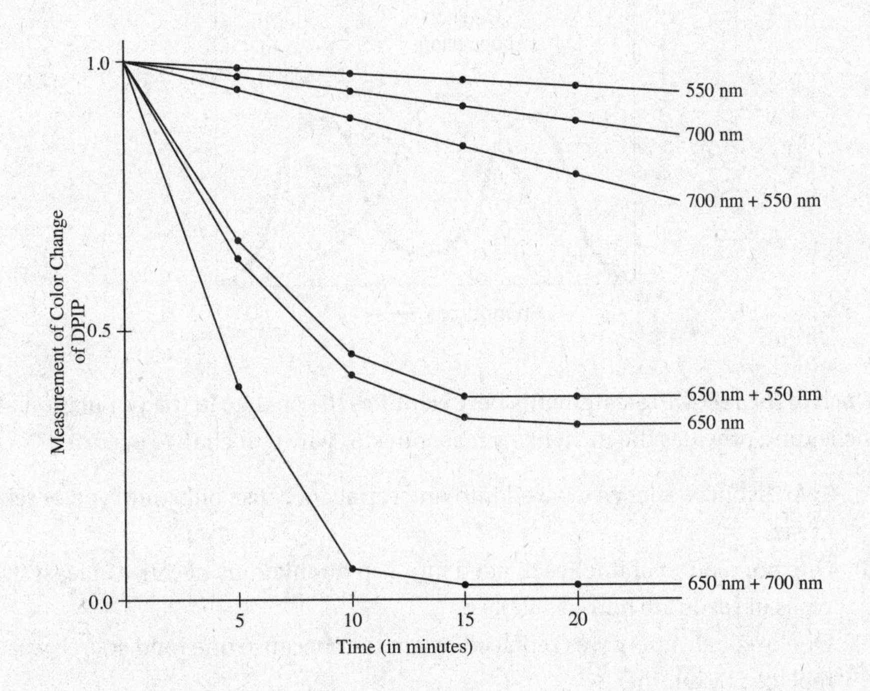

Which statement below best explains the results of this experiment and provides the correct reasoning for why those results occurred?

(A) The rate of photosynthesis is highest when a tube is exposed to light in the 550 nm range because that wavelength of light contains the greatest amount of energy.

(B) The rate of photosynthesis is highest when a tube is exposed to light in the 550 nm range because that wavelength of light contains the least amount of energy.

(C) The rate of photosynthesis is highest when a tube is exposed to combined light in the 650 nm and 700 nm ranges because the combination of the two wavelengths of light contains the greatest amount of energy.

(D) The rate of photosynthesis is highest when a tube is exposed to combined light in the 650 nm and 700 nm ranges because there are two photosystems in chloroplasts, each absorbing a different wavelength of light.

52. Today, two distinctly different beak sizes occur in a single population of finches in an isolated region of West Africa. This finch, the black-bellied seedcracker, is considered a delicacy. The oldest residents of the region remember that all black-bellied seedcrackers used to appear identical. The change in population is shown in the following graph:

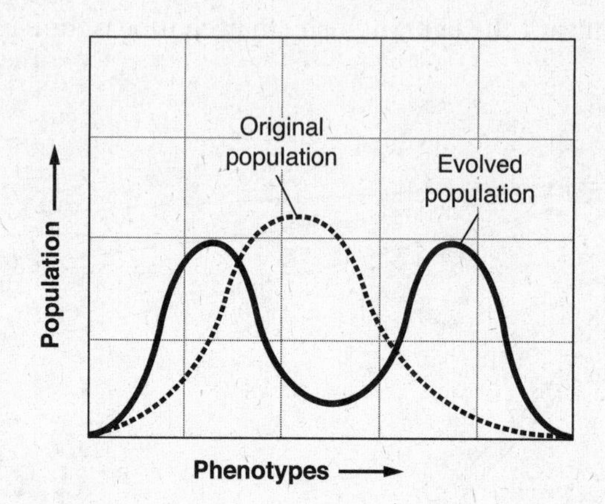

Which of the following statements best identifies the change in the population of finches and provides the most likely reasoning for why that change occurred?

(A) Two distinct varieties evolved into one variety because only one type of seed now exists.
(B) One population of finches divided into two populations because at least two types of seeds are now available.
(C) One original variety was replaced by another because one food source was replaced by another.
(D) The original population died out, leaving only individuals with either a long or a short beak size.

53. Below is a figure that shows the change in one ancestral population of birds over time:

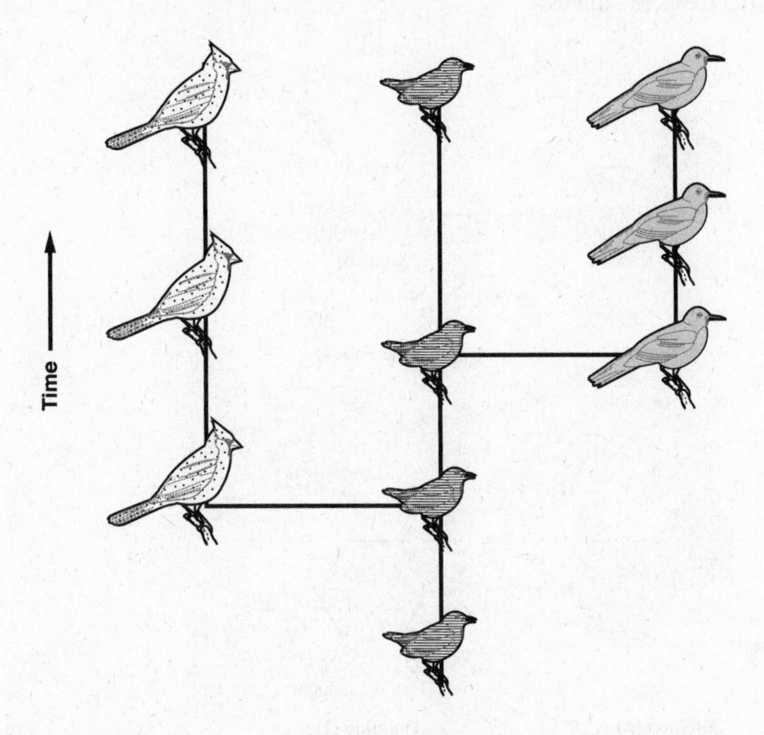

Which of the following statements best supports the evidence presented in the figure?

(A) Evolution was gradual and regular with small changes over a long period of time.
(B) The environment stayed the same for long periods of time.
(C) These birds evolved from a single species that went extinct.
(D) These birds evolved mainly as a result of genetic drift.

54. Which of the following statements is correct about the DNA content of a particular diploid cell just prior to mitosis if the DNA content of the same diploid cell in G_1 is X?

(A) The DNA content of the cell in metaphase is 0.5X.
(B) The DNA content of the cell in metaphase is X.
(C) The DNA content of the cell in metaphase is 2X.
(D) The DNA content of the cell in metaphase is 4X.

55. Which of the following illustrations shows the correct way in which two nucleotides pair up in a molecule of DNA?

(A)

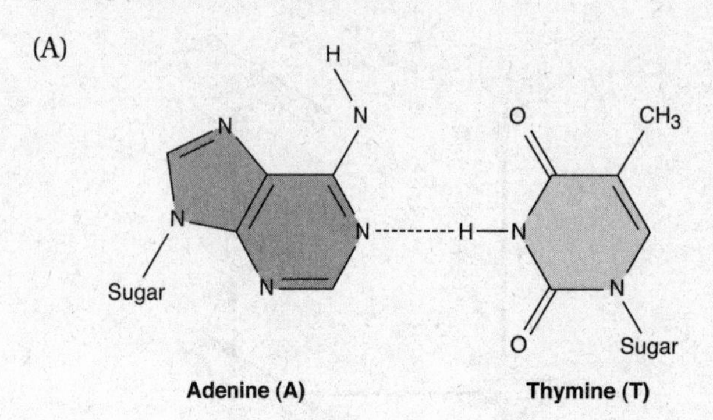

(B)

Adenine (A) Thymine (T)

(C)

Adenine (A) Thymine (T)

(D)

Guanine (G) Cytosine (C)

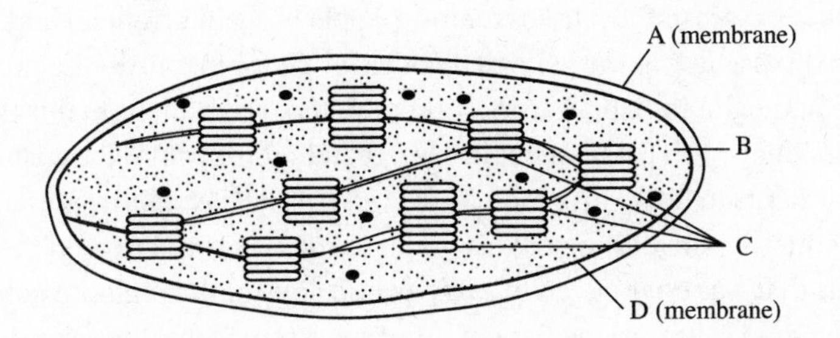

56. Which of the following statements is correct about the plant cell structure shown above?

    (A) A is the site of ATP synthase.
    (B) B is the site where ATP is produced.
    (C) C is the structure from which oxygen is released.
    (D) D is the site of PGAL formation.

57. Except for identical twins, no two people are genetically identical. Which of the following statements most accurately describes the main source of variation in humans?

    (A) The main source of variation in humans is genetic drift.
    (B) Changes in the environment cause rapid changes in different human populations.
    (C) Shuffling of alleles is responsible for the greatest variation in humans.
    (D) Mutations are responsible for the greatest changes in different human populations.

58. β-thalassemia is a disease that strikes mainly people of Mediterranean and Asian descent. It involves a flaw in the gene that codes for β-globin, one of the protein chains that make up hemoglobin (Hb). A trial was conducted. The patient in this trial was typical of individuals with β-thalassemia. He needed monthly blood transfusions and daily treatments to lower his blood iron levels. In 2007, when the patient was 19 years old, scientists removed some of his bone marrow cells and treated them with a modified virus that was engineered to carry a good copy of the β-globin gene. They infused the repaired cells into the patient. After several months, the scientists assessed the patient. Below is a graph of the results of this trial:

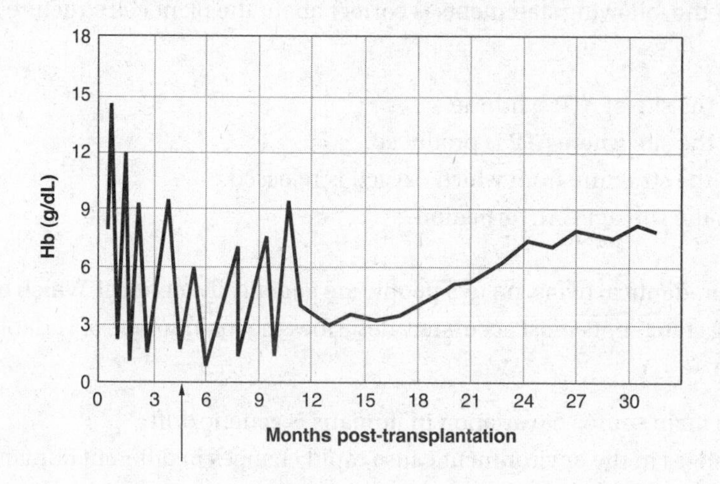

Which of the following statements is supported by the information provided?

(A) The gene therapy trial failed.
(B) The gene therapy trial was a success, but the patient still requires blood transfusions.
(C) The scientists transplanted a gene directly into the patient's red blood cells.
(D) The scientists inserted a gene into a virus vector that carried the normal hemoglobin gene into the patient's cells.

59. In Chesapeake Bay, most oysters reach their prime spawning size by the time they are three years old. An oyster releases millions of eggs into the water, which join with sperm during the fertilization process. Fertilized eggs drift away from the spawning grounds in water currents. Most larvae that hatch from these fertilized eggs die from predation. Those few offspring that survive long enough to attach to a suitable substrate and begin growing a hard shell tend to survive for a relatively long time.

The graph below represents three idealized survivorship curves for different organisms.

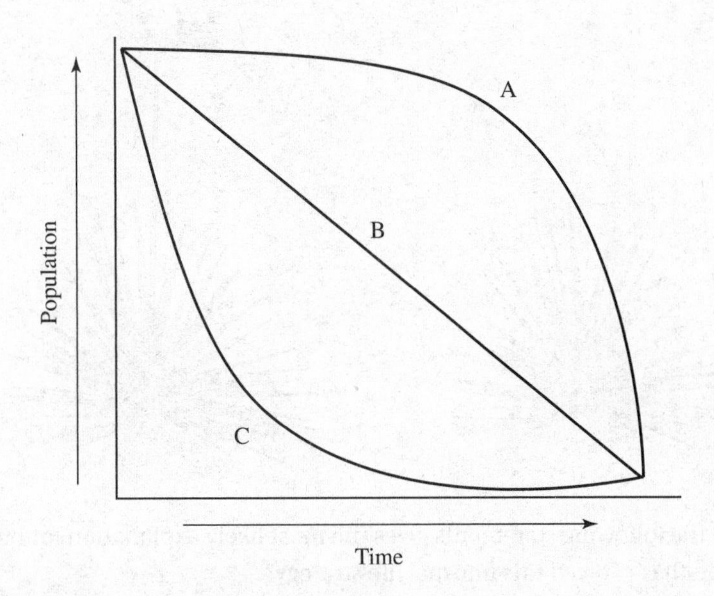

Which line on this graph best fits the life history of the oyster, as described above, and includes the correct justification for that answer?

(A) Line A best fits the life history of the oyster because those individuals that make it to adulthood live a long time.

(B) Line B best fits the life history of the oyster because those individuals that make it to adulthood live a long time

(C) Line C best fits the life history of the oyster because most larvae die from predation.

(D) Lines A and C both best fit the life history of the oyster, depending on the environment the oyster lives in.

PRACTICE TEST 2

60. The agave, or century plant, generally grows in arid climates with unpredictable rainfall and poor soil. The plant exhibits what is commonly called big bang reproduction. An agave grows for years, accumulating nutrients in its tissues until there is an unusually wet year. Then it sends up a large flowering stalk, produces seeds, and dies.

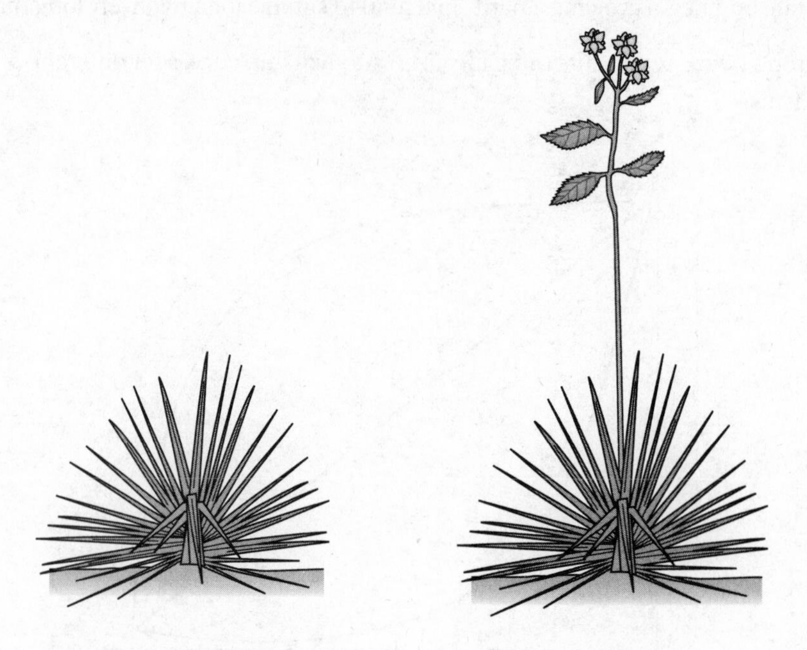

Which of the following statements gives the most likely explanation for how the agave plant has evolved this unusual life strategy?

(A) This strategy is determined by the plant's genes, which are inherited.
(B) This particular life strategy was selected for by the harsh, desert environment.
(C) This strategy is the result of a series of mutations that occurred because of the great heat in the desert.
(D) This strategy is the result of a series of mutations that occurred because of the great heat in addition to the lack of rain in the desert.

SECTION II

Time: 90 minutes

Directions: Questions 1 and 2 are long free-response questions that require about 25 minutes each to answer and are worth 8 to 10 points each. Questions 3, 4, 5, and 6 are short free-response questions that require about 8 to 10 minutes each to answer and are worth 4 points each. You are advised to spend the first 10 minutes of this section as a "reading period" for planning your responses; however, you may begin writing your responses right away. On the actual exam, you must write your responses in the space provided for each question because *only material written in the space provided will be scored.* Your responses must be written out in paragraph form. Outlines, bulleted lists, and/or diagrams alone are not acceptable.

1. Transpiration is the loss of water from a leaf. The theory of transpiration-pull–cohesion tension explains how water and nutrients get "pulled" upward in the xylem tissue of a plant, from the roots, without the plant's cells having to expend any energy. According to this theory, as one molecule of water evaporates from the stomate of a leaf, another molecule of water is pulled into the plant through the roots.

   (a) **Explain** how transpiration is dependent on the structure/properties of water, referencing TWO of water's properties in your response. You may use illustrations in your explanation, but they must be labeled, and they do not substitute for a written explanation. (2 points)

   (b) **Identify** an experimental procedure that would involve the use of green plants and demonstrate differences in the rates of transpiration in TWO different environmental conditions. **Describe** how you would collect quantitative data. (4 points)

   (c) **Analyze** your data. **Explain** how you would decide (based on your data) whether the environmental conditions you studied had an impact on the rate of transpiration in your subject plants. (2 points)

   (d) **Make** a prediction about what the rates of transpiration will be based on the two different environmental conditions in your experiment, and justify your prediction. (2 points)

2. Exercise affects our brains. Studies have shown that regular exercise over time increases the production of neurochemicals in the brain and increases the volume of the hippocampus (a key part of the brain's memory network). In a study that involved exercise and memory, scientists recruited 50 healthy men and women, who were between the ages of 55 and 85 and had no serious memory problems. They were asked to visit a gym on two occasions. On both visits, half the group rested quietly while the other half rode an exercise bike for 30 minutes, a workout the scientists hoped would stimulate the subjects' brains. (Note that both groups included men and women, and, on both visits, the same group exercised or rested.) After the visit to the gym, both groups were given a sports drink that contained radioactive glucose. After allowing enough time for the radioactive glucose to be absorbed by the brain, the subjects were given a PET scan, which detects radioactivity in the brain. While inside the PET scan machine, names were flashed in front of the subject's eyes. Some of the names were the

names of celebrities; others were not. For 30 minutes, the subjects were asked to press one key on a screen when they recognized a celebrity's name and a different key when the name was unfamiliar. During this memory task, the scientists tracked the subjects' brain activity of the portions of the brain that are involved in memory. The results of this experiment are shown in the following table:

Treatment	Memory as Tested by Brain Function in the PET Scan Machine (in Relative Values) (%) (average +/– $2SE\bar{x}$)	
	Visit 1	**Visit 2**
Exercise	0.84 +/– 0.09	0.80 +/– 0.07
Resting	0.70 +/– 0.08	0.58 +/– 0.06

(a) **Explain** the role of the radioactive glucose in determining the level of activity of various areas of the brain in this experiment. (1 point)

(b) On the grid lines below, **construct** an appropriately labeled graph that illustrates the effect that exercise had on memory in this experiment. Use the data provided on the table above to construct this graph. (4 points)

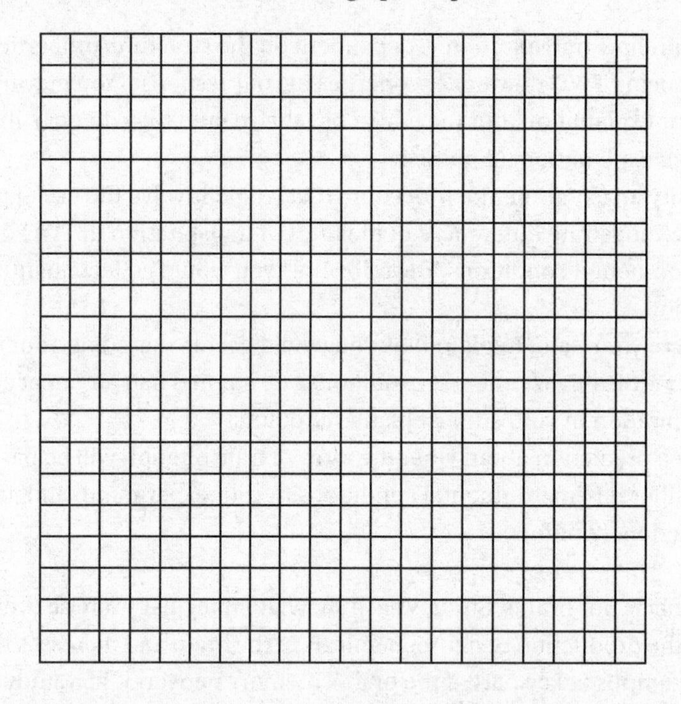

(c) **Analyze** the data from this experiment. Did the data show a positive correlation between exercise and brain function related to memory? (2 points)

(d) **State** a null hypothesis for this experiment. Should you reject the null hypothesis or fail to reject the null hypothesis for each visit? **Justify** your answer. (3 points)

3. Below is an illustration of a living plant cell under normal conditions as it would be seen under a light microscope at 40x magnification.

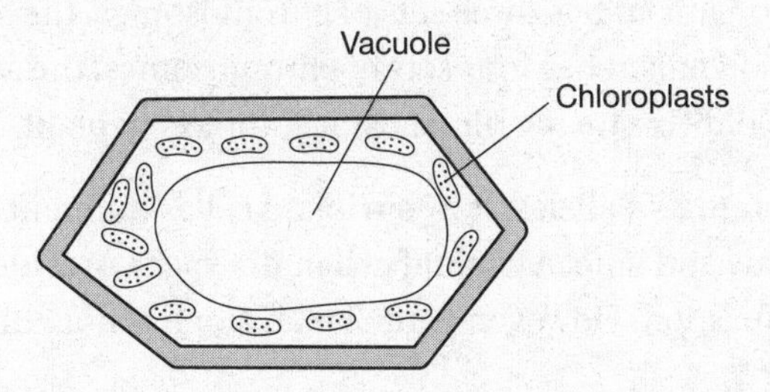

(a) **Describe** what water potential is and how it can cause a plant cell to shrink or swell. (1 point)

(b) **Identify** an experimental procedure that would demonstrate how a change in the water potential could change the integrity and appearance of a plant cell. (1 point)

(c) In the space provided below, **draw** and **label** this plant cell as you **predict** it will appear after the experimental procedure described in part (b). (1 point)

(d) **Justify** your prediction (from part (c)), and **explain** the process that occurred in the cell. (1 point)

PRACTICE TEST 2

4. Some organisms have more than the normal two complete sets of chromosomes (2n) in all their somatic cells. The general term for this chromosomal aberration is polyploidy. If the organism has three sets of chromosomes, the abnormality is known as triploidy. If the organism has four sets of chromosomes, the abnormality is known as tetraploidy. Polyploidy is rare in animals, but common in plants.

   (a) **Describe**, in terms of alleles, how a triploid cell is different from a diploid cell. (You may draw and label a triploid cell in the space provided below to help you answer the question. However, your sketch alone is not adequate as an answer.) (1 point)

   (b) **Explain** how triploidy can arise in a previously diploid organism. (1 point)

   (c) **Predict** ONE reason why a triploid organism might have a problem in terms of reproducing sexually. (1 point)

   (d) **Justify** your prediction (from part (c)). (1 point)

5. The figure below represents the process of expression of the HBA1 gene, which provides instructions for making a protein called alpha-globin (which is a component of a larger protein called hemoglobin). This stretch of DNA is located on chromosome 16.

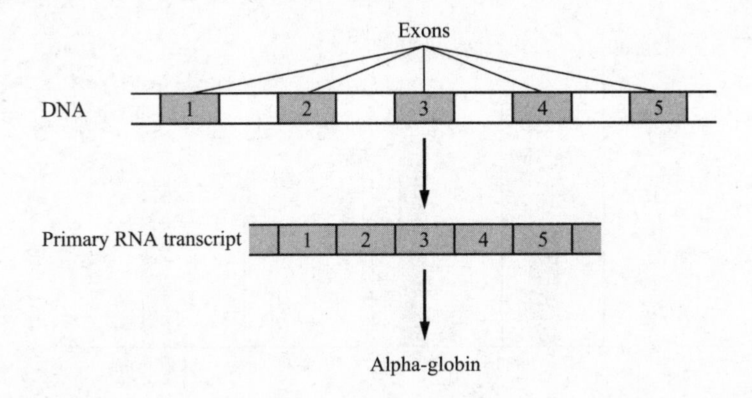

(a) **Describe** the modification(s) that would take place in a cell that would alter pre-RNA (primary transcript) to produce mature mRNA, and **state** where in a cell this process occurs. (1 point)

(b) A different protein, besides alpha-globin, can be made from this same original stretch of DNA. **Explain** how two different proteins can be produced in a cell from the same piece of DNA. (1 point)

(c) **Explain** the relationship between exons and introns and their respective roles in human DNA. (1 point)

(d) Referring to the figure presented at the beginning of this question, **explain** how the process shown in the figure relates to the concept that only 3% of human DNA codes for proteins. (1 point)

6. Two different human cell samples (Cell Type A and Cell Type B) from the same human body were exposed to a chemical that disrupted their plasma membranes. The samples were then sequentially spun in an ultracentrifuge to isolate layers of subcellular components. Below is a histogram that shows the results of this experiment:

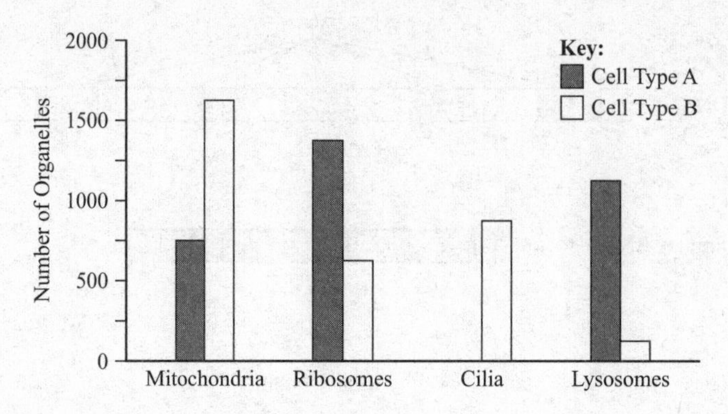

(a) **Describe** the key differences between Cell Type A and Cell Type B, based on this histogram. (1 point)

(b) **Describe** how these data suggest different functions for Cell Type A and Cell Type B. (1 point)

(c) Based on the data provided, **predict** what type of cell each cell (A and B) is, and **justify** your predictions. (1 point)

(d) **Explain** how these two cell types could have identical DNA yet different cellular characteristics. (1 point)

ANSWER KEY

Section I

1. **C**	11. **D**	21. **A**	31. **D**	41. **D**	51. **D**
2. **B**	12. **D**	22. **D**	32. **B**	42. **D**	52. **B**
3. **C**	13. **D**	23. **B**	33. **A**	43. **B**	53. **B**
4. **C**	14. **C**	24. **D**	34. **A**	44. **C**	54. **C**
5. **D**	15. **D**	25. **A**	35. **B**	45. **D**	55. **B**
6. **C**	16. **B**	26. **B**	36. **C**	46. **C**	56. **C**
7. **C**	17. **C**	27. **B**	37. **B**	47. **C**	57. **C**
8. **C**	18. **B**	28. **A**	38. **C**	48. **C**	58. **D**
9. **D**	19. **B**	29. **C**	39. **B**	49. **A**	59. **C**
10. **C**	20. **C**	30. **D**	40. **B**	50. **C**	60. **B**

Section II

See the Answers Explained section.

ANSWERS EXPLAINED

Section I

1. **(C)** This graph is focused on the strength of the bond between Hb and oxygen at pH 7.2 and pH 7.4. At a lower pH, which is more acidic, the attachment is weaker. This makes sense because as cells carry out cellular respiration, they release CO_2, which makes the blood more acidic. The blood pH changes from normal 7.4 to the more acidic 7.2. At pH 7.2, the Hb can release more of its oxygen where it is needed, at the cells.

2. **(B)** The Law of Independent Assortment states that, during gamete formation, the alleles for a gene for a trait, such as height (Tt), segregate independently (randomly) from the alleles for a gene for another trait, such as seed color. Choice A is incorrect because the number of gametes will not increase; the variety of gametes will increase. Choice C is incorrect because the number of offspring is not related to the types of gametes. Choice D is incorrect because the variety of gametes will increase, not decrease.

3. **(C)** Although it is true that during the Krebs cycle pyruvate is completely broken down into CO_2, the most energy from cellular respiration is released during oxidative phosphorylation, when all the energy stored in NADH and $FADH_2$ from previous stages of cellular respiration is released.

4. **(C)** Choice A is not correct because the electron transport chain (ETC) is also located in the grana of chloroplasts. Choice B is incorrect for two reasons. First, energy cannot be produced; only ATP can be produced. Second, the citric acid cycle produces ATP by substrate-level phosphorylation. Choice D is not correct because pyruvate is a product of glycolysis and enters the citric acid cycle.

5. **(D)** The father's genotype is X-Y because he has the disease. The mother's genotype is XX. In this case, the father passes his X-gene to all of his daughters, so they will all be carriers.

PRACTICE TEST 2

6. **(C)** Recombination occurs when homologous pairs bind together during synapsis and crossing-over occurs. Picture 1 is in metaphase. Synapsis and crossing-over have already occurred. The homologous pairs are about to separate. Picture 2 shows prophase. Synapsis is occurring, and crossing-over can occur at this point. Picture 3 shows sister chromatids lined up on the metaphase plate. There can be no crossing-over because there is only one chromosome.

7. **(C)** It is clear from the first cross that the red and smooth traits are dominant. The results of the testcross show a phenotypic ratio of 1:1:1:1 in the offspring. This could only have happened if there was independent assortment of the trait for color and the trait for pod character. Here is the testcross: RrSs × rrss

Test Cross

One Parent (Pure Recessive: White/Wrinkled)

	rs	rs		
RS	RrSs	RrSs	= Red/Smooth	1
Rs	Rrss	Rrss	= Red/Wrinkled	1
rS	rrSs	rrSs	= White/Smooth	1
rs	rrss	rrss	= White/Wrinkled	1

One Parent

8. **(C)** A soil with an acidic pH causes the flowers to be pink; a basic soil gives rise to blue flowers. ABO blood type is an example of codominance because both A and B genes show. Polygenic inheritance is characteristic of traits like skin color, hair color, and height that exist in a complete spectrum.

9. **(D)** By analyzing the F_1 results, you see that there is almost a 50-50 ratio of Gray/Normal to Black/Vestigial. The 50–50 ratio of those traits tells you that the traits are linked. If the body is gray, the wings are normal; if the body is black, the wings are vestigial. However, a small variation from the expected values occurred. Those numbers, 190 and 184, are due to crossing-over. Genes that are very close together usually are inherited together. So these genes must be far apart on the same chromosome.

10. **(C)** The two genes are linked and far apart on the same chromosome.

11. **(D)** The coupling of exergonic and endergonic reactions is an important biochemical strategy. The exergonic flow of electrons in the cristae membrane of mitochondria provides the energy to pump protons across the cristae membrane and maintain a gradient.

12. **(D)** An example of positive feedback occurs during labor and childbirth. The pressure of the baby's head against sensors near the opening of the uterus stimulates uterine contractions, which causes even more pressure against the sensors and more contractions. This continues until the baby is born, and then the process ceases. Choices A and C are incorrect because they describe negative feedback, not positive feedback. Choice B is incorrect because both positive feedback mechanisms and negative feedback mechanisms are universal.

13. **(D)** Diversity results from having the richest environment with the most varied niches. Choice A is incorrect for two reasons. First, food chains never consist of more than 4 or 5 trophic levels anywhere on Earth. Second, the existence of a large animal population does not mean that the population is diverse. On the contrary, it could mean that one invasive species has outcompeted all other organisms and taken over the area. Choice B is incorrect because high temperatures do not necessarily cause greater rates of speciation in a region. What does enhance diversity, however, is having many different niches in which animals can live. Choice C is a nonsense statement. The tropics are spread across the equator worldwide and did not spread from one location inside Pangaea.

14. **(C)** The definition of Hardy-Weinberg equilibrium is that the allelic frequencies will not change. If one trait becomes advantageous and the allelic frequency changes, then the population is no longer in Hardy-Weinberg equilibrium.

15. **(D)** Choice D is correct because painting the surface of most of the leaves with nail polish will cover the stomates and prevent transpiration, which is shown in Line *C*. Green plants reflect green light; they do not absorb it and cannot use it as an energy source. Placing a plastic bag over the leaves will decrease transpiration, resulting in Line *C*. Placing a fan near the plant will increase transpiration, causing Line *A*.

16. **(B)** Choice A is not correct because water moves up a tree in the xylem via passive transport with the help of transpiration pull. Choice C is not correct because the movement of water up a tree can rise hundreds of meters. Choice D is not correct because the higher the humidity, the *slower* the rate of transpiration.

17. **(C)** The importance of internal organelles is that they compartmentalize the cell so that widely different reactions can all occur at the same time, each in their own space.

18. **(B)** Glycolysis reoxidizes NADH to NAD^+ so that glycolysis can continue. Glycolysis would cease if there were not a constant supply of NAD^+ to bind to hydrogen. Choice A is incorrect because glycolysis produces a small amount of ATP by substrate-level phosphorylation. Choice C is incorrect because glycolysis produces pyruvate for the citric acid cycle. Choice D is incorrect because, although it is correct that glycolysis takes place in the cytoplasm, where glycolysis takes place is inconsequential in the broader scheme of energy production compared to the important role described in choice B.

19. **(B)** Earth is about 4.5 billion years old. The planet took about 1 billion years to cool down and be stable enough for life to evolve, and for the first cell—a prokaryotic cell—to develop. About 1.5 billion years ago, the first eukaryotic cell evolved. We have fossil evidence of that. Humans, bananas, and all other eukaryotic cells evolved from some ancestral eukaryotic cell. That is why there are basic similarities among all living things.

20. **(C)** Choice A is not correct because that statement describes punctuated equilibrium, a theory that was put forth by Stephen Jay Gould and Niles Eldredge in 1972. Choice B is not correct because mutations provide new genes for natural selection to "work on." Choice D is incorrect because it describes the discredited theory of "use and disuse" by Lamarck.

21. **(A)** After deforestation, the nitrate runoff into the stream water increased. There is no evidence presented that trees were replanted. Choice D is clearly false.

22. **(D)** Even though a number is mentioned in the question, it is not relevant. There is no way to determine if hawks will begin to eat frogs, which is niche displacement. You can deduce only the obvious.

23. **(B)** Apoptosis is programmed cell death for cells that are irrevocably damaged or infected, or for those cells that have simply come to the end of their life. When such a cell is identified, signals are sent to it to begin the process of self-destruction. Apoptosis evolved early in eukaryotic history, so it is found in all kinds of organisms.

24. **(D)** In general, substances that are polar and/or charged cannot diffuse through a membrane; only nonpolar molecules can.

25. **(A)** As correctly stated in choice A, the sequence of amino acids determines the primary structure of a protein. Once the amino acids are in their ordered place, there are intramolecular interactions (like hydrogen bonding) that cause the protein to fold into its unique shape. Secondary structure, tertiary structure, and quaternary structure are all built on primary structure. So, it is the primary structure that is least affected by hydrogen bonding.

26. **(B)** The drug Oblimersen is an antisense RNA. Therefore, it binds to the overexpressed RNA and prevents it from translating. Theoretically, Oblimersen should work in preventing cancer.

27. **(B)** The extra nitrate does not contribute any more to the algae growth than the initial increase. Phosphate is the main contributing factor to algae growth in this example.

28. **(A)** This is an accurate description of an autoimmune disease.

29. **(C)** Only about 10% of the energy from one trophic level is available to the next trophic level. The remainder is lost as heat, waste products, and uneaten body parts.

30. **(D)** Protons flowing down a gradient through ATP synthase provide energy to phosphorylate ADP into ATP. Energy from the exergonic flow of electrons to oxygen provides the energy to pump protons across the cristae membrane and form a gradient. Choice A is incorrect because this flow of electrons is used to pump protons to the outer mitochondrial compartment and create a proton gradient from which ATP will be produced later. Choice B is incorrect because it describes substrate-level phosphorylation, not oxidative phosphorylation. Choice C is incorrect because the bonding of electrons to oxygen produces water.

31. **(D)** The exergonic flow of electrons in the electron transport chain of mitochondria provides the energy to establish a proton gradient. Most CO_2 that is released from cellular respiration comes from the citric acid cycle. Protons flow through the ATP synthase down a gradient, not by active transport.

32. **(B)** Oxygen is needed to attract electrons down an electron transport chain in the light-dependent reactions. Water entering the plant through the roots ultimately provides hydrogen for the Calvin cycle, but first it is needed in the light-dependent reactions. CO_2 enters the plant through the stomates; it is not released by the light-dependent reactions.

33. **(A)** The only thing this process produces is ATP. No NADPH or oxygen is released. The function of this process is to provide ATP for the Calvin cycle, which uses enormous quantities of ATP.

34. **(A)** All the other choices promote outbreeding, the introduction of new genes into a population.

35. **(B)** First, you must determine the pattern of inheritance for this type of deafness. You can eliminate dominant, recessive, or sex-linked because none of the affected children in the third generation had a parent with the condition. It also cannot be inherited as a sex-linked recessive trait because in order for female #1 to have the condition, her father would have had to have it. He does not. Therefore, the trait must be inherited as an autosomal recessive. That being the case, the affected male is homozygous *dd*, where *d* is the trait for deafness. He has a child with a carrier female (*Dd*). See the cross below. There is a 50% chance that any child will have the condition. It is the same probability for daughters because the trait is not sex-linked.

	d	*d*
D	*Dd*	*Dd*
d	*dd*	*dd*

36. **(C)** In order for transcription of a gene or genes to occur, RNA polymerase must bind to a promoter. In choices A and B, the repressor is bound to the operator, which excludes RNA polymerase from binding to the promoter and prevents transcription. Choice D does not make any sense because RNA polymerase does not bind to the operator. It's a matter of specificity, like a lock and key.

37. **(B)** A mutation in AUG would have disastrous consequences because translation could not begin. Choice C refers to what happens after transcription. RNA processing occurs after transcription.

38. **(C)** The new strand is built 5′ to 3′, and the Okazaki fragments form, moving away from the replication fork.

39. **(B)** RNA processing occurs before the preRNA transcript leaves the nucleus. Introns, which are intervening sequences, are removed. Exons are the expressed sequences, which are the genes or coding regions.

40. **(B)** Restriction enzymes are referred to as "molecular scissors" because they cut DNA at specific *restriction sites*. They are used in all biotechnology labs, but they were discovered in bacteria, where they are part of the bacterial immune system. They cut up attacking viral DNA or RNA and make it non-functional. EcoRI was the first restriction enzyme that was discovered; it was named for the bacteria it was discovered in, *E coli*, which is found in the human large intestine. Choice A is incorrect because that statement describes *ligase*, not EcoRI. Choice C and choice D are both incorrect because EcoRI is not involved in either transcription or replication, even in *E. coli* bacteria.

41. **(D)** RNA processing occurs in the nucleus after transcription. During this process, introns are excised from the pre-RNA, and a 5′ cap and a poly-A tail are added onto the pre-RNA transcript. Choice A is incorrect because splicing out introns occurs in the nucleus (after transcription), not in the cytoplasm. Choice B is incorrect because snRNPs have nothing to do with translation. Choice C is incorrect because exons are transcribed and translated; they are not replaced with intervening sequences (introns).

42. **(D)** Acid rain is caused by the accumulation of sulfuric, sulfurous, nitric, and nitrous acids in the air from the burning of fossil fuels. A pH of 6.0 is 100 times more acidic than a pH of 8.0. The hydrogen ion concentration of a pH of 6.0 is equal to 1×10^{-6} M.

43. **(B)** Without the signal from the gonads to develop into a male, the rabbit remains a female.

44. **(C)** The testosterone would send the signal that the male gonad would have sent. Both the male and female embryos would develop testes.

45. **(D)** The point of meiosis is variation. Daughter cells are different from the parent cells and are different from each other. Homologues pair during prophase I, not prophase II. DNA replication occurs before meiosis I only.

46. **(C)** Pioneer organisms, like lichens, colonize an area that is not habitable by other organisms and alter it over time in a way that makes the area habitable for other organisms.

47. **(C)** Almost by definition, the two species do not share a niche because both populations are surviving. If they shared a niche, the fitter species would outcompete the weaker one. However, the environment has limited resources. In fact, the S-shaped growth curve shown in all three of these graphs is called a logistic growth curve. It is characteristic of some microorganisms under conditions of limited resources.

48. **(C)** The concept stated in choice C is an important idea to remember. Choice A is incorrect because four walking legs first evolved in the frog, not the turtle. Choice B is factually incorrect. All animals with a backbone, or a vertebral column, evolved from a common ancestor. Choice D is incorrect because many animals, besides mammals, have amnions or amniotic sacs. From the table, you can see that the turtle (a reptile) also has an amnion.

49. **(A)** Bacteria can carry out horizontal gene transfer and pass their genes to other bacteria. This is how the epidemic is spreading so rapidly. A spontaneous mutation would not be the cause of the spread of antibiotic resistance because the likelihood of millions of mutations in the same place is nil. Choice C is backward; it is the antibiotic-resistant bacteria that have the advantage. Choice D is not possible because bacteria have only one chromosome and do not undergo recombination of alleles during meiosis.

50. **(C)** This is a fact. Some individuals argue that we are currently in the 6th major extinction—this one caused by humans. Choice A is not correct because there is a lot of scientific evidence that proves that Earth formed 4.6 billion years ago when all the planets in our solar system formed. Choice B is not correct because scientists do not agree on exactly where or how life first developed on Earth because there is inadequate evidence to make a definitive determination. Choice D is not correct because there was no free oxygen on early Earth. In addition, oxygen is such a strong oxidizer that no complex organic molecules would have formed early in Earth's existence if oxygen had been present.

51. **(D)** The amount of energy in light is inversely proportional to its wavelength. The shorter the wavelength of light, the more energy it contains. The lines on the graph that are closest to the *x*-axis represent the greatest amount of reduction and, therefore, the greatest amount of photosynthesis. According to the graph, the most reduction or photosynthesis occurs with two wavelengths of light combined. The reason is that in chloroplasts, two photosystems, P700 and P680, absorb light of different wavelengths. Choice (C) is incorrect because the issue is how much light is absorbed (by the two photosystems), not how much is being emitted. If 10 times the light in the 1,000 nm range were shown on the plant, it would not get absorbed because there would be no photosystem in the plant to absorb it.

52. **(B)** There used to be one population of black-bellied seedcracker; now, there are two. The single population diverged because of a change in food.

53. **(B)** There are three branches of birds that all evolved from an ancestral bird in the middle at the bottom. Each vertical line represents a long period of time when the environment did not change, so, neither did the birds. However, at two different times in the course of this history, two bird lineages evolved suddenly from the ancestral bird lineage. (We can assume that the environment changed suddenly during those times.) One gave rise to a new species on the left; another gave rise to a new species on the right. Once each new species arose, the environment remained the same for a long time and the species did not change. The ancestral species is still alive; it did not go extinct. None of the information leads you to think that evolution occurred because of genetic drift.

54. **(C)** The cell cycle consists of the following stages in this order: G_1 (the cell is growing), S (DNA replicates), G_2 (second gap of growth), and then mitosis (division of the nucleus). Therefore, if the quantity of DNA in G_1 is X (before the S phase), then in mitosis (after the S phase), the DNA content must be double X or 2X.

55. **(B)** There is a double hydrogen bond between adenine and thymine in DNA and a triple bond between cytosine and guanine. Breaks in DNA occur more often between adenine and thymine than between cytosine and guanine.

56. **(C)** This is a chloroplast. The light-dependent reactions, where ATP is synthesized and oxygen is released, occurs in the grana (C). The light-independent reactions, where PGAL is made, occur in the stroma (B). Nothing related to photosynthesis is occurring in A, which is the outer membrane of the chloroplast.

57. **(C)** More than anything else, sexual reproduction with recombination of alleles is the greatest source of variation in humans. Genetic drift and mutations are also an important source of variation in humans, but they do not provide nearly as much variation as does recombination of alleles. Changes in the environment provide pressure for evolution, but they are not a source for variation.

58. **(D)** This was a very successful treatment. The patient's hemoglobin (Hb) levels are approaching normal levels. There are no sudden and severe drops in Hb levels, and the patient no longer requires transfusions.

59. **(C)** Most oyster larvae die from predation, as stated in the information that precedes the question. Even without any other information provided, you would have to choose Line C.

60. **(B)** Choice B is correct because this life strategy evolved by natural selection, and the selecting pressure was the harsh environment. Choice A is not the most likely explanation for this occurrence. Choices C and D have the theory backward. Mutations may provide variety in a population, but they do not occur to provide a variety or a solution to a problem. Mutations just occur. If it is advantageous, that new characteristic may increase in frequency in a population. If it is deleterious, that trait may disappear.

Section II

1. (a) The phenomenon of transpiration is dependent on the fact that water molecules are strongly attracted to each other. The first property of water, known as cohesion tension, is the force of attraction between molecules of water as a result of hydrogen bonding. Hydrogen bonding is the attraction of an oxygen atom from one molecule of water to a hydrogen atom from another molecule of water. The second property of water that is responsible for water molecules "sticking" together is water's strong polarity. The positive side of one water molecule is attracted to the negative side of an adjacent water molecule.

   (b) For this experiment, I would use three geranium plants (that are similar in size, age, and appearance) and place them in 4″ pots. I would weigh each at the beginning of the experiment, and then I would weigh each again at the end of the experiment. Since the mass of the plants will not be exactly the same as at the outset, I will calculate and use percent change in mass for my results. (In all three setups, I will place a heat sink [a large beaker of water] between the light and the plant to absorb the heat that is produced by the light source. This will eliminate heat as an unwanted variable in the experiment.)

   My control will be Plant I, which is placed under a 100-watt light.

   My treatment 1 will be Plant II, which has a clear baggie placed over it and is placed under a 100-watt light.

   My treatment 2 will be Plant III, which will have a fan pointed directly at its leaves while it is exposed to a 100-watt light. This setup will be placed away from the other two plants so that the fan will not affect them.

   I would begin the experiment at time zero and leave all three plants under the same conditions for 24 hours. After 24 hours, I would weigh them all again and determine the percent change in mass for each. Any percent decrease in mass would be due to water loss by transpiration.

   (c) If the mass of the plant after 24 hours was less than it was at time zero, the plant lost mass due to transpiration. The greater the percent loss of mass, the faster was the rate of transpiration. Similarly, the less the percent loss of mass, the slower was the rate of transpiration.

   (d) I predict that Plant I will have lost some mass because it transpired normally over the 24-hour period. I predict that Plant II will have lost less mass than Plant I lost because the baggie over the leaves will have kept the humidity high around the

PRACTICE TEST 2

leaves of the plant, which slowed down the rate of transpiration. I predict that Plant III will have lost the most mass because the fan will have blown the humidity away from the plant, thus increasing the rate of transpiration.

2. (a) Radioactive glucose is absorbed by the body in the same way that nonradioactive, normal glucose is. It is used for cellular respiration to provide ATP for all of a cell's activities, and the amount of glucose absorbed is a function of how active a cell is. Brain cells are always active, even when you are asleep, but they uptake even more glucose when you are engaged in a specific task. In this experiment, the task was identifying names of people, and an increase in glucose absorption indicated increased activity in the brain cells of the hippocampus.

(b) Refer to the graph below:

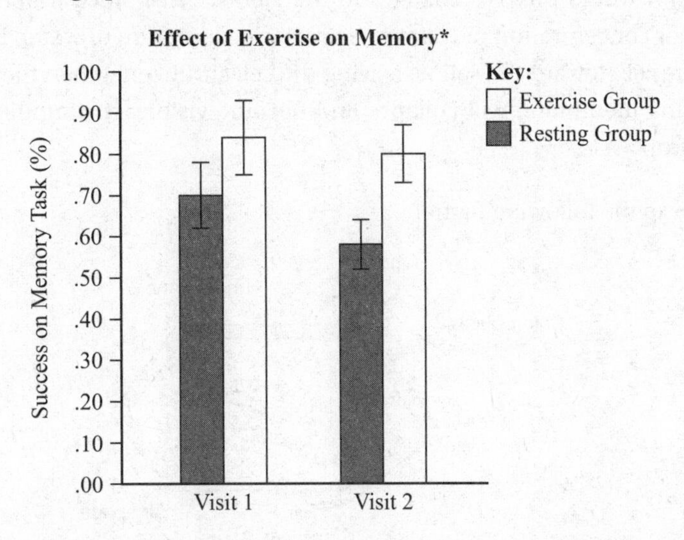

*The data represent the means +/– 2*SE* $\bar{x}$.

(c) From looking at the Exercise Group bars and the Resting Group bars for both Visit 1 and Visit 2 on the graph, it looks like the Exercise Group did better than the Resting Group on the memory task on both visits. However, it is important to look deeper—specifically, at the error bars. With respect to Visit 1, the error bars overlap at the upper limit of the Resting Group bar (0.78) and on the lower limit of the Exercise Group bar (0.75). That means that the difference in the two bars is not statistically significant, and nothing was proven in that experiment. With respect to Visit 2, the Exercise Group did better than the Resting Group on the memory task because the error bars did not overlap. Thus, the difference in the two Visit 2 bars is statistically significant, and the results for that visit demonstrate that exercise helps memory (meaning that there is a positive correlation between exercise and brain function related to memory).

(d) The null hypothesis is that "Exercising before taking a memory test does not increase an individual's results on the memory test." I would fail to reject the null hypothesis for Visit 1 because the differences between the two groups are not statistically significant. In other words, there is no difference on the memory task whether the subject exercised before the task or not. (*It is incorrect to say, "I accept the null hypothesis." You either "reject it" or "fail to reject it."*) I would reject the null hypothesis for Visit 2, however, because there was a statistical difference: memory was improved by exercising before participating in the memory task.

3. (a) Water potential (ψ) is the tendency of water to move across a permeable membrane into a solution. It results from two factors: solute concentration and pressure. If a cell is surrounded by a hypertonic solution, water will tend to leave the cell because water flows from higher water potential to lower water potential (toward the higher solute), and the cell will shrink. If the cell is surrounded by a hypotonic solution, water will tend to flow into the cell and cause it to swell or burst.

   (b) I would create the following experiment. First, I would tear off a thin piece of healthy elodea tissue and prepare a wet mount of it. While observing several single cells under 40x magnification, I would place 2 drops of concentrated (5%) saline solution on the slide next to the cover slip. Next, I would place a piece of paper towel on the opposite side of the cover slip to draw the salt water across the elodea. Then, I would observe changes to the elodea cells. Since water diffuses from a higher concentration of water to a lower concentration of water, it will diffuse out of the cell, toward the saline, leaving the cell shrunken. When the cell shrinks, the plasma membrane will collapse and become visible, surrounding the cluster of chloroplasts.

   (c) Refer to the following figure:

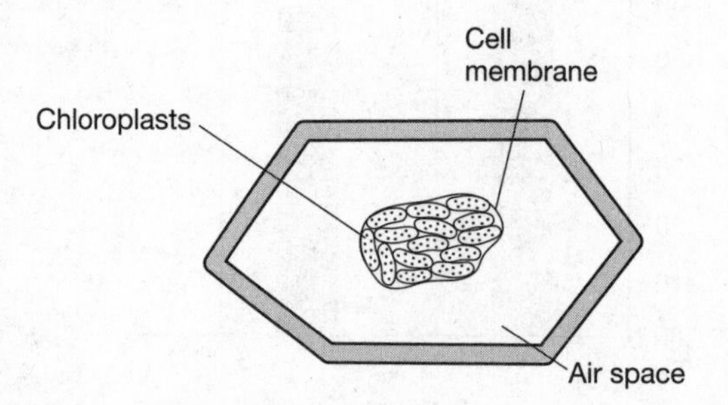

   (d) The cell collapsed as water flowed out of the plant cell toward the hypertonic solution. The cell membrane became visible, surrounding the cell "innards." The cell membrane had not previously been visible in the healthy cell because it was pushed up against the cell wall, and it could not be seen as a separate entity.

4. (a) Whereas a diploid cell has two homologues for each chromosome, a triploid cell has three homologues for each chromosome. This results in a cell with an extra set of chromosomes, as in the figure below:

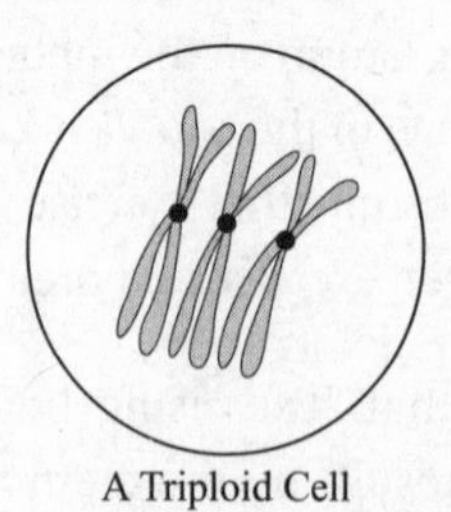

A Triploid Cell

   (b) Triploidy can arise from an error during meiosis when one homologous pair of chromosomes fails to separate as it normally should. Refer to the figure below:

PRACTICE TEST 2

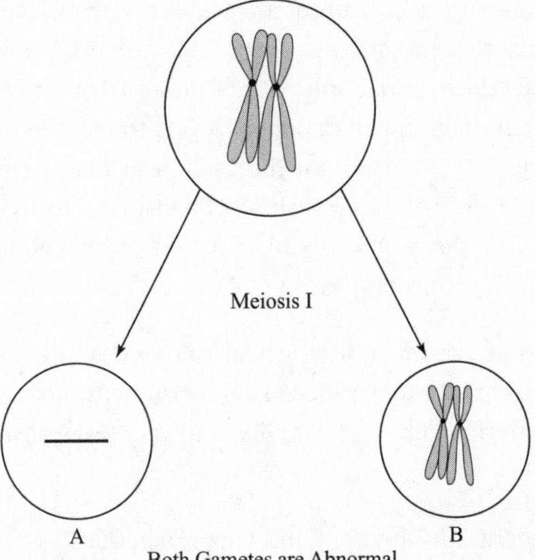

Both Gametes are Abnormal

If that abnormal gamete B combines with a normal gamete during fertilization, the resulting zygote will have three homologues instead of two, and it will be triploid for that chromosome. If gamete A combines with a normal gamete during fertilization, the resulting zygote will lack an entire chromosome and most likely will not survive.

(c) The triploid cell (3n) could have a problem reproducing sexually because it could not produce normal gametes.

(d) The reason why a triploid cell could not produce normal gametes is because triplet homologues could not line up correctly (double file) during metaphase of meiosis I. As a result, homologous pairs would not be able to separate correctly, and normal gametes would not be produced.

5. (a) While the newly formed mRNA is still in the nucleus, it is altered or processed. Editing molecules called snRNPs and spliceosomes splice out introns from the pre-RNA strand. Also, a 5′ cap is added to the 5′ end to help the RNA strand bind to the ribosome. In addition, a poly-A tail is added to the 3′ end to protect from degradation by hydrolytic enzymes in the cytoplasm. The mature mRNA is shorter than the pre-RNA.

(b) If different introns were removed from the primary RNA transcript, different proteins will be formed from the same original stretch of DNA. The process is known as alternative splicing. Referring to the figured provided, if exons 1, 2, 3, 4, and 5 get transcribed, the protein alpha-globin will be produced. If exons 1, 3, and 5 gets transcribed, a different protein will be formed.

(c) Exons are discontinuous sequences of DNA that get transcribed and translated into polypeptides. Genes consist of more than one exon. Introns are intervening sequences of DNA, and they do not code for proteins. They must be spliced out of the pre-RNA transcript before the pre-RNA moves to ribosomes in the cytoplasm. Some introns regulate the expression of exons.

(d) The figure illustrates how different proteins can be created by transcribing the same set of exons in a different order. This process, known as alternative splicing of DNA, is so effective that only 3% of human DNA is needed to code for proteins. In addition, although much of our DNA gets transcribed, it does not get translated into proteins. This transcribed (but not translated) DNA has many important functions in a cell. For example, some of it forms spliceosomes to splice introns out of pre-RNA. The rest of it probably consists of regulatory sequences that control the transcription of other parts of DNA.

6. (a) Cell Type A has mitochondria, lots of ribosomes, no cilia, and many lysosomes. Cell Type B has more than twice as many mitochondria as Cell Type A, about half the number of ribosomes, many cilia, and very few lysosomes compared with Cell Type A.

(b) Structure determines function and vice versa. Cell Type A is a cell whose function is related to digestion. Cell Type B could either be a mobile cell, or it could serve some movement function.

(c) Cell Type A is probably a type of white blood cell (that digests and destroys pathogens) because it has so many lysosomes, which are sacs of digestive enzymes. In addition, digestive enzymes are proteins, which are manufactured in ribosomes. As expected, Cell Type A has a huge number of ribosomes. Cell Type B is probably an epithelial cell that lines the human respiratory system because it has many cilia. Cilia remove dust and germs from the nasal passages, bronchi, and lungs.

(d) These two cells were taken from one human body. That means that they both contain identical DNA in their nuclei. The reason why they are so different in their structures and functions is because each cell expresses different genes and makes different proteins. The same stretch of DNA can produce different proteins by expressing a different combination of exons. Some genes can be suppressed or deactivated entirely or underexpressed (called "downregulated"). Some genes can be upregulated and can produce more of certain proteins than others. Also, genes can be turned on and off at different times in the life of a cell. All of these facts contribute to one person having many different types of body cells, each with the same DNA.

PART 7: APPENDIXES

Once you've worked your way through all of the review and practice this book offers, you may want to review the four **appendixes** that follow. These include:

- Appendix A: Bibliography, which contains a list of textbook resources that you may want to consult for additional information about many of the Biology concepts that were discussed throughout this book.
- Appendix B: Measurements Used in Biology, which provides a list of important quantities, names of units, symbols, and conversions that you should be familiar with for the AP Biology course and exam.
- Appendix C: Reference Tables, which mirrors the types of equations, formulas, and terms you will have access to during your exam.
- Appendix D: Glossary, which defines key terms that are often covered throughout the AP Biology course and exam.

Appendix A: Bibliography

These textbooks are valuable resources for any AP Biology course or any basic college Biology course:

Campbell, Neil A. and Jane Reece. *Biology*, 10th ed., San Francisco, California: Pearson, Benjamin Cummings, 2014

Freeman, Scott. *Biological Science*, 2nd ed., Upper Saddle River, New Jersey: Pearson Prentice Hall, 2005

Hillis, David M. and H. Craig Heller. *Principles of Life, High School Edition*, Sinauer Assoc., W. H. Freeman and Co., 2012

Mader, Sylvia S. *Biology*, 9th ed., Dubuque, Iowa: McGraw-Hill, 2007

Nelson, David L. and Michael M. Cox. *Lehninger Principles of Biochemistry*, 4th ed., New York: W. H. Freeman, 2005

Purves, William K., Gordon H. Orians, and H. Craig Heller. *Life: The Science of Biology*, 6th ed., New York: W. H. Freeman, 2000

Raven, Peter H. and George B. Johnson. *Biology*, 9th ed., Dubuque, Iowa: William C. Brown/ McGraw-Hill, 2011

Starr, Cecie and Ralph Taggart. *Biology: The Diversity of Life*, 12th ed., Belmont, California: Wadsworth Publishing, 2009

Tobin, Allan J. and Jennie Dushek. *Asking About Life*, 2nd ed., Orlando, Florida: Harcourt College Publishers, 2001

Urry, Lisa A. *Campbell Biology in Focus*, Harlow, UK: Pearson Education Inc., 2014

Wallace, Robert A., Gerald P. Sanders, and Robert J. Ferl. *Biology: The Science of Life*, 4th ed., New York: Addison-Wesley Publishing Co., 1996

Appendix B: Measurements Used in Biology

Measurements			
Quantity	**Name of Unit**	**Symbol**	**Conversion**
length	meter	m	1 m = 1,000 mm = .001 km = 100 cm
	kilometer	km	1 km = 1,000 m
	centimeter	cm	1 cm = 1/100 m
	millimeter	mm	1 mm = 1,000 µm
	micrometer	µm	1 µm = 1,000 nm
	nanometer	nm	1 nm = 1/1,000 µm
area	square meter	m^2	area encompassed by a square, each side of which is 1 meter in length
	hectare	ha	1 ha = 10,000 m^2 = 2.47 acres
	square centimeter	cm^2	1 cm^2 = 1/10,000 m^2
volume	liter	L	1 liter = 1/1,000 m^3 = 1.057 quarts
	milliliter	mL	1 mL = 0.001 L
	microliter	µL	1 µL = 0.001 mL
mass	kilogram	kg	1 kg = 1,000 g
	gram	g	1 g = 1,000 mg
	milligram	mg	1 mg = 1,000 µg
	microgram	µg	1 µg = 1/1,000 mg
temperature	Kelvin	K	0 K = absolute zero
	degrees Celsius	°C	°C + 273 = Kelvin
heat, work	calorie	cal	1 calorie = the amount needed to raise 1 gram of pure water 1° Celsius (from 14.5°C to 15.5°C = 4.184 joules)
	kilocalorie	kcal	1 kcal = 1,000 cal
	joule	J	1 cal = 4.184 J

Measurements			
Quantity	**Name of Unit**	**Symbol**	**Conversion**
electric potential	volt	V	a unit of potential difference or electromotive force
	millivolt	mV	1 mV = 1/1,000 V = 10^{-3} V
time	second	s.	1 s. = 1/60 min.
	minute	min.	1 min. = 1/60 hr.
	hour	hr.	1 hr. = 1/24 d.
	day	d.	1 d. = 24 hr.

Appendix C: Reference Tables

Equations, Formulae, and Terms

Standard Error of the Mean	Mean
$SE_{\bar{x}} = \frac{s}{\sqrt{n}}$	$\bar{x} = \frac{1}{n}\sum_{i=1}^{n} x_i$
Standard Deviation	**Chi-Square**
$s = \sqrt{\frac{\sum(x_i - \bar{x})^2}{n-1}}$	$\chi^2 = \sum\frac{(o-e)^2}{e}$

Chi-Square Table

	Degrees of Freedom							
p value	1	2	3	4	5	6	7	8
0.05	3.84	5.99	7.81	9.49	11.07	12.59	14.07	15.51
0.01	6.63	9.21	11.34	13.28	15.09	16.81	18.48	20.09

s = sample standard deviation (i.e., the sample-based estimate of the standard deviation of the population)

$\bar{x}$ = sample mean

n = size of the sample

o = observed results

e = expected results

Σ = sum of all

Degrees of freedom are equal to the number of distinct possible outcomes minus one.

Laws of Probability

If A and B are mutually exclusive, then P (A or B) = P(A) + P(B)

If A and B are independent, then P (A and B) = P(A) × P(B)

Hardy-Weinberg Equations

$p^2 + 2pq + q^2 = 1$	p = frequency of allele 1 in a population
$p + q = 1$	q = frequency of allele 2 in a population

Mode = value that occurs most frequently in a data set
Median = middle value that separates the greater and lesser halves of a data set
Mean = sum of all data points divided by number of data points
Range = value obtained by subtracting the smallest observation (sample minimum) from the greatest (sample maximum)

Metric Prefixes

Factor	Prefix	Symbol	Factor	Prefix	Symbol
10^9	giga	G	10^{-3}	milli	m
10^6	mega	M	10^{-6}	micro	μ
10^3	kilo	k	10^{-9}	nano	n
10^{-1}	deci	d	10^{-12}	pico	p
10^{-2}	centi	c			

Water Potential (ψ)

$\psi = \psi_p + \psi_s$ ψ_p = pressure potential ψ_s = solute potential

The water potential will be equal to the solute potential of a solution in an open container, since the pressure potential of the solution in an open container is zero.

The Solute Potential of a Solution

$$\psi_s = -\ iCRT$$

i = ionization constant (For sucrose this is 1.0 because sucrose does not ionize in water.)
C = molar concentration
R = pressure constant (R = 0.0831 liter bars/mole K)
T = temperature in Kelvin (273 + °C)

Acid and Base: pH
$\mathbf{pH} = -\log[H^+]$

Rate and Growth		
Rate	dY/dt	dY = amount of change dt = change in time B = birth rate D = death rate N = population size K = carrying capacity r_{max} = maximum per capita growth rate of population
Population Growth	$dN/dt = B - D$	
Exponential Growth	$\frac{dN}{dt} = r_{max}N$	
Logistic Growth	$\frac{dN}{dt} = r_{max}N\left(\frac{K-N}{K}\right)$	

Simpson's Diversity Index
$\text{Diversity Index} = 1 - \sum\left(\frac{n}{N}\right)^2$ n = total number of organisms of a particular species N = total number of organisms of all species

Surface Area and Volume		
Volume of a Sphere	$V = \frac{4}{3}\pi r^3$	r = radius l = length h = height w = width s = length of one side of a cube SA = surface area V = volume
Volume of a Cube	$V = s^3$	
Surface Area of a Sphere	$SA = 4\pi r^2$	
Surface Area of a Cube	$SA = 6s^2$	
Surface Area of a Rectangular Solid	$SA = 2lh + 2lw + 2wh$	
Volume of a Rectangular Solid	$V = lwh$	
Surface Area of a Cylinder	$SA = 2\pi rh + 2\pi r^2$	
Volume of a Cylinder	$V = \pi r^2 h$	

Appendix D: Glossary

ABA (abscisic acid) Plant hormone that inhibits growth, closes stomates during times of water stress, and counteracts breaking of dormancy.

Abiotic Nonliving and includes temperature, water, sunlight, wind, rocks, and soil.

Abscission The process of leaves falling off a tree or bush.

Acetylcholine One of many neurotransmitters.

Acid rain Caused by pollutants in the air from combustion of fossil fuels. The pH is less than 5.6.

Actin Thin protein filaments that interact with myosin filaments in the contraction of skeletal muscle.

Action potential A rapid change in the membrane of a nerve or muscle cell when a stimulus causes an impulse to pass.

Active immunity The type of immunity when an individual makes his or her own antibodies after being ill and recovering or after being given an immunization or vaccine.

Adaptive radiation The emergence of numerous species from one common ancestor introduced into an environment.

Adenine A nucleotide that binds to thymine and uracil. It is a purine.

Adipose Fat tissue.

Allopatric speciation The formation of new species caused by separation by geography, such as mountain ranges, canyons, rivers, lakes, glaciers, altitude, or longitude.

Allosteric A type of enzyme that changes its conformation and its function in response to a modifier.

Alu elements The most abundant, transposable elements; the human genome contains over 1 million copies.

Amoebocytes Found in sponges, these cells are mobile and perform numerous functions, including reproduction, transport of food particles to nonfeeding cells, and secretion of material that forms the spicules.

Amphipathic A molecule with both a positive and negative pole.

Anaerobic respiration The anaerobic breakdown of glucose into pyruvic acid with the release of a small amount of ATP.

Analogous structures Structures, such as a bat's wing and a fly's wing, that have the same function, but the similarity is superficial and reflects an adaptation to similar environments, not a common ancestry.

Aneuploidy Any abnormal number of a particular chromosome.

Angiosperms Flowering plants.

Anode The positive pole in an electrolytic cell.

Antenna pigment Accessory photosynthetic pigment that expands the wavelengths of light that can be used to power photosynthesis.

Anterior pituitary Gland in the brain that releases many hormones, including growth hormone, luteinizing hormone, thyroid-stimulating hormone, adrenocorticotropic hormone, and follicle-stimulating hormone.

Anther Part of a flowering plant that produces male gametophytes.

Antheridia Structures in plants that produce male gametes.

Antibodies Produced by B lymphocytes and destroy antigens.

Anticodon The three-base sequence of nucleotides at one end of a tRNA molecule.

Antidiuretic hormone Released by the posterior pituitary, its target is the collecting tube of the nephron.

Apical dominance The preferential growth of a plant upward (toward the sun), rather than laterally.

Apoplast The network of cell walls and intercellular spaces within a plant body that permits extensive extracellular movement of water within a plant.

Apoptosis Programmed cell death.

Aposematic coloration The bright, often red or orange coloration of poisonous animals as a warning that predators should avoid them.

Archegonia Structures in plants that produce female gametes.

Artificial selection The intentional selection of specific individuals with desired traits for breeding.

Associative learning One type of learning in which one stimulus becomes linked, through experience, to another.

ATP synthase channels Located in the cristae of mitochondria and thylakoids of chloroplasts, these are membrane channels that allow protons to diffuse down a gradient in the production of ATP.

Autonomic nervous system The branch of the vertebrate peripheral nervous system that controls involuntary muscles.

Autotrophs Organisms that synthesize their own nutrients.

Auxin A plant hormone that stimulates stem elongation and growth, enhances apical dominance, and is responsible for tropisms.

Backcross *See* testcross.

Bacteriophage A virus that attacks bacteria.

Balanced polymorphism The presence of two or more phenotypically distinct forms of a trait in a single population, such as two varieties of peppered moths, black ones and white ones.

Barr body An inactivated X chromosome seen as a condensed body lying just inside the nuclear envelope.

Batesian mimicry The copycat coloration in which one harmless animal mimics the coloration of one that is poisonous. An example is the viceroy butterfly, which is harmless but looks similar to the monarch butterfly.

Binomial nomenclature A scientific naming system in which every organism has a unique name consisting of two parts: a genus name and a species name.

Biological magnification A trophic process in which substances in the food chain become more concentrated with each link of the food chain.

Bioluminescence Conversion of energy stored in certain organic molecules to light.

Biomes Very large regions of Earth, named for the climatic conditions and for the predominant vegetation. Examples are marine, tropical rain forest, and desert.

Biosphere The global ecosystem.

Biotic potential The maximum rate at which a population could increase under ideal conditions.

Blastocyst The blastula stage of mammalian embryonic development.

Blastula A hollow ball of cells that marks the end of cleavage.

B lymphocyte A lymphocyte that produces antibodies.

Bottleneck effect An example of genetic drift that results from the reduction of a population, typically by natural disaster. The surviving population is no longer genetically representative of the original one.

Botulinum The genus name for the bacterium that produces botulism, a very serious form of food poisoning.

Bryophytes Nonvascular plants like mosses.

Bulk flow The general term for the overall movement of a fluid in one direction in an organism, such as sap flowing in a tree or blood flowing in a human.

Bundle sheath cell A type of photosynthetic plant cell that is tightly packed around the veins in a leaf.

C-3 plant The common type of plant, different from C-4 and CAM plants.

C-4 plant A plant with anatomical and biochemical modifications for a dry environment that differ from C-3 and CAM plants. Examples are sugarcane and corn.

Calvin cycle A cyclical metabolic pathway in the dark reactions of photosynthesis that fixes or incorporates carbon into carbon dioxide and produces phosphoglyceraldehyde (PGAL), a three-carbon sugar.

CAM (crassulacean acid metabolism) A form of photosynthesis that is an adaptation to dry conditions; stomates remain closed during the day and open only at night.

Capsid The protein shell that encloses viral DNA or RNA.

Carbon fixation Carbon becomes fixed or incorporated into a molecule of PGAL. This happens during the Calvin cycle.

Carbonic acid anhydrase An enzyme found in red blood cells that catalyzes the conversion of carbon dioxide and water into carbonic acid as part of the system that maintains blood pH at 7.4.

Carotenoid Accessory photosynthetic pigment that is yellow or orange.

Carrying capacity The limit to the number of individuals that can occupy one area at a particular time.

Catalase An enzyme produced in all cells to decompose hydrogen peroxide, a by-product of cell respiration.

Cathode The negative pole in an electrolytic cell.

CDKs (cyclin-dependent kinases) A kinase whose activity depends on the level of cyclins and that controls the timing of cell division.

Cell plate A double membrane down the midline of a dividing plant cell between which the new cell wall will form.

Centriole One of two structures in animal cells involved with cell division.

Centromere A specialized region in a chromosome that holds the two chromatids together.

Chemiosmosis The process by which ATP is produced from the flow of protons through an ATP synthase channel in the thylakoid membrane of the chloroplast during the light reactions of photosynthesis and in the cristae membrane of the mitochondria during cell respiration.

Chemokines A chemical secreted by blood vessel endothelium and monocytes during an immune response to attract phagocytes to an area.

Chiasma/chiasmata The site at which a crossover and recombination occurs.

Chitin A structural polysaccharide found in the cell walls.

Chlorophyll *a* One type of chlorophyll that participates directly in the light-dependent reactions of photosynthesis.

Chlorophyll *b* One type of chlorophyll that acts as an antenna pigment, expanding the wavelengths of light that can be used to power photosynthesis.

Chloroplast The site of photosynthesis in plant cells.

Choanocytes Collar cells that line the body cavity and have flagella that circulate water in sponges.

Chromatid Either of the two strands of a replicated chromosome joined at the centromere.

Chromatin network The complex of DNA and protein that makes up a eukaryotic chromosome. When the cell is not dividing, chromatin exists as long, thin strands and is not visible with the light microscope.

Cilia Hairlike extensions from the cytoplasm used for cell locomotion.

Citric acid cycle Another name for the Krebs cycle.

Cladogenesis Branching evolution occurs when a new species branches out from a parent species.

Cladogram A diagrammatic reconstruction of phylogenetic history.

Classical conditioning One type of associative learning that is widely accepted because of the ingenious work of Ivan Pavlov associating a novel stimulus with an innately recognized one.

Cleavage furrow A shallow groove in the cell surface in an animal cell where cytokinesis is taking place.

Climax community The final, stable community in an ecosystem.

Cline A variation in some trait of individuals coordinated with some gradual change in temperature or other factor over a geographic range.

Clonal selection A fundamental mechanism in the development of immunity. Antigenic molecules select or bind to specific B or T lymphocytes, activating them. The B cells then differentiate into plasma cells and memory cells.

Cnidocytes Stinging cells in all cnidarians.

Codominance The type of inheritance when there is no trait that dominates over another; both traits show.

Codons The three-base sequence of nucleotides in mRNA.

Coelom The body cavity that arises from within the mesoderm and is completely surrounded by mesoderm tissue.

Coevolution Evolution that is caused by two species that interact and influence each other. All predator-prey relationships are examples.

Cohesion tension Force of attraction between molecules of water due to hydrogen bonding.

Collaboration Two genes interact to produce a novel phenotype.

Collenchyma cells Plant cells with unevenly thickened primary cell walls that are alive at maturity and that function to support the growing stem.

Commensalism A symbiotic relationship in which one organism benefits and one is unaware of the other organism (+/o).

Community All the organisms living in one area.

Companion cell Connected to each sieve tube member in the phloem and nurtures the sieve tube elements.

Complement An important part of the immune system, a group of about twenty proteins that assists in lysing cells.

Complementary genes The expression of two or more genes in which each depends upon the alleles of the other in order for a trait to show.

Conformation The particular three-dimensional shape of a protein molecule.

Conjugation A primitive form of sexual reproduction that is characteristic of bacteria and some algae.

Convergent evolution Evolution that occurs when unrelated species occupy the same environment and are subjected to similar selective pressures and show similar adaptations.

Countercurrent mechanism A mechanism or strategy to maximize the rate of diffusion. This is a major strategy to transport substances across membranes passively, such as in the nephron.

Cristae The internal membranes of mitochondria that are the site of the electron transport chain.

Crop Part of the digestive tract of many animals where food is temporarily stored until it can continue to the gizzard.

Crossing-over The reciprocal exchange of genetic material between nonsister chromatids during synapsis of meiosis I.

Cutin The main component of the waxy cuticle covering leaves to minimize water loss.

Cyclic-AMP (CAMP) Cyclic adenosine monophosphate made from ATP; a common intracellular signaling molecule.

Cyclic photophosphorylation Part of the light-dependent reactions in photosynthesis where electrons travel on a short-circuit pathway to replenish ATP levels only.

Cyclin A regulatory protein whose levels fluctuate cyclically in a cell, in part, related to the timing of cell division.

Cystic fibrosis The most common lethal genetic disease in the United States; characterized by a buildup of extracellular fluid in the lungs and digestive tract.

Cytochrome An iron-containing pigment present in the electron transport chain of all aerobes.

Cytokines Chemicals that stimulate helper T cells, B cells, and killer T cells.

Cytokinesis Division of the cytoplasm.

Cytokinins Plant hormone that stimulates cell division and delays senescence (aging).

Cytosine A nucleotide that binds with guanine. A pyrimidine.

Cytoskeleton A complex network of protein filaments that gives a cell its shape and helps it move.

Cytotoxic T cells A type of lymphocyte that kills infected body cells and cancer cells.

Decomposers Organisms, like bacteria and fungi, that recycle nutrients back to the soil.

Deletion A chromosomal mutation in which a fragment is lost during cell division.

Dendrites The sensory processes of a neuron.

Denitrifying bacteria Convert nitrates (NO_3) into free atmospheric nitrogen.

Density-dependent factors Factors, such as starvation, that increase directly as the population density increases.

Density-dependent inhibition A characteristic of normal cells grown in culture that causes cell division to cease when the culture becomes too crowded.

Density-independent factors Factors, such as earthquakes, whose occurrence is unrelated to the population density.

Depolarization An electrical state in which the inside of an excitable cell is made less negative compared with the outside. If an axon is depolarized, an impulse is passing.

Detritivores Consumers that derive their nutrition from nonliving, organic matter.

Deuterostomes Animals in which the blastopore becomes the anus during early embryonic development.

Dicotyledon A subdivision of flowering plants whose members possess an embryonic seed leaf made of two halves or cotyledons.

Dihybrid cross A cross between individuals that are hybrid for two different traits, such as height and seed color.

Diploblastic An organism whose body is made of only two cell layers, the ectoderm and the endoderm. The two are connected by a noncellular layer called the mesoglea. Animal phyla that are diploblastic are the Porifera (sponges) and the Cnidaria (jellyfish and hydra).

Directional selection Selection in which one phenotype replaces another in the gene pool.

Disruptive selection Selection that increases the extreme types in a population at the expense of intermediate forms.

Divergent evolution Evolution that occurs when a population becomes isolated (for any reason) from the rest of the species, becomes exposed to new selective pressures, and evolves into a new species.

DNA ligase An enzyme that permanently attaches pieces of DNA together.

Dopamine A neurotransmitter.

Down syndrome A genetic condition caused by trisomy 21.

Duodenum The first 12 inches (30 cm) of the human small intestine.

Ecdysone A hormone that helps control metamorphosis in insects.

Ecological niche The sum of a species' use of all the resources in its environment.

Ecological succession The sequential rebuilding of an entire ecosystem after a disaster.

Ecosystem All the organisms in a given area as well as the abiotic (nonliving) factors with which they interact.

Ectoderm The germ layer that gives rise to the skin and nervous system.

Effectors Muscles or glands.

Electron transport chain A sequence of membrane proteins that carry electrons through a series of redox reactions to produce ATP.

Endergonic Any process that absorbs energy.

Endoderm The embryonic germ layer that gives rise to the viscera, the digestive tract, and other internal organs.

Endodermis The tightly packed layer of cells that surrounds the vascular cylinder in the root of a plant.

Endoplasmic reticulum (ER) A system of transport channels inside a eukaryotic cell.

Endosperm The food source for the growing embryo in monocots.

Endosymbiosis This theory states that mitochondria and chloroplasts were once free-living prokaryotes that took up residence inside larger prokaryotic cells in a permanent, symbiotic relationship.

Endotherms Animals that can raise their body temperature, although they cannot maintain a stable body temperature.

Entropy Measure of disorder in a system.

Envelope Cloaks the capsid of a virus and aids the virus in infecting the host. The envelope is derived from membranes of host cells.

Enzyme A protein that serves as a catalyst.

Epicotyl Part of the developing embryo that will become the upper part of the stem and the leaves of a plant.

Epinephrine A neurotransmitter.

Epiphytes Photosynthetic plants that grow on other trees rather than supporting themselves.

Epistasis Two separate genes control one trait, but one gene masks the expression of the other gene.

Epitope A small, accessible portion of an antigen; a single antigen usually has several epitopes.

Esterase An enzyme that breaks down an excess neurotransmitter.

Ethylene A gaseous plant hormone that promotes fruit ripening and opposes auxins in its actions.

Eukaryotes Cells with internal membranes.

Eutrophication Translates as "true feeding." A process begun by the entrance of large amounts of nutrients into a lake, ultimately ending with the death of the lake.

Exaptation A trait used for something other than the one produced by natural selection.

Exergonic A reaction that releases energy.

Exocytosis The process by which cells expel substances.

Exons Stands for expressed sequences of DNA. These are genes.

Exothermic Any process that gives off energy.

Expressivity The range of expression of mutant genes.

Extranuclear genes Genes outside the nucleus, in the mitochondria and chloroplasts.

Facultative anaerobes Organisms that can live without oxygen in the environment.

FAD (flavin adenine dinucleotide) A coenzyme that carries protons or electrons from glycolysis and the Krebs cycle to the electron transport chain.

Fermentation A synonym for anaerobic respiration. The anaerobic breakdown of glucose into pyruvic acid.

Fixed action pattern An innate, highly stereotypic behavior, which when begun, is continued to completion, no matter how useless.

Flagella The tail-like structure that propels some single-celled organisms. Flagella consist of microtubules.

Follicle-stimulating hormone (FSH) A hormone released from the anterior pituitary that stimulates the ovarian follicle.

Food chain The pathway along which food is transferred from one trophic level to the next.

Food pyramid A model of the food chain that demonstrates the interaction of the organisms and the loss of energy.

Food web The interconnected feeding relationships of organisms in an ecosystem.

Founder effect An example of genetic drift, when a small population breaks away from a larger one to colonize a new area; it is most likely not genetically representative of the original larger population.

Frameshift One type of mutation caused by a deletion or addition where the entire reading sequence of DNA is shifted. AAA TTT CCC GGG could become AAT TTC CCG GG.

Frequency-dependent selection A form of selection that acts to decrease the frequency of the more-common phenotypes and increase the frequency of the less-common ones.

Fruit A ripened ovary of a flowering plant.

Fungi The kingdom that consists of heterotrophs that carry out extracellular digestion and have cell walls made of chitin; includes mushrooms and yeast.

GABA (gamma-aminobutyric acid) A neurotransmitter.

Gametangia A protective jacket of cells that prevents some plants' gametes and zygotes from drying out.

Gametophyte The monoploid generation of a plant.

Gastrodermis Cells that line the gastrovascular cavity in cnidarians.

Gastrovascular cavity A digestive cavity with only one opening, characteristic of cnidarians.

Gated ion channel A channel in a plasma membrane for one specific ion, such as sodium or calcium. In the terminal branch of a neuron, it is responsible for the release of a neurotransmitter into the synapse.

Gene flow The movement of alleles into or out of a population.

Genetic drift Change in the gene pool due to chance.

Genetic engineering The technology of manipulating genes for practical purposes.

Genomic imprinting Certain traits whose expression varies, depending on the parent from which they are inherited. Diseases that result from imprinting are Prader-Willi and Angelman syndromes.

Genotype The types of genes an organism has.

Gibberellins Plant hormone that promotes stem elongation.

Gizzard Part of the digestive tract of many animals. It is the site of mechanical digestion.

Glial cells Cells that nourish neurons.

Gluteraldehyde A chemical fixative often used in the preparation of tissue for electron microscopy.

Glycocalyx The external surface of a plasma membrane that is important for cell-to-cell communication.

Glycolysis A nine-step, anaerobic process that breaks down one glucose molecule into two pyruvic acid molecules and four ATP.

Golgi apparatus An organelle in eukaryotes that lies near the nucleus and that packages and secretes substances for the cell.

Gonadotropin-releasing hormone (GgRH) A hormone released by the hypothalamus that stimulates other glands to release their hormones.

Gradualism The theory that organisms descend from a common ancestor gradually, over a long period of time, in a linear or branching fashion.

Grana Membranes in the chloroplast where the light reactions occur.

Greenhouse effect The warming of the planet because of the accumulation of atmospheric carbon dioxide.

Gross primary production (GPP) The total amount of energy from light (or from chemicals in chemosynthetic systems) that is converted into organic molecules per unit of time.

Ground tissue The most common tissue type in a plant, functions mainly in support and consists of parenchyma, collenchyma, and sclerenchyma cells.

GTP (guanosine triphosphate) A molecule closely related to ATP that provides the energy for translation.

Guanine A nucleotide that binds with cytosine. A purine.

Guttation Due to root pressure, droplets of water appear in the morning on the leaf tips of some herbaceous leaves.

Gymnosperms Conifers or cone-bearing plants.

Habitat isolation Separation of two or more organisms of the same species living in the same area but in separate habitats, such as in the water and on land.

Habituation One of the simplest forms of learning in which an animal comes to ignore a persistent stimulus.

Halophiles (halobacteria) Aerobic bacteria that thrive in environments with very high salt concentrations.

Hatch-Slack pathway An alternate biochemical pathway found in C-4 plants; its purpose is to remove CO_2 from the airspace near the stomate.

Head-foot The part of the body of mollusks that contains both sensory and motor organs.

Helicase An enzyme that untwists the double helix at the replication fork.

Helper T cells One type of T lymphocyte that activates B cells and other T lymphocytes.

Hemocoels Blood-filled cavities within the body of arthropods and mollusks with open circulatory systems.

Hemophilia An inherited genetic disease caused by the absence of one or more proteins necessary for normal blood clotting.

Hermaphrodites Organisms possessing both male and female sex organs.

Heterosis *See* hybrid vigor.

Heterosporous A plant that produces two kinds of spores, male and female.

Heterotroph hypothesis This theory states that the first cells on Earth were heterotrophic prokaryotes.

Heterotrophs Organisms that must ingest nutrients rather than synthesize them.

Histamine A chemical released by the body during an inflammatory response that causes blood vessels to dilate.

Homeotherms Organisms that maintain a consistent body temperature.

Homologous structures Structures in different species that are similar because they have a common origin.

Homosporous A plant that produces a single bisexual spore.

Huntington's disease A degenerative, inherited, dominant disease of the nervous system that results in certain and early death.

Hybrid vigor A phenomenon in which the hybrid state is selected because it has greater survival and reproductive success. Also known as heterosis.

Hydrophilic Having an affinity for water.

Hyperpolarized An electrical state in which the inside of the excitable cell is made more negative compared with the outside of the cell and the electric potential of the membrane increases (gets more negative).

Hypertonic Having a greater concentration of solute than another solution.

Hypocotyl Part of the developing embryo that will become the lower part of the stem and roots.

Hypothalamus Gland located in the brain above the pituitary that is the bridge between the endocrine and nervous systems.

Hypotonic Having a lesser concentration of solute than another solution.

Immunoglobulins *See* antibodies.

Immunological memory The capacity of the immune system to generate a secondary immune response against a specific antigen for a lifetime.

Imprinting A type of learning that is responsible for the bonding between mother and offspring. Common in birds, it occurs during a sensitive or critical period in early life.

Incomplete dominance The type of inheritance that is characterized by blending traits. For instance, one gene for red plus one gene for white results in a pink four o'clock flower.

Indoleacetic acid IAA. A naturally occurring auxin.

Inflammatory response A nonspecific defensive reaction of the body to invasion by a foreign substance that is accompanied by the release of histamine, fever, and red, itchy areas.

Interferons A class of chemicals that block viral infections.

Interneuron Also known as association neuron, resides within the spinal cord and receives sensory stimuli and transfers the information directly to a motor neuron or to the brain for processing.

Interphase The longest stage of the life cycle of a cell; it consists of G_1, S, and G_2.

Introns Intervening sequences, the noncoding regions of DNA, that lay between coding regions.

Inversion A chromosome mutation in which a chromosomal fragment reattaches to its original chromosome but in the reverse orientation.

In vitro In the laboratory.

In vivo In the living thing.

Isotonic Two solutions containing equal concentrations of solutes.

Isotope Different forms of the same element; they differ in their number of neutrons.

Karyotype A procedure that analyzes the size, number, and shape of chromosomes.

Kinase An enzyme that transfers phosphate ions from one molecule to another.

Kinetochore A disc-shaped protein on the centromere that attaches the chromatid to the mitotic spindle during cell division.

Klinefelter syndrome A genetic condition in males in which there is an extra X chromosome; the genotype is XXY.

Kranz anatomy Refers to the structure of C-4 leaves and differs from C-3 leaves. In C-4 leaves, the bundle sheath cells lie under the mesophyll cells, tightly wrapping the vein deep within the leaf, where CO_2 is sequestered.

Krebs cycle Also known as the citric acid cycle, it completes the breakdown of pyruvic acid into carbon dioxide, with the release of a small amount of ATP.

Lactic acid fermentation The process by which pyruvate from glycolysis is reduced to form lactic acid or lactate. This is the process that the dairy industry uses to produce yogurt and cheese.

Lateral meristem Growth region of a plant that provides secondary growth, increase in girth.

Law of dominance One of Mendel's laws. It states that when two organisms, each pure for two opposing traits, are crossed, the offspring will be hybrid but will exhibit only the dominant trait.

Law of independent assortment States that each allelic pair separates during gamete formation. Applies when genes for two traits are not on the same chromosome.

Law of segregation During the formation of gametes, allelic pairs for two traits separate.

Learning A sophisticated process in which the responses of the organism are modified as a result of experience.

Linked genes Genes that are on the same chromosome.

Luteinizing hormone Triggers the ovulation of the secondary oocyte from the ovary.

Lysosomes Sacs of hydrolytic enzymes and the principal site of intracellular digestion.

Lytic cycle A type of viral infection that results in the lysing of the host cell and the release of new phages that will infect other cells.

Macroevolution The development of an entirely new species.

Macrophage While acting as an antigen-presenting cell, it engulfs bacteria by phagocytosis and presents a fragment of the bacteria on the cell surface by an MHC II molecule.

Malpighian tubules The organ of excretion in insects.

Mantle The part of the body of mollusks that contains specialized tissue that surrounds the visceral mass and secretes the shell.

Map unit The distance on a chromosome within which recombination occurs 1 percent of the time.

Marsupials Animals whose young are born very early in embryonic development and where the joey completes its development nursing in the mother's pouch. Includes kangaroos.

Matrix The inner region of a mitochondrion, where the Krebs cycle occurs.

Medusa The free-swimming, upside-down, bowl-shaped stage in the life cycle of the cnidarians. An example is jellyfish.

Megaspores In flowering plants, these produce the ova.

Meiosis Occurs in sexually reproducing organisms and results in cells with half the chromosome number of the parent cell.

Membrane potential A measurable difference in electrical charge between the cytoplasm (negative ions) and extracellular fluid (positive ions).

Memory cells A long-lived form of a lymphocyte that bear receptors to a specific antigen and that remains circulating in the blood in small numbers for a lifetime.

Meristem Actively dividing cells that give rise to other cells such as xylem and phloem.

Mesoderm The germ layer that gives rise to the blood, bones, and muscles.

Methanogens Prokaryotes that synthesize methane from carbon dioxide and hydrogen gas.

MHC (major histocompatibility complex) A collection of cell surface markers that identify the cells as self. No two people, except identical twins, have the same MHC markers. Also known as HLA (human leukocyte antigens).

Microevolution Refers to the changes in one gene pool of a population.

Microfilaments Solid rods of the protein actin that make up part of the cytoskeleton.

Micropyle The opening to the ovule in a flowering plant.

Microspores In flowering plants, these produce sperm.

Microtubules A hollow rod of the protein tubulin in the cytoplasm of all eukaryote cells that make up cilia, flagella, spindle fibers, and other cytoskeletal structures of cells.

Middle lamella A distinct layer of adhesive polysaccharides that cements adjacent plant cells together.

Minority advantage *See* frequency-dependent selection.

Mitchell hypothesis An attempt to explain how energy is produced during the electron transport chain by oxidative phosphorylation.

Mitochondria The site of cell respiration and ATP synthesis in all eukaryotic cells.

Mitosis Produces two genetically identical daughter cells and conserves the chromosome number ($2n$).

Molarity Number of moles of solute per liter of solution.

Monera No longer used as the name of the kingdom that contains all the prokaryotes, including bacteria.

Monoclonal antibodies Antibodies produced by a single B cell that produces a single antigen in huge quantities. They are important in research and in treating and diagnosing certain diseases.

Monocotyledon A subdivision of flowering plants whose members possess one embryonic seed leaf or cotyledon.

Monocytes A type of white blood cell that transforms into macrophages, extends pseudopods, and engulfs huge numbers of microbes over a long period of time.

Monohybrid cross This is the cross between two organisms that are each hybrid for one trait.

Monotremes Egg-laying mammals whose embryos derive nutrition from the yolk, like the duck-billed platypus.

Motor neuron A neuron that stimulates effectors (muscles or glands).

Mucosa The innermost layer of the human digestive tract. In some parts of the digestive system, it contains mucus-secreting cells and glands that secrete digestive enzymes.

Müllerian mimicry Two or more poisonous species resemble each other and gain an advantage from their combined numbers. Predators learn more quickly to avoid any prey with that appearance.

Multiple alleles More than two allelic forms of a gene.

Mutagenic agents Substances that cause mutations.

Mutation Changes in DNA.

Mutualism Asymbiotic relationship in which both organisms benefit (+/+).

Mycorrhizae The symbiotic structures consisting of the plant's roots intermingled with the hyphae (filaments) of a fungus that greatly increase the quantity of nutrients that a plant can absorb; enabled plants to move to land.

Myosin Thick protein filaments that interact with actin filaments in the contraction of skeletal muscle.

NAD (nicotinamide adenine dinucleotide) A coenzyme that carries protons or electrons from glycolysis and the Krebs cycle to the electron transport chain.

NADP (nicotinamide nucleotide phosphate) Carries hydrogen from the light reactions to the Calvin cycle in the dark reactions of photosynthesis.

Natural killer (NK) cells Part of the nonspecific immune response that destroys virus-infected body cells (as well as cancerous cells).

Natural selection A theory that explains how populations evolve and how new species develop.

Net primary production (NPP) The NPP is equal to the GPP minus the energy used by primary producers for their own respiration.

Neuromuscular junction The place where a neuron synapses on a muscle.

Neurotransmitter The chemical held in presynaptic vesicles of the terminal branch of the axon that are released into a synapse and that excite the postsynaptic membrane.

Neutrophils A type of white blood cell that engulfs microbes by phagocytosis.

Niche Organisms that live in the same area and use the same resources.

Nitric oxide Acts as a local signaling molecule.

Nitrifying bacteria Convert the ammonium ion into nitrites and then into nitrates.

Nitrogen-fixing bacteria Convert free nitrogen into the ammonium ion.

Nondisjunction Homologous chromosomes fail to separate as they should during meiosis.

Norepinephrine A neurotransmitter.

Notochord A rod that extends the length of the body and serves as a flexible axis in all chordates.

Nucleoid Nuclear region in prokaryotes.

Nucleolus Located in the nucleus and is the site of protein synthesis.

Nucleotides The building blocks of nucleic acids. They consist of a five-carbon sugar, a phosphate, and a nitrogenous base: adenine, thymine (in DNA), cytosine, guanine, or uracil (in RNA).

Obligate anaerobes Prokaryotes that cannot live in the presence of oxygen.

Okazaki fragments In DNA replication, the segments in which the 3′ to 5′ lagging strand of DNA is synthesized.

Omnivores Organisms, like humans, that eat both plants and animals.

Operant conditioning A type of associative learning in which an animal learns to associate one of its own behaviors with a reward or punishment and then repeats or avoids that behavior. Also called trial-and-error learning.

Operator In an operon, the binding site for the repressor.

Operon Functional genes and their switches that are found in bacteria.

Osmotic potential The tendency of water to move across a permeable membrane into a solution; also called *solute potential.*

Outbreeding Mating of organisms that are not closely related; it is a major mechanism of maintaining variation within a species.

Oxidation The loss of hydrogen (H^+) or electrons (e^-).

Oxidative phosphorylation The production of ATP using energy derived from the electron transport chain.

Oxytocin Hormone released by the posterior pituitary that stimulates labor and the production of milk from mammary glands.

Parallel evolution Evolution that occurs when two related species have made similar evolutionary adaptations after their divergence from a common ancestor.

Parasitism A symbiotic relationship (+/–) where one organism, the parasite, benefits while the other organism, the host, is harmed.

Parasympathetic One of two branches of the autonomic nervous system that has a relaxing effect.

Parenchyma cells Traditional plant cells with primary cell walls that are thin and flexible and that lack secondary cell walls.

Passive immunity Immunity is transferred to an individual from someone else.

Pathogens Organisms that cause disease.

Pedigree A family tree that indicates the phenotype of one trait being studied for every member of a family and will help determine how a particular trait is inherited.

Peroxisomes Organelles in both plants and animals that break down peroxide, a toxic byproduct of cell respiration.

Phage Short form of bacteriophage, the virus that attacks bacteria.

Phagocytes A type of white blood cell that ingests invading microbes.

Phagocytosis The process by which a cell engulfs large particles using pseudopods.

Phenotype The appearance of an organism.

Phenylketonuria An inborn inability to break down the amino acid phenylalanine. It requires elimination of phenylalanine from the diet, otherwise serious mental retardation will result.

Phloem Transport vessels in plants that carry sugars from the photosynthetic leaves to the rest of the plant by active transport.

Phosphodiester linkages The bonds that join the nucleotides in DNA.

Phosphofructokinase (PFK) An allosteric enzyme important in glycolysis.

Phosphoglyceraldehyde (PGAL) A three-carbon sugar, the first stable carbohydrate to be produced by photosynthesis.

Photolysis The process of splitting water, providing electrons to replace those lost from chlorophyll *a* in P680. This is powered by the light energy absorbed during the light-dependent reactions.

Photoperiod The environmental stimulus a plant uses to detect the time of year and the relative lengths of day and night.

Photophosphorylation The process of generating ATP by means of a proton motive force during the light reactions of photosynthesis.

Photorespiration A process that occurs when rubisco binds with O_2 instead of CO_2. It is a dead-end process because no ATP is produced and no sugar is formed.

Photosynthesis The process by which light energy is converted to chemical bond energy.

Photosystem I (P700) Energy, with an average wavelength of 700 nm, is absorbed in this photosystem and transferred to electrons that move to a higher energy level.

Photosystem II (P680) Energy, with an average wavelength of 680 nm, is absorbed in this photosystem and transferred to electrons that move to a higher energy level.

Photosystems Light-harvesting complexes in the thylakoid membranes of chloroplasts. They consist of a reaction center containing chlorophyll *a* and a region containing several hundred antenna pigment molecules that funnel energy into chlorophyll *a*.

Phycobilin A photosynthetic antenna pigment common in red and blue-green algae.

Phylogeny Evolutionary history of a species or group of related species.

Phytochrome The photoreceptor responsible for keeping track of the length of day and night. There are two forms of phytochrome, Pr (red light absorbing) and Pfr (infrared light absorbing).

Pili Cytoplasmic bridges that connect one cell to another and that allow DNA to move from one cell to another in a form of primitive sexual reproduction called conjugation.

Pinocytosis A type of endocytosis in which a cell ingests large, dissolved molecules.

Pioneer organisms The first organisms, such as lichens and mosses, to inhabit a barren area.

Pistil Part of a flowering plant that produces female gametes.

Placental mammals Animals whose embryos develop internally in a uterus connected to the mother by a placenta where nutrients diffuse from mother to embryo. Also called eutherians.

Plasma cells A short-lived form of a lymphocyte that secretes antibodies.

Plasmid Foreign, small, circular, self-replicating DNA molecule that inhabits a bacterium and imparts characteristics to the bacterium such as resistance to antibiotics.

Plasmodesmata An open channel in the cell walls of plant cells allowing for connections between the cytoplasm of adjacent cells.

Plasmolysis Cell shrinking.

Plastids Organelles in plant cells, including chloroplasts, chromoplasts, and leucoplasts.

Plastoquinone A proton and electron carrier in the electron transport chain during the light reactions of photosynthesis.

Pleiotropy One single gene affects an organism in several or many ways.

Poikilotherms Cold-blooded animals.

Point mutation A change in one nucleotide in DNA.

Polarized membrane An axon membrane at rest where the inside of the cell is negative compared with the outside of the cell.

Pollen One pollen grain contains three monoploid nuclei, one tube nucleus, and two sperm nuclei.

Pollination The transfer of pollen from the stamen to the pistil.

Polygenic Genes that vary along a continuum, like skin color or height.

Polymerase chain reaction (PCR) A cell-free, automated technique by which a piece of DNA can be rapidly copied or amplified.

Polyp A vase-shaped body or the sessile phase in the life cycle of cnidarians.

Polyploid A chromosome mutation in which the organism possesses extra sets of chromosomes; the cell becomes $3n$, $4n$, $5n$, and so on.

Population A group of individuals of one species living in one area.

Predation One animal eating another animal, or it can also refer to animals eating plants.

Primary consumer The animal that eats the producer.

Primary immune response The initial immune response to an antigen.

Primary transcript Also called pre-RNA; the initial RNA transcript from a gene.

Primase An enzyme that joins RNA nucleotides to make the primer.

Prions Infectious proteins that cause several brain diseases: scrapie in sheep, mad cow disease in cattle, and Creutzfeldt-Jakob disease in humans.

Producer Those photosynthetic organisms at the bottom of any food chain.

Prokaryotes Cells with no internal membranes. Bacteria are one example.

Promoter The binding site of RNA polymerase in an operon.

Prophage A phage genome that has been inserted into a specific site in a bacterial chromosome.

Prostaglandin A hormone that promotes blood supply to an area.

Protista The kingdom that consists of single-celled and primitive multicelled organisms, such as paramecium and ameba.

Proton pump A mechanism in cells that uses ATP to pump protons across a membrane to generate a membrane or electric potential.

Protostome An animal in which the blastopore becomes the mouth during early embryonic development. Literally, first opening.

Pseudocoelomate A body cavity with mesoderm on only one side, characteristic of nematodes.

Pseudogenes Former genes that have accumulated mutations over eons and no longer produce functional proteins.

Pseudopods Cellular extensions of amoeboid cells used in moving and feeding.

Punctuated equilibrium A theory that proposes that new species appear suddenly after long periods of stasis.

Purines A class of nucleotides that includes adenine and guanine.

Pyrimidines A class of nucleotides that includes cytosine, thymine, and uracil.

Pyrogens A chemical released by certain leukocytes that increases body temperature to speed up the immune system and make it more difficult for microbes to function.

Pyruvate A variant of pyruvic acid.

Pyruvic acid A three-carbon molecule that is the product of glycolysis and is the raw material for the Krebs cycle.

Quorum sensing The ability of bacteria to sense and respond to changes in their population density.

Radicle In the embryonic root, the first organ to emerge from the germinating seed.

Radiometric dating A method of determining the absolute age of rocks and fossils.

Radula A movable, tooth-bearing structure that acts like a tongue in mollusks.

Receptor-mediated endocytosis The uptake of specific molecules by a cell's receptors.

Recessive trait The trait that remains hidden in the hybrid state.

Recognition sequence A specific sequence of nucleotides at which a restriction enzyme cleaves a DNA molecule.

Recombinant chromosomes Chromosomes that combine genes from both parents due to crossing-over.

Recombination The result of a crossover.

Reduction The gain of hydrogen (H^+) or electrons (e^-)

Reflex arc The simplest nerve response; it is inborn, automatic, and protective.

Refractory period The period of time during which a neuron cannot respond to another stimulus because the membrane is returning to its polarized state.

Replication bubbles There are thousands of replication bubbles along the DNA molecule that speed up the process of replication.

Replication fork A Y-shaped region where the new strands of DNA are elongating.

Repressor Binds to the operator of an operon and prevents RNA polymerase from binding to the promoter, thus blocking transcription.

Resolution A measure of the clarity of an image; the ability to see two objects as separate.

Resource partitioning The exploitation of environmental resources by organisms living in the same area so that each group of organisms can occupy a different niche.

Restriction enzymes Enzymes, naturally occurring in bacteria, that cut DNA at certain specific recognition sites.

Restriction fragment length polymorphisms (RFLPs) Noncoding regions of human DNA that vary from person to person. They can be used to identify a single individual. Pronounced "riflips."

Restriction fragments Fragments of DNA that result from the cuts made by restriction enzymes.

Reverse transcriptase An enzyme found in retroviruses that facilitates the production of DNA from RNA.

Rhizobium A symbiotic bacterium that lives in the nodules on roots of specific legumes and that incorporates nitrogen gas from the air into a form of nitrogen the plant requires.

Ribosomes The site of protein synthesis in the cytoplasm.

RNA polymerase The enzyme that binds to the promoter in DNA and that begins transcription.

RNA primer An already existing chain of RNA attached to DNA to which DNA polymerase adds nucleotides during DNA synthesis.

Rubisco (ribulose biphosphate carboxylase) The enzyme that catalyzes the first step in the Calvin cycle: the addition of RuBP (ribulose biphosphate) to CO_2.

Sarcolemma The modified plasma membrane surrounding a skeletal muscle cell and that can propagate an action potential.

Sarcomere The basic functional unit of skeletal muscle.

Sarcoplasmic reticulum Modified endoplasmic reticulum in skeletal muscle cells.

Satellite DNA Short sequences of DNA that are tandemly repeated as many as 10 million times in the DNA. Much of it is located at the telomeres.

Schwann cells Glial cells that are located in the peripheral nervous system and that form the myelin sheath around the axon of a neuron.

Sclerenchyma cells Plant cells with very thick primary and secondary cell walls fortified with lignin.

Secondary consumer The animal that eats the primary consumer.

Seed After fertilization, the ovule becomes the seed.

Semiconservative replication The way DNA replicates, each double helix separates and forms two new strands of DNA. Each new molecule of DNA consists of one old strand and one new strand.

Senescence Aging.

Sertoli cells The cells found in the mammalian testes that nourish the developing sperm cells, which contain no cytoplasm.

Sessile Nonmotile.

Sex-influenced trait The inheritance of a trait influenced by the sex of the individual carrying the trait.

Sexual selection Selection based on variation in secondary sexual characteristics related to competing for and attracting mates.

Sieve tube members Along with companion cells, these make up the phloem.

Signal transduction pathway A series of steps that converts a mechanical, chemical, or electrical stimulus to a cellular response.

Single-stranded binding proteins Proteins that act as scaffolding, holding two DNA strands apart during replication.

snRNPs (small nuclear ribonucleoproteins) Help to process RNA after it is formed and before it moves to the ribosome.

Sodium-potassium pump A protein pump within a plasma membrane of an axon that restores the membrane to its original polarized condition by pumping sodium and potassium ions across the membrane.

Solute The substance dissolved.

Solvent The substance doing the dissolving.

Somatic cell A body cell.

Somatic nervous system The branch of the vertebrate peripheral nervous system that controls skeletal (voluntary) muscles.

Sori Structures on the underside of the fern leaves that are clusters of sporangia containing monoploid spores.

Specific heat The amount of heat a substance must absorb to increase 1 gram of the substance by 1°C.

Spicules Found in sponges, these consist of inorganic materials and support the animal.

Spindle fibers Made of microtubules that connect centrioles to kinetochores of chromosomes and that separate sister or homologous chromosomes during cell division.

Spiracles Openings in the exoskeleton of arthropods, such as the grasshopper, that connect to internal cavities called hemocoels where respiratory gases are exchanged.

Spliceosomes Enzymes that (along with snRNPs) help process RNA after it is formed and before it moves to the ribosome.

Spongocoel Found in sponges, it is the central cavity into which water is drawn to filter nutrients.

Sporopollenin A tough polymer that protects plants in a harsh terrestrial environment.

Stabilizing selection Selection that eliminates the extremes and favors the more common intermediate forms.

Stele The vascular cylinder of the root, consisting of vascular tissue.

Stroma The site of the light-independent (dark) reactions in chloroplasts.

Submucosa A layer of the human digestive system that contains nerves, blood vessels, and lymph vessels.

Substrate-level phosphorylation The direct enzymatic transfer of phosphate to ADP to produce ATP.

Survivorship curves Show the size and composition of a population.

Sympathetic nervous system One of two branches of the autonomic nervous system that is generally excitatory.

Sympatric speciation The formation of new species without geographic isolation; such as polyploidy or behavioral isolation.

Symplast A continuous system of cytoplasm of cells interconnected by plasmodesmata.

Synapsis The process of pairing replicated homologous chromosomes during prophase I of meiosis.

Systematics Scientific study of the classification of organisms and their relationships to one another.

Taq polymerase A heat-stable form of DNA polymerase extracted from bacteria that live in hot environments, such as hot springs, that is used during PCR technique.

Taxa A particular group at a category level; such as kingdom or genus.

Taxonomy The study of classification of organisms.

Tay-Sachs disease An inherited genetic disease that is caused by lack of an enzyme necessary to break down lipids necessary for normal brain function and results in seizures, blindness, and early death. Common in Ashkenazi Jews.

Telomerase An enzyme that catalyzes the lengthening of the telomeres at the ends of eukaryotic chromosomes.

Telomeres The protective ends of eukaryotic chromosomes.

Tertiary consumer The third trophic level of consumer in a food chain.

Testcross A cross done to determine whether an individual plant or animal showing the dominant trait is homozygous dominant (*B/B*) or heterozygous (*B/b*). The individual in question (*B/_*) is crossed with a homozygous recessive individual.

Tetanus The smooth, sustained contraction of a skeletal muscle.

Thermophiles Prokaryotes that thrive in very high temperatures.

Theta replication The way in which prokaryotes replicate their DNA.

Thylakoids Membranes in chloroplasts that make up the grana, the site of the light reactions.

Thymine A nucleotide that binds with adenine. It is a pyrimidine and is not present in RNA.

T lymphocytes One type of lymphocyte that fights pathogens by cell-mediated response.

Tracheids Long, thin cells that overlap and are tapered at the ends and that, along with vessel elements, make up xylem in a plant.

Tracheophytes Plants that have transport vessels, xylem and phloem.

Transcription The process by which DNA makes RNA.

Transduction Transfer of bacterial DNA by phages from one bacterium to another.

Transformation The transfer of genes from one bacterium into another.

Translation The process by which the codons of an mRNA sequence are changed into an amino acid sequence.

Translocation A chromosome mutation in which a fragment of a chromosome becomes attached to a nonhomologous chromosome; the transport of sugar in a plant from source to sink.

Transpiration Loss of water from stomates in leaves.

Transpirational pull–cohesion tension theory This theory describes the passive transport of water up a tree. For each molecule of water that evaporates from a leaf by transpiration, another molecule of water is drawn in at the root to replace it.

Transposons Transposable genetic elements, sometimes called jumping genes.

Triploblastic Having three cell layers: ectoderm, mesoderm, and endoderm.

Triploid A chromosome mutation in which an organism has three sets of chromosomes ($3n$) instead of two ($2n$).

Trisomy A chromosome condition in which a cell has an extra copy of one chromosome. The cell has three of that chromosome, instead of two.

Trophic level Any level of a food chain based on nutritional source.

Tropic hormones Hormones released by one endocrine gland that stimulate other endocrine glands to release their hormones.

Tropism The growth of a plant toward or away from a stimulus, for example, phototropism.

T system A set of tubules that traverse the skeletal muscle, conduct the action potential deep into the cell, and stimulate the sarcoplasmic reticulum to release calcium ions.

Turgid Firm. Plant cells swollen because they have absorbed water.

Turner syndrome A genetic condition in females caused by a deletion of one of the two X chromosomes.

Typhlosole A large fold in the upper surface of the intestine of the earthworm that increases surface area to increase absorption.

Ultramicrotome An instrument used to cut very thin sections of tissue for use in the transmission electron microscope.

Uracil A nucleotide in only RNA that binds with adenine. It is a pyrimidine.

Vacuoles A membrane-enclosed sac for storage in all cells, particularly in plant cells.

Vasodilation The enlargement of blood vessels to increase blood supply.

Vegetative propagation Plants can clone themselves or reproduce asexually from any vegetative part of the plant; the root, stem, or leaf.

Vesicles Small vacuoles.

Vessel elements Wide, short tubes that, along with tracheids, make up the xylem.

Vestigial structures Structures of no importance, such as the appendix, that were once important to ancestors.

Visceral mass The part of the body of mollusks that contains the organs of digestion, excretion, and reproduction.

Water potential The tendency of water to move across a semipermeable membrane, ψ. The water potential of pure water is zero. Any solution has a water potential less than zero.

Wave of depolarization The wavelike reversal of the polarity of the membrane when an impulse passes.

Wobble Refers to the translation of mRNA to protein. The relaxation of base pairing rules in which the pairing rules for the third base of a codon are not as strict as they are for the first two bases. UUU and UUA both code for the amino acid phenylalanine.

Xanthophyll A photosynthetic antenna pigment common in algae that is a structural variant of a carotenoid.

Xylem Transport vessels in plants that carry water and minerals from the soil to the leaves.

Z lines These define the edges of the sarcomere in a muscle cell.

Zone of cell division The region of a plant's root with actively dividing cells that grow down into the soil.

Zone of differentiation The region of root tip where cells undergo specialization.

Zone of elongation The region of root tip where cells elongate and that are responsible for pushing the root cap downward, deeper into the soil.

Index